Quickfinder Tools & Methods

Creative activity... innovation! ... and performance metrics, long considered to be polar opposites, turn out to be essential complimenters. Lewrick's *Design Thinking and Innovation Metrics* is a must-read for anyone needing to add analytical strength to design thinking; it is an excellent treatment of both. Flying blind is no way to go into the future!

— Bill Fischer, Professor Emeritus of Innovation Management, IMD Business School, and Senior Lecturer, Sloan School of Management, MIT

Peter Drucker famously said that if you can't measure it, you can't manage it. This latest book of Lewrick's excellent series on design thinking is a highly practical guide for mastering the complex world of innovation.

— Roland Deiser, Chairman of the Center for the Future of Organization at the Drucker School of Management

This book is an excellent enhanced toolkit and extremely practical resource for any organization, regardless of size or industry.

— Elvin Turner, Best-selling innovation author, *Be Less Zombie*

This book is a game changer on the corporate world's view. It goes far beyond any approach to innovation accounting I have seen before.

— Jean-Paul Thommen, Professor Business and Management Studies, University of Zurich

More books by Michael Lewrick on the subject of design in the business context as well as for personal life and career planning.

DESIGN THINKING FOR BUSINESS GROWTH
HOW TO DESIGN AND SCALE BUSINESS MODELS AND BUSINESS ECOSYSTEMS
MICHAEL LEWRICK
ILLUSTRATIONS DONIKA PALAJ
WILEY
THINKING TOOLBOX
A GUIDE TO MASTERING THE MOST POPULAR AND VALUABLE INNOVATION METHODS
Michael Lewrick
Patrick Link
Larry Leifer
Achim Schmidt
WILEY
THE DESIGN THINKING PLAYBOOK
MINDFUL DIGITAL TRANSFORMATION OF TEAMS, PRODUCTS, SERVICES, BUSINESSES AND ECOSYSTEMS
MICHAEL LEWRICK
PATRICK LINK
LARRY LEIFER
DESIGN: NADIA LANGENSAND
WILEY
THE DESIGN THINKING LIFE PLAYBOOK
EMPOWER YOURSELF, EMBRACE CHANGE, AND VISUALIZE A JOYFUL LIFE
MICHAEL LEWRICK
LARRY LEIFER
DESIGN ACHIM SCHMIDT
WILEY

Lewrick
Design Thinking For Business Growth

How to Design and Scale Business Models and Business Ecosystems
ISBN: 978-1119815150

Lewrick, Link, Leifer
The Design Thinking Toolbox

A Guide to Mastering the Most Popular and Valuable Innovation Methods
ISBN: 978-1119629191

Lewrick, Link, Leifer
The Design Thinking Playbook

Mindful Digital Transformation of Teams, Products, Services, Businesses and Ecosystems
ISBN: 978-1119467472

Lewrick, Thommen, Leifer
The Design Thinking Life Playbook

Empower Yourself, Embrace Change, and Visualize a Joyful Life
ISBN: 978-1119682240

DESIGN THINKING AND INNOVATION METRICS

Published by John Wiley & Sons, Inc., Hoboken, New Jersey.
Published simultaneously in Canada.

Library of Congress Cataloging-in-Publication Data is Available:

ISBN 9781119983651 (Paperback)
ISBN 9781119983668 (ePub)
ISBN 9781119983804 (ePDF)
ISBN 9781119983675 (print replica)

Cover Design and Illustrations: Rukaiya Karim

SKY10041967_022323

DESIGN THINKING AND INNOVATION METRICS

POWERFUL TOOLS TO MANAGE CREATIVITY, OKRs, PRODUCT, AND BUSINESS SUCCESS

MICHAEL LEWRICK

ILLUSTRATIONS
RUKAIYA KARIM

WILEY

THE PROBLEM TO GROWTH AND SCALE FRAMEWORK

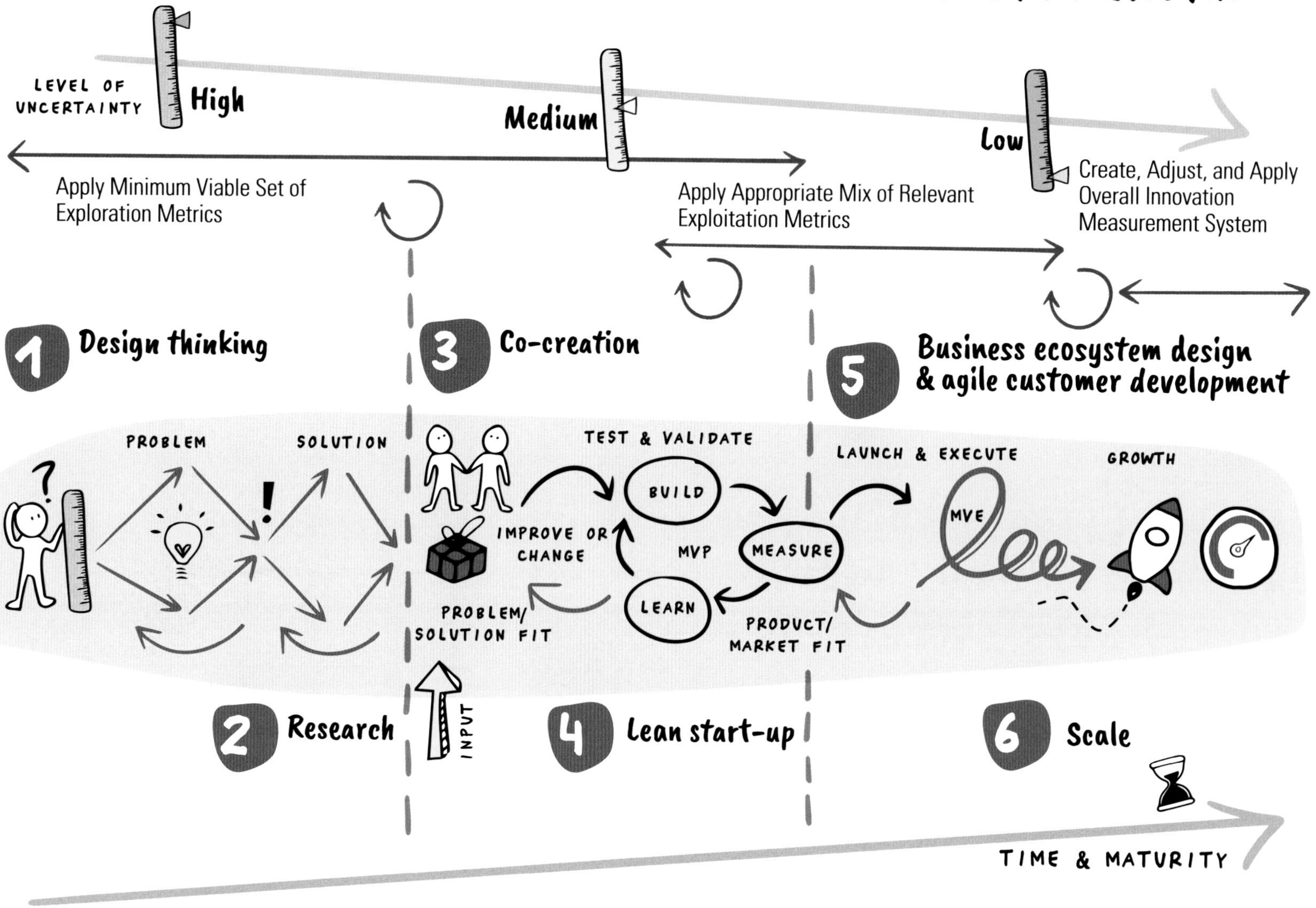

LEVEL OF UNCERTAINTY
High
Medium
Low
Apply Minimum Viable Set of Exploration Metrics
Apply Appropriate Mix of Relevant Exploitation Metrics
Create, Adjust, and Apply Overall Innovation Measurement System
1 Design thinking
3 Co-creation
5 Business ecosystem design & agile customer development
PROBLEM
SOLUTION
TEST & VALIDATE
BUILD
IMPROVE OR CHANGE
MVP
MEASURE
LEARN
PROBLEM/ SOLUTION FIT
PRODUCT/ MARKET FIT
LAUNCH & EXECUTE
MVE
GROWTH
INPUT
2 Research
4 Lean start-up
6 Scale
TIME & MATURITY

MVP – minimum viable product
MVE – minimum viable ecosystem

Metrics and meaningful measurements along the problem to growth and scale framework.

The Design Thinking Playbook introduces the overall context of design thinking, from problem definition to scalable solutions. It is complemented by the methods discussed in ***The Design Thinking Toolbox***. ***Design Thinking for Business Growth*** presents a paradigm shift that many companies will face in the upcoming years in terms of business model innovation, value stream definition, and business growth. This book, ***Design Thinking and Innovation Metrics***, complements the other books of the Wiley design thinking series and focuses on specific approaches for measuring, selecting, and predicting the respective activities from problem definition to scaling solutions.

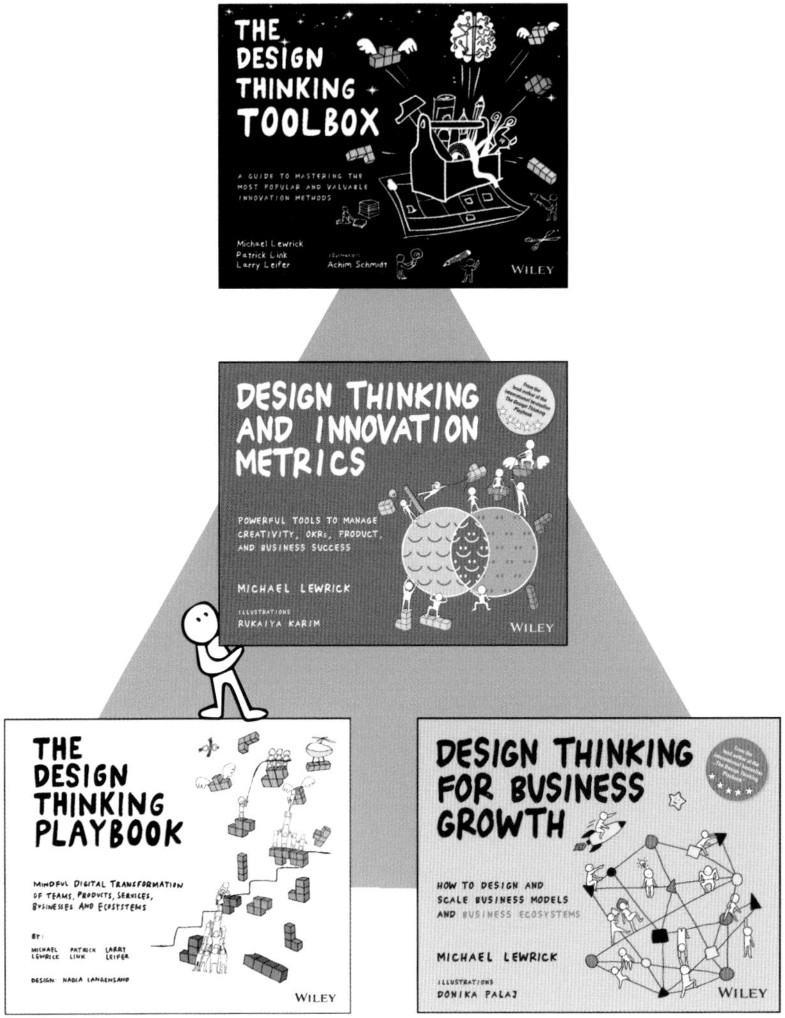

THE DESIGN THINKING TOOLBOX
A GUIDE TO MASTERING THE MOST POPULAR AND VALUABLE INNOVATION METHODS
Michael Lewrick
Patrick Link
Larry Leifer
Achim Schmidt
WILEY
DESIGN THINKING AND INNOVATION METRICS
POWERFUL TOOLS TO MANAGE CREATIVITY, OKRs, PRODUCT, AND BUSINESS SUCCESS
MICHAEL LEWRICK
ILLUSTRATIONS
RUKAIYA KARIM
WILEY
THE DESIGN THINKING PLAYBOOK
MINDFUL DIGITAL TRANSFORMATION OF TEAMS, PRODUCTS, SERVICES, BUSINESSES AND ECOSYSTEMS
BY:
MICHAEL LEWRICK PATRICK LINK LARRY LEIFER
DESIGN: NADIA LANGENSAND
WILEY
DESIGN THINKING FOR BUSINESS GROWTH
HOW TO DESIGN AND SCALE BUSINESS MODELS AND BUSINESS ECOSYSTEMS
MICHAEL LEWRICK
ILLUSTRATIONS
DONIKA PALAJ
WILEY

WWW.DESIGN-METRICS.COM

To the
Galaxy

This book should be on the desk of every employee who has an active role in shaping the future.

Preface

Happy customers get you paid, and doing this repeatedly and efficiently is the goal of every business.

ASH MAURYA
Best-selling Author & Creator of *Lean Canvas*

Design thinking has become one of the most promising mindsets to address wicked problems in different contexts and from various disciplinary standpoints. However, the measurement of creativity, innovation, and business success is still challenging—from identifying the right problems worth solving to driving ideas to product/market fit.

My approach for addressing these challenges is embracing a continuous innovation framework that marries concepts from innovation accounting, running lean, agile, and customer development. The conceptual models and descriptions in this book perfectly fit this approach.

Michael Lewrick has done a great job of providing an entire toolkit of measurement techniques that can be used depending on the organization's objectives, culture, and maturity level.

In my view, this book has four crucial characteristics that make it an indispensable companion, especially for business practitioners:

1. Making decisions over the entire design cycle and at different points of EXPLORE and EXPLOIT
2. Tracking and measuring the success of design challenges, innovation projects, and cultural and organization change
3. Assessing the impact that innovation has on the business, employees, and the strategy level
4. Hands-on tools and governance models for implementing innovation measurement systems as well as objectives and key results

The ability to measure our innovation activities, mitigate risks, and create real value are key skills of start-up, design thinking, and innovation teams.

I wish you much success in applying design thinking and innovation metrics throughout the entire design cycle and beyond.

Ash Maurya

THE DESIGN THINKING AND MEASURING MINDSET

Focus on the current and future customers' needs

- Place the customer/user at the center of all activities, break internal silos, and remove barriers to external partners so you can create winning value propositions.
- Apply and use design thinking tools in solving customer, strategic and business problems, and designing physical, digital, and virtual customer experiences, products, and services.

Define a minimum viable set of exploration metrics

- Introduce the measurements for EXPLORE step by step and in iterations, but with the same professionalism as the measurements for EXPLOIT.
- De-risk development by continually employing evidence across desirability, viability, and feasibility dimensions to evaluate highly uncertain, transformative innovation initiatives.

Grow with the team beyond the problem and solution space

- Build capabilities in design thinking and hold all relevant teams accountable for defining their objectives and key results.
- Establish interdisciplinary design thinking and innovation teams that follow the North Star of exploratory innovation initiatives.

Transform traditional metrics into meaningful measurement systems

- Promote the right behavior and appropriate mindset in order to foster both radical and incremental innovation.
- Provide strategic direction by indicating shifts in priorities.
- Direct the (re)allocation of resources and assess the effectiveness of innovation spending
- Analyze and enhance innovation performance and culture.

NEW MINDSET.
NEW PARADIGM.
BETTER SOLUTIONS.

WWW.DESIGN-METRICS.COM

Apply Moonshot and Everest thinking for innovation teams

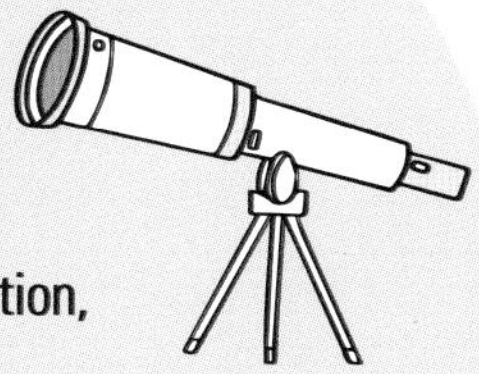

- Start to link objectives to real-world projects and initiatives.
- Align the teams based on impact and not on activities.
- Define audacious and inspirational objectives and measurable key results within a pre-defined scoring system.
- Turn it into culture where failure is a part of success, in exploration, if the team strives to achieve remarkable results.

Create impact with the interplay of big data analytics, artificial intelligence (AI), and neurodesign across the entire design cycle

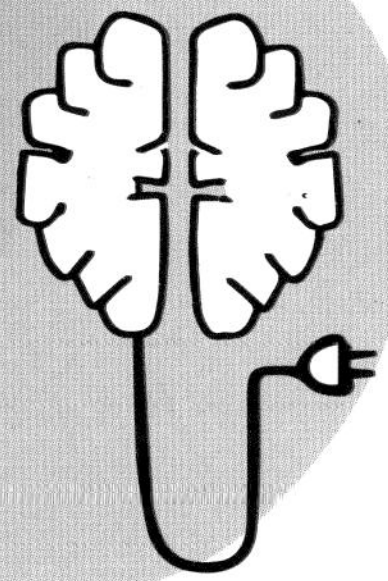

- Leverage the potential of data and AI automation to extend data insights and deliver "unknown unknowns" based on the ability to explore millions of hypotheses in a very short time.
- Reveal the truth of design thinking with neurodesign or the simple satisfaction of formal expectations about human behavior that are still not completely understood.

How to Get the Most out of the Book

The following elements make it easier to find your way in the book:

Example

Examples of applying design thinking or the use of specific metrics will be described and presented.

Toolkit

Various known and new methods, tools, and procedural models will be presented. You'll find an overview of the most frequently applied tools and methods on the first page of this book. Each logical section has a toolkit with selected tools.

To the Point!

At the end of each logical section, the content is reflected upon and summarized.

Template Download

Selected templates are available for download as PDF files. Premium templates for the professional workshop facilitators and training purposes, including a detailed description of the application shown on the template, are available in the online store:

www.dt-toolbook.com/shop

Adapt

The traditionally rigid management framework has been obsolete at least since the turn of the millennium. So please adapt the procedural models in this book to your specific situation.

Each enterprise, organization, or team has a different purpose. Besides financial indicators, even the way a defined success is achieved might also be different. The primary goal of this book is to inspire you to adapt the current innovation measurement system and understand how it can be set up for innovation and design thinking teams.

When defining and applying performance measurement systems for innovation teams, it is especially important to establish a system that fits the organization, the current transformation, and the desired ambition.

In addition, the snippets on data analytics, AI, and neuroscience in combination with design thinking provide insights into how future data and measurement points can enrich your work on the upcoming market opportunities. However, the approaches presented are also adaptable to the situation at hand.

It cannot be stressed enough that the tools, methods, and procedural models presented in this book must always be adapted to your specific situation.

Contents

For an easy entry into the world of design thinking and measurements, the essentials are provided in two comprehensive 101s. This is complemented by the definition of performance measurement systems for innovation teams (201). Data analytics, artificial intelligence, and neurodesign provide insights for applying data-driven innovation (301). Discussions are complemented by hands-on tools, methods, and suggestions for current and future metrics and measurements.

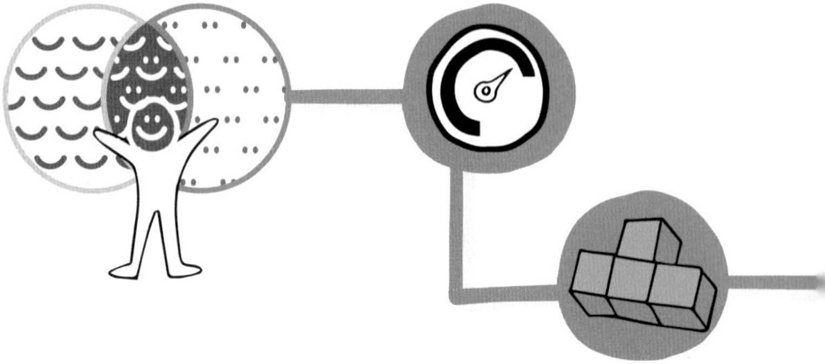

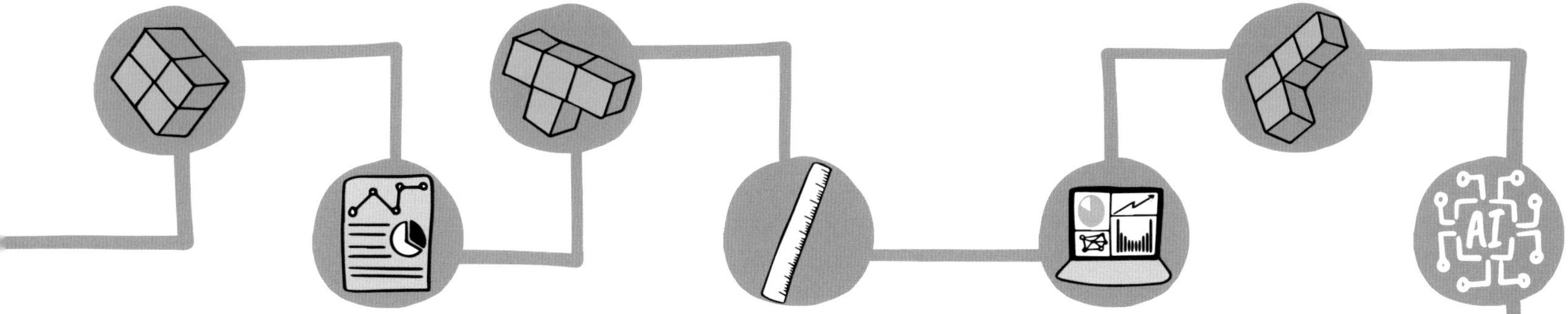
AI

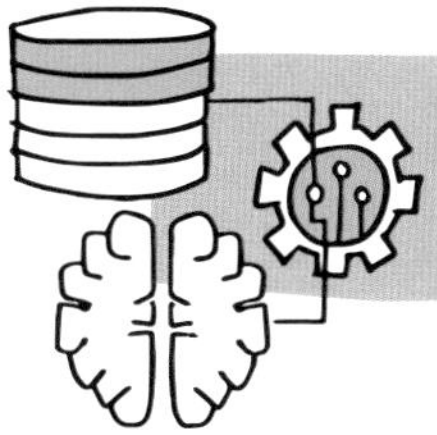

The Future of Design Thinking & Data-Driven Innovation 312

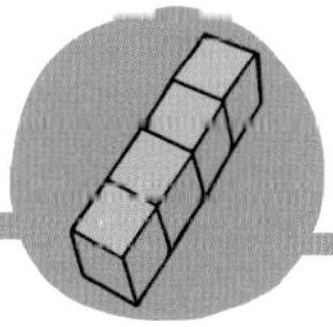

Motivation for this book

Michael Lewrick, PhD | MBA, has worked very intensively in recent years on the mindset that enables us to solve different types of problems. Michael is a best-selling author, award-winning design thinking and business ecosystem design thought leader, business entrepreneur, and visiting professor at various universities globally. His ideas, books, and company, Lewrick & Company, help mobilize people around the world to better lead innovation, digital transformation, and business growth in an era of increasingly rapid change. He is the author of the international best-sellers *Design Thinking for Business Growth*, *The Design Thinking Toolbox,* and *The Design Thinking Playbook*, in which he describes the mindful transformation of people, teams, ecosystems, and organizations. He works intensively with universities and companies, and places the self-efficacy of people in personal and organizational change projects at the center of his activities. In recent years, he has expanded his toolbox for measuring creativity, innovation, and business success. As an internationally recognized expert in the field of digital transformation and the management of innovation, Michael has helped numerous companies to develop and scale growth strategies.

Core statements with regard to managing creativity, OKRs, product and business success

> *"A **balanced portfolio of EXPLORE and EXPLOIT** leads to the realization of **new market opportunities** and sustaining the existing business at the same time."*

> *"In the **current discussion** and application of **innovation measurement systems**, it is believed that the work of design thinking and innovation teams **is subject to constant change.**"*

"**Growth-leading companies** align their metrics and incentives to a minimum viable set of exploration metrics and an appropriate mix of relevant exploitation metrics **while avoiding measurement traps.**"

"A state-of-the-art innovation measurement system should allow **prudent risk taking**, embedded in a **culture and mindset** that is tolerant of **risk and failure.**"

"**Advanced AI** will make metrics for innovation and business success **more forward-looking** rather than retrospective."

"Understanding **neuroscience** is an integral part of the **future of design thinking and innovation metrics**. It will transform the way how creative teams work and how new product, service, and experiences are designed."

"**A modern measurement system** should align the **North Star with objectives** that need to be achieved by **teams and networks of teams.**"

Why Is Design Thinking & Measurement a Relevant and Central Topic for Me?

The title of this book alone triggers an already controversial discussion. Chief innovation officers, innovation and creative teams, as well as intrapreneurs are fighting every day for more freedom to explore and experiment so they can eventually achieve radical innovations that were previously unknown to customers and to the market.

To justify this important innovation work, nice-sounding key performance indicators (KPIs) often have to be invented to justify the budgets for exploration. On the other hand, in companies, we have a whole faction of executives and decision makers who follow the principle "What you can't measure, you can't manage."

There is a need for a sensible approach toward metrics so that measurements do not take on a life of their own and become instruments of justification, defense, and power, rather than fulfilling their intended value of self-reflective direction setting.

I work with both groups in optimizing the existing business (EXPLOIT), as well as in realizing new market opportunities and disruptive business ecosystems (EXPLORE). Proper measurement ensures that a company's available resources are used in the best possible and most profitable way.

This book aims to bridge the gaps between creativity, innovation, and measurement, and it offers a comprehensive collection of powerful tools to manage creativity, objectives and key results (OKRs), products, and business success. I have seen an increasing demand for this in many projects, and all of them reached a point in the design cycle, from finding the right problem to scaling the solution, where questions are asked about the appropriate metrics.

In some organizations, questions about metrics come very early in the design cycle, and in others, they come later, for example, when it comes time to invest to realize initial prototypes. The fact is that data points provide certainty for decisions, and even small surveys that provide a marginal reduction of uncertainty can be extremely valuable and have a big impact. Thus, the goal should be to apply the appropriate metrics, measurement systems, and frameworks that help to provide true indicators for measuring progress and success. In both cases, they should support deriving the right actions from them.

In many of my engagements with companies over the last 20 years the focus has usually been on two objectives: building new capabilities (learning) and, at the same time, delivering high-impact outcomes in the project. This approach, better known as project-based learning, focuses on the people who are trained to strengthen their creative confidence and to apply the appropriate tools and methods at the right time, as well as to deliver convincing solutions at the end of the innovation journey.

Along with the people, the path from problem definition to solution is equally important in the development of any idea to market. This transformational journey is about networking with other teams, leveraging the strengths of each team member to the best of their ability, and allowing teams to grow beyond the problem and solution space that exists in the heads of individual team members. This way, the networking and interaction usually go through a positive reinforcement that empowers everyone to make a valuable contribution to the organization and its strategic ambition.

Measuring transformation, networking, and building new capabilities are ongoing and related to activities. Thus, not all metrics are the same, and their nature can range from lagging to leading, and financial to nonfinancial.

One possible way to set up a meaningful performance measurement system is to use objectives and key results that pay attention to the larger strategy and help the teams to focus time and energy on the appropriate customer problem. This book not only explains how to design and apply metrics for **EXPLORE and EXPLOIT**, but presents hands-on tools to align between company objectives and team objectives in establishing performance measurement systems.

Applying traditional KPIs in every step, before even starting to explore and test new market opportunities, is destroying creativity before it begins.

In addition, this book builds on design thinking, which gives us the mindset to think from the customer's point of view, to collaborate radically, and finally to better master the dance with ambiguity. If you've read the other books in which I have been the lead author, you know that the current mindset includes the combination of big data analytics, and the application of varying mindsets from systems thinking to design thinking, for example in the realization of business ecosystems.

This book provides a framework to better deal with metrics and also serves as a deepening of and complement to the books *The Design Thinking Playbook*, *The Design Thinking Toolbox*, and *Design Thinking for Business Growth*. In particular, it presents additional methods, tools, and approaches for the selection of ideas and the measurement of success across the problem of growth and scale framework, facilitating decision making, and comparison of options and opportunities. In short, this book is an advanced toolbox that positively supports our daily innovation work. Professional facilitators and innovation experts will use it as a playbook to get inspired.

How many new ideas did we come up with this year?
ONE. A new way of counting ideas.

However, the book is also intended to encourage people to rethink existing views and innovation measurement systems and not to focus on meaningless KPIs, such as the well-known vanity metrics that measure, for example, the number of new ideas generated by teams over a year.

I welcome your feedback on the application of these tools, methodologies, and metrics and wish you good luck in your current and future activities as you realize new market opportunities and business success.

Michael Lewrick

I welcome direct feedback on the book and an exchange of ideas on the application of design thinking and the ways to measure creativity, innovation, and business success.

www.linkedin.com/in/michael-lewrick

The Roots of Design Thinking

The mindset, tools, and methods presented in this book relate mostly to the principles of design thinking applied at Stanford University. Stanford University's involvement with design-oriented programs began in the 1950s, and it has been an influencer of many companies globally ever since. At the same time Stanford has been influenced by new industry movements and developments in applying design thinking for products, services, processes, and business ecosystems. The research and education programs at Stanford have become a global standard for design thinking and applied design. From the early days, Stanford's engineering design was influenced by studies on creativity and human-centered design. One of the postgraduate program's first students was Professor Larry Leifer, who for decades drove Stanford's academic vision of design in engineering, eventually founding the university's Center of Design Research (CDR) in 1984. The CDR has produced gigantic amounts of design thinking–related research throughout the years, mostly focusing on the effectiveness of collaboration, teamwork, and, lately, the application of neurodesign. Since the late 1980s, Larry Leifer has led the iconic and radical engineering design course ME310, which celebrated its 55th year in 2022. The course displays many of the aspects that today are associated with design thinking. For almost 20 years, ME310 has been taught in partnership with universities and clients from around the world, becoming a breeding ground for hundreds of followers of design thinking the Stanford way. In 2005, the d.school was founded by David Kelley (founder of IDEO), a former PhD student of Larry Leifer, who attended the ME310 course back in the 1970s. The d.school and CDR remain Stanford's leading institutions for a more holistic view on design thinking.

Inspired by the research, tools, and methods of Stanford University, many frameworks, like the Business Model Canvas and Value Proposition Design, have been derived over the years. In addition, CX/UX/UI designers use the design thinking process to discover problems and come up with creative solutions by thoroughly understanding the customer/user needs, pain points, and jobs to be done.

The methods presented in this book have also been profoundly influenced by my mentor, friend, and colleague Larry Leifer over recent decades. Completely new research, practical models, and approaches have emerged from this collaboration, such as business ecosystem design, which is based on a skillful combination of systems thinking and design thinking to create the basis for scalable solutions and exponential growth. In the future, design thinking will become even more important and will accompany and co-exist with AI solutions in solving the biggest and most complex questions of mankind, from stopping global warming and dealing with pandemics, to the design and realignment of the political order.

THINK
OUTSIDE THE
BOX

Prof. Larry Leifer

The Evolution of Design Thinking

DESIGN THINKING

[Design thinking is a mindset for creative problem solving. The approach leads to breakthrough innovation.]

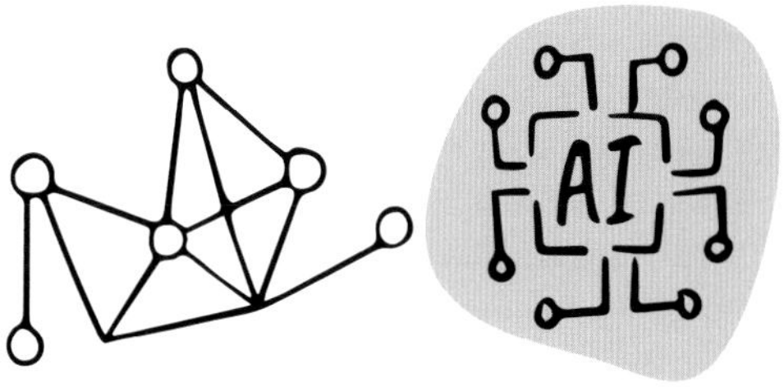
AI

EARLY DAYS

TODAY

FUTURE

Focus on (re)design of products, services, processes, business models, and experiences

Enable digital platforms, business ecosystems, new organizational behavior, and user experiences

Design of immersive experiences, convergence with AI, automation, and mass customization; relevant mindset for tackling the most wicked problems of mankind

1960 1980 2000 2020 2040 TIME

The History of Innovation Metrics

As with design thinking, innovation measurement systems have also evolved over the past decades. Innovation measurement originated in the 1960s, with the establishment of international standards to measure the research and development (R&D) efforts by companies. However, it was soon realized that this measure alone did not cover the diversity of innovation and that adaptable models were needed. In the 1970s and 1980s there were several attempts to overcome the linear models of innovation and new theories of economics of industrial innovation were established. This was followed by a period in which Schumpeter's theories were revived in order to lay down the basis for a new economics of innovation, the classification of innovation typologies has to be given a special attention in the innovation and measurement community. As a result, measurement activities, either quantitative or qualitative, focused on specific business functions and any improvement in these functions was seen as evidence of innovation. At this stage, the measurement of innovation output was seen as a priority. The identification of an innovation process through a few potential outputs, such as a new business model, new product, or new process, was proven to be very effective for the measurement purposes at this specific period of time. With the digital transformation, frameworks with different metrics looking at innovation and investment in intangible assets as two highly correlated phenomena became established. These mostly included business activities related to technology, digitization, environmental and social sustainability, customer experience and branding, internal innovation networks, purpose, and external innovation ecosystems.

In the current discussion and application of innovation measurement systems, it is believed that the work of design thinking and innovation teams is subject to constant change. For this reason, on the one hand, the full range of statistical tools should be used to keep pace with this change. On the other hand, measurements should be made on two levels: a carefully described object and productivity as a measure of innovation. Thus, besides the qualitative indicators, the quantitative output Indicators are an important element in understanding whether individual measures, from building new capabilities to implementing a strategy, are effective and successful.

Thus, many more approaches to various assessment and measurement methods will be added in the future to account for the multiple impacts of innovation, including the quantitative and qualitative dimensions of the performance and experiences of individuals, teams, and organizations.

Likewise, new names are invented for the respective measurement systems to emphasize the importance of certain elements, such as the aspiration aspects of OKRs for innovation teams based on customer insights, so called AKIs (Aspirations and Key Insights).

ADAPT THE MEASUREMENT SYSTEM
Joseph A. Schumpeter

The Evolution of Innovation Metrics

INNOVATION MEASUREMENT SYSTEM

[Today organizations build custom-made innovation and performance measurement systems based on a set of minimum viable exploration metrics.]

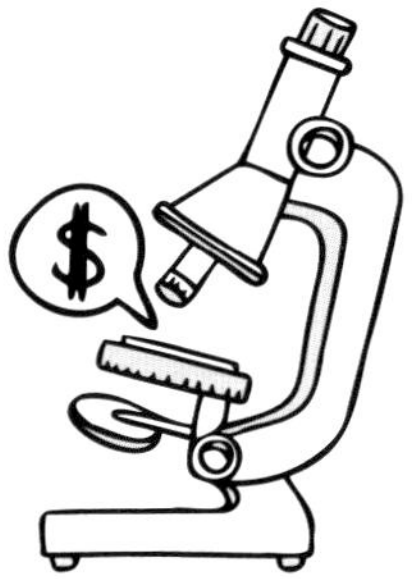

$

EARLY DAYS

TODAY

FUTURE

Focus on R&D expenditures, capital, tech intensity, patents, number of publications, and number of new products

Focus on outcomes, processes, portfolios, risk/return, clusters, network effects, design thinking and systems thinking capabilities, and team performance

Extension toward system dynamics, collaboration, future capabilities, supported by AI, big data analytics, and emerging measurements about communities and ecosystem capital

1960 1980 2000 2020 2040

TIME

To the Point!

In an era of increasingly rapid change, wicked problems to solve, and extraordinary opportunities for business growth and positive change, design thinking provides the appropriate mindset.

Measuring creative work, business, and innovation success remains a challenge for individuals, teams, organizations, and entire ecosystems.

In many instances vanity metrics are applied, for example, in counting the number of randomly collected ideas without meaning. However, most of those traditional measurements are not related to the customers' needs and do not help make better decisions and take effective action.

This book is an advanced toolbox that positively supports the daily innovation work and deepens the mindset that is needed to create purpose, action, and impact for customers, employees, people, and society.

INTRODUCTION TO THE TOPIC

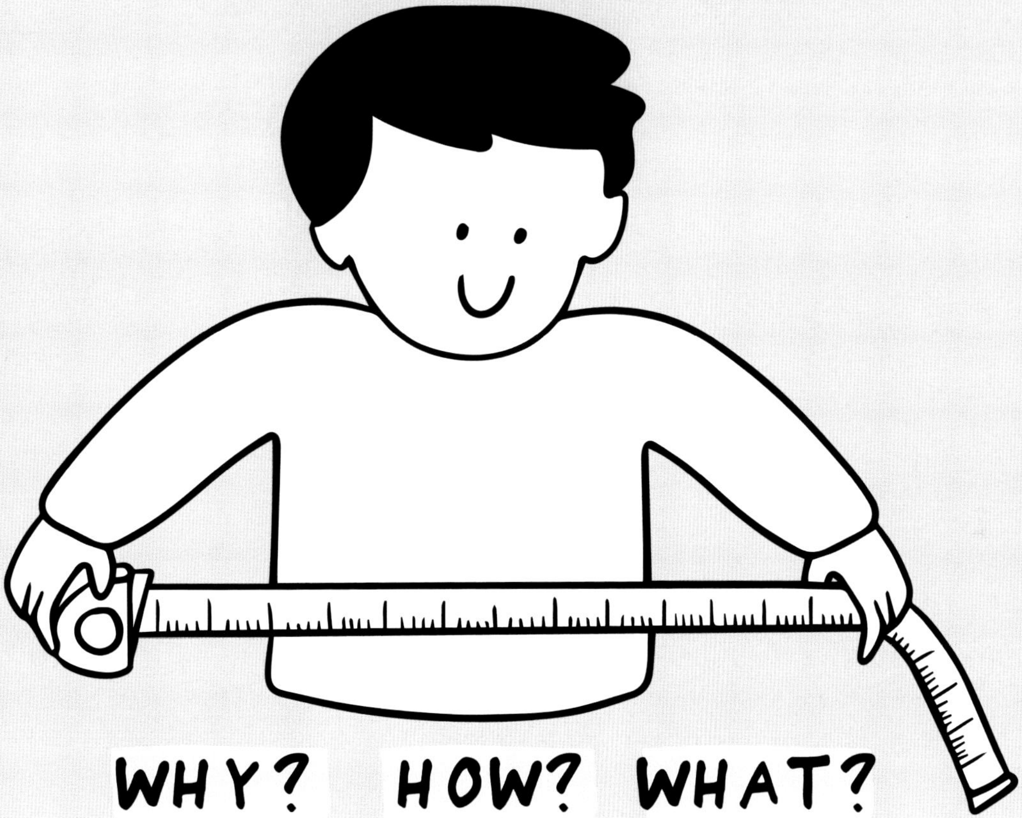

WHY? HOW? WHAT?

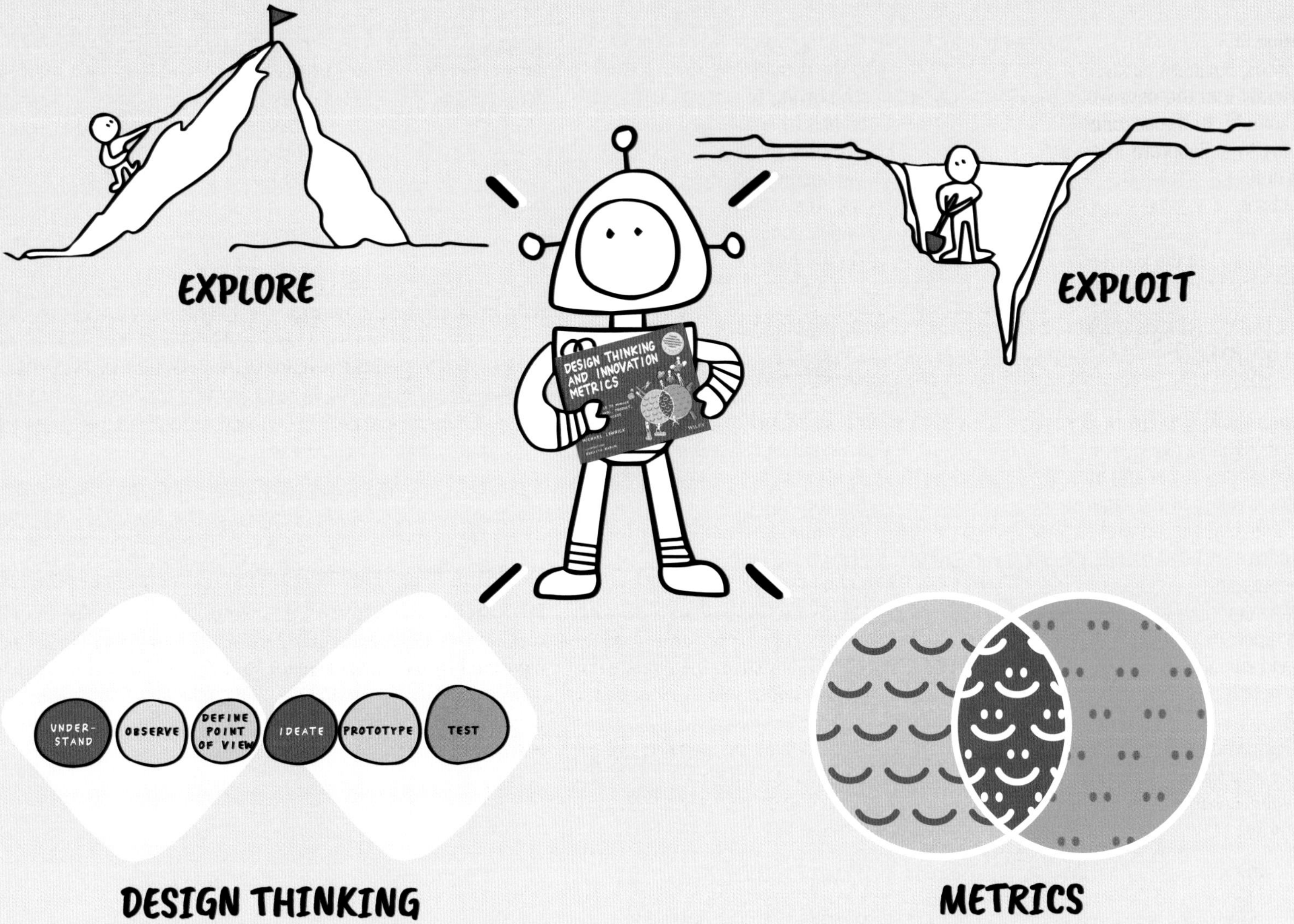
EXPLORE
EXPLOIT
DESIGN THINKING AND INNOVATION METRICS
WILEY
UNDER-STAND
OBSERVE
DEFINE POINT OF VIEW
IDEATE
PROTOTYPE
TEST
DESIGN THINKING
METRICS

WHY Are Design Thinking & Innovation Metrics Important?

Design thinking and meaningful metrics are key for creating new and viable products, services, business ecosystems and business models. The design thinking mindset fits perfectly into the new world in which companies are striving to become even more meaningful to the customer. It is also ultimately intended to enable interdisciplinary teams to collaborate in an ad hoc and radical way across the established silos of the enterprise. Design thinking offers the best starting position especially for wicked problems and the development of radical innovations. The design thinking mindset helps the respective teams today in the most innovative companies worldwide not only to explore the problem space and develop initial prototypes, but also to respond to constantly changing customer needs in the setting of agile customer development.

> *Change requires innovation, and innovation leads to progress, which can be measured at different levels.*

Design thinking helps organizations, teams, and each individual employee to define success and to work out the critical functions and experiences that are important for the customer today and in the future. Design thinking also helps establish a common understanding of the organization's ambitions and the North Star, so it is not surprising that building up and applying design thinking as well as measuring innovation success are skills that are in high demand. Both are becoming increasingly important for companies to establish a mindset and the development of appropriate innovation measurement systems. Together, they help not only to realize value capture, but also to support and realize value creation and delivery in the best possible way. However, individuals, teams, and decision makers underestimate in many cases their innate desire to optimize their behaviors to meet the objectives with which they are presented.

In many companies, the objectives are still designed to support this behavior, which leads employees to design their activities around it to optimize numbers, but not focus on value creation, value delivery, or value capture depending on the growth phase or maturity of the initiative the individual or team is working on.

> *Value creation, value delivery, and value capture become more important than optimizing numbers.*

In organizations and environments where vanity metrics have a long tradition, employees tend to make them their individual and team objectives. An especially hard-to-understand strategic vision can lead teams to lose sight of their North Star and instead only optimize themselves for the short term. Often such behavior can also be observed in agile teams, which are required to develop solutions based on ideas and do not get the opportunity to explore the problem space first to know what the ultimate purpose of the solution should be for the customers or other stakeholders. As a result, the organization, the team, or each individual misses the wrongly set targets, and leaders criticize not the system of measurement and how objectives have been defined, but the individual employees and teams that have been subjected to this system. In most cases, failure occurs because the organization measures efficiency, but they actually want to know more about effectiveness.

> *If the measurement doesn't make sense, you can't expect the behaviors of each individual and the team to be logical, either.*

Practical experience shows that teams are more willing to engage in design thinking activities when they feel they are not measured by their achievement hard requirements, but by their achievement of great outcomes that create value. In exploration, outcomes are often related to behavior change that drives business results.

Increasing Interest in the Design Thinking Mindset...

Design thinking as a mindset has become very popular in the last decade. More and more companies have realized that exploring the problem space more deeply before finding a solution helps them to discover better solutions, and thus increases the success of market opportunities. Traditional companies from the pharmaceutical, banking, insurance, retail, and engineering industries are increasingly recognizing the need to build design thinking capabilities and empower their employees with new tools, methods, and approaches.

The Google Trends radar chart for the search term "design thinking" confirms that there is a very great global interest in design thinking. This also brings us a statistic on the interest in the design thinking mindset, which correlates very well with the number of requests training and consulting companies are receiving to make design thinking a central mindset across organizations.

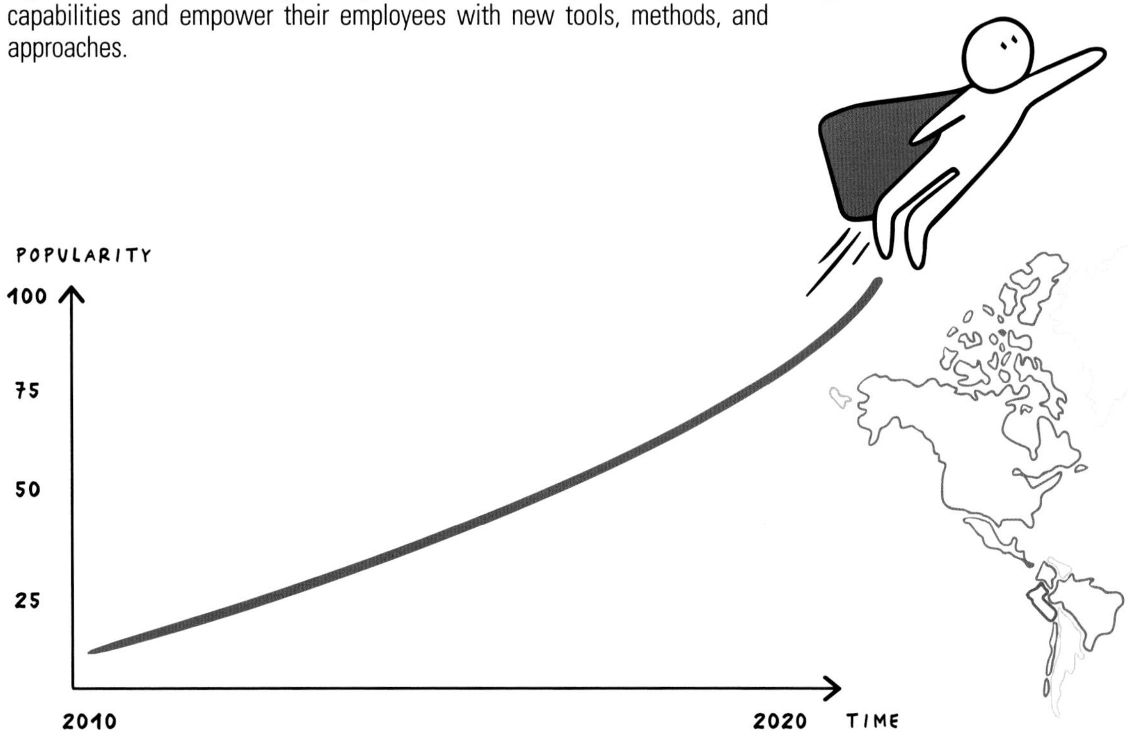
POPULARITY
100
75
50
25
2010
2020
TIME

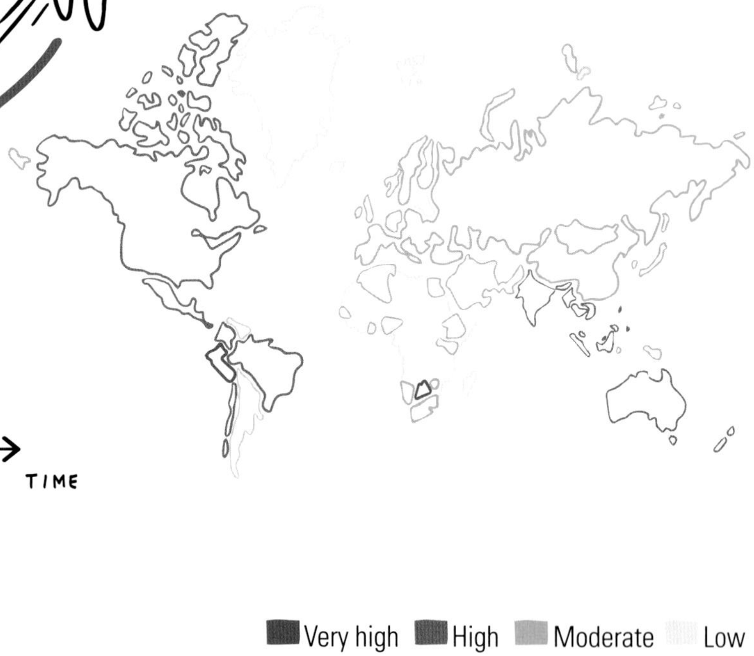
Very high
High
Moderate
Low

Source: Google Trends "Design Thinking" (2010–2020)

... and Even More Potential by Applying a Broader Design Lens

Today, design thinking is much broader than the (re)design of products, services, and processes. It is used for a wide range of problem statements. Design thinking teams radically design new value propositions for business ecosystems. The mindset is used to improve business models or to combine completely new value streams in such a way that exponential growth can be realized. Design thinking is also used in combination with big data analytics to design better features and experiences for customers/users. Furthermore, it is applied to set up conversational AI solutions, such as responding to already known customer needs and frequent inquiries, so customers receive fast and promising answers.

Since the onset of the digital transformation of products, services, organizations, and teams, the design thinking mindset has become indispensable for companies, start-ups, and non-profit organizations. Business development and strategy teams base their strategy considerations on customer insights and iteratively develop new product, service, and growth strategies. For this and many other reasons, the interest in design thinking and innovation metrics is also increasing. In the end, what counts is the impact, transformation, and customer delight regarding all activities delivered.

Evolving design thinking from pure product / service development into a wider design thinking lens with transformational digital and business growth opportunities

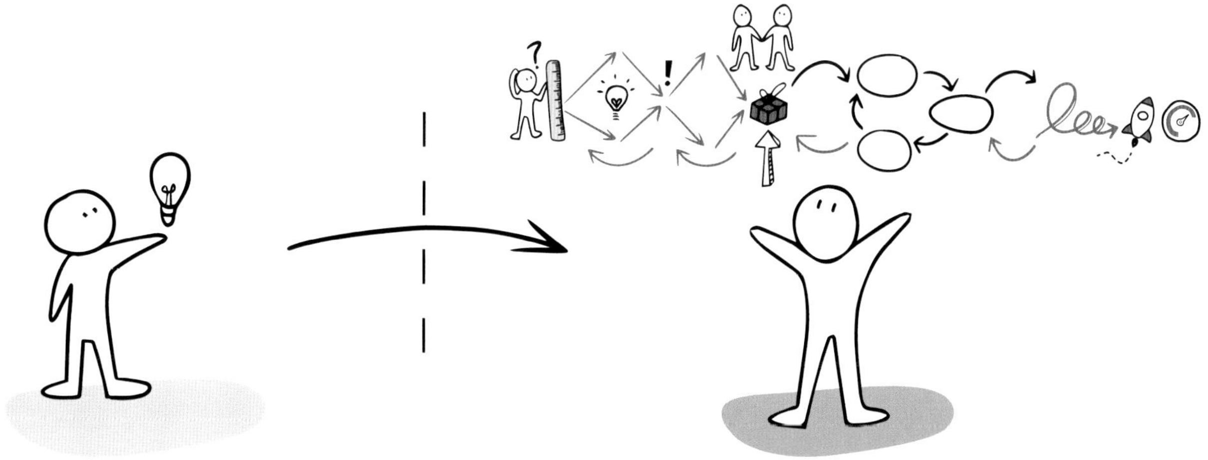

Design thinking has evolved over the last 50 years. Today, the design thinking mindset is used for the (re)design of products, services, business models, and even complex business ecosystems. The examination of the problem and solution space is the basis for strategy work, the definition of OKRs, and initiating new organizational models.

Increased Interest in Measuring Innovation ...

There is also a great interest in measuring innovation. One reason for this is that business models have changed in recent years and, at the same time, dynamics have increased, which means that companies have to react to changes much more quickly than in the past. Companies use big data analytics for agile product and service development and act proactively on current and future customer needs with tailored offerings. However, the measurement of innovation often lags behind the described developments, although the data, information, and inferences would positively support decision making and business management.

It is not surprising that the Google Trends search on "Innovation Index" is also gaining in importance, even if the global distribution shows that India, Australia, Europe, and North America have so far shown the greatest interest. A quick check of the search trends on Baidu shows that the map in Asia continues to fill up for the keywords "Design Thinking" and "Innovation Index" as well.

POPULARITY
100
75
50
25
2010
2020
TIME

Very high
High
Moderate
Low

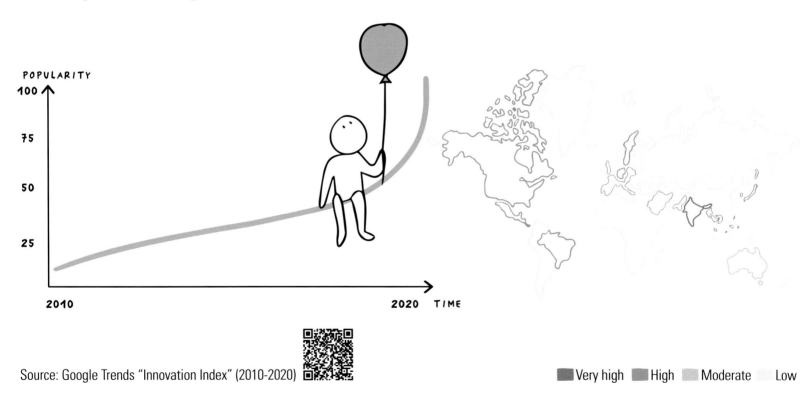

Source: Google Trends "Innovation Index" (2010-2020)

... But Still a High Popularity of Vanity Metrics

Vanity metrics can be found in many innovation frameworks of organizations. Especially in large companies, such indicators have a high longevity, as they are an easy way to quantify the claimed innovation power in an impressive way. In most cases, however, they only provide an indication of the current impact and neglect the benefits that will be generated by the respective activities in the future. This is another reason to broaden the view and present innovation metrics that are actionable, informative, and measurable.

It is also important to differentiate between what your organization wants to achieve currently and in the future regarding EXPLORE and EXPLOIT and how other key objectives relate to transforming and developing new capabilities. This includes the establishment of new organizational models, in addition to the pure realization of market opportunities.

Thus, in this book a rough distinction is made between:

- North Star metrics
- Minimum viable set of exploration metrics
- Appropriate set of exploitation metrics
- Performance measurements (OKRs)

The overall motivation is to provide individuals, teams, decision makers, and organizations with a choice of tools and methods for creating actionable measurements and objectives, which allows them to reveal possible directions for change, motivates them to create impact with purpose, and provides for more frequent and time-bound evaluations. However, for many metrics it is also necessary to add external data points for measuring innovation and defined success.

From vanity metrics to meaningful metrics

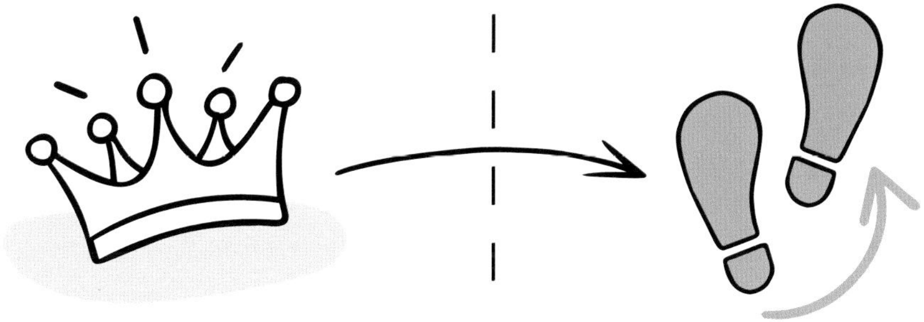

Meaningful measurements for creativity, innovation, and business success range from a minimum viable set of exploration metrics embedded in a powerful innovation measurement and performance measurement system, to the appropriate set of exploitation metrics aiming to steer, reflect, and adjust activities toward the desired future outcomes.

The Symbiosis of Design Thinking and Measurements

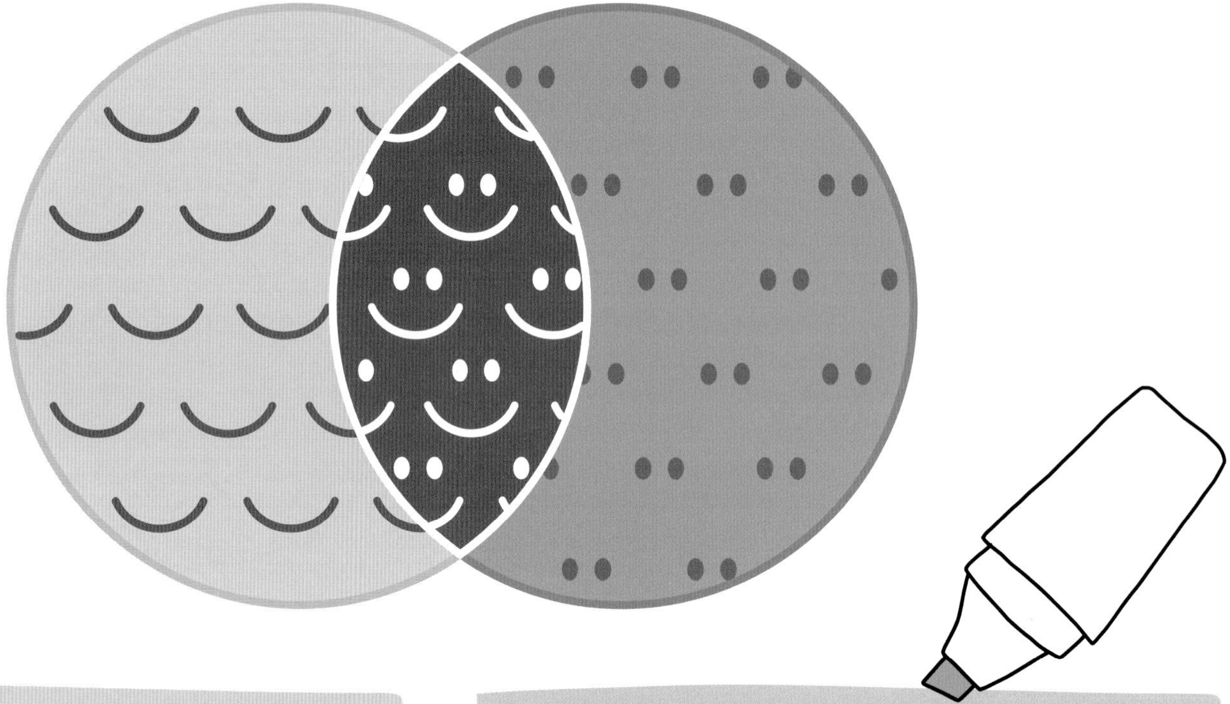

- Does the mindset of your organization consider the pains, gains, and jobs to be done for your customers?
- Does the organizational setup allow radical collaboration and a balanced portfolio of EXPLORE and EXPLOIT?

- Do the metric and the measurement system of your organization matter to your customer?
- Does the metric lead to taking actions and making better decisions?

HOW to Apply It to Unlock Value

As an extended toolbox for design thinkers and business practitioners, this book presents the different levels of defining metrics and the new and known metrics for daily work. This book takes a broader view of the design lens from problem identification to solution scaling. In addition, approaches are presented that help in applying the design thinking mindset in defining objectives and key results.

The subject of metrics is roughly divided into EXPLORE and EXPLOIT because the goals, the underlying mindset, and the measurement are very different between these two categories. EXPLORE usually has a focus on creativity and experimentation up to the establishment of new capabilities and organizational structures. By contrast, EXPLOIT focuses on achieving the best possible efficiencies, optimizing costs, and increasing output. These are measures that help to optimize ROI and achieve the targets of business plans.

Metrics based on the one-size-fits-all pattern for EXPLORE and EXPLOIT often do not achieve the desired goal.

Depending on the organization, there are different levels for the definition of metrics. The framework presented in this book includes elements that are usually relevant for established companies and illustrates how different the basis for the respective definition of metrics can be, depending on the focus and purpose: EXPLORE or EXPLOIT.

In contrast, start-up companies usually deal with EXPLORE at the beginning and can start with a greenfield approach in terms of structure, culture, and the targeted portfolio. Often OMTM (one metric that matters) is applied in the early incubation phase. However, the start-up philosophy is often also found in established companies in the form of intrapreneur units, which are assigned the task of disrupting the existing business. At the same time, other established companies have neglected their EXPLORE capabilities and directed all activities toward EXPLOIT.

Focusing on the team/culture and project/business spheres of measuring EXPLORE and EXPLOIT over time and iterations can create a superior innovation measurement system. This unlocks value and taps into the full potential of design thinking and measurements.

Innovation Measurement System

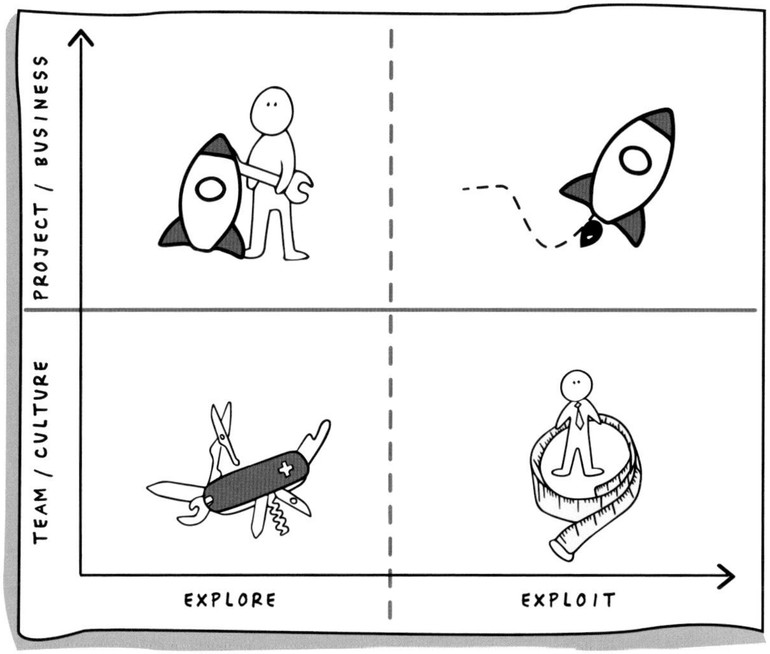
PROJECT / BUSINESS
TEAM / CULTURE
EXPLORE
EXPLOIT

EXPLORE needs minimum viable metrics that allow an organization to measure progress and learning while reducing risk from problem definition to scale.

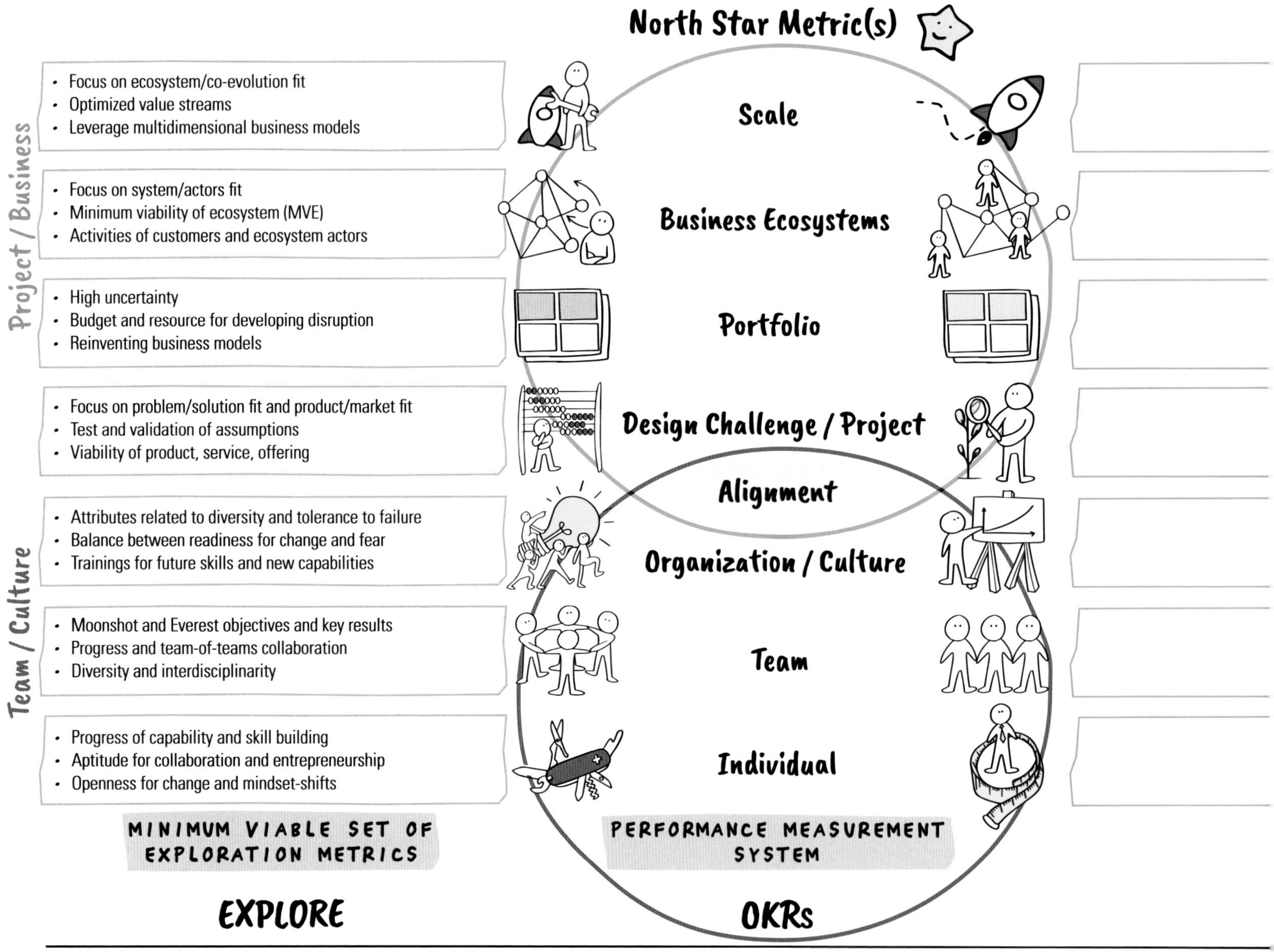
North Star Metric(s)
Project / Business
Focus on ecosystem/co-evolution fit
Optimized value streams
Leverage multidimensional business models
Scale
Focus on system/actors fit
Minimum viability of ecosystem (MVE)
Activities of customers and ecosystem actors
Business Ecosystems
High uncertainty
Budget and resource for developing disruption
Reinventing business models
Portfolio
Focus on problem/solution fit and product/market fit
Test and validation of assumptions
Viability of product, service, offering
Design Challenge / Project
Alignment
Team / Culture
Attributes related to diversity and tolerance to failure
Balance between readiness for change and fear
Trainings for future skills and new capabilities
Organization / Culture
Moonshot and Everest objectives and key results
Progress and team-of-teams collaboration
Diversity and interdisciplinarity
Team
Progress of capability and skill building
Aptitude for collaboration and entrepreneurship
Openness for change and mindset-shifts
Individual
MINIMUM VIABLE SET OF EXPLORATION METRICS
PERFORMANCE MEASUREMENT SYSTEM
EXPLORE
OKRs

- Network effects and exponential growth
- Leverage of digital, physical, and hybrid touchpoints
- Scalable processes, IT, data analytics

- Adaption of ecosystem and co-evolution of actors
- Building up and leveraging of ecosystem capital
- Enhancement and expansion of core value proposition

- Low uncertainty
- Budget and resources for developing the core
- Sustaining business models

- Focus on the systematic improvement
- Based on forecasts and more detailed plans
- Achievement of growth targets or efficiency gains

- Driven by automation and process efficiency
- Balance between agility and discipline
- Culture of continuous improvement

- Roofshot objectives and key results
- Efficiency gains and reorganization
- Standardization and specialization

- Individual contribution to the team
- Aptitude for efficiency and automation
- Openness for exponential growth

APPROPRIATE MIX OF RELEVANT EXPLOITATION METRICS

EXPLOIT

Building Blocks of the Innovation Measurement System (IMS)

Levels for Defining Metrics and Examples

The different levels in the illustration show how complex and comprehensive the management and measurement of creativity, products, and business success are. The focus in this book is on the largely ill-defined measurements, which represent the greatest challenge in the context of EXPLORE. Likewise, different approaches are presented to appropriately select the more often well-defined measurements, in the context of EXPLOIT. Another central element relates to the tools and methods that are suitable for the implementation of performance measurement systems for innovation and design thinking teams. Performance measurement systems, minimum viable exploration metrics, and exploitation metrics are briefly defined next.

Performance Measurement Systems

In terms of measuring and breaking down strategic goals and priorities, there are different approaches. One way is to work with objectives and key results. This approach is grounded in having a concrete North Star, clearly defined ambitions, or the corresponding mandate to the teams to consciously radically rethink things. Good performance measurement systems support the company in achieving the objectives in both EXPLORE and EXPLOIT. At the team and individual levels, objectives are defined that are action oriented, significant, concrete, and, especially in EXPLORE, inspirational. The respective key results act as benchmarks and monitoring for the teams. Most of the key results are measurable and verifiable. This book places a special focus on the OKRs for innovation teams, which often work with stretch goals for the achievement of Moonshot and Everest objectives. However, the applied system for performance objectives must also fit the culture of the company. With the appropriate performance measurement system, design thinking and innovation teams are able to cascade and align goals to different levels of an organization, defining outcome-based key results that help verify the innovation success of the objective. Moreover, a wider acceptance guides daily work and connects all employees to a larger purpose. Many companies add powerful performance measurement systems to the measurement of EXPLORE and EXPLOIT (see the next page).

Minimum Viable Set of Exploration Metrics

It is not easy to measure EXPLORE activities and impacts, but it is a necessity that the metrics applied by decision makers, practitioners, and design teams on a daily basis should have more impact. The upshot is to start with a set of minimum viable exploration metrics that are known to immediately derive initial indications for concrete action. Some parts of these experiments with measurements will succeed and others will fail. However, the approach is appropriate because it fits well into the current belief that experimentation is part of creating value in a dynamic environment.

As the metrics are dynamically adjusted over time, the measurement system becomes more robust and can be expanded based on initial data points. Metrics that do not generate valuable information and do not contribute to innovation success are discarded. In this way, the measurement system can be continuously developed to suit each company, organization, or team. Of course, this always needs to take into account the respective objectives, from the long-term increase in the value of the company to the development of new capabilities or the transformation of new forms of collaboration.

A minimum viable set of exploration metrics is an efficient core piece of a potential innovation measurement system. Initially, it should be assumed that stakeholders, decision makers, and teams have a certain need, and every metric, measurement, and data point will help them to make better decisions in the short and long term.

A first minimum viable set of exploration metrics, co-created with or exposed to teams, might begin to measure the validated lessons learned and provide crucial feedback on individual items, as well as capture combined values derived from the metrics. Based on the information, the metric(s) will either remain, be adapted, or be discarded. Over time the data points will help to convert the most powerful metrics and measurements into more sophisticated systems.

Appropriate Mix of Relevant Exploitation Metrics

Measuring EXPLOIT seems to be easier because many data points already exist from customer interactions, sales, and operational activities. However, the challenge remains to select the appropriate mix of relevant exploitation metrics and include lagging with leading indicators.

In the context of EXPLOIT (i.e., after the launch phase of product-/market fit or systems-/actors fit in the case of business ecosystem initiatives), an appropriate mix of relevant exploitation metrics is increasingly applied, as these metrics are common indicators for business growth, quantity, and the basis for scaling and exponential growth.

Common quantifiable metrics include ratios indicating the percentage of successfully met financial ambitions; the percentage of sales from new products in the past n-years; the return on investment in design thinking and innovation activities; the percentage of profits from new users, customers or categories; and customer satisfaction/net promoter scores. In addition, all metrics related to the scalability of IT or the performance of real-time data analytics are of interest.

For an appropriate mix of exploitation metrics, it is in many cases valuable to question classic KPIs to see if health metrics are not the better option of measurement to derive the desired action. As the name implies, health metrics describe a business's vital signs, much like physical vital signs. Especially in turbulent times, health metrics can be the better choice, as they help to find out how resources and time are used. It is important to select the appropriate metrics, which are meaningful and really help to improve organizational efficiency, scalability, and processes.

The input variables for exploitation metrics often come from financial controlling, sales, HR, IT, operational excellence process monitoring, or external data sources.

In the (re)definition of the metrics, it is recommended to avoid using vanity metrics, even if they are the easiest way to indicate results. Unfortunately, vanity metrics still appear in innovation accounting and controlling books and in wider management literature. This results in vanity metrics still being used for innovation in many companies.

Recommended Procedure:

- Explain the purpose and goals: Is the focus of the defining and measuring to develop specific measures for EXPLORE or EXPLOIT?
- Be clear about what is to be measured: Measure at various levels as appropriate, reaching out from individual, team, culture, project portfolio, or ecosystem to scale.
- Work with minimum viable exploration metrics before metrics are fixed in a rigid measurement system: Apply the cycle of measuring, evaluating, and operationalizing
- Find alignment between business and teams: Very vague objectives and key results without alignment to a North Star can lead to wrong actions.
- Measurement can be done at different levels: Often, even an initial measurement at the tool level helps to derive initial insights and actions.

Avoid the Typical Vanity Metrics

- Number of ideas
- Number of patents
- R&D spending
- Number of employees in R&D

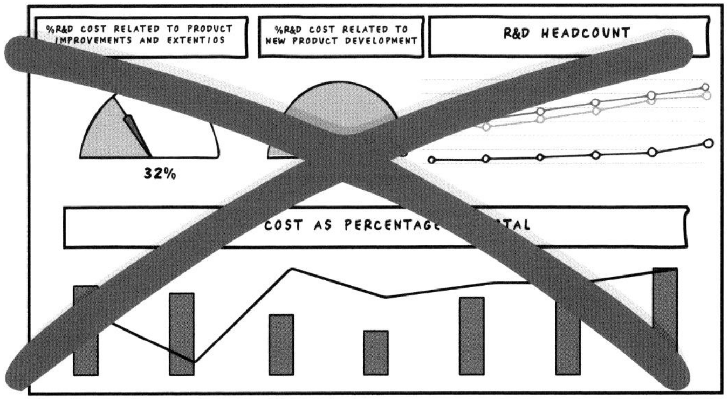

%R&D COST RELATED TO PRODUCT IMPROVEMENTS AND EXTENTIOS
%R&D COST RELATED TO NEW PRODUCT DEVELOPMENT
R&D HEADCOUNT
32%

Vanity metrics feel good, are great for marketing, and are easy to visualize. In large organizations, these are often called activity metrics to demonstrate how busy the teams are and to justify the innovation theater. However, for most cases they do not give a clear indication of concrete actions.

WHAT Tools, Methods, and Mindsets Are Required to Become Successful?

The introduction to design thinking (see pages 61–72) and measuring (see pages 146–156) is deliberately kept short in this book to give more space and room for the applications, tools, and measurement instruments that help to question or expand existing approaches. For a better understanding and a common language, the basics regarding the design thinking process, mindset, and the most important tools are briefly addressed in the first part of this book. For the same reasons, an overview of the concept of measurements, metrics, and statistical models will be given, since we know but rarely use many models from Moore's law to Monte Carlo simulations for modeling uncertainty in decisions. Especially in the context of measuring innovation, the ultimate outcome must be to support decisions that result in guidance, actions, and transparency.

The performance and innovation objective approach in this book reflects on well-known concepts ranging from the OKR movement to FAST goals and provides a selection of hands-on tools for defining objectives and breaking them down into missions and tasks.

All proposed concepts and tools have been tested and implemented in various initiatives in recent years, often in a combination of building new capabilities, initiating transformation, and creating tangible outcomes for business growth. However, it is important to be mindful while implementing any of them because every organization, culture, and team is different. Not all organizations and settings are ready to implement the same mindset and measurement approach that most innovative companies have been cultivating for decades. In addition, the ways we collaborate, define success, and impact the world and society are also constantly changing.

A state-of-the-art innovation measurement system should allow prudent risk-taking, embedded in a culture and mindset that is tolerant of risk and failure.

In this decade, new goals such as carbon neutrality, sustainability, geo-political challenges, and responsible growth are on the corporate agenda more often than ever before. In parallel, the demands, capabilities, and desires of employees with regard to the working environment have changed radically, impacting not only how innovation and objectives are measured, but also how companies align and prepare themselves for the future.

Employees, teams, and entire organizations are taking the next step in embracing value-creating activities and contemporary skills.

Key interpersonal skills include capabilities that help strengthen relationships with customers and colleagues. This includes many elements that are part of the design thinking mindset, such as the concepts of building empathy, trusting in the work of other teams, and putting people and their needs at the center of thinking. In the area of collaboration with other teams and the realization of the team-of-teams approach, empowerment through radical collaboration is just as important as basing team composition on different T-shape profiles and different thinking references. This promotes integration and motivates the collaboration of different personalities.

From the point of problem definition all the way to the first prototypes and design of an entire business ecosystems, cognitive abilities are also of great importance. These range from a new mental flexibility, which is needed to apply design thinking, systems thinking, and data analytics in varying states of mind throughout the entire design cycle, to the ability to imagine things and to leave the predefined creative frame.

The ability to question existing systems becomes central, not only to define the right metrics but also to be able to solve wicked problems and understand biases. The ability to communicate results is also becoming increasingly important, in the form of storytelling and visualizations based on data analytics and transporting core statements from in-depth interviews or tests of initial prototypes with potential customers. However, this also requires the ability to formulate powerful questions, to set up in-depth interviews, and finally to listen actively.

In the context of state-of-the-art employees, target and performance management where teams and individual employees set their own objectives, it is increasingly important that employees master self-leadership and self-efficacy. This requires self-confidence and a good understanding of one's own strengths, as well as the ability for self-control and regulation. For activities in the EXPLORE area, it is also important that employees work with energy, optimism, and passion on the objectives and missions. In this context, it must also be allowed that accepted beliefs are broken, change is actively shaped, and risks are taken within the defined design principles agreed upon by the team and project sponsors. For the self-imposed objectives, however, the employees and teams must then also assume the corresponding ownership and learn to deal with the uncertainty and, in case of doubt, to manage this uncertainty by making decisions using the appropriate methods (qualitative and quantitative).

A good understanding of the associated enabler technologies, digital fluency, and an understanding of digital systems is self-evident, especially because there will be few innovations in the future that do not have a connection to digital components or that innovate or predict based on data.

Design thinking and innovation teams are encouraged to expand their capabilities, adapt their ways of working, and create value supported by automation and artificial intelligence.

The design thinking mindset and the future skills of employees are just as important as the methods and tools they apply for finding the appropriate problem to solve, identifying appropriate metrics for steering decisions, and applying varying design lenses.

Approaches such as a hybrid model that consists of a combination of data analytics and design thinking can create new combined data from human insights and data insights and support tapping into a collective intelligence. Such a hybrid model includes powerful design thinking elements which drive actions through compelling "show, do not tell" approaches with data visualization and tangible prototypes at the same time. However, curiosity remains the main driver for both, tapping into data and human insights.

The skills and mindset described are helpful in a variety of situations, but they are essential in times of massive uncertainty and change.

Along with the application of design thinking in combination with data analytics, completely new ways of applying the design thinking mindset have emerged over time. Today, it is for example the basis for the development of AI solutions. The breadth and depth of designing AI solutions is increasingly complex; it requires tools, methods, and a mindset that design thinking can provide. In the design of the human-machine relationships and their functioning as a team, design thinking plays an especially important role. Interdisciplinary teams with data scientists, design thinking experts, and software developers continuously deal with interaction, change, and increased automation. In addition to functionality and efficiency, the most important elements, the customer focus and experience, remain at the center of all considerations.

What will the future bring?

Superior AI systems can achieve higher-quality outcomes faster than humanly possible.

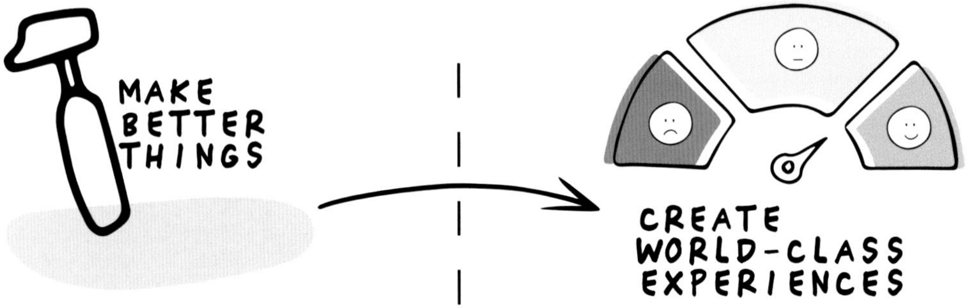
MAKE BETTER THINGS
CREATE WORLD-CLASS EXPERIENCES

In addition to big data analytics, neuroscience and, in connection with innovations, neurodesign are playing an increasingly important role. Neuroscience can also be considered the root of today's artificial intelligence. It is becoming increasingly evident that neuroscience can enrich many topics related to AI and the work of design thinking and innovation teams.

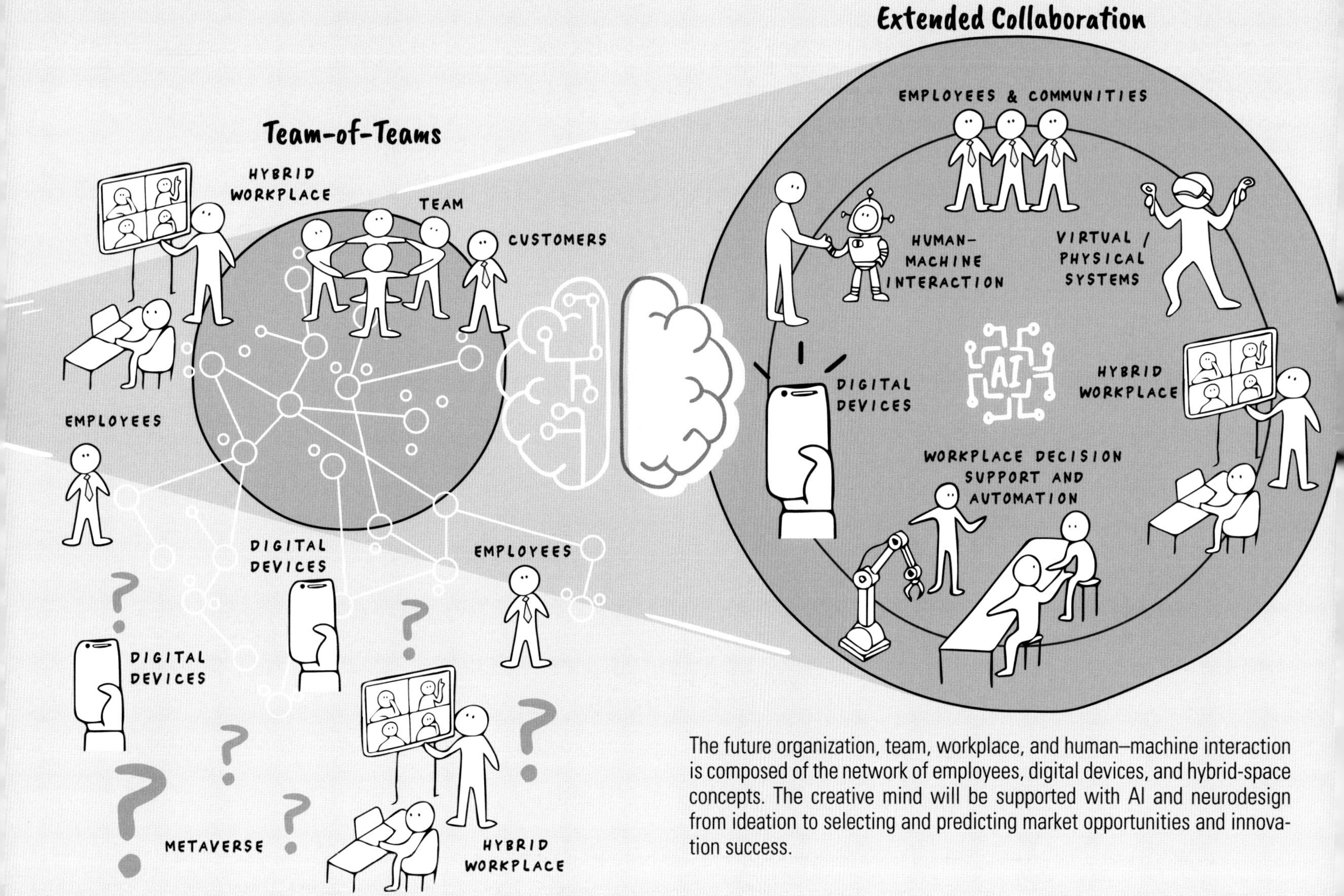
Extended Collaboration
Team-of-Teams
HYBRID WORKPLACE
TEAM
CUSTOMERS
EMPLOYEES
DIGITAL DEVICES
EMPLOYEES
DIGITAL DEVICES
METAVERSE
HYBRID WORKPLACE
EMPLOYEES & COMMUNITIES
HUMAN-MACHINE INTERACTION
VIRTUAL / PHYSICAL SYSTEMS
AI
DIGITAL DEVICES
HYBRID WORKPLACE
WORKPLACE DECISION SUPPORT AND AUTOMATION

The future organization, team, workplace, and human–machine interaction is composed of the network of employees, digital devices, and hybrid-space concepts. The creative mind will be supported with AI and neurodesign from ideation to selecting and predicting market opportunities and innovation success.

To the Point!

Competitive advantages evolving rapidly across all industries and sectors, design thinking, and meaningful measurements are key for creating new and viable products, services, business ecosystems, and models. Innovation needs purpose, objectives, and key results to create impact for customers, people, and society.

Many levels and parameters have to be considered in defining a solid measurement system. Metrics based on the one-size-fits-all pattern often do not achieve the desired goal. In contrast, powerful measurement systems are aiming to steer, reflect, and adjust activities toward the desired future outcomes.

However, the design thinking mindset and the future skills of employees are just as important as the methods and tools they apply over the entire design cycle.

A QUICK READ TO GET INSPIRED

A Journey of Thought ....
into the Jungle and the Superhighways of Corporations

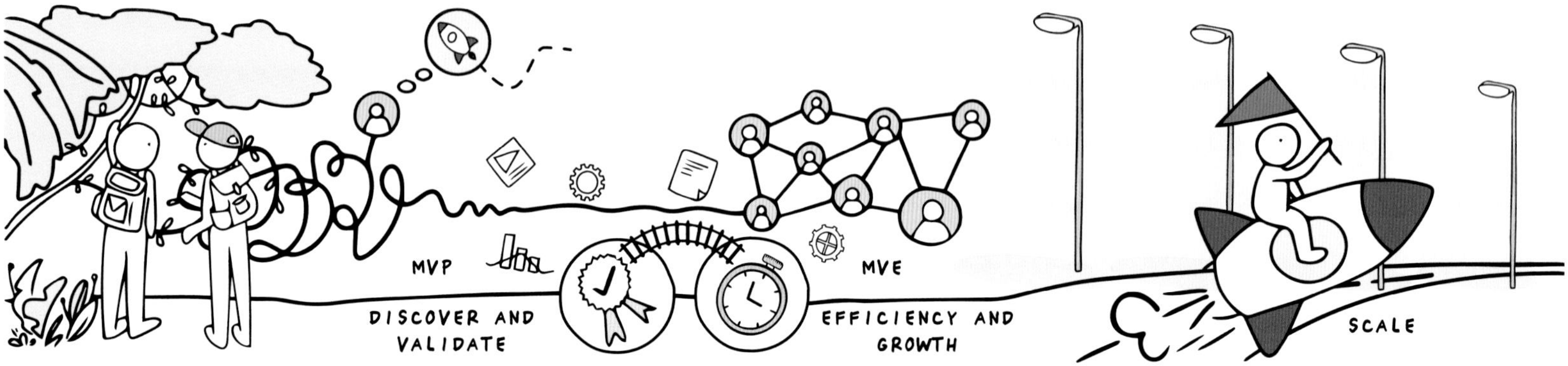
MVP
DISCOVER AND VALIDATE
MVE
EFFICIENCY AND GROWTH
SCALE

This short thought journey provides a better understanding of design thinking and innovation metrics. It briefly summarizes the most important characteristics and typical activities to get a quick overview into the topic. We will use the jungle as an analogy for the ambiguity of design thinking and the superhighway as an analogy for efficiency-driven organizations. It seems particularly important to mention that we need both, but should be very mindful of which metrics we apply and when. In addition, a moderated transition from EXPLORE to EXPLOIT is needed to ensure the right measure of the innovation work, later implementation, and scaling. Furthermore, this quick read contains seven tips and tricks to create, build, and gradually establish an innovation engine and measurement system, including an example of how the result of a traditional measurement system can be transferred to a more modern approach. In addition, the importance of North Star metrics in defining OKRs is introduced as well as how measurement systems might evolve over time.

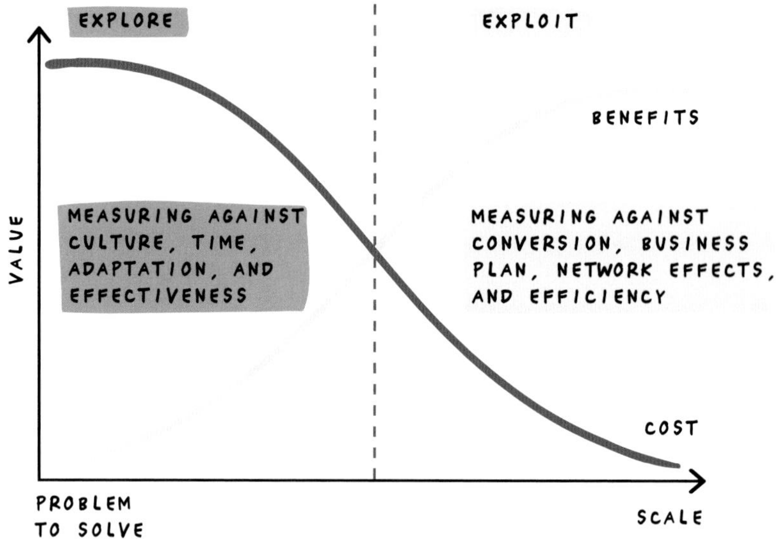
EXPLORE
EXPLOIT
BENEFITS
MEASURING AGAINST CULTURE, TIME, ADAPTATION, AND EFFECTIVENESS
MEASURING AGAINST CONVERSION, BUSINESS PLAN, NETWORK EFFECTS, AND EFFICIENCY
VALUE
COST
PROBLEM TO SOLVE
SCALE

This quick read should inspire you to dive deeper into the topic of design thinking and innovation metrics and at the same time offer a quick access to the broader topic.

EXPLORE and EXPLOIT are based on different values and activities, so the innovation and scaling engines need different metrics to match their respective mindset and activities. EXPLOIT feels like driving a car on a well-lit "superhighway," with cruise control and adaptive driving programs, thanks to intelligent assistance systems.

EXPLORE, on the other hand, feels more like being exposed to the jungle, far away from the known civilization and without a path to get to the destination, where many things are unknown, and an ad hoc reaction to the respective situations is necessary.

This is exactly where the challenge lies in measuring the activities around problem solving, creativity, and the impact of innovation work. The measurement is situationally different and a comparison with the activities on the superhighway is not adequate.

Only much later, after countless deep interviews with customers/users, shared insights, formulated point of views, built prototypes, and iterations of individual functions to the complete value proposition, there is a transition to the measurement systems and efficiency-driven superhighways that allows team and organizations to scale the market opportunity.

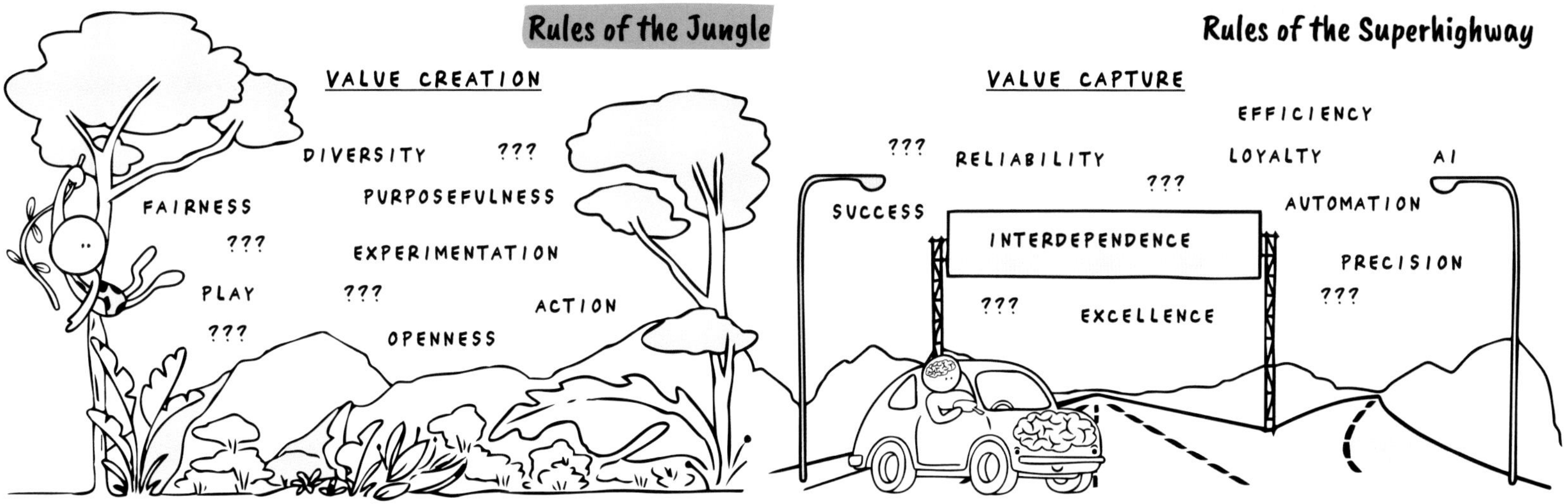
Rules of the Jungle
VALUE CREATION
DIVERSITY
???
PURPOSEFULNESS
FAIRNESS
???
EXPERIMENTATION
PLAY
???
ACTION
???
OPENNESS
Rules of the Superhighway
VALUE CAPTURE
EFFICIENCY
???
RELIABILITY
LOYALTY
AI
???
SUCCESS
AUTOMATION
INTERDEPENDENCE
PRECISION
???
EXCELLENCE
???

TYPICAL ACTIVITIES IN EXPLORE

- Breaking rules and dreaming
- Opening doors and listening
- Trusting and being trustworthy
- Seeking fairness, not advantage
- Experimenting and co-creating together
- Making mistakes, failing, and persevering
- Asking if it is worth it

TYPICAL ACTIVITIES IN EXPLOIT

- Outstanding in your job
- Being loyal to your team
- Working with people you can rely on
- Seeking competitive advantages
- Getting the job right the first time
- Striving for perfection
- Returning favors

Top Five Challenges for Companies and Leadership Teams Aiming to Perform in EXPLORE and EXPLOIT

The analogies to the jungle and the superhighway are reflected in the top five challenges that can be observed in many companies. This is one of the many reasons that there is a need for a mindset, approaches, and metrics that can address these dynamics.

1 Jumping into solutions without understanding the problem!

Beginning at the beginning for both creating new market opportunities and measuring innovation

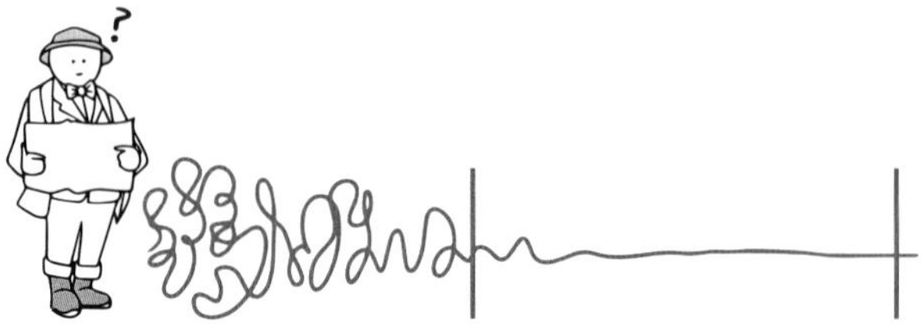

2 Having one culture, measurement system, and mindset across all functions and teams!

Building awareness for different settings and activities in EXPLORE and EXPLOIT

3 Outdated measurement systems bump into iterative, customer-centric, and agile customer development!

Accepting the concept of minimum viable exploration metrics

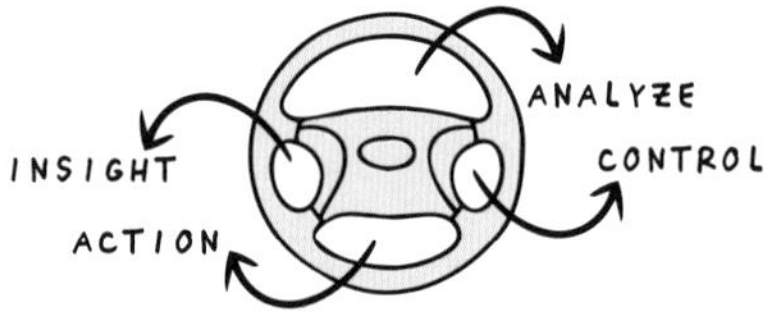
ANALYZE
CONTROL
INSIGHT
ACTION

4 Implementers and scalers are not involved early in the EXPLORE process to make the transition part of how to bring it home.

Forming a transition from EXPLORE to EXPLOIT and the relevant measurements

5 No freedom, budgets, and time for teams to run through the entire macro-cycle in realizing new market opportunities.

Guiding the teams to optimize the core and declare and empower breakthrough thinking

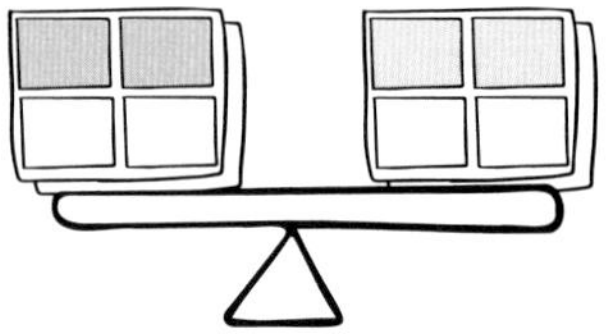

The Path from Status Quo to Superior Innovation Measurement Systems

It is recommended to consistently apply the iterative approach based on the design thinking mindset for initial development as well as for the continuous further development of the innovation measurement system (IMS). Design thinking is not only suitable for the development of products and services but can also be used for the design of processes and entire systems. The terms used, such as the use of minimum viable exploration metrics, suggest that typical iterations such as build, measure, and learn can be adapted and applied to the design of metrics (see page 175). In addition, there are well-selected EXPLOIT metrics (see page 187), which, for example, monitor the general "health" of the enterprise and are therefore also referred to as "health metrics" in this book about design thinking and innovation metrics.

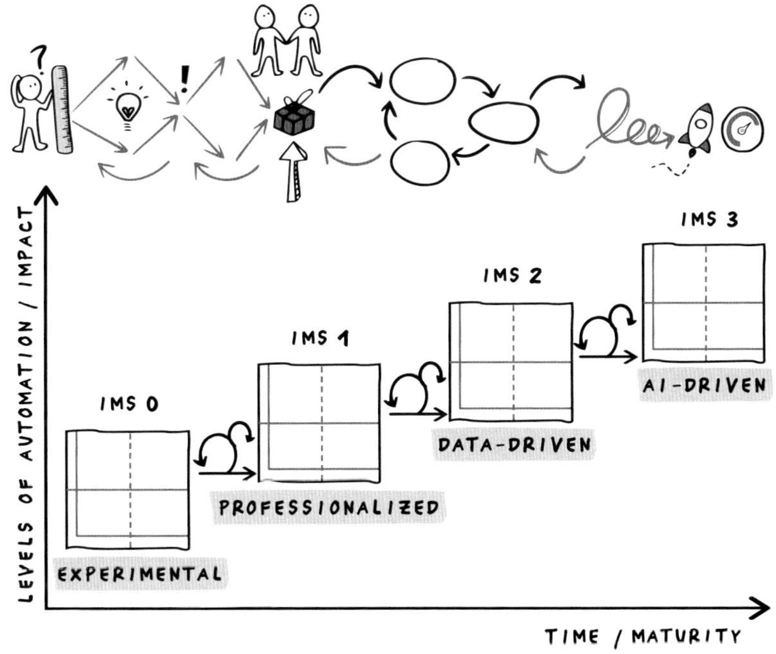

LEVELS OF AUTOMATION / IMPACT
IMS 0
EXPERIMENTAL
IMS 1
PROFESSIONALIZED
IMS 2
DATA-DRIVEN
IMS 3
AI-DRIVEN
TIME / MATURITY

The North Star

Important inflection points are anticipated via sigmoidal curves (S-curves) and automatically trigger new design cycles of the IMS (see page 167). The strategic inflection points are influenced by new/changing customer behavior, business cycles, capabilities, and technologies. Based on this, the north star also forms the guiding principles for the design thinking and innovation teams, the application of OKRs, the definition of team and individual objectives, and key results within the performance measurement system.

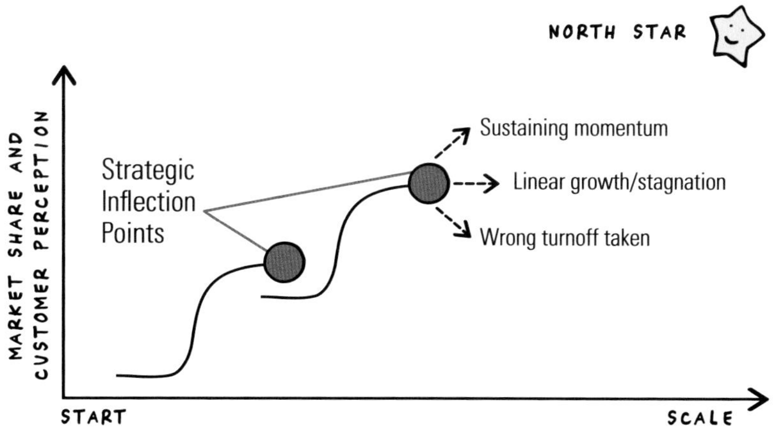

NORTH STAR
MARKET SHARE AND CUSTOMER PERCEPTION
Strategic Inflection Points
Sustaining momentum
Linear growth/stagnation
Wrong turnoff taken
START
SCALE

Performance Measurements

Modern performance measurement systems (see page 226), such as the ones that work with OKRs or FAST goals, allow an alignment of the objectives on the company level to the respective team and individual objectives. The key results serve as team-defined metrics that are measured on a short cycle (for example, quarterly) and when applied properly are an essential part of a dynamic innovation measurement system.

Objectives & Key Results

The respective objectives and key results can be equipped with different ambition levels within the framework of the performance measurement system. For innovation and design thinking teams, bold Moonshot or so-called Everest objectives are often defined. As the name suggests, these are larger projects with a rather radical innovation character, which require the teams to formulate bold and ambitious mid- and long-term objectives. So, it is not surprising that true Moonshots in terms of agreed scoring are usually not 100 percent achieved; scores between 60 and 70 percent are very decent results. Everest objectives are applied, for example, in conjunction with innovative business ecosystem initiatives, with the overall objective of realizing new value propositions, which are new to market and to the customer. Objectives related to Everest adventures provide, in general, more room for creative solutions, and, most of the time, they offer a very exciting, stimulating, and impassioned journey for the teams. In contrast, many operational and short-term improvements might be defined as Roofshot objectives, which are usually 100 percent achievable. In dealing with Moonshot objectives, it is good to give the teams the possibility to prioritize the initiatives based on missions or to equip them with kill metrics (see pages 182–183), which leads to stopping larger innovation initiatives or design challenges when certain requirements, feasibilities, or budget limitations come into play.

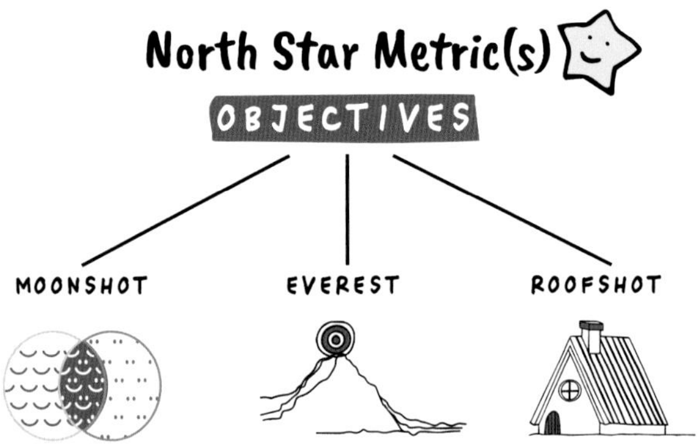
North Star Metric(s)
OBJECTIVES
MOONSHOT
EVEREST
ROOFSHOT

Evolution of Innovation Measurement Systems Metrics into Meaningful Measurement Systems

The individual building blocks (North Star, minimum viable exploration metrics, appropriate set of exploitation metrics, and the use of a performance measurement system) help in the continuous development of the IMS and, at the same time, support efforts toward high-performing teams up to the realization of a team-of-teams organization. In addition, the maturity of the teams and the organization in terms of new capabilities is gradually increased, which is supported by the right activities to increase the innovation and business performance.

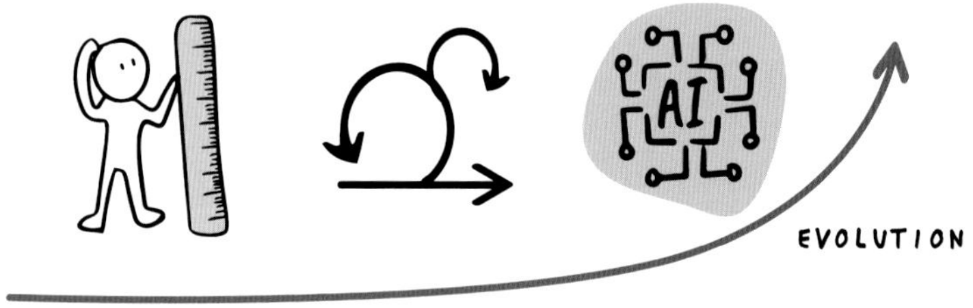
AI
EVOLUTION

By establishing useful measurement points at different levels, meaningful measurements can be operationalized and automated over time. Measurements related to customer interaction and gathering insights, such as the use of neuroscience (see page 369), provide high-value data that helps enrich design thinking activities with insights from data analytics activities (see page 332). The utilization of AI for innovation work and measurement of innovation may currently seem a long way on the journey from IMS 0, to IMS 2 or 3, but the respective iterations and steps are the basis for a superior innovation measurement system and data-driven innovation possibilities.

The book is structured so that the individual building blocks can be used independently, starting with the first experiences with design thinking to reduce uncertainties, including first measurements on the tool level, and up to detailed instructions for building a performance measurement system. Regarding metrics, examples of EXPLORE and EXPLOIT metrics are presented, as well as approaches for measuring the creativity of teams and individuals.

Begin at the Beginning and Go on Until the Exploitation Starts

It is important to identify a suitable metric for each phase and the goals to be achieved. They should be linked to actions that help to improve the activities, measure the impact, or optimize the processes. In most cases, it is not enough to simply copy a predefined measurement system. However, individual metrics can serve as inspiration to experiment with.

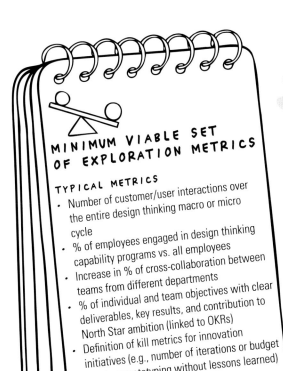

MINIMUM VIABLE SET OF EXPLORATION METRICS

TYPICAL METRICS

- Number of customer/user interactions over the entire design thinking macro or micro cycle
- % of employees engaged in design thinking capability programs vs. all employees
- Increase in % of cross-collaboration between teams from different departments
- % of individual and team objectives with clear deliverables, key results, and contribution to North Star ambition (linked to OKRs)
- Definition of kill metrics for innovation initiatives (e.g., number of iterations or budget spent on prototyping without lessons learned)

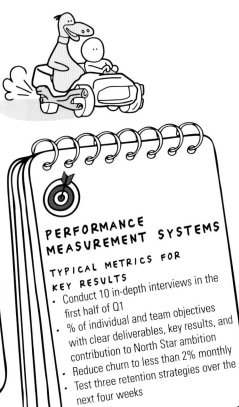

PERFORMANCE MEASUREMENT SYSTEMS

TYPICAL METRICS FOR KEY RESULTS

- Conduct 10 in-depth interviews in the first half of Q1
- % of individual and team objectives with clear deliverables, key results, and contribution to North Star ambition
- Reduce churn to less than 2% monthly
- Test three retention strategies over the next four weeks

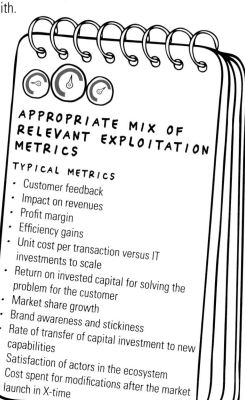

APPROPRIATE MIX OF RELEVANT EXPLOITATION METRICS

TYPICAL METRICS

- Customer feedback
- Impact on revenues
- Profit margin
- Efficiency gains
- Unit cost per transaction versus IT investments to scale
- Return on invested capital for solving the problem for the customer
- Market share growth
- Brand awareness and stickiness
- Rate of transfer of capital investment to new capabilities
- Satisfaction of actors in the ecosystem
- Cost spent for modifications after the market launch in X-time

GOLDEN RULE: START APPLYING THE METRIC WHEN IT MATTERS, AND NOT TOO SOON!

Tips and Tricks to Create, Build, and Gradually Establish an Innovation Engine and Measurement System

On the journey to a suitable measurement system, principles and tips can be helpful. The outlined seven tips and tricks from the corporate practice help to make the start with measurement as easy as possible.

1: It is hard to measure what hasn't happened – focus on the team and capabilities

- Start with a team that has the right attitude, mindset, and diverse skill set to help solve wicked problems.
- Focus on a minimum viable set of exploration metrics with the team, for example, focusing first on capability building and training innovation skills to survive in the jungle together.

2: One size does not fit all – distinguish between EXPLORE and EXPLOIT

- Metrics used to evaluate the existing core business usually have no relevance to the design thinking and innovation work in EXPLORE.
- Experiment and create a map of exploration metrics and measurements that fits the culture, maturity level, objectives, and ambition of the organization.

3: Keep it simple - create acceptance in the team / organization

- In EXPLORE, avoid collecting all kinds of data and measuring just to have measured something.
- Use metrics that are simple, meaningful, and intuitive in the context of the design challenge at hand or that focus on collaboration efforts.
- Metrics have the greatest impact when they are understood and accepted by the team.

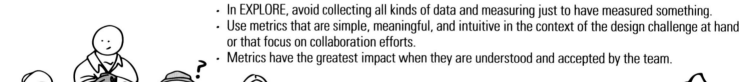

?

#4: Leverage existing measurement systems – focus on the relevant metrics

- Historically, organizations have often grown measurement systems with a strong focus on Exploit.
- Constructive reflection and step-by-step adjustment should be aligned with the development of new capabilities, cultural change, and transformation.
- Fewer and modified metrics often find more acceptance in an early phase than radically new or additional metrics

#5: Don't be afraid to modify or adapt a metric – love it, change it, or leave it

- Gradually establish a culture where metrics are perceived as something vibrant.
- For teams that are evolving to a higher design thinking maturity, metrics should reflect that maturity.
- If a particular metric does not provide the insight in the thick of the jungle or the guidance hoped for, it should be adjusted or replaced.

#6: Focus on the customer with the metric – create actions from measurements

- Measure also aspects of customer orientation, such as sales of new products, to supplement the internally oriented metrics. If the market opportunity has no significance for the customer, it must be realigned.
- An initial simple dashboard with the most important and meaningful metrics contributes significantly to building a changed culture.
- Use the attention of both individual team members and leaders to focus on outcomes from which appropriate activities can be derived.

#7: Don't let the wrong metrics stifle innovation – create transition into exploitation

- Let everyone know that the team and the organization are on the right track, and make sure that there is a transformation from the jungle to the superhighway, and that the measurement system will change accordingly.

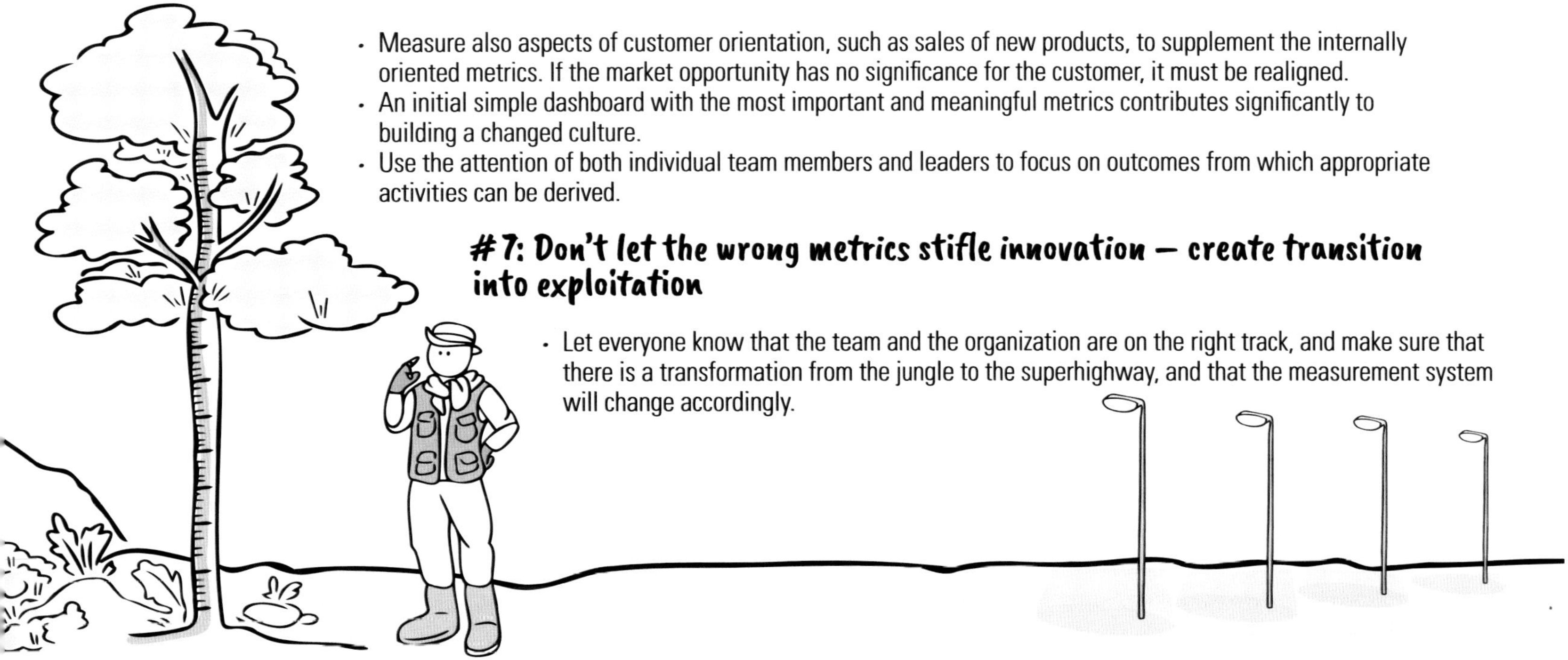

Transforming Traditional Metrics into Meaningful Measurement Systems

Most traditional measurement systems have a strong focus on performance outcomes per gate or market launch because they are relative straightforward to track. However, those may not always be the most useful or relevant metrics in the realization of radical new market opportunities and the objective to change behavior and culture as well. This means that neither important elements of innovation work, such as cross-functional collaboration, nor learning are taken into account. State-of-the-art measurement systems usually follow the logic of the Problem to Growth and Scale Framework, which mitigates risks from customers and market validation. Stage-gate processes, on the other hand, are usually based on classical input variables and a collection of many ideas (based on assumptions without customer validation), which are processed on the basis of criteria per gate.

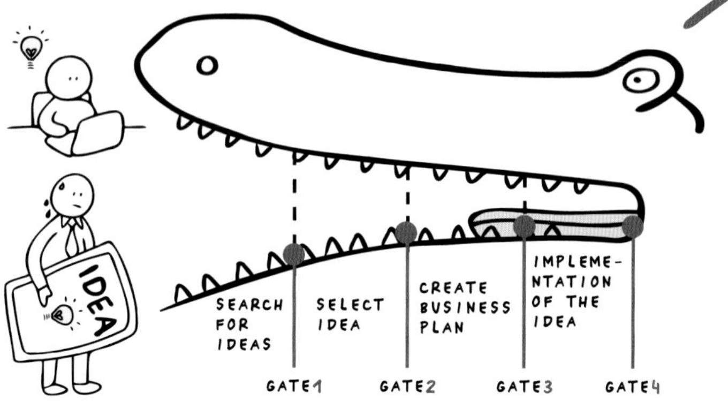
IDEA
SEARCH FOR IDEAS
SELECT IDEA
CREATE BUSINESS PLAN
IMPLEMENTATION OF THE IDEA
GATE 1
GATE 2
GATE 3
GATE 4

Example of a typical traditional innovation measurement system

INPUT

- Spending (percentage of sales)
- Human resources devoted to innovation
- Pipeline of ideas/concepts
- Number of projects in active development
- Percent of ideas/concepts from outside the firm
- Ratio of ideas from inside/outside

> A stage-gate process is lethal for designing radical new market opportunities: promising options are discarded before they are properly explored.

PROCESS EFFECTIVENESS

1. Development activities
- % hitting gates on time
- % meeting quality guidelines

2. Patenting activity
- # patents filed
- # patents commercialized
- % ideas covered by patents

3. Budget versus actual
- Time
- Cost/investment

4. Average time to market
- # of new products launched
- % of projects that are major improvements

PERFORMANCE OUTCOMES

- % of sales from new products in the past N years
- Success ratio (percentage of meeting financial goals)
- Revenue growth
- Return on investment in innovation (ROIC)
- % of profits from new customers (or occasions)
- % of profits from new categories
- Average time to break-even/cash
- Customer satisfaction
- Profit growth due to new products/services
- Percent of profits from new products in a given period
- NPV of portfolio
- Potential of portfolio to meet growth targets

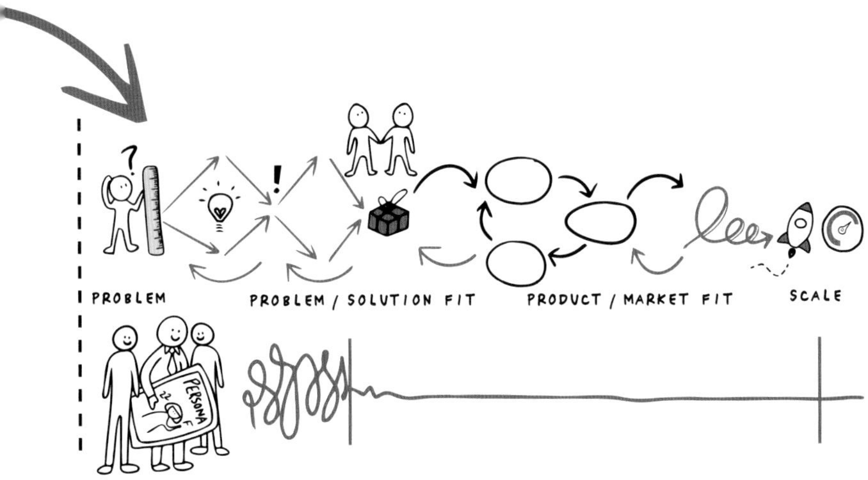
PROBLEM
PROBLEM / SOLUTION FIT
PRODUCT / MARKET FIT
SCALE
PERSONA

A modified innovation measurement system replaces or adds a set of minimum viable exploration metrics. Even the transformation of shifting toward actionable metrics might be a very strong metric for a certain time because it indicates the speed of transforming the organization into a new mindset. Many of the measures below have the objective of promoting the right behavior in order to further innovation over the entire design cycle. Some incentives even diffuse through every level of the organization, so that fresh eyes and minds are involved and can use their subject matter expertise to achieve innovation. These stand in stark contrast to the previously outlined traditional measures, which mainly focused on what had been achieved and provided almost no framework for actions or improving the culture, procedures, and processes accordingly.

Example of a modified innovation measurement system

DESIGN CHALLENGE / PROJECT

- % of new customer problems solved versus identified in X-time logged
- Tested versus untested assumptions the design cycle
- Cost spent for modifications after the market launch in X-time

SCALE

- # of customer adoption in X-time versus market size
- Customer stickiness versus all customers in X-time
- Unit cost per transaction versus IT investments to scale

PORTFOLIO

- Innovation portfolio balance (projects related to EXPLORE versus EXPLOIT)
- % of product/service Innovations to market share in X-time
- Average time spent for adopting to unexpected strategic changes

BUSINESS ECOSYSTEMS

- Real Options: market opportunities versus relevant opportunities
- Engagement level ecosystem actors contributing to proposition
- Actual versus realization of ecosystem capital in X-time

ORGANIZATION / CULTURE

- % of employees engaged in capability programs versus all employees
- % of intrapreneurs versus innovation projects per business unit
- # of employees using free-time-to-experiment
- % of employees motivated by intrinsic motivation

TEAM

- Degree of trust between teams and top management
- Number of conflicts within the team that were resolved constructively

INDIVIDUAL

- Tracking of individual's confidence level [...]
- .....

To the Point!

Build awareness that EXPLORE and EXPLOIT are based on different values and activities, and therefore the innovation and scaling engine requires different metrics in each case.

Start to experiment and apply each metric when it matters and not too early in the design cycle. Create an appropriate and dynamic way of measuring innovation for your team, organization, or company.

A good way to start is to work with a minimum viable set of exploration metrics, for example, focusing first on the training and development of design thinking capabilities or on measurement and data points at the tool level.

Do not be afraid of adjusting the existing measurement system. Most existing systems are not corresponding to the way innovation work should be performed nowadays.

THE 101s

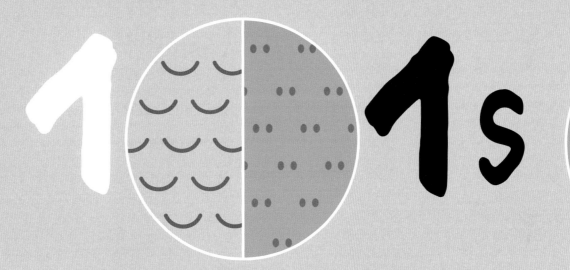

ESSENTIALS OF DESIGN THINKING & MEASURING

ONE-oh-ONE

The two 101s, Design Thinking 101 and Measuring 101, are must-reads into the topics of design thinking and innovation metrics. They provide the most important concepts everyone working nowadays in the arena of innovation and business growth should understand. The 101s have all the basic principles and concepts that are expected for professional and extended design thinking teams.

Even for very experienced design thinking facilitators and innovation accounting cracks, the 101s will give new impulses or broaden views. This initial collection of introductory materials to the topic will also be very helpful for fully understanding the concepts presented later in the 201 and 301 sections.

The design thinking mindset has become a mainstay in many teams, organizations, and companies around the globe. Centered around empathy with the customer/user, creativity, and a strong focus on finding the appropriate problem before jumping into solutions, design thinking is seen today as a paramount paradigm for driving innovation, digital transformation, organizational design, and sustainable business growth.

At the same time, there is a lack of consistent and appropriate approaches for helping novices learn to speak metrics. There is also a reflection that hands-on interaction with data is often more valuable than the repetition of statistical concepts. In this book, data literacy in relation to innovation is more about the continuous learning process that creates the ability to recognize, understand, interpret, create, communicate, and compute information. This leads to better insights that help to define the right actions. For this reason, this book is only going to touch very lightly on the subjects of mathematics and theorems.

However, the applied knowledge and the ability to design minimum viable metrics and select the appropriate metrics for exploiting and scaling should be part of the necessary set of basic knowledge of any employee working in an agile environment in an extended innovation team setting. This should be a part of the work and reflection of the design, implementation, and scaling team.

In addition, many companies are still facing data challenges in managing their next steps in digitization. These data challenges are related to people and how they interact with data analytics and related technologies, including leadership, talent management, and decision makers. That's why this book also aims to close the gap of applying design thinking and data analytics in a hybrid model of design that is human led and data driven at the same time.

Shall we fall in love?
Let's pivot first!

The 101s aim to provide a basic understanding of the design thinking paradigm and the impact of applying it. In addition, they provide guidance in how to measure and create appropriate metrics and innovation measurement systems.

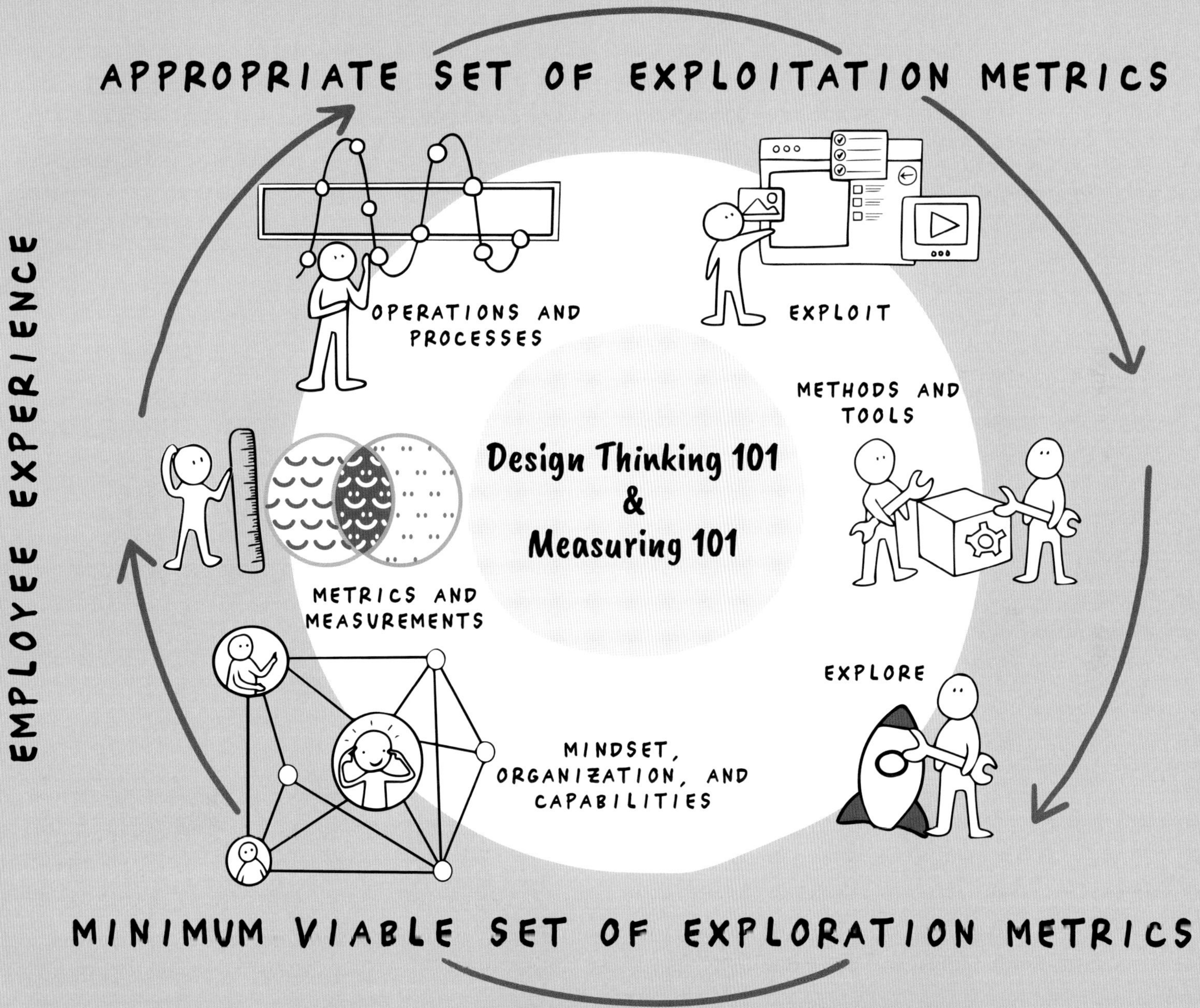
APPROPRIATE SET OF EXPLOITATION METRICS
OPERATIONS AND PROCESSES
EXPLOIT
METHODS AND TOOLS
EMPLOYEE EXPERIENCE
CUSTOMER EXPERIENCE
Design Thinking 101 & Measuring 101
METRICS AND MEASUREMENTS
EXPLORE
MINDSET, ORGANIZATION, AND CAPABILITIES
MINIMUM VIABLE SET OF EXPLORATION METRICS

DESIGN THINKING 101

The Design Thinking Paradigm

Begin at the Beginning

Design thinking has evolved in recent decades. Originally, it was specifically focused on the re(design) of products, services, and processes. Today design thinking is a mindset that is the basis for business model innovation, value proposition design, and the design of entire business ecosystems within the framework of design thinking for business growth.

However, design thinking also supports digital transformation and testing new organizational models in which radical collaboration takes place across the known silos in the company, ultimately establishing the team-of-teams concept. Design thinking is increasingly influencing strategy work in which strategic directions and ambitions are iteratively developed on the basis of customer needs.

Especially where offers that are new for the customer and new for the market are defined, Design Thinking can unleash its full potential. In addition, many companies have realized that agile alone is not enough. Teams need to understand where a problem originates, what the customer needs are, and what the purpose of the potential solution is. This means that beginning at the beginning becomes the credo for the successful realization of market opportunities and exponential growth.

This is one of the reasons why design thinking has become an incredibly popular approach in small and large organizations. However, it requires a certain type of mindset. Everyone who wants to re(design) their work and create better products, services, or new experiences for their users/customers might benefit from applying certain attitudes that can empower their thinking and creativity.

Too often, design thinking is assumed to be part of the agile toolbox, and companies realize too late that their teams need additional skills in exploring the problem space.

Agile
This consultant is here to install agile.
Ah, finally!
...9 MONTHS LATER
Agile
The agile methodology is now installed!
Unfortunately, it won't work until your teams understand the customer problems they have to solve and apply a design thinking mindset.
AGILE TOOLS
MINDSET
13%
87%

The design thinking mindset is based on the core beliefs that the teams in the exploration of the problem space and solution space rely on an overarching collaboration and focus on thinking from the perspective of the customer's/user's needs. Based on the gains, pains, and jobs to be done of a potential customer/user, new solutions are iterated. As already noted, it is important to first explore the appropriate problem before solutions are developed.

In the best case, interdisciplinary teams (see page 82) that know each other's strengths, skills, and thinking preferences work on the respective problem statements. The relevant skills and thinking preferences are used depending on the situation, for example to build empathy for the customer/user, to cluster insights, to create a vision prototype, or to carry out in-depth number crunching and testing during the validation of economic viability.

The goal is to build different prototypes to get important feedback for validating assumptions and to iteratively improve the solution through interactions with the potential customer/user. It usually takes many solution approaches, interactions, and feedback loops in the design of outstanding experiences, products, and services. It is important that all team members know where the team stands in the overall design cycle. Thus, it has become a principle in design thinking to announce the current behavior. In other words, it is important to openly state whether the team is hunting for the next opportunity or whether insights, results, or observations are being shared with the team in order to define corresponding point-of-views, which in turn form the basis for further experiments (see page 91).

Collaboration in design thinking is based on a bias toward action. The team comes together in "doing" mode to either gather new insights, build a prototype, or test it. Many design challenges in the business context today have digital components or target growth opportunities in business ecosystems. However, this also means more complexity in the projects, which has to be accepted by all involved. Such projects also require varying mental states, usually involving a combination of design thinking, lean start-up, systems thinking, and scale methodology. An approach that was also highlighted by Ash Maurya in the preface of this book as an innovation framework that marries various concepts in the realization of innovations.

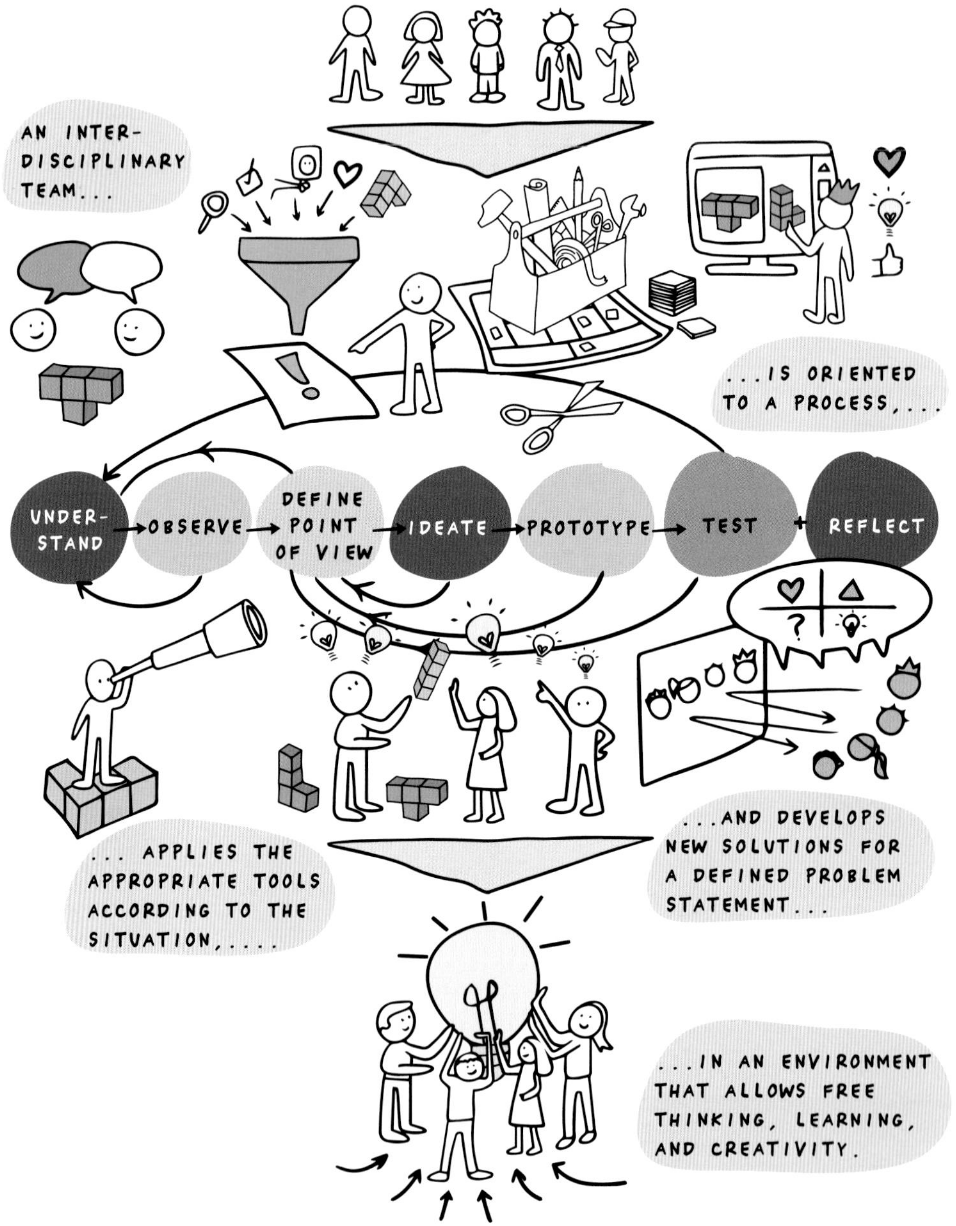
AN INTER-DISCIPLINARY TEAM...
...IS ORIENTED TO A PROCESS,...
UNDER-STAND
OBSERVE
DEFINE POINT OF VIEW
IDEATE
PROTOTYPE
TEST
+
REFLECT
... APPLIES THE APPROPRIATE TOOLS ACCORDING TO THE SITUATION,...
...AND DEVELOPS NEW SOLUTIONS FOR A DEFINED PROBLEM STATEMENT...
...IN AN ENVIRONMENT THAT ALLOWS FREE THINKING, LEARNING, AND CREATIVITY.

The Innovator's Essentials

This 101 on design thinking presents an initial overview of the design thinking mindset, process, tools, and appropriate metrics. The metrics are part of a minimum viable set of exploration metrics, which can help, especially in the EXPLORE phase to select ideas, to measure the creativity of teams, and to steer a design challenge and an innovation project at the project and meta levels. The metrics have the purpose of making better decisions and deriving the next actions and iterations. They should not limit creativity, nor should they obscure the target to finish design challenges and projects at an early phase.

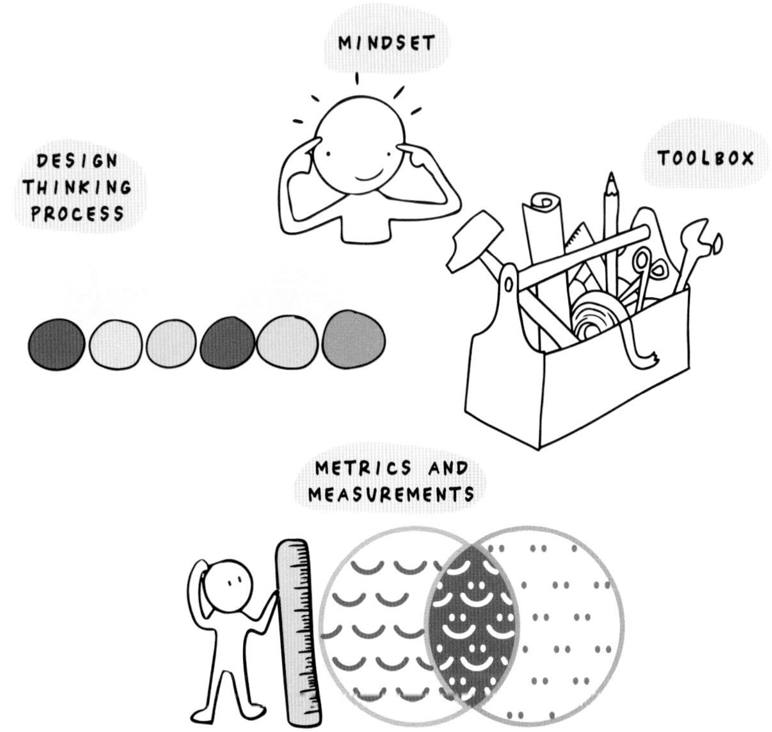
MINDSET
DESIGN THINKING PROCESS
TOOLBOX
METRICS AND MEASUREMENTS

The tools and methods presented are a means to an end, meaning the metrics as well as the tools are always to be adapted to the respective situation.

The central element in design thinking, however, is the mindset (see page 62). It is the linchpin of agile customer development, the way teams work together, and how solutions are iteratively developed starting from a problem statement.

For many of us, no matter our age or years on the job, the ability to explore and learn from our respective experiments has often been forgotten. And in most cases, our education at schools and universities discourages the detailed questioning and examination of facts, but solutions are quickly sought and implemented.

This is one of the reasons why in design thinking there is usually no team of experts working iteratively on solutions to wicked problems. A group of experts usually tends to solve the problem in a known way due to their experience, which often only leads to incremental improvement and innovation.

Experience and research shows that in interdisciplinary teams there is a better mix of many skills and domain know how, which helps to take different perspectives on the problem and potential solutions.

The best performing design thinking teams are those in which the team members have attitudes that are:

- **Open to a world of possibilities** and "what if…?" questions
- **Free of preconceptions** about how something will work
- **Filled with curiosity** to understand things more deeply
- **Free of expectations** about what will happen
- **Open to failing early and often,** while learning quickly

The Design Thinking Mindset

The Stanford d.school design thinking mindset presented here is the basis for many manifestations that are applied in the field of design thinking today. In the context of digitalization, design thinking for business growth, and an expanded design lens, elements such as the acceptance of complexity and the application of variable thinking states have been added to elements presented below.

FOCUS ON HUMAN VALUES

Build up empathy for the customer/user or other stakeholder the team is designing for. Explore the needs, pains, gains, and jobs to be done, and obtain feedback.

SHOW, DO NOT TELL

Communicate the vision and prototypes in a meaningful way by creating an experience, applying illustrative visuals, and telling a powerful and inspiring story.

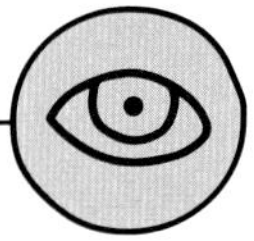

CRAFT CLARITY

Develop a coherent vision out of a wicked problem to solve. Frame it in a way to inspire the team and to fuel ideation.

BIAS TOWARDS ACTION

Move the idea forward together with the team. Maintain a bias toward doing and making over thinking and meeting.

EMBRACE EXPERIMENTATION

Prototyping is not just a way to validate an idea with the customer/user, it is an integral part of the design thinking process. Ideation, prototypes, and testing are to be understood as one entity to think and learn about a potential market opportunity.

RADICAL COLLABORATION

Create teams with different T-shapes, thinking preferences, backgrounds, and points of views. Breakthrough solutions and innovations emerge from diversity and ad hoc collaboration.

BE MINDFUL OF THE PROCESS

Know and communicate where in the design thinking process the team is right now. Decide about the appropriate tools and methods to use and remember what the objectives are.

Design thinking is often incorrectly categorized as an innovation method, but it is actually a mindset of finding the appropriate problem as a starting point for developing solutions that matter for customers/users.

Extract Long-Term Value

Design thinking is the basis for long-term value. It provides the foundation of the appropriate topics that allow us to become more meaningful to the customer. Moreover, design thinking has a transformative character that allows us to establish a design-led organization that uses the design thinking paradigm from the elaboration of critical functions and experiences to the definition of growth strategies. The agile toolbox uses elements, tools, and methods from design thinking in many respects, but primarily has the task of developing the validated prototypes quickly and purposefully and testing them on the market. For agile and the related tasks other skills are needed and metrics are also changing. Finally, lean thinking (EXPLOIT) focuses on efficient implementation and forms the basis for later scaling, which allows the customer base to grow exponentially and costs to remain moderate. However, applying the well-established exploitation metrics for increasing operational efficiency and reducing waste in an effort to measure the ability of extracting long-term value will remove any chance of breakthrough discovery, invention, and transformation.

DESIGN THINKING → EXTRACT LONG-TERM VALUE

BY APPLYING DISTINCT CAPABILITIES AND METRICS...

- to understand users and potential customers, their pains, gains, and jobs to be done, preferences, behavior, and needs
- to establish a design-led organization moving from product-centric to more personalized and customized services, products, and experiences
- to react closer and faster to improve and react to new customer behaviors and market changes

AGILE → DEVELOP & TEST RAPID

BY APPLYING DISTINCT CAPABILITIES AND METRICS...

- to drive project speed and predictability with small, frequent delivery increments
- to establish a collaborative business and IT management framework
- to establish a new risk management engineered directly into an agile execution life cycle

EXPLOIT → RUN CHEAP

BY APPLYING DISTINCT CAPABILITIES AND METRICS...

- to drive realization, knowing when to turn and when to persevere and grow a business with maximum acceleration
- to allow a principled approach to phases of exploiting, sustaining, and retiring portfolio elements

No Shortcuts – Do Not Start with Just Building Solutions Agile

In our efficiency-driven organizations, we often try to take shortcuts because the exploratory work is time-consuming, and many teams have to leave their comfort zone for the interaction with the potential customer/user. However, the respective phases of the design thinking micro cycle and macro cycle (see page 69) are central to the subsequent success of problem/solution and product/market fit. Solutions based on our own assumptions without customer validation often die quickly, because customer behaviors, needs, pains, gains, and jobs to be done have not been considered.

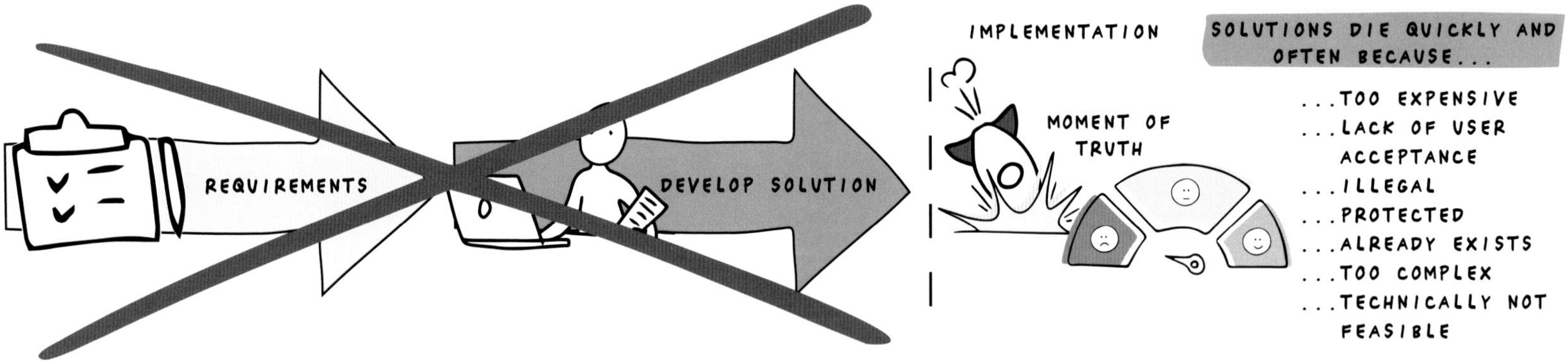
REQUIREMENTS
DEVELOP SOLUTION
IMPLEMENTATION
MOMENT OF TRUTH
SOLUTIONS DIE QUICKLY AND OFTEN BECAUSE...
...TOO EXPENSIVE
...LACK OF USER ACCEPTANCE
...ILLEGAL
...PROTECTED
...ALREADY EXISTS
...TOO COMPLEX
...TECHNICALLY NOT FEASIBLE

NEEDFINDING
PROTOTYPE
TESTING
ERROR
IMPLEMENTATION
BETTER TO SEARCH FOR...
...BEHAVIOR
...NEED
...PAINS
...GAINS
...JOB TO BE DONE
...USE CASES

Why Start with Design Thinking

Impact of Benefits, Cost, Customer, and Employee Satisfaction

The importance of design thinking in the search for the appropriate problem and in the development of solutions that meet the needs of the customer is best illustrated by looking at the risks and costs of changes, resistance, or customer feedback over the course of a typical innovation project. The cost of changing requirements increases over the course of the project. Major changes toward the end, or even in the implementation phase of an innovation project, are much higher than doing the design thinking homework early in the process.

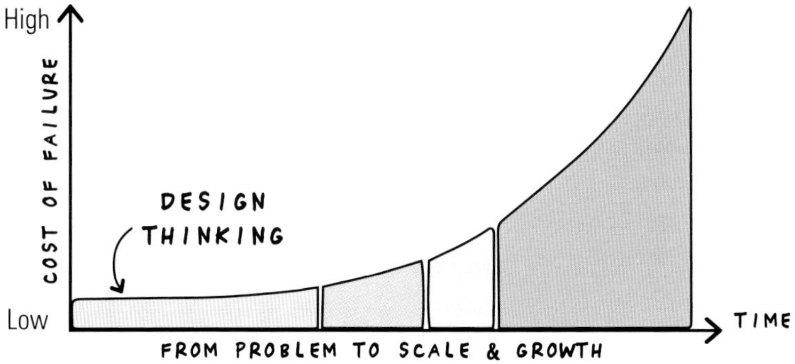
High
COST OF FAILURE
DESIGN THINKING
Low
TIME
FROM PROBLEM TO SCALE & GROWTH

The impact on customers and the associated bottom line is self-evident. Therefore, organizations and companies are well advised to equip their teams with the appropriate skills and create structures in which iterative work on the problems can be done early on, and a pivot can be carried out by the design thinking team, so that expensive delays can be avoided, competitive advantages can be realized, and exceptional customer experiences can be created. See the typical impact on various factors on the next page. In addition, morale and employee satisfaction can be kept high by applying design thinking principles. Thus, the design thinking mindset becomes the basis for agile delivery models, which are better at dealing with change, and thus helps significantly reduce costs and resistance in later phases of project work for all project types.

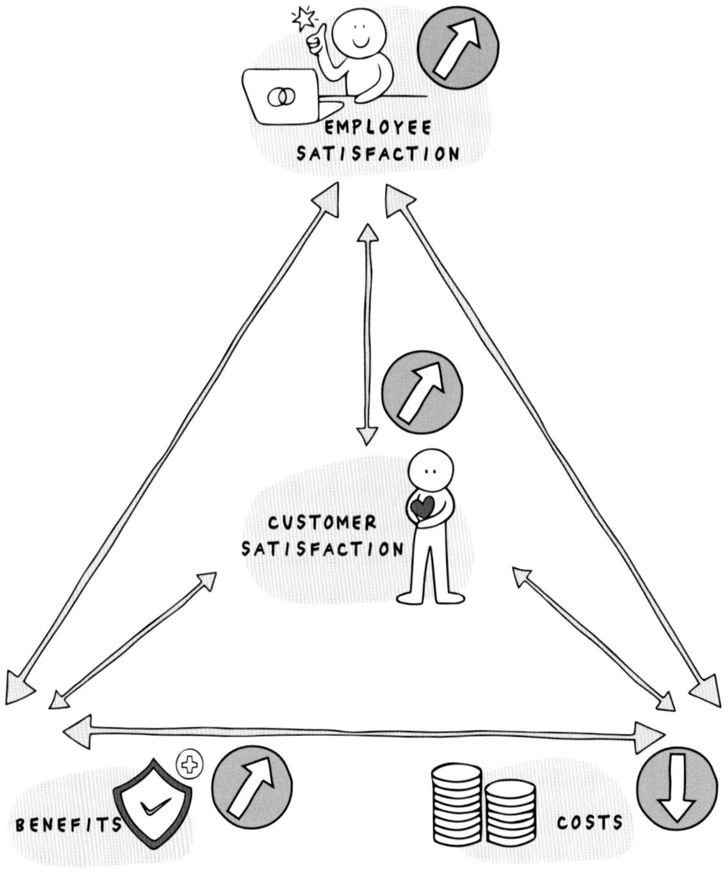
EMPLOYEE SATISFACTION
CUSTOMER SATISFACTION
BENEFITS
COSTS

Design thinking has a positive effect on several levels and also saves massive costs for later changes to the product, service, or customer experience.

Typical Impact on (Re)design, Innovation Success, and Employee and Customer Satisfactions as Phases of Innovation Projects Elapse

	NEEDFINDING →	PROTOTYPING →	TESTING →	IMPLEMENTING
PROCESS COSTS AND COSTS FOR CHANGES	$	$$	$$$	$$$+
TIME REQUIRED FOR CHANGES	• Hours to days	• Days to weeks	• Weeks to months	• Months to years (sometimes never)
RESISTANCE TO CHANGE	• Low	• Medium	• Medium to high	• High to very high
AGILITY AND EFFORT TO PIVOT	• Observing and understanding create many points of view, which provide the basis for ideation and quick feedback from potential customers.	• Depending on the resolution of a prototype, changes might be accommodated with low to medium costs and resistance to change becomes more likely.	• Solution testing and tests on finished prototypes operate in many cases with already defined functions and experiences. Major changes are expensive and need extended approval.	• Pivoting after implementing is in many circumstances a new innovation project, which needs additional resources or major re-engineering or development of solution.
INDIVIDUAL AND TEAM MOTIVATION	• A new design challenge is energizing the teams, awakening curiosity, and inspiring them to be creative.	• Enthusiastic teams retain their creativity, think through problems, and design radical new solutions.	• Many iterations and rework frustrate teams, and often, motivation is lost if this state continues for too long.	• Pessimism is widespread as employees resist realigning entire projects or the subsequent changes are very costly and time-consuming.
IMPACT ON CUSTOMER SATISFACTION AND BOTTOM LINE	• Discovering of new or changing customer/user needs.	• De-risk potential of failure.	• Major changes in evolving and higher maturity.	• Market flop has impact on revenues, brand image, retention, and acquisition.
RECOMMENDATION AND MINDSET TO BE APPLIED TO AVOID NEGATIVE IMPACTS	✓ **Problem exploration and definition are key.**	✓ **Fail fast, fail cheap.**	✓ **Test frequently and in every iteration.**	✓ **Plan implementation while hunting for opportunities.**

The Design Thinking Micro Process

The design thinking process exists as a process representation in countless variants. In the end, however, it always has the same purpose: to serve as a basis for knowing where the team currently stands in order to plan the next appropriate interactions with the customer, team, or other stakeholders. In all books of this design thinking series, the process is taken as a basis, which is also used by the pioneers and the active design thinking community at Stanford University, the d.school, and in many of the leading companies worldwide. The connecting lines of the Double Diamond make it clear that work is done in several iterations in the process. They also show that the understanding and observing steps form one entity, while the ideation, prototyping, and testing steps form another entity. In both cases, new insights are collected, which results in a new point of view. The design thinking process is usually started with a problem statement or an already formulated "How might we...?" (HMW) question, which already suggests an opening for new solutions and at the same time contains the problem and the already known customer/user needs.

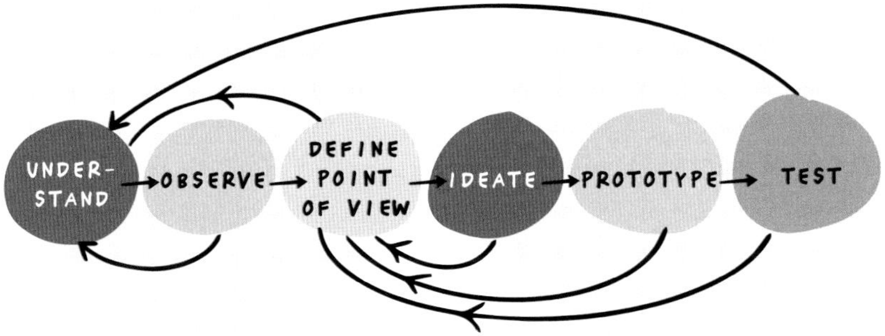
UNDER-
STAND
OBSERVE
DEFINE
POINT
OF VIEW
IDEATE
PROTOTYPE
TEST

In addition, the Double Diamond provides orientation. It divides the design thinking process into the problem space and the solution space. The individual phases are briefly described below.

Understand

In the first phase of the microcycle, we want to learn more about the potential user, their wishes, and the tasks to be fulfilled. At the same time, we sharpen the creative framework for which we want to design solutions. To define the design challenge, we use *why* and *how* questions, for example, to further open or narrow the scope (see page 108). Tools such as the empathy map and viable metrics to track the constant interactions with customers/users (see page 110) or categorization of critical assumption (see page 109) support this phase. Likewise, different jobs to be done can be compared and evaluated (see page 112). The subsequent phases and tools help the team to successively learn more about the potential customer/user. Professional design thinking teams measure how confident the team is to work on the problem space.

Observe

Only reality shows whether the assumptions about customers/users (e.g., as represented in a persona or several juxtaposed personas [see page 114]) are validated, and therefore it is best to go where potential users are. Various tools help in observing customers/users in the real environment or context of the problem at hand. Criteria for comparing personas might be formulated, quantified, and measured. The findings from the observe phase will help in the next phase (i.e., developing or improving the persona and in creating a new point of view). When the team talks to potential customers/users to learn more about their needs, open-ended questions should be asked if possible, and a question map should be used. A structured interview guide can also be helpful, but that usually only confirms the assumptions that prevail in the team.

Define Point of View

In this phase, the team focuses on evaluating, interpreting, and weighing the collected findings. It is essential to understand that analysis and synthesis often occur at the same time in design thinking (see page 119). Synthesis creates the bigger picture and forms the basis for experimenting and applying concepts. By structuring the analysis, the problem is framed, while synthesis is emergent, helping to create connections that identify breakthrough ideas and opportunities (see page 118).

This eventually culminates in the synthesis of the results (the point of view). The point of view is usually formulated as "How might we" question, to make a statement based on the findings according to the following pattern:

> **Name of the user/ persona: (who)** ______________________
>
> **needs: (what is needed/desired)**______________________
>
> **in order to: (job-to-be-done/outcomes)**______________________
>
> **because: (insight/finding)**______________________

Ideate

After the team has defined the point of view, the IDEATE phase takes place. Ideation serves to find solutions to the defined problem. Usually, different forms of brainstorming or specific creativity techniques, such as working with analogies, are used for this purpose (see *The Design Thinking Toolbox*, pages 151-184). Simple dot-voting tools; scoring of ideas in terms of desirability, feasibility, and viability (see page 121); and the application of business value versus the urgency matrix help to cluster and finally select the ideas. The assumptions of ideas are later tracked with each experimental prototype (see page 122).

Prototype

Building prototypes helps the team to test the ideas or solutions quickly, and without risk, with the potential customers/users. Digital solutions in particular can be prototyped with simple paper prototypes or mockups. The tools are as easy as can be: craft materials, paper, aluminum foil, strings, glue, and scotch tape are often enough to make our ideas tangible and experienceable. Prototypes range from critical-experience prototypes to final prototypes. Ideation, building, and testing are each to be considered as a joint and interlinked activity. They cover the so-called solution space.

Test

Testing should take place after each prototype is built, or even when individual functions, experiences, or characteristics have been developed. The most important aspects of testing are that there is interaction with the potential user/customer and that the results are documented accordingly. Here, for example, the number of prototype iterations per feature can be measured. The measurement per feature gives the impression of a vanity metric, but it is still important because it enables a comparison between projects, regardless of the size of the project or the feature set. In addition to conventional user testing, it is possible to use digital solutions for testing nowadays (e.g., online tools in the context of A/B testing). This allows prototypes or individual functionalities to be tested quickly and with a large number of users/customers. From the tests, the team receives feedback that helps them to improve the prototypes. Competing prototypes with regard to novelty might be compared with the exploration grid (see page 125). The design thinking team should continue to learn from this process and develop prototypes further until the potential customers/users are fully convinced of these ideas. If that fails, the team should abandon or change the prototype.

Reflect

Reflection is a constant companion in design thinking, as it helps the team to learn and to improve the project work. The so-called retrospective sailboat or feedback rules based on "I like, I wish, I wonder" support this mindset.

The Design Thinking Macro Cycle

In many design thinking 101 discussions, only the design thinking micro process is presented. In some cases, this leads to the assumption that the micro process is only completed once and that the design thinking activities are completed at the end of the process. However, in most cases, the micro process is repeated several times, and, depending on the novelty of the solution, feedback from customers/users, and corresponding new insights, it is decided which elements require more attention or demand more extensive activities from the team in order to increase creativity.

The goal in the divergent phase of the macro cycle is, for example, to generate as many and as wild ideas as possible, and, ultimately, prototypes that help sharpen the vision. In the converging phase, the prototypes become more concrete and high fidelity. The functional prototype, for example, helps to test the problem/solution fit for individual elements before it matures into a final prototype. This phase is usually followed by implementation and market launch. However, many products and services today also require a well-thought-out business ecosystem design, which should be created on the basis of an MVP or after the completion of the final prototype.

The Groan Zone is the most challenging area for most teams, as they have to select the features, experiences, and solutions that will later make it into the high-resolution and final prototype. Tools like the Exploration Map (see page 72) help to document the respective experiments. In addition, the exploration map measures the expectations of an experiment and its effect on the target group. It also expresses and visualizes the radicality of the potential solutions. If the documentation of all outcomes in each phase of the micro and macro process is required, the *Design Thinking Canvas* is the perfect tool to capture the most important activities (see page 129).

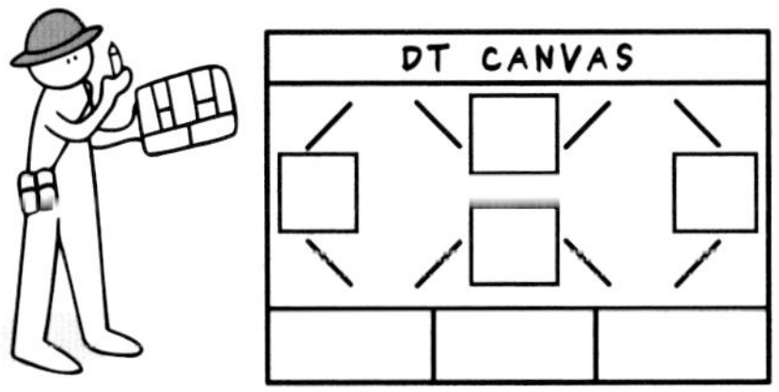
DT CANVAS

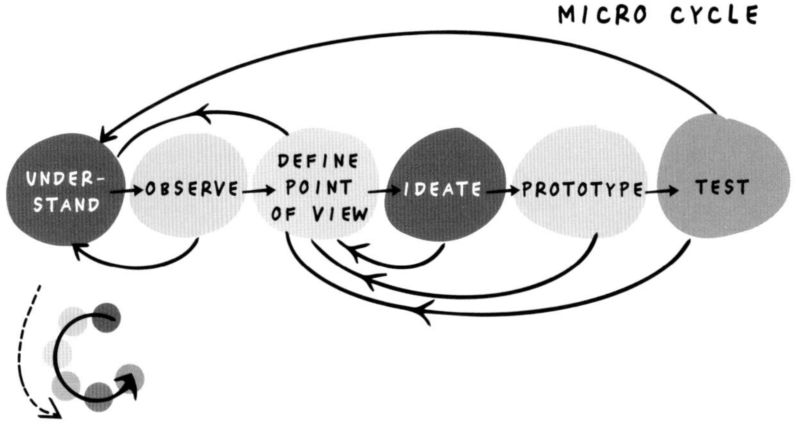
MICRO CYCLE
UNDER-STAND
OBSERVE
DEFINE POINT OF VIEW
IDEATE
PROTOTYPE
TEST

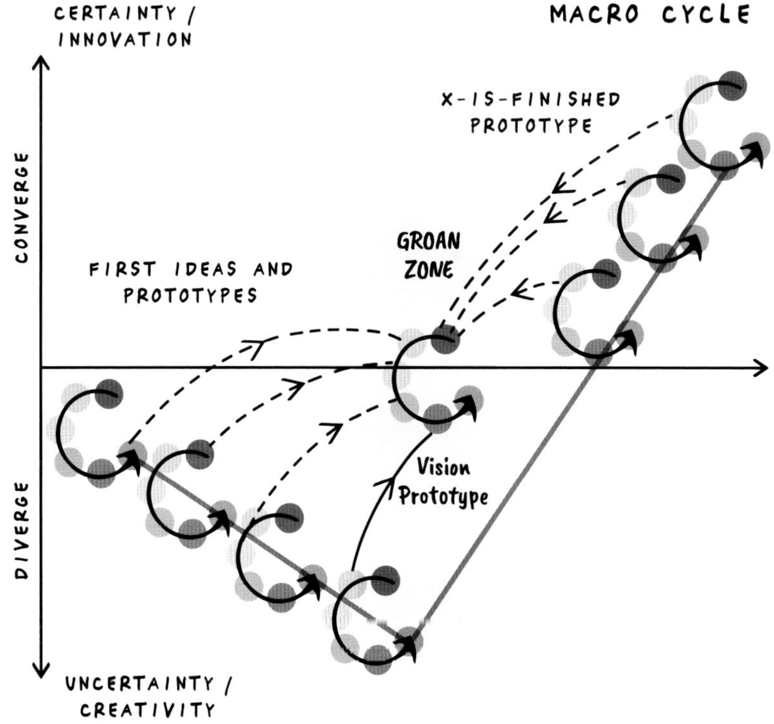
CERTAINTY / INNOVATION
MACRO CYCLE
X-IS-FINISHED PROTOTYPE
CONVERGE
GROAN ZONE
FIRST IDEAS AND PROTOTYPES
Vision Prototype
DIVERGE
UNCERTAINTY / CREATIVITY

The Exploration Map

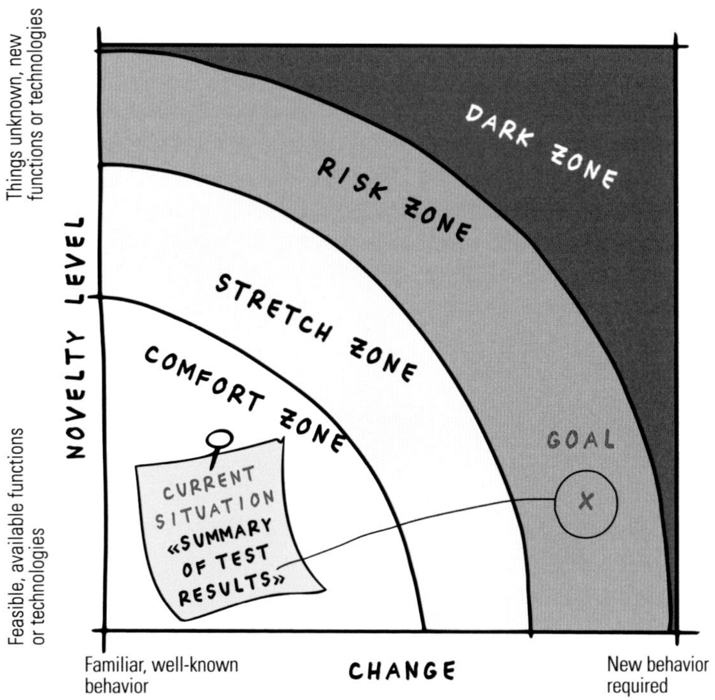
Things unknown, new functions or technologies
NOVELTY LEVEL
Feasible, available functions or technologies
DARK ZONE
RISK ZONE
STRETCH ZONE
COMFORT ZONE
GOAL
X
CURRENT SITUATION «SUMMARY OF TEST RESULTS»
Familiar, well-known behavior
CHANGE
New behavior required

The exploration map is documenting the novelty and the degree of change in every experiment performed over the entire macro design cycle. After each iteration in the solution space, outcomes of the experiment with a summary of test results can be documented on the exploration map. It provides guidance to the team about an experiment performed and at the same time the basis for the next planned one. The exploration map supports the critical discussion about experiments performed. Another great advantage of using an exploration map is that it shows, at the end of the entire design cycle, the path the team took to reach the ultimate solution. Metrics and measurements, in conjunction with exploration maps and how to estimate the probability of innovation success, are described on page 124.

Innovation Success

In the run-through of the micro and macro design cycles, the team repeatedly asks itself the central questions of desirability, feasibility, and viability. These three dimensions help to reduce the risks associated with the launch of new solutions, and the company and teams learning faster based on the "fail fast, fail early" approach. In addition, design thinking leads (if applied appropriately) to solutions that are innovative, and not only incremental.

> **Design thinking is a human-centered approach to innovation that draws from the designer's toolkit to integrate the needs of people, the possibilities of technology, and the requirements for business success.**
> **–Tim Brown**

What Is Desirability, Feasibility, and Viability?

Desirability by the potential customer must be understood and acknowledged by the design thinking team in order for customer-centric innovation to be possible at all. Even if it is tedious and takes effort to approach potential customers and users, the exploration of the problem space and the associated needs of the customer must be clearly understood and accepted to ensure a good basis for the subsequent work in the solution space. **Feasibility** means that a solution is technically possible, for example. It is the simple reality of the possibility that a design challenge can be realized with reasonable effort and cost. For many digital solutions today, feasibility is not a major challenge. **Viability** is the long-term understanding that the project is worth the time and cost, taking into account all the effort and potential returns or benefits. At best, the solution is not only of great value for the customer, but also for society or the environment by supporting a more sustainable way of life. Subsequent considerations or doubts are usually very costly and should therefore be taken into account in the design thinking activities (see costs of change on page 66). Other elements to consider might relate to environment, social, and governance (ESG) initiatives (see page 126).

The Design Thinking Lens

[Innovation success evolves from a combination of the following: the needs of the customer user (desirability), a solution that is profitable (viability), and technical implementability (feasibility).]

DESIRABILITY

- Who has the problem?
- How is it solved today?
- Which tasks of the customer/user (job to be done) are to be solved?
- Which customer/user pains and gains are known?
- Which customer needs and emotions are relevant?
- What critical functions and experiences for the customer/user should the solution include?
- How does the solution approach differ from known solutions?
- To what extent do customer/user benefits match the values of the service/product delivery?

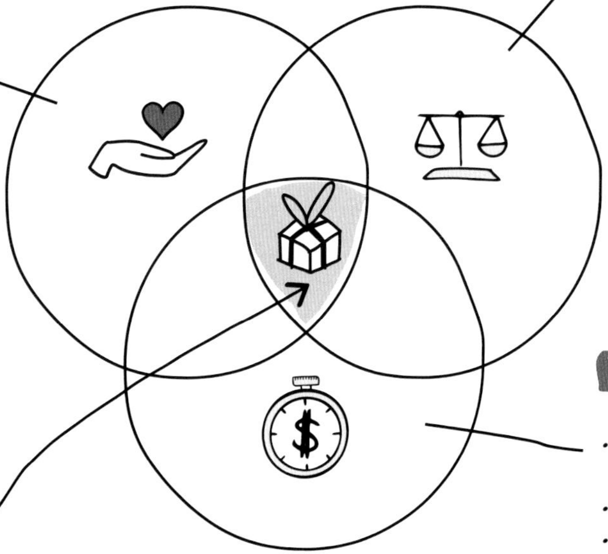

FEASIBILITY

- What are important components for implementation?
- What kind of technology is needed for this?
- What skills are needed for design, construction, and operation?
- Which systems are affected?
- What workflows are needed?
- How can "fail fast, fail often" be lived without destroying market opportunity?
- How is success defined, measured, and communicated to stakeholders?

VIABILITY

- How are revenues generated with the solution?
- How can costs be saved with it?
- Where does added value arise from for the organization, company, or a potential business ecosystem?
- How can the investments be justified?

INNOVATION SUCCESS

FACILITATING WORKSHOPS

Performing Design Thinking in Physical, Hybrid, and Virtual Settings

Design thinking exists in the interaction both within the team and with the potential customers/users. The interactive space and its respective environment are and remain an important component in design thinking (see tips and tricks for setting up creative space and environment in *The Design Thinking Playbook*, pages 132–143). New technologies and virtual environments, such as online whiteboards, virtual realities, conferences, or meeting rooms, have gained popularity in recent years.

Each format of interaction, from physical, to hybrid, to virtual, has its advantages and disadvantages. Reflecting on the neuroscience research during design thinking workshops in these different settings, it quickly becomes apparent that the interaction of a team in front of a physical whiteboard triggers more activity in different areas of the brain than the interaction using an online whiteboard. Likewise, there are only limited possibilities to build and experience prototypes in front of virtual space. Even quick sketches of ideas on a sheet of paper are currently faster in physical space than in hybrid or virtual spaces.

However, interaction with virtual reality has evolved dramatically over the last 10 years, and powerful VR goggles and modern applications can now provide excellent eye-tracking and full-body tracking. In many of the new virtual rooms, documents and 3D models can be loaded, and the room can be configured with meeting tables, whiteboards, or lecture hall seating.

Regardless of setting, some rules for a well-performed design thinking workshop still remain.

Golden Rules of Performing Design Thinking Workshops

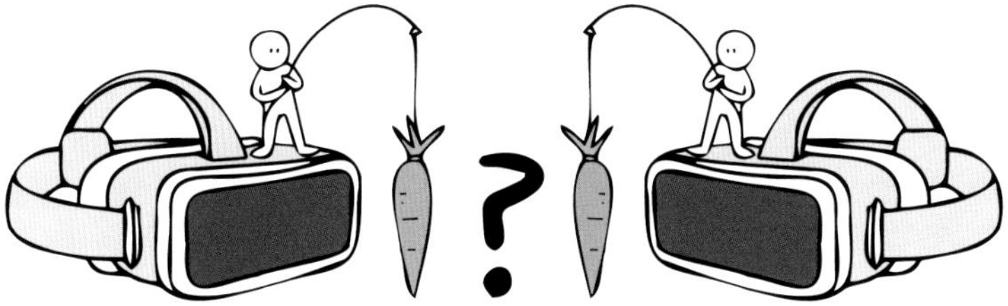
?

- Professional preparation and planning
- Information about the design brief/topic of the problem
- Invitation with agenda, directions, and access code
- Positive atmosphere for participants and sufficient space for collaboration
- An experienced workshop facilitator with the ability to apply the tools and methods as needed
- Use of appropriate warm-ups that bring out the desired mood state of the participants
- Material or tools to build physical/virtual prototypes
- Time and access to customers/users for testing initial features, experiences, and solutions
- Transparent communication about the intended mindset, the (virtual) etiquette of collaboration, and design principles
- Assembling teams based on individuals' T-shapes and thinking preferences
- Documentation of results, customer feedback, and experiments
- Measurement via a set of exploration metrics
- Embedding of objectives and key results (OKRs) from problem identification to implementation and scaling of solutions

A design thinking workshop is a hands-on, activity-based session that is based on the design thinking mindset. Facilitating workshops in physical, hybrid, and virtual spaces is an important skill that helps the team to navigate through divergent and convergent thinking.

Choose the appropriate setting... ...for the suitable collaboration and the desired results!

PROS

+ Real human interactions
+ Faster realization of moods, feelings, and work progress
+ Creation of tangible product and service prototypes

CONS

- Limited space and room
- Cost and time to travel to location
- Limited amount of participants

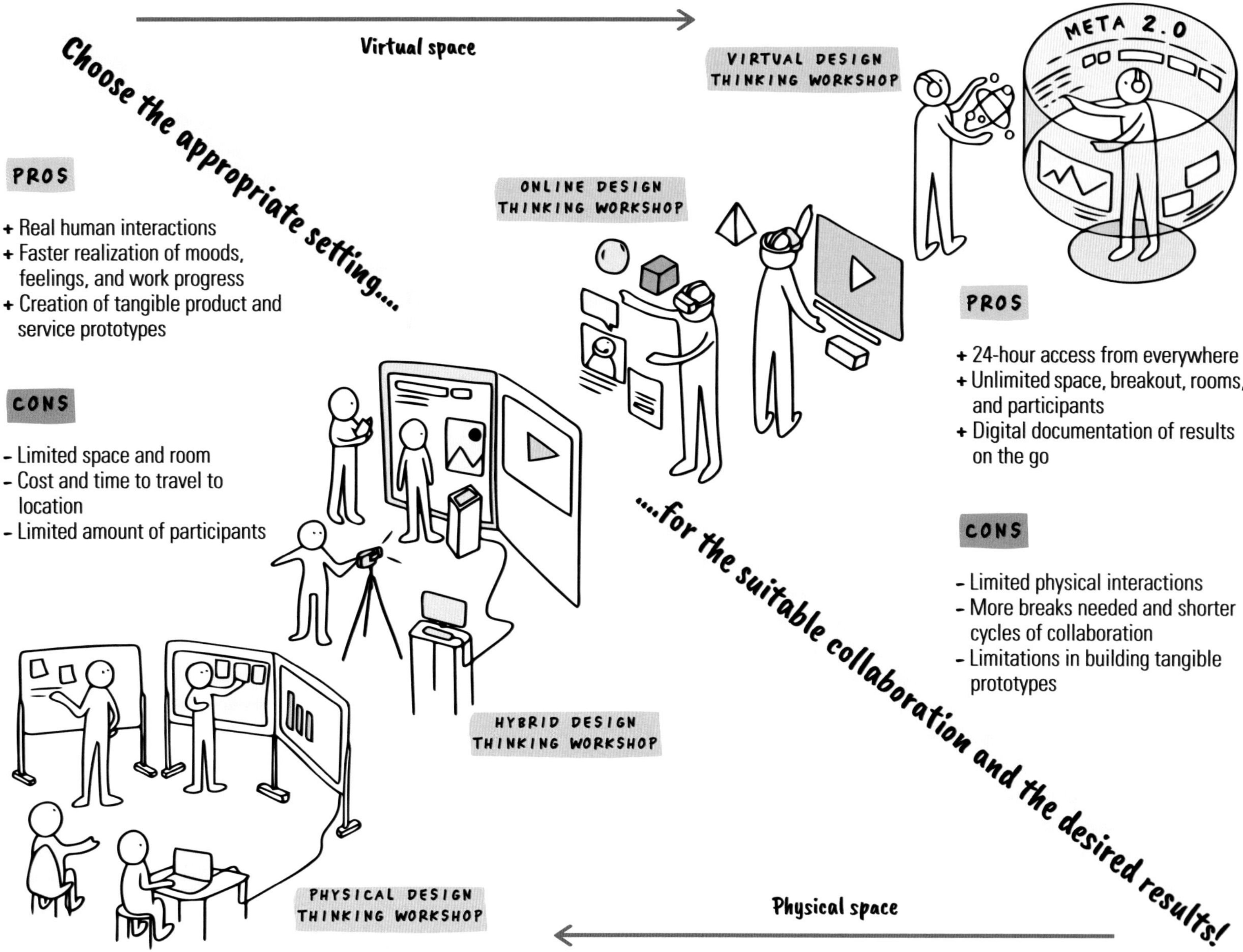

Virtual space
VIRTUAL DESIGN THINKING WORKSHOP
META 2.0
ONLINE DESIGN THINKING WORKSHOP
HYBRID DESIGN THINKING WORKSHOP
PHYSICAL DESIGN THINKING WORKSHOP
Physical space

PROS

+ 24-hour access from everywhere
+ Unlimited space, breakout, rooms, and participants
+ Digital documentation of results on the go

CONS

- Limited physical interactions
- More breaks needed and shorter cycles of collaboration
- Limitations in building tangible prototypes

Maximizing the Human Design Thinking Workshop Experience in the Metaverse

New forms of collaboration have brought more efficiency to some of the innovation work, but at the same time let important elements, such as building physical prototypes, take a back seat. Digital and hybrid interaction have became the basis of daily collaboration. The interaction and execution of design thinking workshops in the metaverse is the next level of immersion. However, this requires a radical realignment of the approach and design of the workshops. The good news is that the design thinking mindset and tools help for designing such an experience, as one of the basic principles in the metaverse is to create a human-centered experience. From this perspective, the needs and urges of the workshop participants, in the virtual setting to explore new experiences, are the focus.

In many ways, workshop participants become a new dimension of design thinking players who are challenged to solve a wicked problem. Unlike digital whiteboards, which only had the function of supporting communication, the metaverse provides the possibility to build a closer relationship with the other design thinking players, the space, and experiences. At the same time, a balance between reality and the virtual world has to be kept, as well as between the existing 2D and the new 3D worlds, so that everyone can participate in the most human way.

This means that, in the best case, the design thinking players should have the possibility to participate from any place and with all their senses.

Next-generation virtual ecosystems are a reality to consider, not just as a conceptual futuristic vista. The design thinking community should consider a metaverse world and remain curious for the possibilities

Simply copying existing experiences, procedures, and known workshop elements into the virtual environment has proven to be unpromising. The metaverse offers more. It gives the design thinking teams and each individual player the opportunity to see things that were previously impossible to grasp, to have access to knowledge that was previously unknown, and to experience things that the real world cannot offer.

There is also the possibility to create new communities, to express belonging, and to get involved in new ways. This applies to the work on the design challenge and, of course, beyond. The metaverse provides the opportunity to think beyond the existing boundaries of the company and to work together on the big challenges of tomorrow.

The teams have the opportunity to work together virtually, to playfully shape the design cycle, to treat intellectual property (IP) and ideas as digital assets, and not only to design new products and services for the real world, but also to design, test, and ultimately implement things, buildings, and experiences for environments in the metaverse.

The possibilities of interaction among teams, stakeholders, and communities are enormous and are on the journey to perfecting the experience. The use of design thinking tools and interaction will once again require an iterative approach and experimentation to inspire the design thinking players and to open up the appropriate worlds that allow for divergence and convergence. To what extent and how quickly the metaverse will make all assets available as a digital mirror, and how the augmented realities blend for each use case, depend on the speed of adaptation and the quality of the experience, functionality, and, ultimately, a compelling value proposition that speaks to the new interactions.

However, design thinking and innovation teams now have a unique opportunity to co-create their new environment of work and, ultimately, to create environments that are not limiting, but rather extend existing realities.

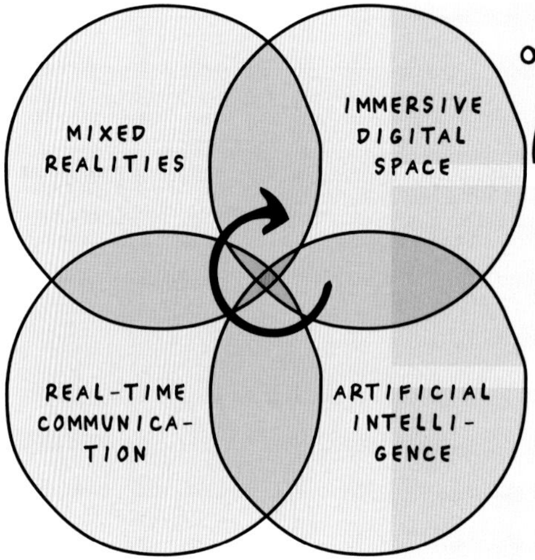
MIXED REALITIES
IMMERSIVE DIGITAL SPACE
REAL-TIME COMMUNICA-TION
ARTIFICIAL INTELLI-GENCE

Procedure for Creating Workshop Experiences in the Metaverse:

1. Define the main objectives of the interaction with the design thinking players (e.g., team and community building, exploring the problem space, working with analogies and benchmarks)
2. Determine the roles of the design thinking players and other stakeholders (e.g., active participant, providing briefing, engaging for testing)
3. Launch the activities and tap into intuitive immersive experiences with the design thinking players. Do not overwhelm the participants with information and options to apply tools and to navigate along the design process
4. Measure results and analyze the behavior of participants in different thinking modes and activities (e.g., eye tracking and facial attributes data collection via XR technologies)
5. Constantly adapt and improve the experience for design thinking players
6. Embed the metaverse activities in physical interactions or extended realities
7. Share best practices with the wider design thinking and innovation community

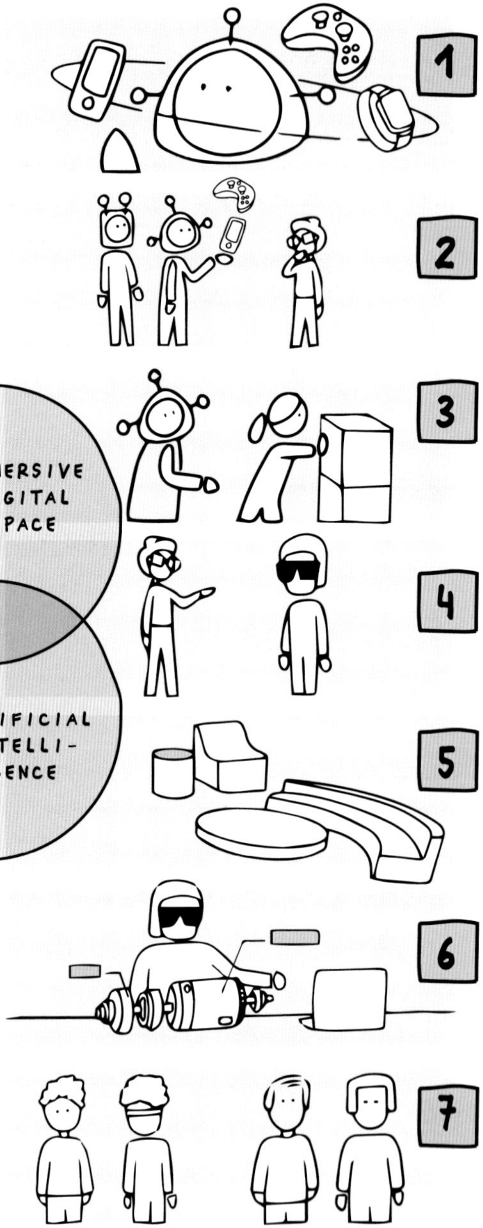
1
2
3
4
5
6
7

Leading and Facilitating Design Thinking Workshops

Leading and facilitating design thinking workshops, regardless of the setting, require experience in facilitation and, at the same time, knowledge of design thinking tools, methods, and processes. The cognitive process in design thinking is based on design abduction, which means that, in many cases, two unknowns lead to a process of creative exploration. The dance with ambiguity is challenging for both the facilitator and the team because, in a more traditional setting, the cognitive process is based on reasoning, deduction, and induction. Dorst (2017) provided a simple equation of the cognitive process in design thinking to illustrate the differences.

MOM!!

REASONING

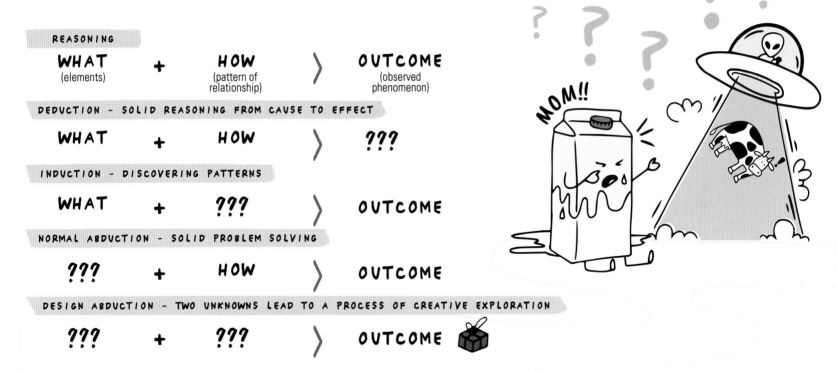

WHAT (elements) + HOW (pattern of relationship) 〉 OUTCOME (observed phenomenon)

DEDUCTION – SOLID REASONING FROM CAUSE TO EFFECT

WHAT + HOW 〉 ???

INDUCTION – DISCOVERING PATTERNS

WHAT + ??? 〉 OUTCOME

NORMAL ABDUCTION – SOLID PROBLEM SOLVING

??? + HOW 〉 OUTCOME

DESIGN ABDUCTION – TWO UNKNOWNS LEAD TO A PROCESS OF CREATIVE EXPLORATION

??? + ??? 〉 OUTCOME

Micro and Macro Process Objectives

As was clear with the introduction of the design thinking mindset on page 62, it is important to know where one stands in the design thinking process and which tools and methods should be applied. Especially in the role of a facilitator, it is important to understand what the respective objectives are in the design thinking micro and macro processes. Experienced facilitators have a very good sense of which activities, tools, and thinking states are needed both in the moment and in the next step in guiding the team to ultimately achieve the best results. When going through the design thinking micro process several times, different tasks, thinking states, and activities are performed, depending on the situation. In early phases these elements can be described mainly as collecting, analytical, and observant. In later phases, the elements relate more to selecting (e.g., the appropriate functions and experiences) as well as experimental and observant (i.e., the activity that experiments are built and tested).

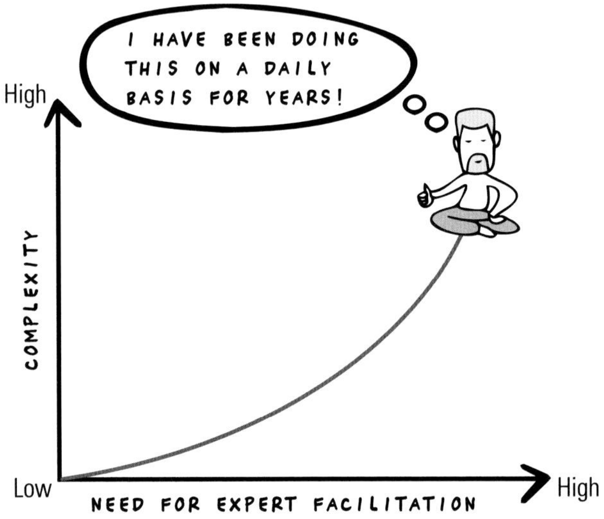
I HAVE BEEN DOING THIS ON A DAILY BASIS FOR YEARS!
High
Low
COMPLEXITY
NEED FOR EXPERT FACILITATION
High

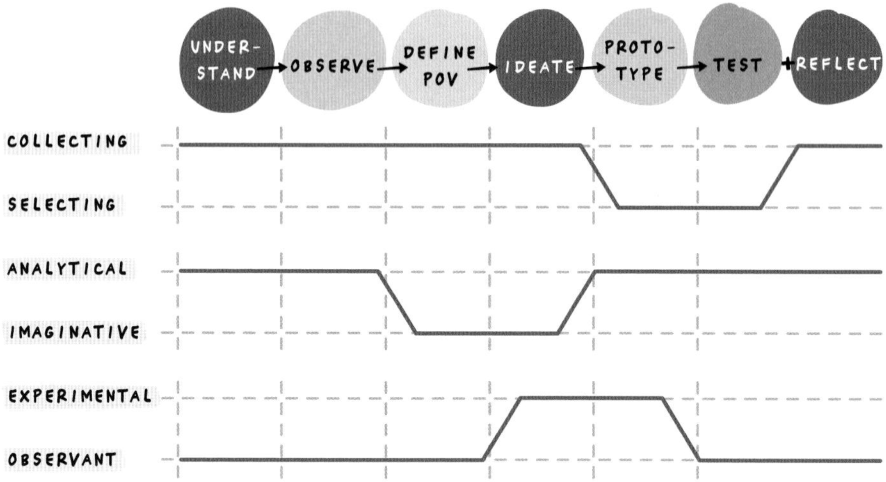
UNDER-STAND
OBSERVE
DEFINE POV
IDEATE
PROTO-TYPE
TEST
REFLECT
COLLECTING
SELECTING
ANALYTICAL
IMAGINATIVE
EXPERIMENTAL
OBSERVANT

Key Expert Facilitator Behaviors:

- Manages complicated engagements that span multiple customer needs, places, times, technologies, and actors
- Works successfully with all levels of the organization, including executives and high-impact stakeholders
- Iterates complex design challenges in real time to adapt to emergent group and project needs
- Innovates designs thinking tools, methods, frameworks, as well as facilitation activities

Highly experienced facilitators are adept at solving all possible problems. They have the ability to explore, understand, and predict the complicated needs of teams and guide them to outcomes that meet those needs. Such facilitators create a coherent web of interactions on an ad hoc basis, combining warm-ups, tools, thinking states, dialogue, and fluidity in such a way that teams achieve outstanding results, even for complex problems and situations.

The Role of the Facilitator

The role of the design thinking facilitator should be clear to everyone. It is to plan and guide activities and instructions throughout the design thinking micro and macro cycles to help the respective teams develop a common understanding with varying states of thinking. The task of a facilitator is not to develop an outstanding creative idea or to make the right decisions.

It is important that a facilitator not take command or dictate the outcome. The key is to allow each team member to contribute fully and equally, and to facilitate a collaborative team outcome. Here, objectivity is key. In addition to a degree of neutrality, their own contributions should be kept to a minimum whenever possible. The focus is on supporting the respective behavior (e.g., hunt or transport) as best as possible (see page 88). More specifically, the role of the facilitator is to help the teams in the beginning to identify the appropriate problem, to formulate a problem statement, to build empathy with the potential customer/user, and to get the teams to develop and test the best possible ideas. Different tools and methods can be offered by the facilitator to help create a dialogue among the teams and make the best decisions.

Purpose: Design thinking facilitation is the means to an end. It allows for inclusive and collaborative decision making, mutual understanding, shared responsibility, and equal participation of all team members.

Principles: Design thinking facilitation is more art than science. The setup and style might vary, but some core principles, like being authentic, embracing constructive conflicts, active listening, and the creation of an inviting physical, hybrid, or virtual space and environment, always stay true.

Toolbox: Design thinking facilitation applies the appropriate tools and methods to start activities, trigger dialog, or to achieve certain results. In addition, the materials for building first prototypes, templates to work with, and workshop supplies are the responsibility of the facilitator.

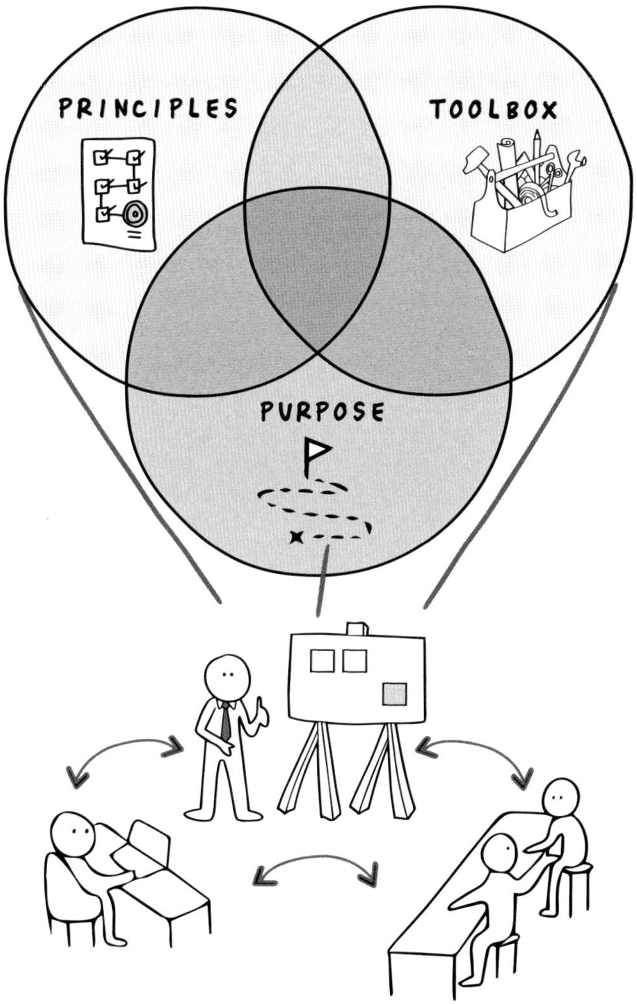
PRINCIPLES
TOOLBOX
PURPOSE

Highly skilled facilitators have a very extensive toolbox of methods and techniques that they can use during the workshop in an improvisational or planned way.

WORKING IN TEAMS

How to Expand Toward a Circle of WE

How to Shift the Culture

- Upshift leadership from commanding to building a shared consciousness
- Expect and empower decision making by whoever is closest to the content
- Work directly together to build trust
- Know your purpose
- Be willing to learn

Lessons from Working with Team-of-Teams

- Teams outperform individuals in complex contexts.
- Any one individual cannot comprehend the whole or have all required skills.
- Most organizations as a whole have yet to even begin to understand how to gain the advantages of teams at scale.

Typical Metrics and Measurements

- Team member performance related to helpfulness
- Team member performance related to initiatives
- Overall team readiness based on feelings and maturity with regard to mindset, tools, and methods
- Overall team commitment for achieving joint objectives and key results
- Overall team collaboration and processes for decision making, communication, conflict resolution, problem handling

Radical Team Collaboration and Organizational Design

The collaboration of interdisciplinary teams is in many cases of high importance for the success of innovation and for the work in the solution and problem space. In many explanations of design thinking, the focus is on team formation, how teams should be composed, and what dynamics should be considered. These elements (thinking preferences, team composition according to growth phases, and T-shaped teams) are important, but the biggest challenge for companies today is to choose the appropriate organizational set-up so that they can work together in the best possible way in a global, diverse, and increasingly virtual and hybrid environment. The requirement for the organizational design and the environment is to realize a seamless collaboration of teams. This often transcends corporate boundaries, as products and services are increasingly delivered in ecosystems of business rather than vertically integrated through one company, with traditional supplier relationships.

Many companies are also leveraging the external talent ecosystem to quickly realize capabilities for realizing projects with new skills and technologies. Thus, multiple teams are configured ad hoc to collaborate on a common core value proposition and set ambition. Measuring the collaboration of teams and the holistic team-of-teams design of an organization is possible on different levels. For example, it is possible to perceive the extent to which teams align their activities with a defined North Star, their respective actions are autonomous, and communication is open and transparent (e.g., via objectives and key results). The central element in such approaches is to establish a bowl of trust as a foundation. A supportive team community is the be-all and end-all for a culture in which the WE stands above disparate team interests, and conflicts are solved profitably, as the lessons learned help to move all teams forward. The relationship of the teams with each other and with individuals is important. It is about understanding what the others do, where their strengths are, and in which environment and under which conditions they work. Central to this is understanding how the other teams can help as one moves forward together and vice versa.

Team-of-teams is based on the fact that teams outperform individuals in complex contexts because individuals cannot comprehend the whole or possess all the required skills.

What Characterizes an Interdisciplinary Team?

In very general terms, interdisciplinary means comprising several disciplines. In interdisciplinary teams, ideas are produced on a collective basis. In the end, everybody feels responsible for the overall solution. A methodological and conceptual exchange of ideas takes place on the way to the overall solution.

What Characterizes a Multidisciplinary Team?

A multidisciplinary team has the advantage in that, at the end, everybody stands behind the commonly created product or service. This is a factor of success that interdisciplinary teams, for example, cannot afford. On a multidisciplinary team, every member is an expert who advocates his or her specialization. The solution is often a compromise.

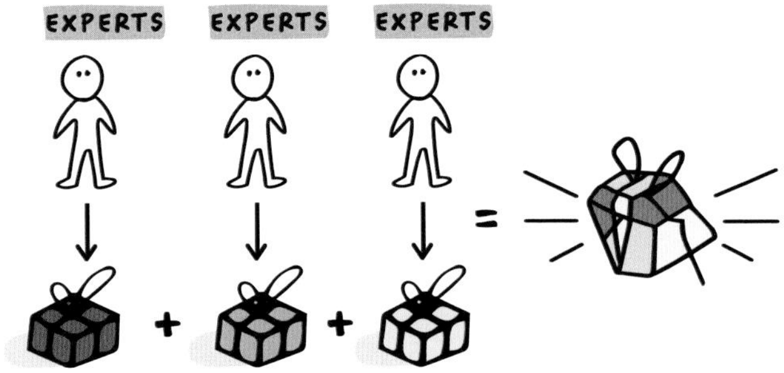
EXPERTS
EXPERTS
EXPERTS

What Characterizes T-shaped Profiles?

The vertical bar of the T profile stands for the respective specialist skills that a team member has acquired, usually through vocational training or specialized academic studies. Most of the time, this is mirrored by the job descriptions for specific functions of experts in organizations to analyze and solve a specific problem.

The horizontal bar of a T-shaped team member is defined by many cultures, disciplines, and systems. These skills and capabilities are related to experiences and learnings, which are usually not apparent from the job descriptions. These include, for example, experience of living in and understanding different cultures, to skills and achievements that help to analyze problems from a different perspective, or to re-frame the problem.

T-shaped team members (with their unique skills and capabilities) are open, interested in other perspectives and topics, and curious about other people, environments, and disciplines. The better the understanding is for the way others think and work in a team, the faster and greater the common progress and success of the design thinking work is.

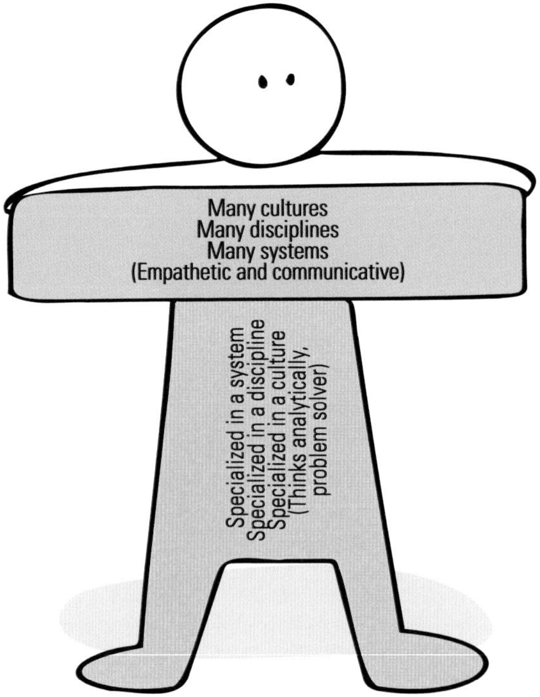
Many cultures
Many disciplines
Many systems
(Empathetic and communicative)
Specialized in a system
Specialized in a discipline
Specialized in a culture
(Thinks analytically,
problem solver)

Team-of-Teams Evolution

In an ideal world of team-of-teams, employees have a constant exchange and dialog. They review their previous actions to learn from mistakes. They compare new methods with proven ones to find the most effective practices. They proactively acquire new information and skills and coach each other across boundaries so that teams succeed together. However, such a transformation does not happen overnight. It takes time, skills, and the right leadership to respond fluidly to complex environments before an organization becomes adaptable and resilient. This in turn requires that all team members adopt core elements of the design thinking mindset (see page 62):

Dance with Ambiguity: Learn to expect the unexpected

Radical Collaboration: Connect in a way that allows rapid self-configuration to respond to new threats or opportunities

Leaders must be able to link organizational elements both vertically (up and down the command chain) and horizontally (across units and divisions within the organization, as well as across external partner organizations).

How to Achieve it

Working with different companies has shown that a good starting point is to make sure each team member has a systemic understanding of the big picture and knows how their work is interdependent.

It is also paramount for the teams to understand the strategy and to learn how to link the North Star to the team objectives and key results. As mentioned earlier, the strong lateral connectivity between teams through personal relationships between individual team members is essential. Everyone benefits if T-shapes and thinking preferences of each team member are made transparent. Knowing each other better and tapping into vertical AND horizontal skills helps in this process. Equally important, however, is that for a team-of-teams to be born, the leadership team must be willing to let go and to share power. A popular approach is to empower teams with new techniques and ways of thinking (e.g., design thinking, systems thinking, critical thinking).

The non-structured mesh of individual connections is what makes team-of-teams successful.

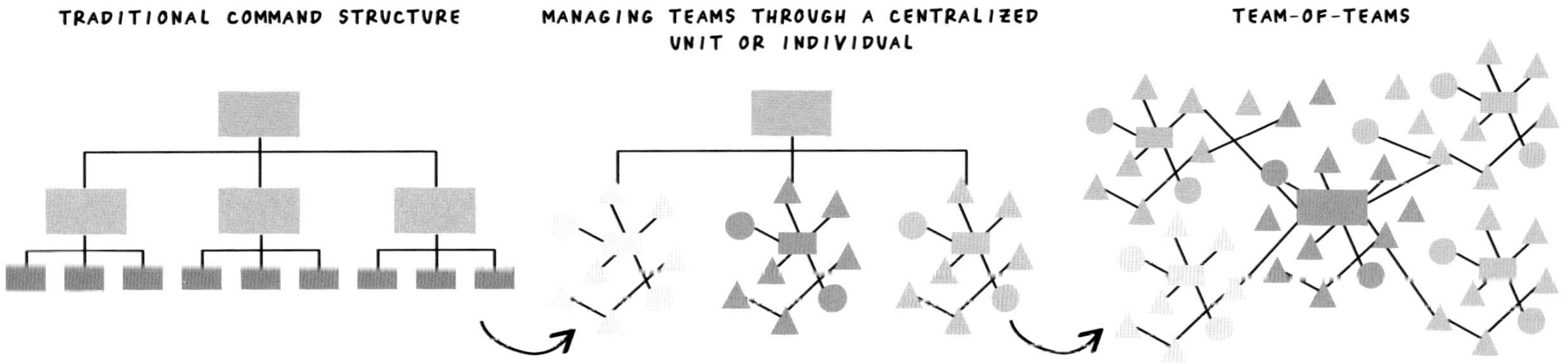
TRADITIONAL COMMAND STRUCTURE
MANAGING TEAMS THROUGH A CENTRALIZED UNIT OR INDIVIDUAL
TEAM-OF-TEAMS

Design Thinking & Organizational Network Analysis (ONA) for Organizational (Re)design

By applying design thinking to redesign organizations, organizational network analysis (ONA) helps provide insights that go beyond shifting through job titles, sweeping organizational charts, and unclear role descriptions. As presented in the team-of-teams approach, the best team constellations emerge based on thinking preferences and different T-Shape profiles (see page 84). This is how natural networks form when teams come together that make their work better, easier, or even just more satisfying.

Supporting ONA can create further insights into how, for example, information and knowledge are actually retained and shared through employee and team networks. Visualizing and analyzing the formal and informal relationships in the respective organizations helps to develop forms of collaboration that promote cooperation and improve the effectiveness of tasks from EXPLORE to EXPLOIT. However, ONA is a structured method for visualizing the flow of communication, information, and decisions in an organization. Organizational networks consist of nodes and links that form the basis for understanding how information is flowing between individuals and teams.

Utilizing ONA as the Basis for the Organization's Design Helps:

- to transform organizations in savvier ways by identifying formal and informal organizational leaders who facilitate change and help realize the benefits of transformation more expeditiously.
- to develop more effectively and focused talent by minimizing role confusion and redundancy, as well as identifying interests in specific projects, initiatives, and skill development programs of talents within the teams and organizations.
- to increase operational efficiency and effectiveness by building a structure that is organized dynamically in a way that fosters collaboration and information sharing among the appropriate employees and teams, which contributes to the achievement of the defined objectives.

More advanced application of ONA can be found in the 301 section, which provides enhanced metrics including, for example, culture based on an AI-enabled network analysis (see pages 355–356).

Visualizing and analyzing the formal and informal relationships of teams help to design organizational structures that make the organization more sustainable and effective.

In every organization and in every team, there are employees ("nodes") who serve as important channels for the exchange of ideas, prototypes, and information. A connection is beneficial when the needed information is exchanged and knowledge is shared.

- Central connectors are the team members who seem to know everyone. These central nodes share a lot of information and have a quick influence on all teams. Central nodes can be located in any team, and are often popular and highly involved in company news and developments.
- Knowledge brokers are team members who build bridges between teams. Without knowledge brokers, the exchange of information and ideas stalls.
- Shadow players are easily overlooked and are not connected to the rest of the teams, so they can pose a risk to companies. For example, outstanding data scientists who don't collaborate with design thinking teams not only stagnate in innovation development, but are also easily convinced to take their talents elsewhere.
- Ties are the formal and informal relationships between nodes. Creating optimal relationships between central connectors and knowledge brokers helps ensure that useful information is easily shared between and within teams.

WHO HOLDS THE NETWORK OF TEAMS TOGETHER?

WHO HAS THE POTENTIAL TO BECOME A BOTTLENECK IN EXECUTION?

WHO IS A CRITICAL CONNECTION TO EXTERNAL KNOWLEDGE?

WHO NEEDS TO BE CONNECTED INTERNALLY TO BE EFFECTIVE?

WHO IS TOTALLY DISCONNECTED AND AT RISK OF EXIT?

WHO IS TALENTED, BUT NOT CONNECTED TO RELEVANT FUTURE PROGRAMS AND PROJECTS?

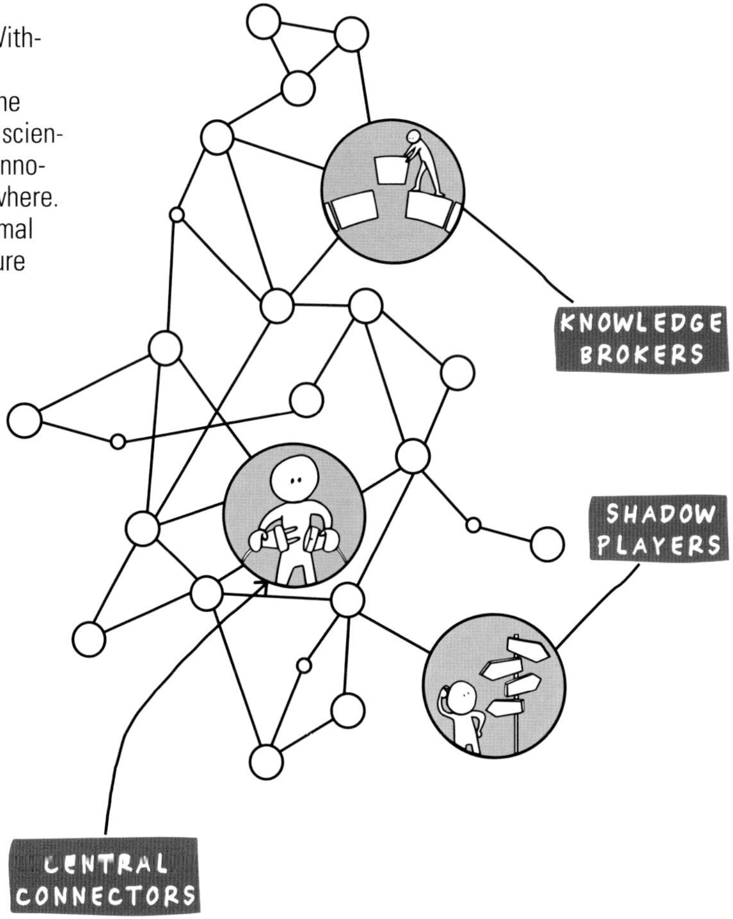
KNOWLEDGE BROKERS
SHADOW PLAYERS
CENTRAL CONNECTORS

BUILDING CAPABILITIES

Building Design Thinking Capabilities and Measuring Maturity

If a company really wants change and aims to strengthen innovation, it cannot expect it to just happen. Rather, an environment must be created in which new market opportunities can thrive. That takes a culture of embracing creative confidence and the appropriate skills to allow teams to collaborate radically.

Building design thinking capabilities is an investment to future-proof the company. Corporate practice shows that there are two paths that work best when done in parallel: capability-building programs and hiring talent. The targeted search for and hiring of talents and high performers who already have the desired mindset and future skills is the quickest way but is usually only feasible to a limited extent. The new joiners are employees who have the power and the will to commit to positive change, who are able to think from the customer's point of view and at the same time connect all the dots (design thinking and systems thinking), and who are able to absorb information at lightning speed and turn it into action with the teams.

In addition to talent, a new system agility (see page 90) is needed at all levels from the work environment to support the measurement of objectives and key results to organizational design, work processes, and structures that allow for design thinking and new thinking. The biggest challenge here is for management teams to break away from traditional command-and-control hierarchies and, for example, through the team-of-teams approach, to focus on more autonomy and accountability of employees, while at the same time strengthening the self-efficacy of each individual.

Empowered employees who have been gradually introduced to the design thinking mindset through project-based learning, for example, are better able to initiate change. Rather than endlessly seeking approval and encountering resistance, they can make decisions and act quickly.

By changing the structures and the organization, the decision makers in the company can also reconnect with each other, which in many cases helps to create better alignment and flexibility, and ultimately a better vision for the future.

By infusing design thinking as a mindset into every aspect of a team, organization, or company, a new culture is created. This influences the strategic thinking about growth initiatives, the attitudes of the workforce, the organizational structure, and the approach to the problems to be solved.

Teams that work close to EXPLORE or that support these activities will feel increasingly confident in exploring new customer problems, developing ideas, and implementing them. Especially in a time where the world is changing faster than ever before–which everyone has felt in the last few years–the companies that have built up the appropriate capabilities and skills will be able to deal with the future better.

Tweaks won't do it. Design teams need to get things off of the whiteboard and into the real world!

Design Thinking Behavior Can Be Trained to Deliver Value

Through project-based learning and tailored capability programs, design thinking behavior can also be trained to create business value and accelerate transformation. It supports teams not only in solving customer problems but also in defining business strategy and growth options to invest in the most promising opportunities, while reducing the risk of making inappropriate investments. In terms of culture, design thinking serves to establish a positive and affirming culture, while encouraging and empowering each team member to think creatively and collaborate without fear of failure or retribution across the well-known silos in the organization. In addition, it can be observed that the planning and execution of projects is significantly accelerated, and overhead can be reduced in the medium term. However, acceptance and transformation need time, constant reflection, and a capability-building program that fits the organization. In this way, the maturity of design thinking can be increased over time, and real business impact can be generated.

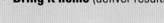

Accept, Learn, and Apply the Hunting and Transport Philosophy

- **Hunting is not wandering** (have a purpose)
- **Never go hunting alone** (multi-capability teams)
- **Don't give up too early** (patience with failure)
- **Don't confuse transport with hunting** (declare the behavior)
- **Bring it home** (deliver results)

The best way to describe the design thinking behavior is the hunting and transport model, which reflects the way true design thinking teams work and has become the core of activities. The model has been shaped by Larry Leifer at Stanford University in the context of ME 310. The hunt for the next big market opportunity starts after the exploration of the problem space, with divergent thinking and a first experimentation with known solution approaches. The goal is identified, first prototypes are built, and then they are tested against what is known to the potential customer/user. For the unknown, the feedback from the customer/user provides information. The respective insights, the feedback, and the discussion in the team contribute to learning and help in the planning and execution of the next experiment, which is usually focused on a target that is situated somewhere else. By frequently alternating inductive and deductive thinking, the unknown can be revealed over time (see page 88). To make this happen, it takes pragmatic abduction to produce a discovery. Transport is in many cases not straightforward. The way home also has ups and downs and road blocks to overcome. It is best to overcome these challenges by involving key stakeholders who are already in the hunt.

The biggest challenge in design thinking training is to teach both individual team members and the team as a whole how to hunt again.

HUNTING

It's best to go hunting as a small and agile interdisciplinary team with a maximum diversity of skills. This includes a good hunter; a gathering specialist; a realist who keeps track of time, equipment, and conditions; and someone who pays attention to team dynamics, emotions, and communication.

A
B0
B1

TRANSPORT

After the big idea is discovered, it has to be made tangible, and the question of how to bring it home is in focus. Now is the time for the team to fix the requirements, make plans for the implementation, and optimize the resources.

Measuring Design Thinking Maturity on Different Levels

The maturity model of applying the design thinking mindset for companies, organizations, and teams could be extended with countless dimensions to achieve more system agility and increase innovation maturity. In addition to the general future skills described on pages 40–41, there is usually a trigger reason why companies consider design thinking to be an important mindset and capability, and use it specifically to achieve progress on different levels. Often, in large organizations, the situation is that almost all teams have a good understanding of agile working, but the structures, processes, and mindset, as well as the overarching collaboration in the form of team-of-teams, are not yet very well developed. Likewise, the measurement of employees has so far been based on a classic management by objective approach. The step-by-step development of design thinking capabilities, within the framework of a project-based learning program, helps to realize new market opportunities, and at the same time to transform the organization from a push culture to one based on a new understanding of ownership and personal responsibility. The parallel introduction of OKRs supports this transformation (see page 269).

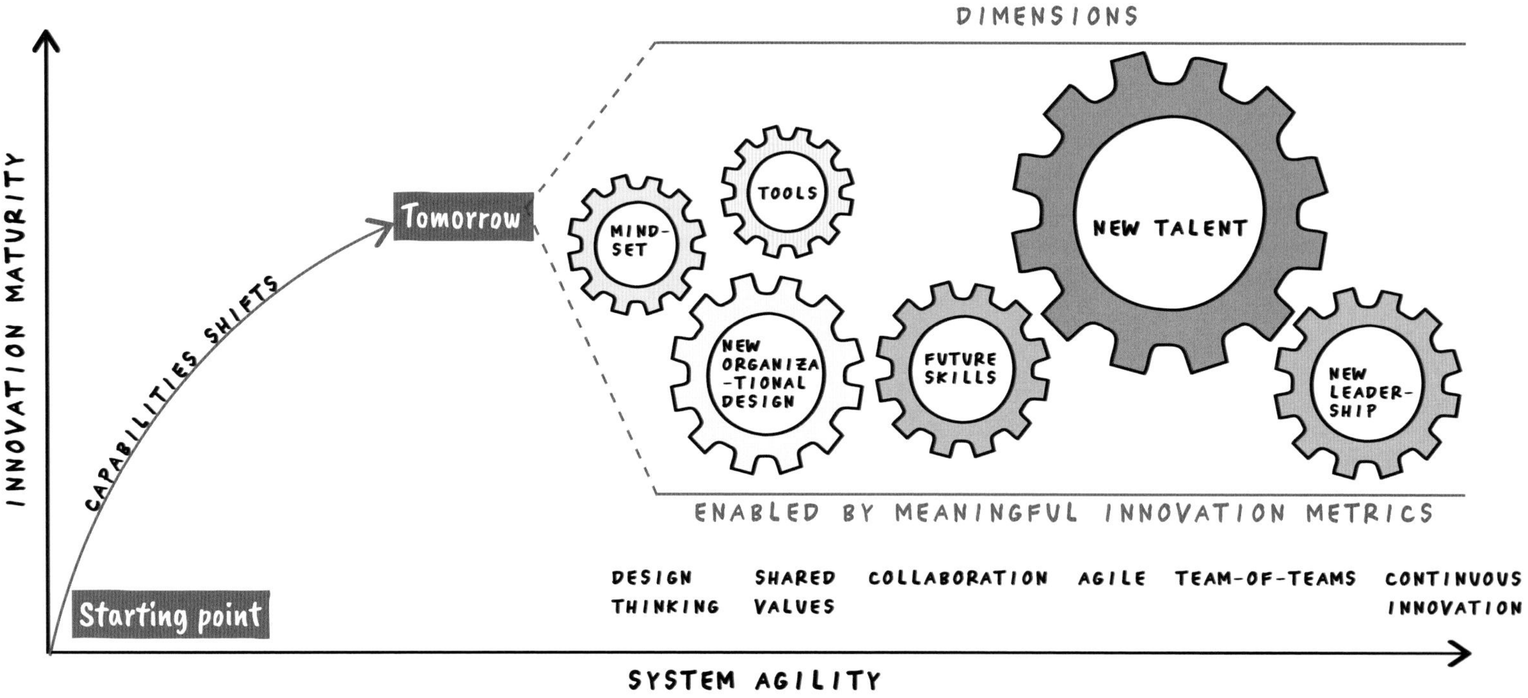
DIMENSIONS
INNOVATION MATURITY
Tomorrow
CAPABILITIES SHIFTS
Starting point
MIND-SET
TOOLS
NEW TALENT
NEW ORGANIZA-TIONAL DESIGN
FUTURE SKILLS
NEW LEADER-SHIP
ENABLED BY MEANINGFUL INNOVATION METRICS
DESIGN THINKING
SHARED VALUES
COLLABORATION
AGILE
TEAM-OF-TEAMS
CONTINUOUS INNOVATION
SYSTEM AGILITY

Applying the Design Thinking Maturity Model

An initial measurement of maturity in terms of the applied design thinking mindset very often shows that the organization and the teams in question have limited capabilities in terms of design thinking and the desired future system agility. Based on an initial assessment of the existing capabilities, targeted capability-building measures, experiments in organizational design, and new ways of designing and measuring innovation and performance, measurements can be initialized. Usually, it takes several capability shifts over a time horizon of 2 to 3 years until the teams master all capabilities. Design thinking and the underlying mindset help to build many areas of future skills, and in some areas these go beyond the toolbox of design thinking.

	LEVEL 1 (limited capabilities)	LEVEL 2 (evolving capabilities)	LEVEL 3 (value driven)	LEVEL 4 (mastering all capabilities)
NEW TALENT / FUTURE SKILLS	• Low mastery • Competency-based teams • Limited design thinking training exposure • Limited shared ownership or team accountability	• T-shaped teams • Teams supported with facilitation and design thinking training • Challenged • Still need for strong guidance	• Team-of-teams established • Strong contribution • Self-organizing • Strong purpose and customer focus • Attracting best talent	• Mindfulness and courtesy • Circle of "we" established • Servant leaders • Belonging • Engaged teams
MINDSET / TOOLS	• Good understanding of agile, basic understanding of new ways of working & design cycle • Limited transparency • Fixed scope • Limited value-driven projects	• Empirical design thinking process established • First value-driven design challenges accomplished • Mastering basic design thinking tools and methods	• Ownership counts more than process • Multiple team cadences • Design thinking scaled across organization • Continuous delivery of projects over macro and micro design thinking cycles	• Driven by team objectives and results • Mastering individual objectives • Delivering tangible results and managing multi-design thinking projects and implementation (i.e., ring it home)
ORGANIZATION / LEADERSHIP / CULTURE	• Competency-based structure • Closed groups and organizational silos • Push culture	• Team-based structure • Learning organization • Push and pull culture	• Self-organizing teams • Strong team contribution • Pushing boundaries	• Innovative • Exploring and high curiosity • Ownership

START → YEAR 1 → YEAR 2 → FUTURE → TIME

How to Start Project-Based Learning and Scale New Capabilities across the Organization

For many organizations, there is also the question of how design thinking experience and skills can be scaled to the entire organization. One possibility is to make design thinking an integral part of the company-wide academy and at the same time offer interested employees the opportunity to expand their design thinking toolbox and, in turn, expand their skills as a facilitator on a project basis. However, this is usually only successful if the framework conditions fit, the gradual build of the innovation and performance management systems are initiated, and the strategy is clear and understandable to the employees. If these conditions are met, far-reaching benefits for transformation and future business success can be realized.

SET A STRONG NORTH STAR
Create a customer-centric strategy and allow teams to create strong missions based on objectives and key results.

ALLOW PROJECT-BASED LEARNING IN EACH MISSION TO COMPLETE
Embed the missions to complete in a design thinking capability program and guide teams to begin at the beginning.

SET CLEAR PRIORITIES AND AMBITION FOR THE DESIGN CHELLENGES
Initially support creating the design principles and how much risk your organization will take driving it.

MANAGE EXPLORE AND EXPLOIT
Actively review, measure, and manage the EXPLORE and EXPLOIT portfolio. Provide feedback and clear governance to teams.

DEVELOP THE BASIS FOR A DESIGN THINKING ACADEMY AND PROFESSIONALIZE FACILITATION
The academy provides structures, approaches, tools, and the development pathways to scale transformation to all relevant parts of the organization.

FUEL AND FUND INITIATIVES
Allow teams to on-board appropriate internal and external resources to accelerate and amplify initiatives—funding, talent, tools, sponsors, and more.

BENEFITS

- **Attracting, developing, and deploying individuals to innovation and the mindset for radical collaboration and operating in a networked organization.**
- **Building connections to internal and external resources who can augment and amplify initiatives.**
- **Establishing a clear understanding in the organization about mindsets, methods, and approaches for understanding the problem and solution space.**
- **Sensing constantly organizational maturity for methodology, measurements, and organizational design and teamwork.**

The applied metrics and measurements depend on the maturity and overall ambition to initiate change. The following simple example shows minimum viable exploration metrics that are predominantly measuring the adoption of the new mindset and team maturity through a three-part defined acceptance criteria.

Example

1 CAPABILITY DEVELOPMENT

Successful completion of the training modules and project-based learning sprints, with each of the teams reaching their learning goals

2 PROJECT DELIVERY

Positive feedback from project customer on project delivery (e.g., time, improved or new products and services, product/market fit)

3 BEHAVIORAL

Positive feedback from the project sponsor and select team members on future skills and mindset displayed on the job

Example of Validating the New Innovation Capabilities through Improved Project Delivery

Success is:

- More projects successfully delivered
- Projects delivered faster
- Reduced change requests late into the process
- Failing earlier and learning from it
- New mindset with better ways of working (measured, for example, in team surveys)

Every single member of a design thinking team in an organization plays a role in spreading design thinking mindsets by looking for opportunities to openly share and pass on their methods and tools. All of them should be guides to human-centered design, not its guardians.

The new capabilities also include the ability to apply the appropriate quantifications, measurements, and metrics. Starting from design thinking, the toolbox expands to innovation metrics. These new capabilities to work with data open up many new opportunities to not only measure innovation, but also to generally bring data-driven innovation together with design thinking to ultimately be faster to market with innovations while reducing uncertainty. The table below summarizes the characteristics of design thinking. The attributes of design thinking and innovation metrics are to be understood as additional elements to design thinking.

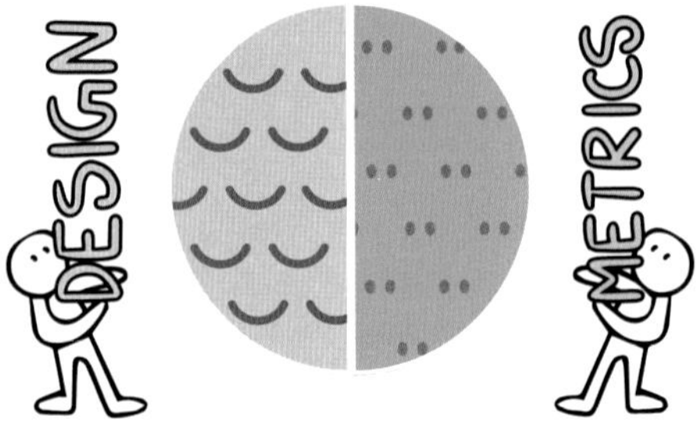
DESIGN
METRICS

What	Design Thinking	Design Thinking & Innovation Metrics
OBJECTIVES	• Product, service, process innovation • Realizing new market opportunities • Transformation/new mindset	• Customer-led and data-led innovations • Reducing uncertainty in dynamic environments • Professionalizing/scaling transformation
FOCUS	• Providing insights rich in strategic information • Understanding the problem space • Building empathy with customers • Creating tangible prototypes	• Providing standardization in comparison • Improving evidence through measurement • Establishing the basis for AI-supported innovation • Creating a better understanding of effectiveness
QUALITATIVE METHODS	• Ethnographic research, synthesis, definition of point of view, brainstorming, idea selection, prototyping, solution tests	• Understanding the need for measurements and metrics and consequences of possible actions • Asking the right questions
QUANTITATIVE METHODS	• Applying mostly descriptive analytics • Application of criteria score, define success criteria, and digital enabled A/B testing	• Applying descriptive, predictive, and prescriptive analytics • Operationalization of measurements and metrics
TYPICAL METRICS	• Problem/customer fit, problem/solution fit, product/market fit • Testing assumptions (viability, desirability, feasibility)	• North Star metrics, minimum viable exploration metrics, kill metrics, OKRs for innovation teams • Transition toward exploitation metrics and health metrics
TEAMS	• Interdisciplinary teams, T-shaped participants, mix of diverse thinking preferences	• Expanded team with data scientist and AI (future)
TYPICAL APPLICATION SCENARIOS	• Solving wicked problems • Creating a team-of-teams culture • Changing behavior	• Anticipating before the next S-curve • Measuring organizational/cultural change • Prove effectiveness of teams and procedures

MEASURING IMPACT

Measuring the Impact of Design Thinking in Business Performance

The burning questions at the end of Design Thinking 101 are whether the impact of design thinking and the activities described can be measured. And the good answer is yes, the business impact from the combination of building design thinking capabilities and value creation can be measured, and evidence of this exists from various design thinking initiatives. In various projects, this approach has made it possible to place the customer at the center of considerations, to offer teams new opportunities for radical collaboration, and to significantly increase work efficiency. Especially when the design thinking mindset is applied across the company's portfolio of products and services, costs can be saved, the development speed increased, and better solutions for the customers can be created. Moreover, it is not a one-time effect, as, due to the increasing maturity of the capabilities in the teams and the accelerated transformation, the potentials for the financial impact for both innovation projects and a successful organizational transformation are enormous.

An evaluation of various design thinking capability programs over a decade shows the most significant outcomes. The improvements, activities, and anticipated change range from improved collaboration to a more sophisticated understanding of how to design value propositions. The fact that customer interaction becomes more intensive through design thinking is due to the nature of a human-centered design approach.

From the observations, however, also emerges a list of the most challenging issues that have been witnessed over the same period of time. In many cases what is needed is a strong-willed decision maker, who knows the potential of design thinking and is willing to lead the mindful transformation of teams, organizations, and organizational structures over a longer period of time. In addition, the aspect of time is a critical element, since working attitudes and routines cannot be changed from one day to the next. As already described on page 88, design thinking behavior can be trained to deliver value, but it needs space for reflection and a speed of change adapted to the pace of employees. It also needs to be accepted that not all employees will embrace the change. It is up to leaders to find appropriate roles and tasks for these employees that better fit their mindset, attitude, and personality. In a normal business cycle, there are also phases of focus, cost savings, and an over-weighing of EXPLOIT. In these phases, there is a danger that, in a very short time, the capabilities and talent that have been painstakingly built up will leave the company or be taken over by the new objectives. The best approach is to embrace design thinking as a mindset that can add value to all areas of an organization, starting with the exploration of customer needs and continuing up to strategy work, which is supported by the iterative approach.

Top Five Results from Building Design Thinking Capabilities

- Improved team collaboration
- Increased customer interaction
- More efficient processes and workflows
- Acceleration of time to market
- More clearly defined value propositions

Top Five Challenges in Disseminating the Design Thinking Mindset

- Perseverance and support from top management
- Time and space for change and transformation
- Appropriate tasks and roles in a changing culture
- Healthy balance between EXPLORE and EXPLOIT
- Perceive design thinking as a mindset

Quantifying the Financial Impact of Design Thinking Projects

Example

The challenge remains to quantify the previously listed benefits of design thinking and to establish the basis for measurement. On page 66, we have already described the impact on costs, customers, and employees that can be observed in typical innovation projects. As a first measure, the reduced costs from the initialization of the projects to successful implementation can be quantified for a defined period of time.*

Metrics Based on Meta Level

- Overall reduced cost from project start to implementation in USD

Metrics Based on Phases

- Reduced cost for project selection in USD
- Reduced cost of prototyping and testing in USD
- Savings for modification in late development state in USD
- Increased profit from better customer/solution fit in USD

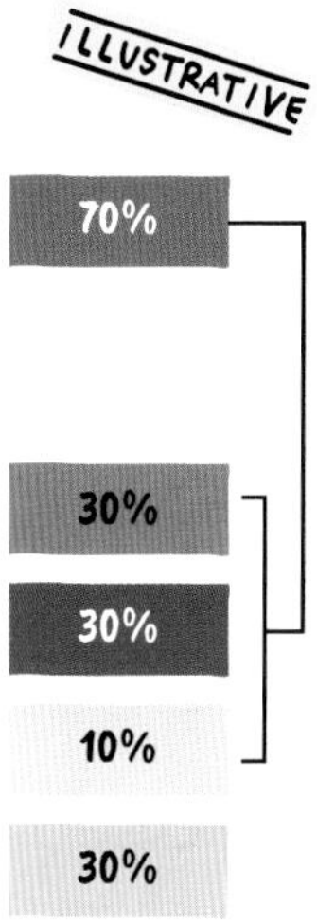

ILLUSTRATIVE
70%
30%
30%
10%
30%

Ratio shift for time spent in problem definition to implementation refinements in % / USD

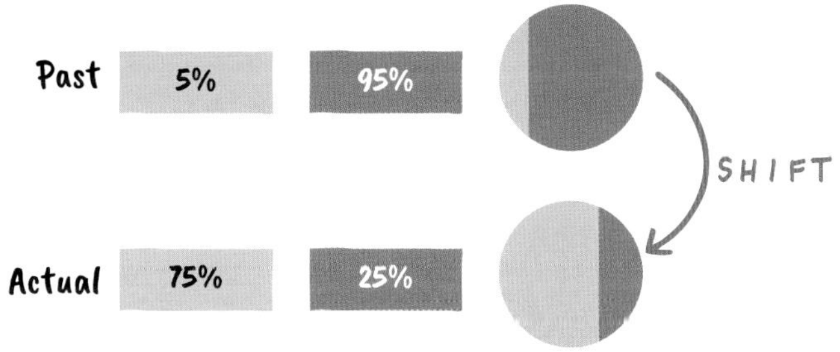

Past
5%
95%
SHIFT
Actual
75%
25%

Innovation Project Portfolio Analysis

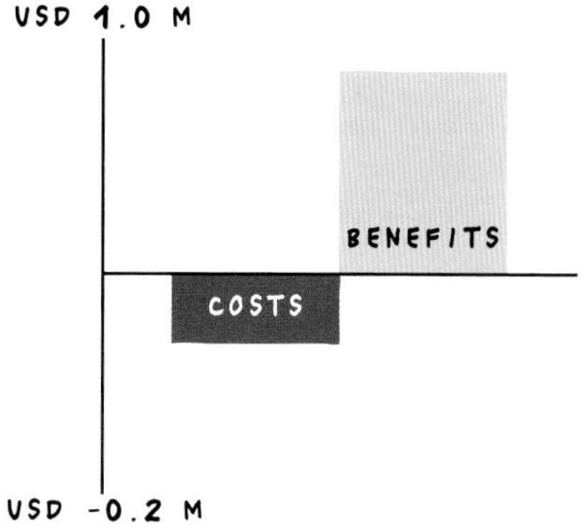

USD 1.0 M
BENEFITS
COSTS
USD -0.2 M

Benefit Breakdown

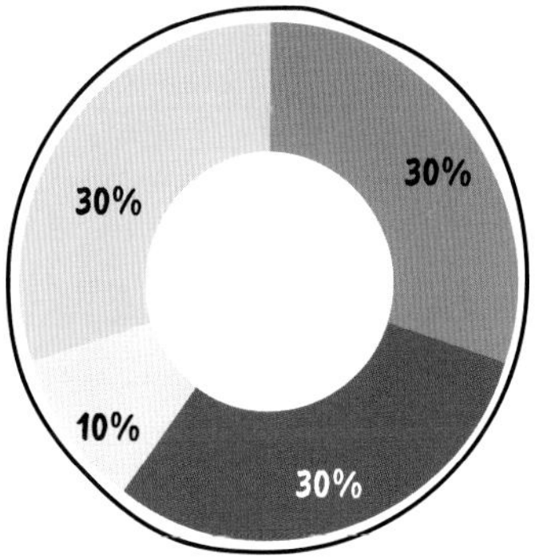

30%
30%
10%
30%

Many organizations realize significant gains because design thinking internally addresses the problem up front, whereas without design thinking capabilities, late changes, throwaway efforts, and rework are the norm.

* The defined period of time might vary for different industries (e.g., for pharma 5–10 years in comparison to digital services 1–2 years).

Calculating Costs / Benefits for Training and Capability-Building Programs

There are various options for building up design thinking capabilities. These range from the establishment of an internal academy, to the use of specialists, to the involvement of large consulting firms, which have purchased the corresponding skills over the last decade through the takeover of smaller design studios and agencies. The costs/benefits vary accordingly.

However, for rapid scaling of capabilities, it has proven to be good to have some of the workforce become design thinking coaches early in their respective training. This approach has a significant impact on the proliferation and number of projects and teams executing projects with the design thinking mindset. It is also important to not only focus on design thinking skills, but also to continuously expand the toolbox to include, for example, business ecosystem design skills in order to achieve a higher skill density.

Furthermore, it is strongly recommended to build on known tools, methods, and design thinking approaches and not to waste budgets on activities such as re-branding and re-naming of methods, tools, and frameworks. Usually that results in more confusion than a positive effect. The best results and acceptance can be achieved by building on existing templates, literature, and process models that are known worldwide and beyond the organization's own boundaries.

Many organizations start capability-building programs with a smaller amount of pilot teams, usually no more than five teams, that gain experience through project-based learning. Mostly, the design challenges used have a manageable time frame of 4–12 months, depending on the complexity, the available time budgets and the desired integration with, for example, the defined objectives and key results.

Based on the first experiences and success stories, the programs can be quickly scaled to 20–30 teams. In parallel, programs for trainers can be initialized and the first teams can work autonomously on the respective design challenges without support. Accompanying measures and communication via town halls and leadership boards increase the spread of relevance.

Calculation of the costs depends on how much time employees spend for the training and how much for the actual application. Real costs are incurred for external support of the teams, surveying the intimate capabilities, conducting trainings, workshop facilitation conducted by design thinking experts, and supporting the measurement of impact from individual activities to overarching outcomes.

Usually, the cost for the training and project-based learning is not more than 40 percent of the total design thinking project costs

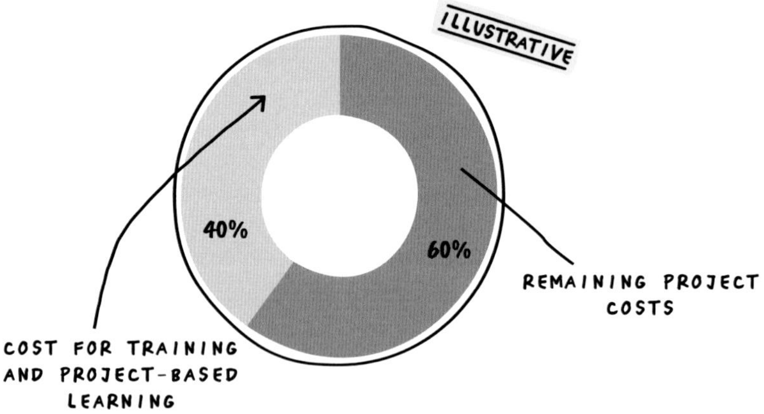
ILLUSTRATIVE
40%
60%
REMAINING PROJECT COSTS
COST FOR TRAINING AND PROJECT-BASED LEARNING

Financial Bravado Exists – Deeper Benefits Are Making the Difference

Example

Metrics / cash flow view on design thinking, capability building, and project-based learning

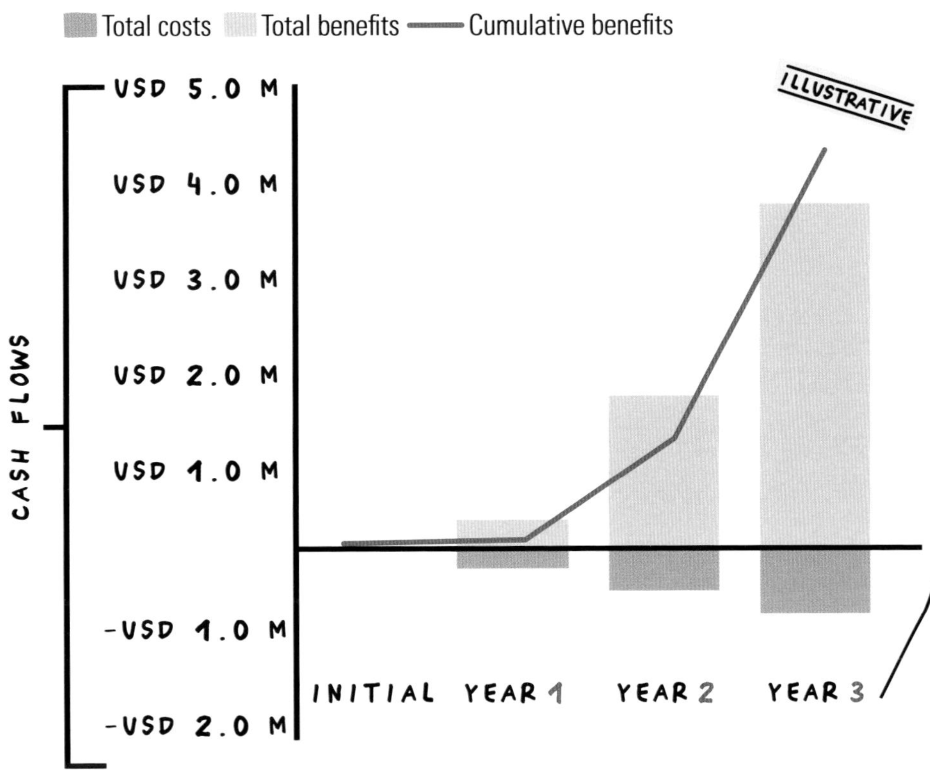

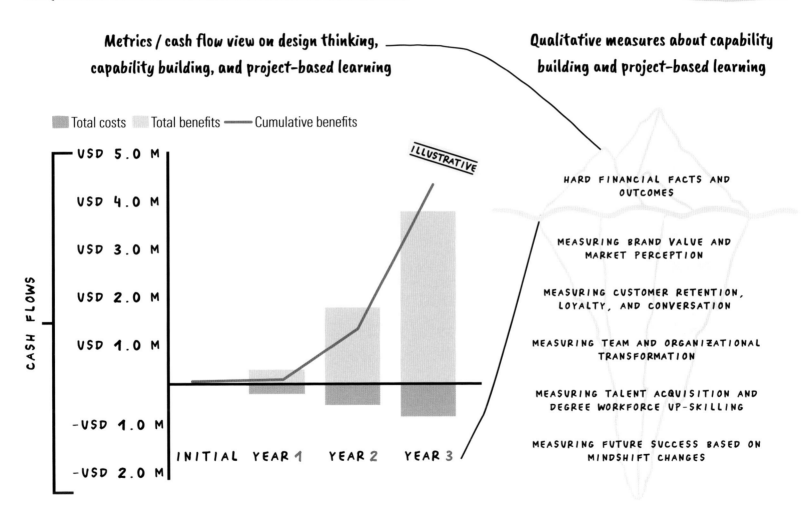

Total costs
Total benefits
Cumulative benefits
ILLUSTRATIVE
CASH FLOWS
USD 5.0 M
USD 4.0 M
USD 3.0 M
USD 2.0 M
USD 1.0 M
-USD 1.0 M
-USD 2.0 M
INITIAL
YEAR 1
YEAR 2
YEAR 3

Qualitative measures about capability building and project-based learning

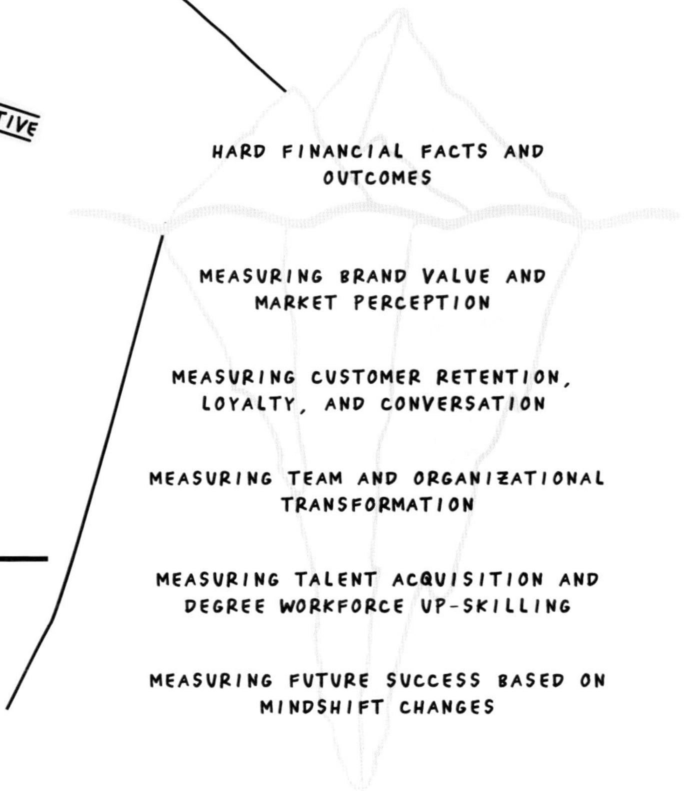

HARD FINANCIAL FACTS AND OUTCOMES
MEASURING BRAND VALUE AND MARKET PERCEPTION
MEASURING CUSTOMER RETENTION, LOYALTY, AND CONVERSATION
MEASURING TEAM AND ORGANIZATIONAL TRANSFORMATION
MEASURING TALENT ACQUISITION AND DEGREE WORKFORCE UP-SKILLING
MEASURING FUTURE SUCCESS BASED ON MINDSHIFT CHANGES

To the Point!

Design thinking is more than a creativity method. Today it is part of the skills of the future and a mindset for the realization of new market opportunities. At the same time, it has a transformative character in the way teams collaborate across the traditional organizational silos.

Design thinking starts the exploration for new market opportunities with a lot of uncertainty. The underlying mindset, tools, and iterative process allow teams to validate the needs of the customer/user, profitability, and technical implementability. As a result, design thinking provides teams a natural way to minimize the risks and to make predictions about market success.

Design thinking is the foundation for long-term value creation and forms the basis for the agile development of products and services, up to the design of complex business ecosystems.

DESIGN THINKING TOOLKIT

Extended Selection of Design Thinking Tools

The entire design thinking toolbox comprises more than 500 methods and tools, which are beyond the scope of this book. The most popular and valuable tools can be found in the book *The Design Thinking Toolbox*, which is highly recommended to everyone aiming to apply design thinking or take on the role of facilitating design thinking workshops. This introduction to design thinking includes a selection of prudent methods to reduce risk from the problem/customer fit all the way to the product/market fit, including the systems/actors fit (if a business ecosystem is required). Most of the presented tools focus on simple tips and tricks that can be used within EXPLORE to compare ideas, provide insights, or make better decisions through data collection and measurements over the entire design cycle. For a quick and easy mapping of the tools and metrics, the tools follow the problem to scale and growth framework. From the previous introduction of design thinking, it becomes already obvious that design thinking means consistently moving from a narrow technical and business perspective to a truly interdisciplinary and collaborative culture of thinking and doing. For many organizations, however, the expectation of design thinking is still mostly linked to the metrics of business success based on traditional financial indicators. This fulfilling viability is a key element, but it should not be viewed in isolation from the other benefits of applying the design thinking mindsets. It can already be guessed that design thinking cannot be measured as a single concept. The presented examples of measures on a tool level are therefore to be understood as supporting components for reducing risks. In addition, the applied metrics and measurements are very often the starting point for creating a larger innovation measurement system. It is highly recommended to adapt the quantifications, measurements, and metrics to the specific situation, objective, and available resources. A step-by-step guide for how to create an innovation measurement system and how causality of two more metrics help in providing evidence is outlined in the Measuring 101 section (see pages 145–255).

Examples and ideas for quantification and measurement at the tool level are included throughout the toolkit.

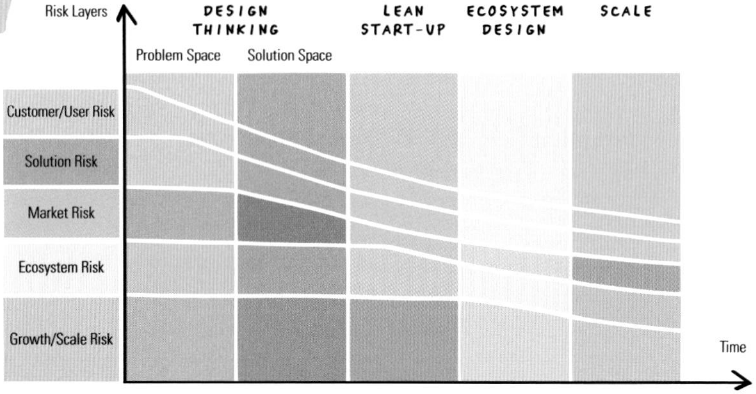

Risk Layers
DESIGN THINKING
LEAN START-UP
ECOSYSTEM DESIGN
SCALE
Problem Space
Solution Space
Customer/User Risk
Solution Risk
Market Risk
Ecosystem Risk
Growth/Scale Risk
Time

Preview of Suitable Methods and Tools for Design Thinking

Toolkit

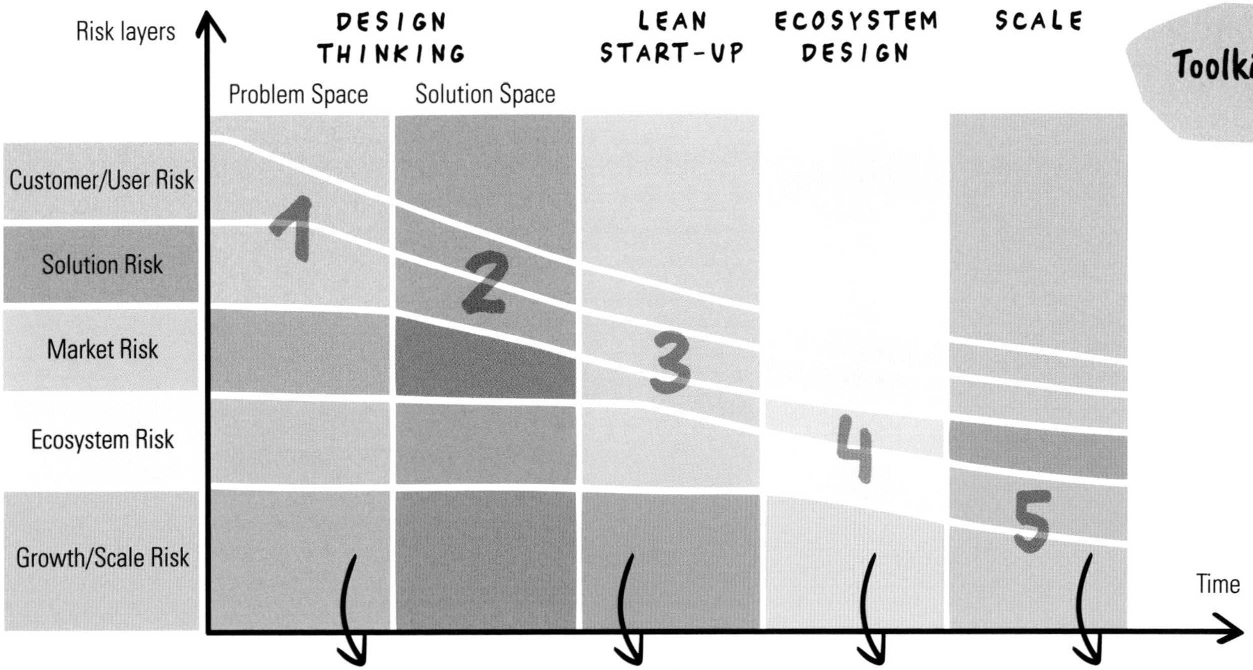

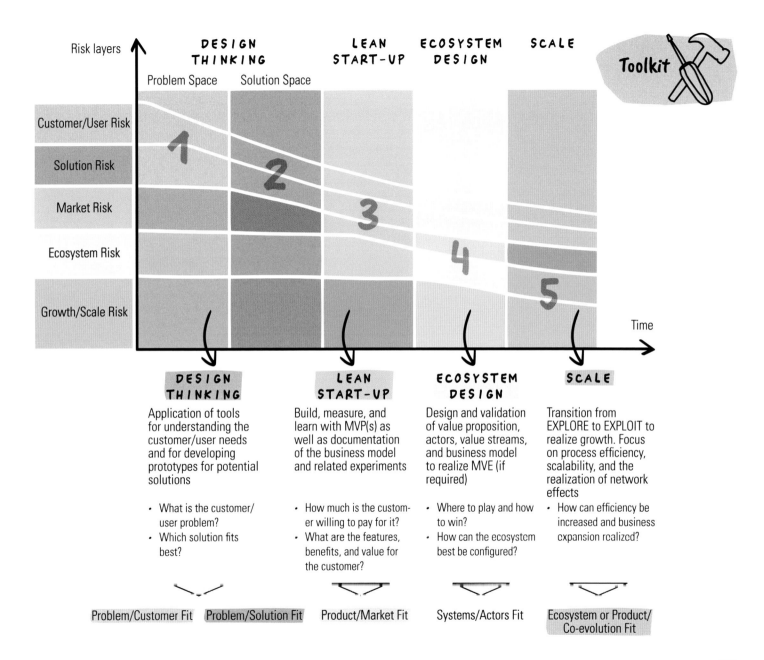

Risk layers
DESIGN THINKING
LEAN START-UP
ECOSYSTEM DESIGN
SCALE
Problem Space
Solution Space
Customer/User Risk
Solution Risk
Market Risk
Ecosystem Risk
Growth/Scale Risk
1
2
3
4
5
Time

DESIGN THINKING

Application of tools for understanding the customer/user needs and for developing prototypes for potential solutions

- What is the customer/user problem?
- Which solution fits best?

LEAN START-UP

Build, measure, and learn with MVP(s) as well as documentation of the business model and related experiments

- How much is the customer willing to pay for it?
- What are the features, benefits, and value for the customer?

ECOSYSTEM DESIGN

Design and validation of value proposition, actors, value streams, and business model to realize MVE (if required)

- Where to play and how to win?
- How can the ecosystem best be configured?

SCALE

Transition from EXPLORE to EXPLOIT to realize growth. Focus on process efficiency, scalability, and the realization of network effects

- How can efficiency be increased and business expansion realized?

Problem/Customer Fit | Problem/Solution Fit | Product/Market Fit | Systems/Actors Fit | Ecosystem or Product/Co-evolution Fit

Design Thinking (Problem Space): Problem / Customer Fit

Design thinking primarily focuses on the customer/user and solution risk. Proficient design thinking and innovation teams ideally spend 80 percent of their time trying to better understand the customer problem (problem space). Without a deep understanding of the customer needs, the efforts in the solution space quickly fizzle out. However, observations and measurements in companies across all industries and sizes unfortunately show a different picture. Many teams spend the lion's share (up to 70 percent) of activities and time on ideation without having understood the customer's problem. Measurements also reveal that teams sometimes spend less than 5% of their available time on the first two quarters of the double diamond. It is not surprising that, on this basis, design thinking generates only output but no results and impact. For this reason, the double diamond should be presented and applied in an asymmetric form to provide the appropriate perspective on the design work that is expected and will be performed.

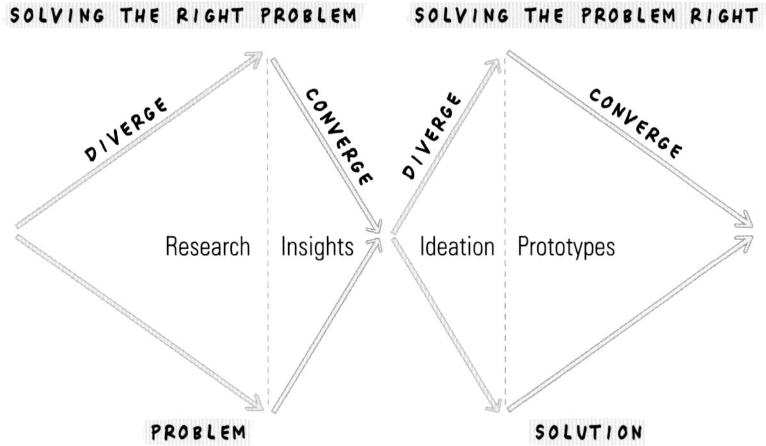

SOLVING THE RIGHT PROBLEM
SOLVING THE PROBLEM RIGHT
DIVERGE
CONVERGE
DIVERGE
CONVERGE
Research
Insights
Ideation
Prototypes
PROBLEM
SOLUTION

The *Design Thinking Canvas* (see page 129) supports the team in documenting the activities from the problem definition to the final prototype. A focused application of the design thinking mindset and tools helps to create long-lasting value based on the problem/customer fit and problem/solution fit.

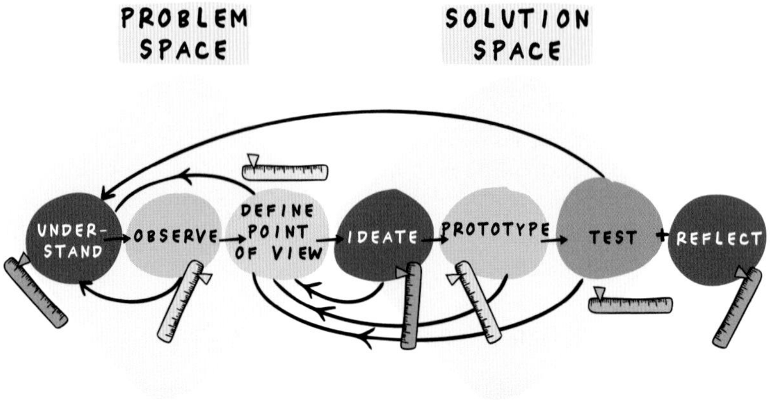

PROBLEM SPACE
SOLUTION SPACE
UNDER-STAND
OBSERVE
DEFINE POINT OF VIEW
IDEATE
PROTOTYPE
TEST
REFLECT

De-risk development by continuously assessing the dimensions of desirability, feasibility, and viability to evaluate high-uncertainty, transformative innovation initiatives.

Examples of Indicators Related to Design Thinking:

- Confidence level of the team regarding problem worth solving
- Overlapping characteristics, needs, behaviors in a set of personas
- Number of new insights collected per deep interview of customers
- Scoring of ideas on desirability, feasibility, and viability
- Number of critical and unknown assumptions identified
- Percentage of change from unknown to known in each iteration of validating critical assumptions
- Number of interactions across the organization to collaborate
- Reduced cost from start to implementation
- Design thinking maturity level of team/organization
- Hard financial facts, outcomes, and impact on band value and customer satisfaction

Define Success

Template download

For measurement purposes over the design thinking cycle, it is possible to initially define success at various levels and applications. What constitutes success on a very high level may already be roughly anchored in a design brief, derived from the design principles or, in many cases, initially defined by the design thinking team itself. However, defining success can be done in different phases of the design cycle. Most teams start before the project. Defining success can also be done based on the point of view, after the problem space has been observed for the first time, or much later in the process close to implementation. Mostly, the type of project determines the appropriate time for defining success.

Defining Success Helps the Design Thinking Team:

- to get alignment from the team members on what success should be.
- to ensure the team understands the organization, management, users/customers, and other stakeholders' requirements (so it is easy to get buy-in later or prepare the transition from EXPLORE to EXPLOIT).
- to select and prioritize throughout the project.
- to create a concrete start for measuring with clear outcomes.

Procedure and Template:

1. The principle is to do an ideation, first considering all the elements that may be defining success, then narrow them down to the critical ones.
2. Typical questions are related to the team, organization, company, customers/users, financial impact, or other stakeholder relationships:
   - What does success mean for the potential customer?
   - What does success mean for the company?
   - What may be the value of the project for the organization?
3. Once all these elements have been shared, the team may need to go through a converging exercise that will help pick the key elements for success. For example, clustering the key areas for defining success, then voting on main areas.
4. As the project progresses it is useful to review the criteria for success several times, as internal or external requirements may change.

EVALUATION AND SELECTION TEMPLATE FOR SUCCESS

Evaluate the answers to relevance. Define the values for the example.

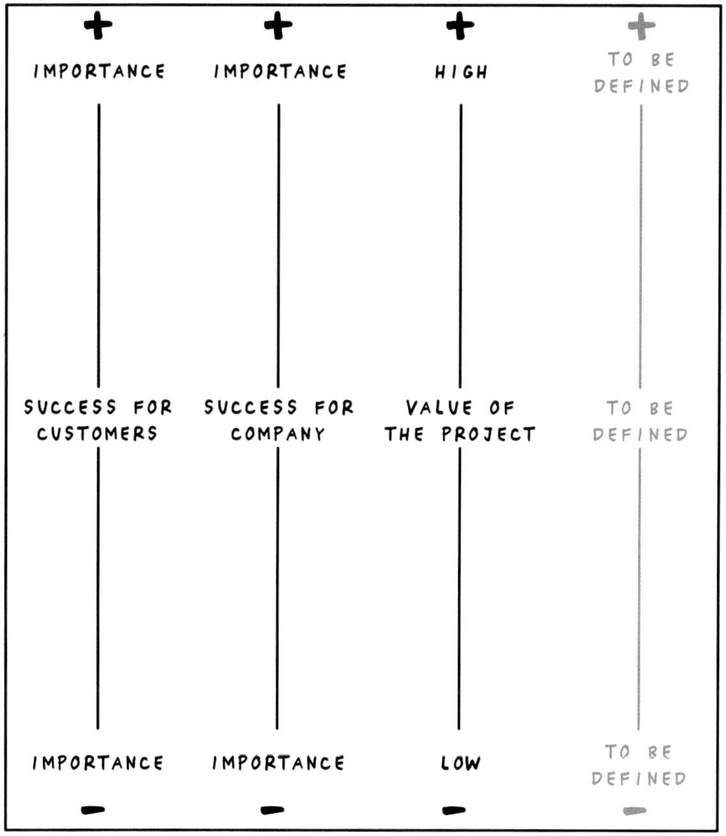
+ IMPORTANCE
SUCCESS FOR CUSTOMERS
IMPORTANCE –
+ IMPORTANCE
SUCCESS FOR COMPANY
IMPORTANCE –
+ HIGH
VALUE OF THE PROJECT
LOW –
+ TO BE DEFINED
TO BE DEFINED
TO BE DEFINED –

DOWNLOAD TOOL
www.design-metrics.com/en/define-success

Problem Statement

The problem statement is, in many cases, based on initial research about the current and future customer needs and number of customers (rough estimate of the size of the market). In some cases, trend reports or pictures of the future are supporting the reasoning. Besides early evidence that the problem matters, many of the customer pains, gains, or jobs to be done are based on assumptions. Without starting the design thinking process with *observing* and *understanding,* most of the assumptions will not be validated. The problem statement outlines which user/customer problem might be solved. In other words, what is the design challenge? The problem statement is part of the empathy phase. Different immersion and observation techniques support the empathy phase (see for example the empathy map on page 110).

The Problem Statement Helps the Design Thinking Team:

- to develop a common understanding of the problem.
- to formulate the collective findings from initial research in a design challenge.
- to outline the direction and next activities (e.g., validating assumptions).
- to create and support a reference for measuring success (see page 107).

Procedure, Template, and Examples of Measurements:

1. Ask key questions to formulate the problem statement: What is the problem? Why is it a problem? Who has the problem? Who has a need? When and where does the problem occur? How is it solved today?

2. Based on the analysis, formulate a problem statement. Use, for example, the following form:

How might we (re)design..(WHAT)
for..(WHOM),
so that.. (THEIR NEED) is satisfied?

Template download
1
2
3
4
5

WH QUESTIONS

WHO	WHAT	WHEN	WHERE	WHY	HOW
Who is involved?	What do we know already about the problem?	When did the problem start?	Where does the problem occur?	Why is the problem important?	How could the problem be an opportunity?
WHO...	WHAT...	WHEN...	WHERE...	WHY...	HOW...

PROBLEM STATEMENT

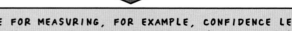

BASELINE FOR MEASURING, FOR EXAMPLE, CONFIDENCE LEVEL AND OUTLINING CRITICAL ASSUMPTIONS (SEE PAGE 109)

DOWNLOAD TOOL
www.design-metrics.com/en/problem-statement

Measuring Confidence Level and Critical Assumptions

Example

For measuring the relevance of a **problem statement**, two methods are helpful to create indicators in two perspectives:

- Assessing the confidence level of the team with periodic check-ins
- Discussion and evaluation of the critical and unknown assumptions

Examples of questions to assess the confidence level of the team with regard to a problem worth solving

How confident is the team in working on the problem?

1. How confident are we with the targeted user profile (persona)?

Low		Not sure		High

2. How confident are we about the deeper needs of the user/customer addressed in the problem statement?

Low		Not sure		High

3. How confident are we about the estimated size of the market (total and addressable)?

Low		Not sure		High

4. How confident are we with any limitating factors from the design principles?

Low		Not sure		High

Categorization of critical assumptions related to initial design phases (e.g., desirability, feasibility, and viability)

What is the team assuming with regard to customer behavior (desirability), monetization (viability), and technology (feasibility)?

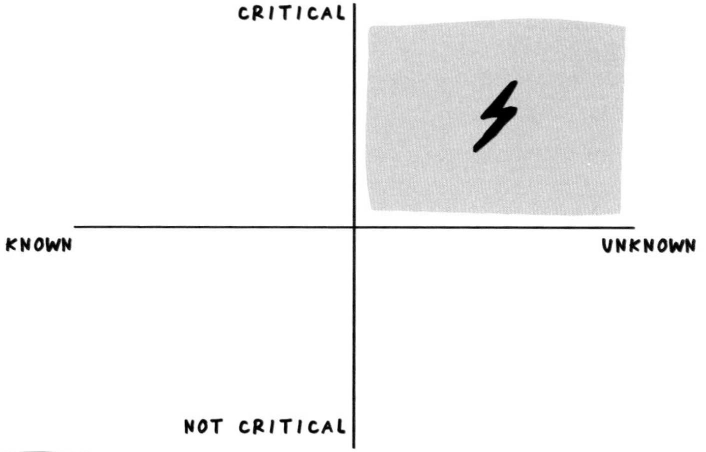
CRITICAL
KNOWN
UNKNOWN
NOT CRITICAL

Initial confidence level of team versus confidence level in every micro design thinking cycle:
The measurement indicates the ambiguity at the start and compares the confidence level with prototypes and experiments performed until the final solution is performed. The metrics also help the team to overcome the "groan zone," switching from diverging to converging.

# of critical and unknown assumptions identified:
The measurement indicates how much uncertainty is related at the start of project and helps to estimate the time and resources needed to validate assumptions.

% of change from unknown to known in each iteration of the design thinking micro cycle of validating critical assumptions:
The metric indicates the continued process of knowledge gain along the activities.

Empathy Map

Template
download

The empathy map is great tool for capturing and synthesizing the conversation informed by the design thinking team's observations. The visualizing of customers/user attitudes and behaviors helps to create a deep understanding of customers/users. The mapping process and discussion with the design team also reveal any discrepancies about the real and assumed behaviors of customers/users. Any of unexpected insights are especially of value for the further development of personas and the creation of jobs to be done (JTBD).

The Empathy Map Helps the Design Thinking Team to Answer the Following Questions:

- WHAT are some quotes and defining words the customer/user said? (**SAY**)
- WHAT actions and behaviors are performed by the customer/user? (**DO**)
- WHAT might the customer/user be thinking? What does this tell you about their beliefs? (**THINK**)
- WHAT emotions might the customer/user be feeling? (**FEEL**)

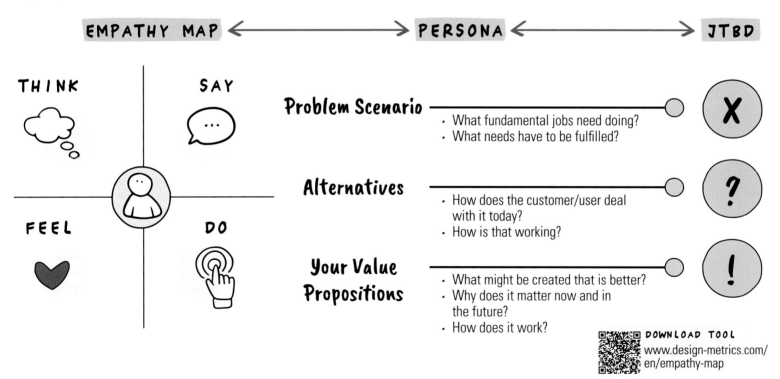

EMPATHY MAP ⟷ PERSONA ⟷ JTBD

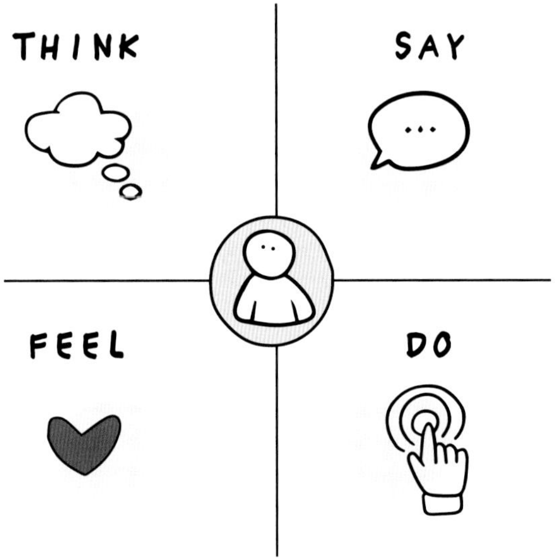
THINK
SAY
FEEL
DO

Problem Scenario

- What fundamental jobs need doing?
- What needs have to be fulfilled?

Alternatives

- How does the customer/user deal with it today?
- How is that working?

Your Value Propositions

- What might be created that is better?
- Why does it matter now and in the future?
- How does it work?

X
?
!

DOWNLOAD TOOL
www.design-metrics.com/en/empathy-map

Measuring Activities Related to Customer/User Interactions

Building empathy with the customer/user is an important part of the design thinking mindset. At the same time, it is also an integral design thinking process step that should be given special attention by the design thinking team. That is why, especially for those situations where the empathy of a design thinking team with the customer/user is central or frequent, appropriate measurements should be applied to evaluate these activities.

1. **Time between respective observations and deep interviews of customers/users:** This measurement can be initially collected for typical innovation projects, and activities can be taken to reduce the time between interactions.

2. **Number of customer/user interactions over the entire design thinking macro or micro cycle:** The ultimate goal is to keep the number of interactions constant throughout the entire design cycle (i.e., from need-finding to solution testing).

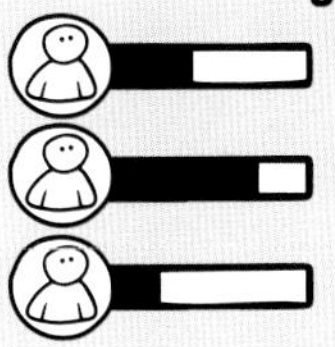

3. **Application of a lagging indicator on the variety of interactions with customers/users:** This measurement compares with the persona used to solve the customer problem with the actual customers/users. The metric allows you to analyze in retrospect to which kinds of customers and needs an interaction has taken place over the whole design cycle.

How to Measure Empathy

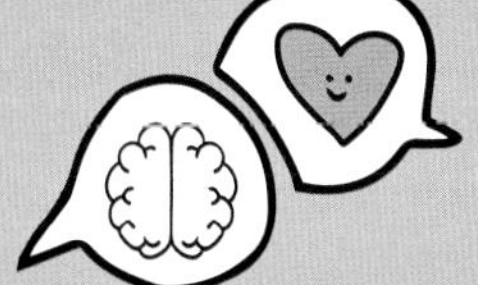

Experienced design thinking teams know that measuring empathy is more difficult than other activities. The future possibilities of neurodesign are discussed in 301, which can fundamentally help to better explore the *why* in the context of the problem space (see page 364), and thus reduce some challenges in the application of the currently used empathy tools. In general, measuring empathy is problematic because meaningful comparisons about how to define and measure empathy are lacking. Hopefully, the ambiguity of the term *empathy* will soon be overcome, and consistent measures will be developed. Research in social cognitive neuroscience suggests that four different neural networks should be activated for a human to have a complete experience of empathy. These neural networks should be the basis for future measurement and are important components of the application of the empathy map for design thinking and innovation teams:

1. Effective sharing: Subjective reflection by the design thinking team on another person's observable experience.

2. Self-awareness: Differentiation between the experiences of the design thinking team and those of the customer/user being observed.

3. Mental flexibility: The ability of the design thinking team to effectively imagine what it would be like to experience the world from the customer's/user's position.

4. Emotion regulation: The design thinking team's ability to regulate the volume of their own feelings as they arise from mirroring the customer/user experience.

Jobs to Be Done

Jobs to be done (JTBDs) is a conceptual tool that is often used in connection with the development of a persona and the reflection on gains and pains in order to convert observational findings into strong statements that the customer/user implicitly or explicitly tries to achieve. Often there are countless JTBDs that a customer/user has to complete, and these JTBDs can contain both emotional and functional aspects, depending on the problem to solve. Thus, the selection of the appropriate JTBD is not an easy task for design thinking teams. Finally, the best possible scenario is to focus on the JTBD that has the greatest potential to realize a market opportunity or is the starting point for radical new business ecosystem considerations. For the formulation of a JTBD, a simple pattern based on an outline of the situation, motivation, and expected result can be applied.

When ____________, I want to ____________ so I can ____________.

(Situation) (Motivation) (Expected result)

Example of Measuring JTBD with the Application of Two Dimensions:

- **How important is the task to be accomplished for the customer/user?**
- **How satisfied is the customer/user with the solutions that already exist?**

For the measurement of JTBDs, the described dimensions can be used to map the customer's viewpoint on the one hand and on the other hand to carry out a valuation of the opportunity. It has proven useful to survey the importance for the customer/user on a Likert scale (degree of importance). Satisfaction with existing solutions can also be surveyed in this way.

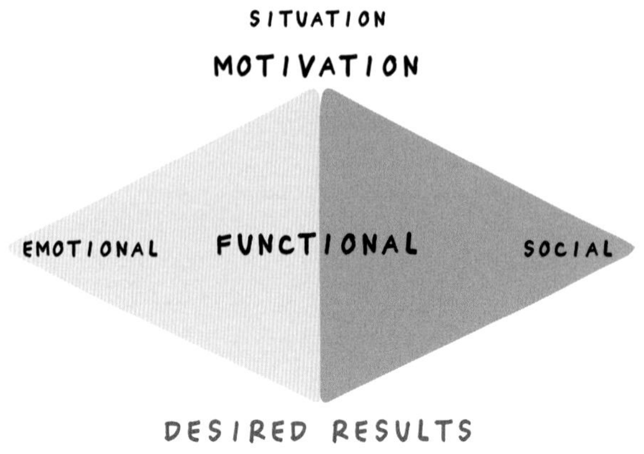
SITUATION
MOTIVATION
EMOTIONAL
FUNCTIONAL
SOCIAL
DESIRED RESULTS

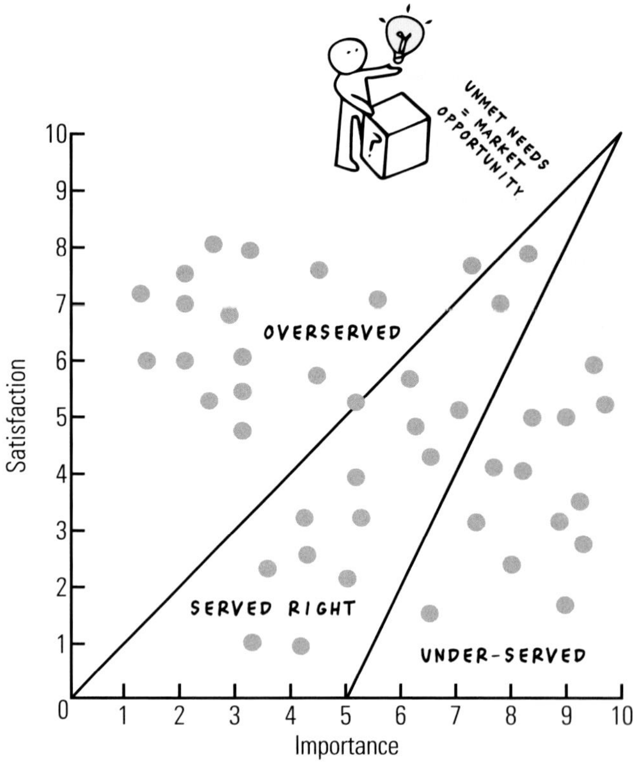
UNMET NEEDS = MARKET OPPORTUNITY
OVERSERVED
SERVED RIGHT
UNDER-SERVED
Satisfaction
Importance
0
1
2
3
4
5
6
7
8
9
10

The Power of Game-changing JTBDs

Example

The specific JTBDs are the basis for many design thinking teams in determining the metrics that customers use to measure success in completing each step of the job to be done. These metrics, based on desired outcomes, give predictability to a potential market opportunity. A desired outcome is a specifically constructed statement that has a set of unique characteristics related to the outcome. Characteristics range from solution-free, stable over time, measurable, controllable, and/or tied to a process (or the task) that the customer/user is trying to complete.

Many of the JTBD statements are created within the first two design thinking activities (understand and observe customers/users) of the design thinking micro cycle. A great example of applying the combination of design thinking and data analytics (see the hybrid model on page 320) for detecting powerful JTBD is the transformation of TikTok (formerly musical.ly) into an engaging consumer platform. TikTok has become the number-one downloaded app by understanding new customers' needs, motivations, and behavior.

Musical.ly is an app that focuses on adding background musics and videos.

The TV show ***Lip Sync Battle*** becomes famous on Thursday nights.

TikTok realizes customer/user needs for lip sync and the JTBD for cool video features.

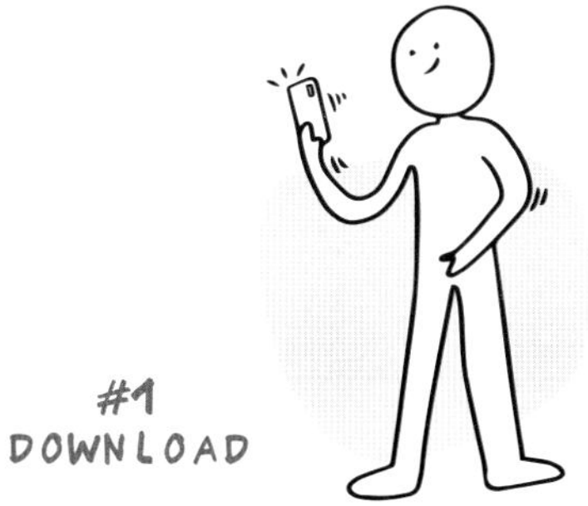
#1
DOWNLOAD

Downloads of the Musical.ly app increase after the TV show

EVIDENCE BASED

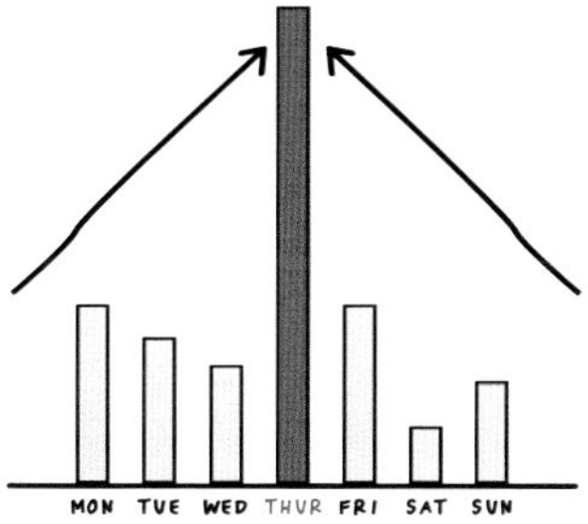
MON
TUE
WED
THUR
FRI
SAT
SUN

WHY MORE DOWNLOADS?

Fans search in App Stores for "lip sync app" after the show. Musical.ly shows up in the top search result, which generates massive downloads.

NEW CUSTOMER NEEDS

A new value proposition based on user motivation, advanced features, and JTBD leads to better ad hoc videos.

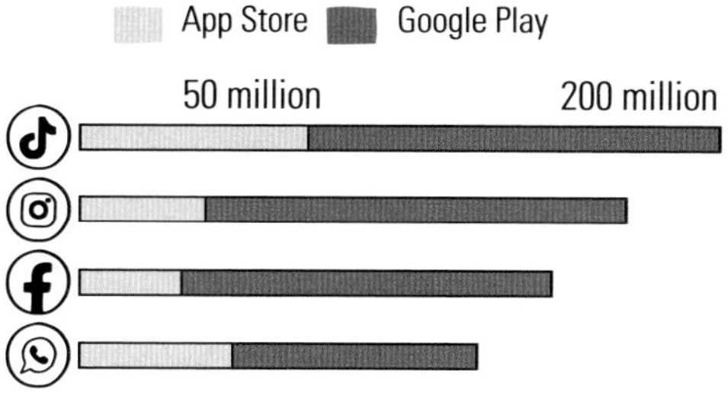
App Store
Google Play
50 million
200 million

2015 — 2018 — 2022 → TIME

Personas/User Profiles

Personas/user profiles are a great way either to define initial assumptions about user behavior and needs (which are later validated) or to document the collected insights from observations and in-depth interviews. For the representation of personas, there are different possibilities, and for the comparison of different personas, it is recommended to choose criteria that can be derived from the design brief. For example, if the final solution has to be provided digitally, it makes sense to compare the personas in terms of their maturity in the use of digital media or to express it in a temporal dimension in terms of hours per day that the persona has active screen time.

Typical Elements to Describe a Persona:

Location:	Where is my persona from?
Gender:	What is the gender of my persona?
Age:	How old is my persona?
Education level:	What education level does my persona have?
Hobbies/Interests:	How does my persona spend leisure time?
Income:	What is the income of my persona?
Marital status:	Does my persona have a family?
Occupation:	What is the professional sphere my persona works in? What are the responsibilities of my persona?
Channels:	What websites or social media does my persona use?
Motivation:	Why does persona buy my products? What is my persona's goal?
Pains:	What are the concerns of my persona? What problems exist?
Gains:	What possibilities does the persona have? What benefits is the persona already experiencing?
JTBD:	What are the jobs to be done (see page 112)?

Template download
$

Typical Criteria for Comparing Personas:

Motivational Criteria:
Achievements, personal growth, efficiency, convenience

Personality Criteria:
Introvert versus extrovert; thinking versus feeling; sensing versus intuition; judging versus perceiving; loyal versus fickle; passive versus active

Customized Criteria Based on Problem Statement:
For example, confidence in digital, level of independence, planning horizon, buying behavior, preferred channels (e.g., multi-, opti-, omnichannel)

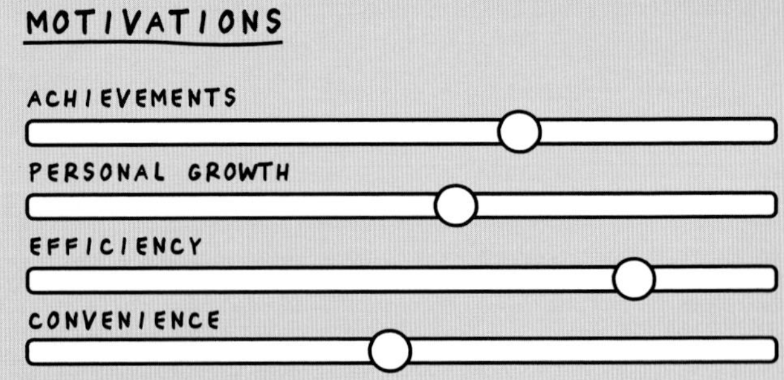
MOTIVATIONS
ACHIEVEMENTS
PERSONAL GROWTH
EFFICIENCY
CONVENIENCE

DOWNLOAD TOOL
www.design-metrics.com/en/persona-comparison

Quantitative Specifications of Personas

Example

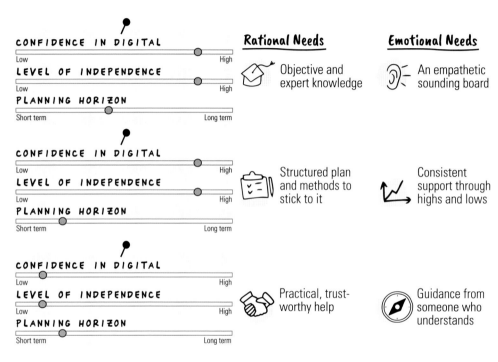

Example of the personas comparison of 65+ segments based on three customized criteria with regard to confidence in digital, level of independence, and (financial) planning horizon

Personas of 65+ Segments

Rational Needs | **Emotional Needs**

Independent Rosemarie

- Leads a financially comfortable life and is confident in managing money
- Proactive planner who thinks about the future

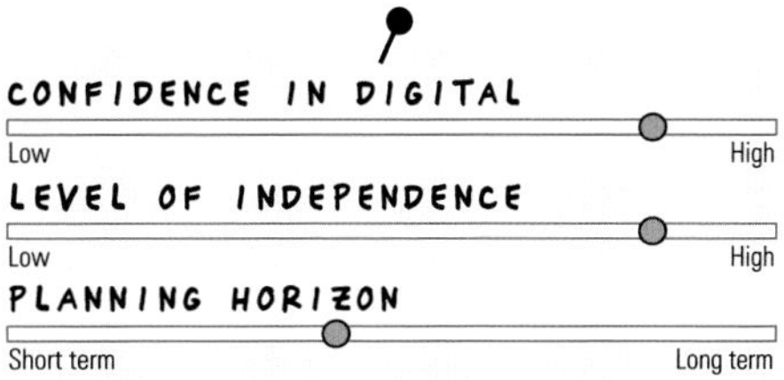
CONFIDENCE IN DIGITAL
Low
High
LEVEL OF INDEPENDENCE
Low
High
PLANNING HORIZON
Short term
Long term

Rational Needs: Objective and expert knowledge

Emotional Needs: An empathetic sounding board

In-the-Moment Bill

- Lives day-to-day and makes the most of life with the money he has
- Deals with problems when they occur

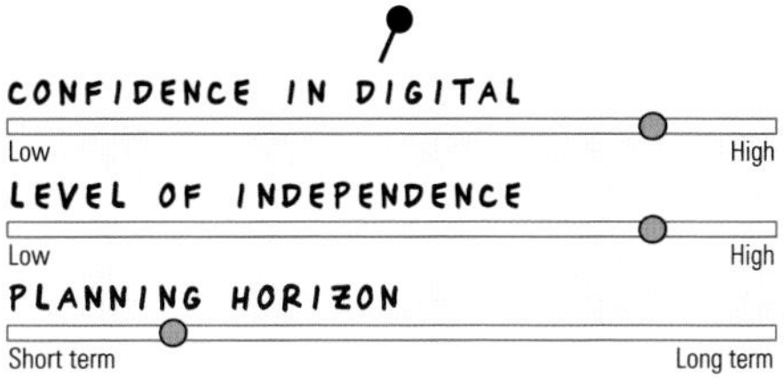
CONFIDENCE IN DIGITAL
Low
High
LEVEL OF INDEPENDENCE
Low
High
PLANNING HORIZON
Short term
Long term

Rational Needs: Structured plan and methods to stick to it

Emotional Needs: Consistent support through highs and lows

Dependent Anna

- Has physical and cognitive disability
- Relies on support in managing finances day-to-day

CONFIDENCE IN DIGITAL
Low
High
LEVEL OF INDEPENDENCE
Low
High
PLANNING HORIZON
Short term
Long term

Rational Needs: Practical, trust-worthy help

Emotional Needs: Guidance from someone who understands

Examples of Measuring the Perceived Value of Potential Solutions

Real customers/users corresponding to the persona can be used in measuring later in the design thinking process in measuring the perceived value of specific features, potential products and services, or the defined value proposition. Key questions to ask and to collect data points might refer to the following three questions and validations:

- **Feature Value:** What features and aspects of the product do potential customers/users care most about? What do they care least about?
- **Quantified Value Propositions:** What value messaging drives them to sticking to a specific proposition?
- **Price Sensitivity and Willingness to Pay:** How much are the customers/users willing to pay to solve their problem?

Critical Items Diagram

The critical items diagram is a frequently applied tool for designing physical, digital, and hybrid experiences, services, and products. It helps the design thinking and innovation team to agree on the critical success elements for the target group based on initial findings, the definition of a point of view (see page 118), or validating a persona. Many of the elements elaborated in the critical items diagram are the ones that are most likely included later at the stage of creating the final prototype to fulfill customer/user needs. The described elements in the critical items diagram can either describe the expected experience or present expected functions. It is recommended to challenge the elements of the diagram with each iteration. However, some of the critical items will necessarily have relevance to a critical experience or critical function up to the final prototype.

The Critical Items Diagram Helps the Design Thinking Team:

- to appraise the results from the understand and observe phases and filter out the crucial elements.
- to prepare the ideate and prototype phases, later the realize-MVP phase, to establish a good starting position.
- to figure out the things that are essential to the project and agree on what might be important to success.

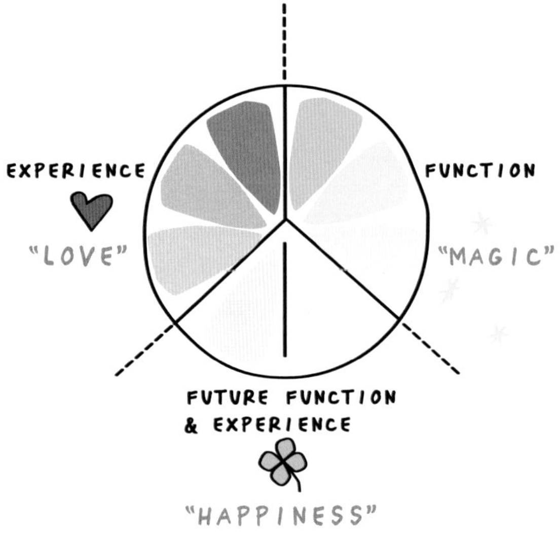
EXPERIENCE
"LOVE"
FUNCTION
"MAGIC"
FUTURE FUNCTION
& EXPERIENCE
"HAPPINESS"

Procedure, Template, and Examples of Measurements:

1. At the beginning of this step, ponder the question: "What is crucial for a successful solution for the customer/user in the context of the problem?" This is based on the findings from the understand and observe phases.
2. Sketch a critical items diagram on the whiteboard or a large sheet of paper and discuss in the team which experiences the user must have/which functions are critical for the user/customer.
3. The team brainstorms and writes elements that are critical for the user on Post-its.
4. The team selects and names four experiences and four functions, one of which focuses on completely new or future expectations.
5. Discuss the results and agree in the team about the final eight critical items. The critical items diagram supports the definition of the "How might we..." questions that are thought provoking enough to launch the ideate phase successfully.

The items (functions and experiences) are best placed by applying the concept of the daisy flower. Each petal has the same importance as the others.

Metrics for Measuring Critical Items

One way of measuring critical items is the extent to which individual critical items develop over time in terms of their relevance and time-related importance. Such measurements have led to insights into how quickly individual functions, such as voice interaction in vehicles, become highly relevant for customers over different design challenges.

CLUSTER OF CRITICAL FUNCTIONS AND EXPERIENCES RELEVANT IN THE FUTURE [YEAR 0-10]

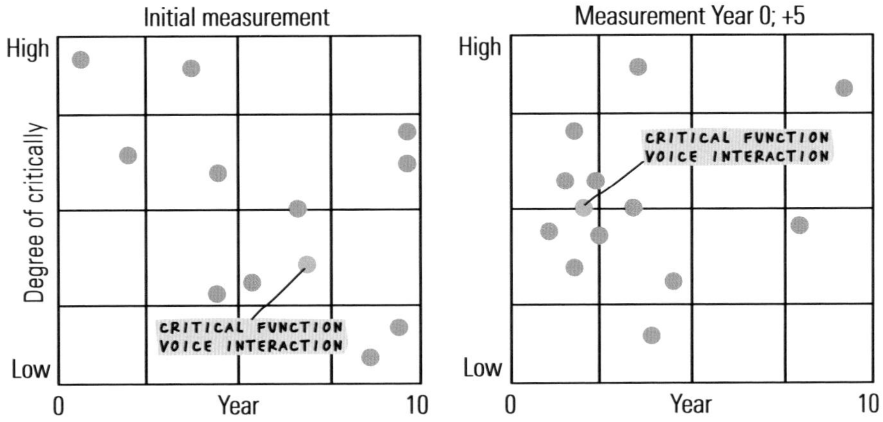
Initial measurement
High
Degree of critically
Low
0
Year
10
CRITICAL FUNCTION VOICE INTERACTION
Measurement Year 0; +5
High
Low
0
Year
10
CRITICAL FUNCTION VOICE INTERACTION

Importance of Future Functions and Experiences for User:

Measurement of the extent to which individual critical items develop over time in terms of their relevance and time-related importance.

In many cases the functions are related to the experiences, future use cases, and JTBD of the user/customer. Some mega and technology trends accelerate the user/customer adaption and experiences from the living environment (e.g., Alexa or other conversational AI technology) becoming a necessity in the application of mobility.

Example

Science fiction and foresight in the 1980s

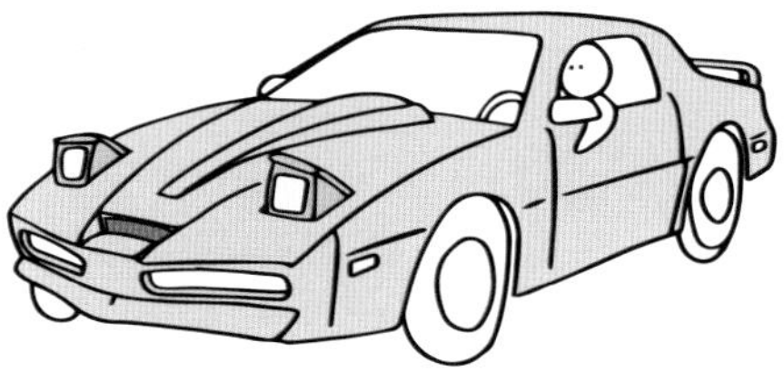

Customer awareness in the 2000s

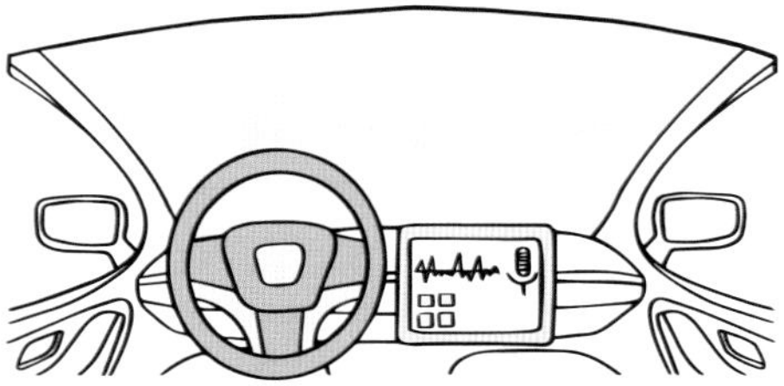

Integration with Siri, Alexa, and Google for all driver and passenger functions +AR experience to be connected to the external world.

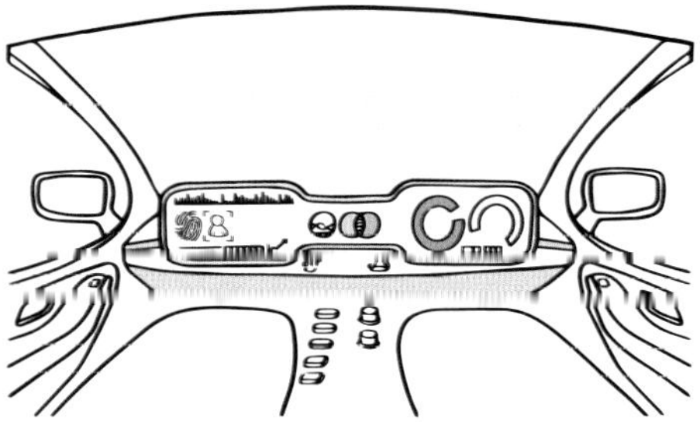

Point of View

The point of view (PoV) is the reframing of the design challenge into an actionable problem statement. The PoV is based on the first two phases of the design thinking process (understand and observe). The analysis of the customer/user behavior and deeper needs is the first step of breaking down complex concepts and problems into constituents which are better and easier to understand. For the creation of the PoV, the design thinking team focus is on organizing, interpreting, and making sense of the collected data and insight. The next step of the design thinking process, ideate, is based on the previously mentioned actionable problem statement based on HMW questions.

1. **Insights:** Collect and organize information. Identify what every team member sees.
2. **Clustering:** Sort gathered information and manipulate or reframe the collected information, as necessary.
3. **Drill down into the design challenge/problem:** Build HMW questions on different PoVs.
4. **Prioritization:** Decide which PoVs and related HMW questions might have the biggest impact and highest urgency.

Applying an analysis and synthesis approach provides design thinking and innovation teams with a path to identify true, core insights on which to build the HMW questions.

How might we redesign ....(WHAT) for ....(WHOM), so that.. (HIS NEED) is satisfied, by considering .... (INSIGHT/REQUIREMENT/LIMITATION)?

HOW TO CREATE POWERFUL HMW QUESTIONS

- Adjust the original problem statement with focused HMW questions.
- Present the identified challenge as a problem statement.
- Use the HMW question to evaluate competing ideas.
- Phrase the question in a way that makes it actionable and tangible.
- Remember always that the HMW questions guide the innovation effort and serve as inspiration to the team.
- For different ways of exploring, expand the HMW questions into "What would happen if ..." or "What is stopping us from ..." once the team feels comfortable.

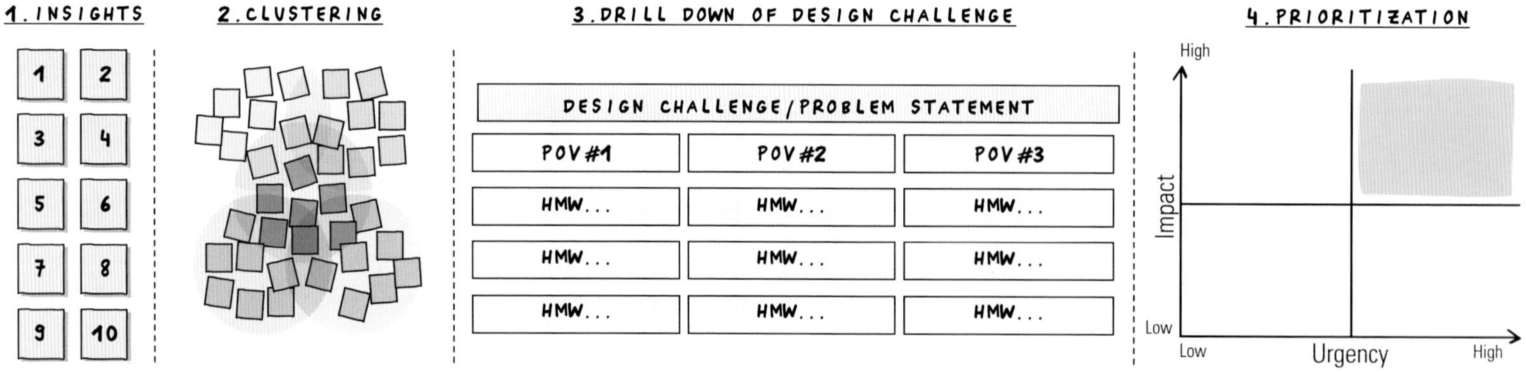

1. INSIGHTS
1
2
3
4
5
6
7
8
9
10
2. CLUSTERING
3. DRILL DOWN OF DESIGN CHALLENGE
DESIGN CHALLENGE/PROBLEM STATEMENT
POV #1
POV #2
POV #3
HMW...
4. PRIORITIZATION
High
Impact
Low
Low
Urgency
High

Understanding the Relationship between Analysis and Synthesis

The analysis of findings from the observe and understand phrases involves disintegrating the fragments. Synthesis creates the bigger picture and forms the basis for experimenting and applying. Analysis and synthesis often occur at the same time. By structuring the analysis, the problem is framed, while synthesis is emergently helping to create connections that identify breakthrough ideas and opportunities. Insights derived from the smart use of data and collected information are hugely powerful. Design thinking teams who are able to develop big insights from any level of data or collected information are usually more successful in creating breakthrough innovation.

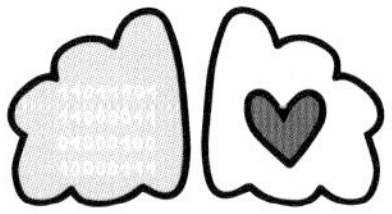

The hybrid model (see pages 320–334), which combines design thinking and data analytics, is ideal for gathering valuable insights at all levels and synthesizing the appropriate conclusions.

	Analysis	Synthesis
DEFINITION	The analysis aims to fragment a complex customer problem into micro sections to acquire more comprehension from it.	Synthesis aims to collect the micro sections and connect them together, in conjunction, to acquire a broader view of the problem.
CORRELATION	Without synthesis, analysis will not get justified.	Without analysis, synthesis will not take place.
KEY FEATURE	It involves research, observation, and theories of customer behavior and needs.	It involves practical experiments and outcomes that are new and are freshly available.
IMPACT	It is easy to understand and execute.	It is complex to understand and execute.
DEPENDENCY	Analysis is an independent process.	The synthesis process is dependent on analysis.
PROCESS STEPS	It is the process of thinking, observing, and then trying.	It is the process that includes applying, experimenting, and then studying and reflecting on the outcomes.
BENEFITS	It provides a detailed investigation of specific topics.	It provides an abstract overview of the whole topic.
TYPES	Qualitative, quantitative, predictive, descriptive, prescriptive, and diagnostic are the different types of analysis.	Sample-based, vector, frequency modulation, granular, wave table, additive, subtractive, physical modeling, spectral, and West Coast are the different types of synthesis.

Design Thinking (Solution Space): Problem / Solution Fit

The point of view (PoV) is the starting point for ideation (solution space). The confidence level of the team is usually much higher than the starting point, since the team has more information about the potential customer/user and his/her needs, and many of the critical assumptions have been verified. At this point confidence indicates how sure the team is about impact, and to some degree also about ease of implementation. The initial chaos is cleared and there is more clarity about the framework of a potential solution. Through different brainstorming variants and the construction of different prototypes, different solutions are made tangible and then tested with the customer/user. However, it is not advisable to measure the number of ideas in this context, as it is not a useful measure. It seems more important to check and evaluate the potential prototypes and solutions for their desirability, feasibility, and viability (see the introduction on page 73). A detailed guide to different creativity techniques, brainstorming variants, and prototype building (e.g., Dark Horse Prototype, Funky Prototype, Functional, X is Finished Prototype, and Final Prototype) can be found in the complementary tools and methods book *The Design Thinking Toolbox*.

Every prototype created helps in the validation of desirability and feasibility. In addition, the question of viability is becoming increasingly important in the solution space.

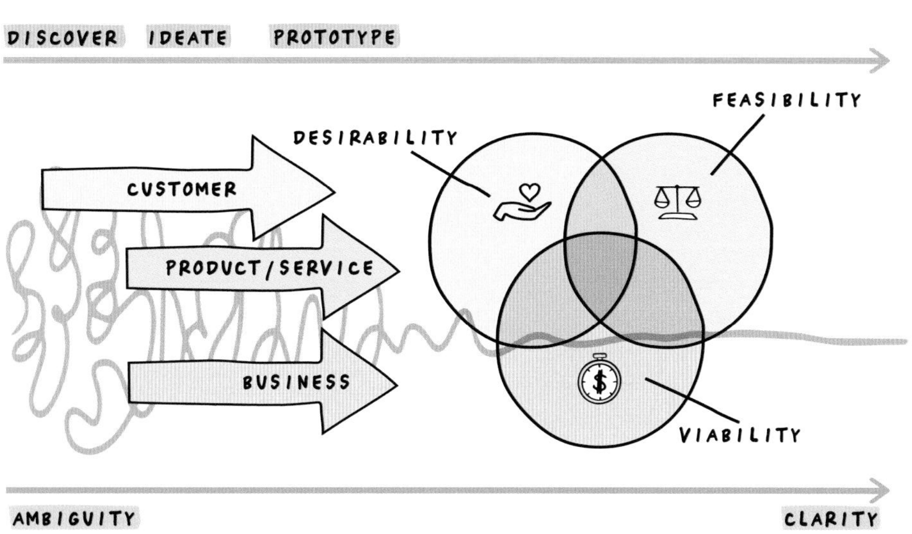
DISCOVER
IDEATE
PROTOTYPE
DESIRABILITY
FEASIBILITY
CUSTOMER
PRODUCT/SERVICE
BUSINESS
VIABILITY
AMBIGUITY
CLARITY

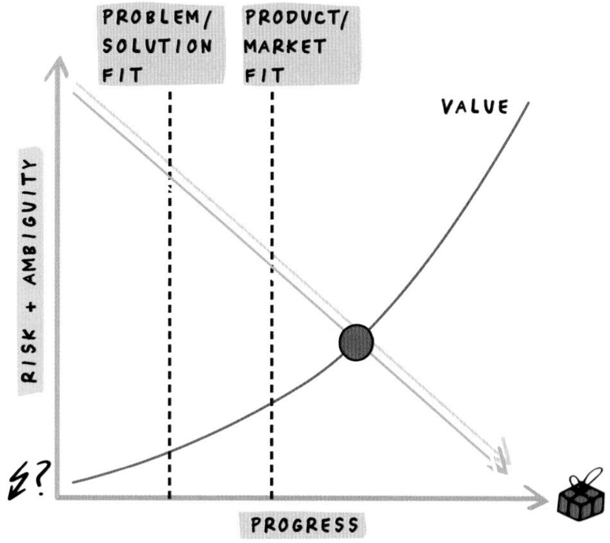
PROBLEM/ SOLUTION FIT
PRODUCT/ MARKET FIT
VALUE
RISK + AMBIGUITY
PROGRESS

Scoring Ideas

One possibility for measuring ideas is based on the dimensions of the previously mentioned design thinking lens. This approach is the foundation for many decisions the design thinking team makes over the entire design thinking cycle. This means that an idea can be reviewed in different phases of the design thinking macro cycle with regard to the central questions of how the idea scores against **desirability, feasibility, and viability**. Whereas in the early design phases the team usually concentrates mainly on desirability, the question of feasibility and viability becomes more important as the solution matures. Finally, all ideas, experiments, and solutions can be evaluated with respect to the three dimensions of the design thinking lens.

Scoring Ideas Helps the Design Thinking Team:

- to find out how much the idea, experiment, or solution is wanted by customers/users.
- to evaluate the possibility that the idea will become a reality.
- to understand if the necessary capabilities, technologies, and resources exist to bring it home.
- to evaluate the economical and sustainable model for creating, building, and maintaining the solution.

Procedure and Template:

1. Create a three-dimensional diagram for the ideas, experiments, or solutions based on the three dimensions (desirability, feasibility, and viability).
Tip: If you operate in only two dimensions, it is recommended to use the size of the idea as the indicator for desirability.
2. Find appropriate parameters to compare ideas on each dimension.
Tip: A good starting point is to compare new approaches with already existing solutions.
3. Discuss with the team the dimensions and constraints and find innovative solutions to overcome limitations' feasibility and viability.

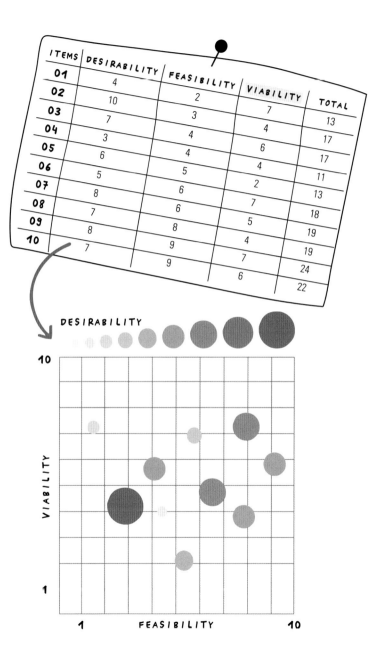

ITEMS	DESIRABILITY	FEASIBILITY	VIABILITY	TOTAL
01	4	2	7	13
02	10	3	4	17
03	7	4	6	17
04	3	4	4	11
05	6	5	2	13
06	5	6	7	18
07	8	6	5	19
08	7	8	4	19
09	8	9	7	24
10	7	9	6	22

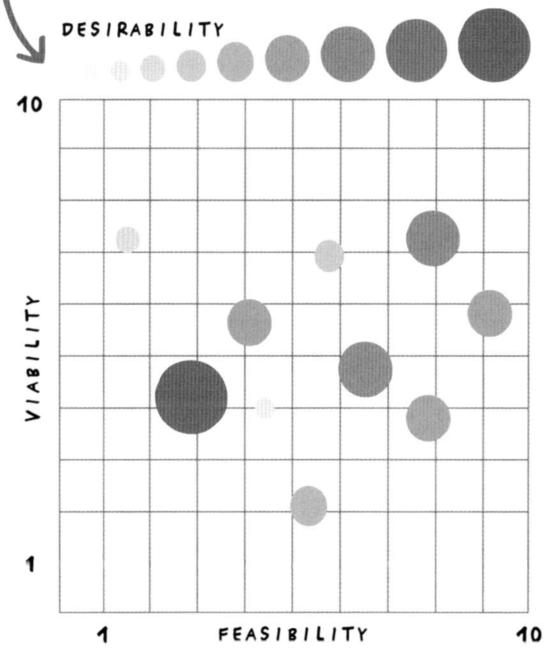
DESIRABILITY
10
VIABILITY
1
1
FEASIBILITY
10

Tracking Experiments / Prototypes

Similar to the previous tool, the individual experiments can be broken down even further. The *Experiments/Prototypes Tracking Canvas* helps to systematically question the assumptions in the dimensions of desirability, feasibility, and viability. On this basis, the confidence level of the teams can also be validated. The actions taken are related to the well-known design thinking phrase: **love it, leave it, or change it!**

The Experiment/Prototypes Tracking Canvas Helps the Design Thinking Team:

- to document all the assumptions for one experiment based on desirability, feasibility, and viability.
- to document what has been learned from the evidence to support or refute a specific assumption.
- to calculate the confidence level based on how strong the insights (I) are and the strength of the evidence (E), Likert: 1 = weak | 3 = strong, with the confidence level calculated as I x E = C (max. 9) (min. 1).
- to decide on action to take based on love it, leave it, or change it.

Procedure and Template:

1. The *Experiments/Prototypes Tracking Canvas* helps before starting an experiment, planning an experiment, capturing the outcomes of an experiment, and creating actions

2. Define the experiment/prototype testing in three phases:

- Before the experiment: Reflect on critical assumptions.
- Planning the experiment: Decide what to test and how.
- After the experiment: Validate insights, check the strength of evidence, and calculate the confidence level.

Compare the confidence level with the general feeling of the team about prototypes tested/experiments conducted.

3. Define actions based on outcomes of testing the *Experiments/Prototypes Tracking Canvas.*

What Actions Derive from Experiment/Prototype Measurements?

- **Love it:** This is typically when the hypothesis is confirmed, and the design thinking team is very confident about the strength of the evidence.
- **Change it:** This is a typical pivot when the design thinking team has gathered enough evidence to change parts of a prototype and to test it again afterwards.
- **Leave it:** This radical move is not an easy decision for design thinking teams. However, if there are multiple pieces of supporting evidence that disprove the hypothesis, the prototype will not make sense in most cases.

Examples of Measuring Activities Related to Experiments:

- # of experiments/prototypes which contribute to significant learning outcomes related to "love it, change it, or leave it"
- # of experiments/prototypes that were independently initialized, planned, and executed by the design thinking team

Template download

THE EXPERIMENTS / PROTOTYPES TRACKING CANVAS

NAME OF EXPERIMENT/PROTOTYPE TESTED:	MAIN OBJECTIVE OF THE EXPERIMENT/PROTOTYPE TESTING:

	CRITICAL ASSUMPTIONS	Confirmed or rejected?	LEARNINGS: Insights are true? (1-3)	Strength of the evidence? (1-3)	= Calculated Confidence Level I x E = C	ACTIONS	Love it, leave it, or change it?
DESIRABILITY							
FEASIBILITY							
VIABILITY							
SCORE							

DOWNLOAD TOOL
www.design-metrics.com/en/tracking

Probability of Innovation Success

Different statistical models (see more in Measuring 101, starting on page 146) can be applied by the design thinking team throughout the entire design cycle to establish measurements and metrics step by step. For example, the application of Bayes' theorem for determining the probability of success for innovations is helpful in determining a potential solution's likely success based on the data from the design thinking team's previously conducted experiments based on the exploration map (see page 70). An experiment is a success if the actual results at least correspond to the expected results. The expected results could, for example, be formulated in the form of the so-called XYZ hypothesis: "At least X% of target market Y will Z," where Z describes how this percentage of the target market will react to the solution idea.

Bayes' theorem – a classic of probability theory – is a calculation rule, which is used to calculate the (conditional) probability of an event.

SUCCESS
COST
RISK
UNKNOWN
TIME
FAILURE
ATTENTION
EMBARRASSMENT

Especially in the execution of different experiments within the comfort, stretch, risk, and dark zones, the advantage of Bayes' theorem is that, after each performed experiment, the learning progress can be evaluated by an updated calculation of the probability of success. The objective is to achieve a predefined minimum value for the probability of success by conducting experiments, from which point there is sufficient certainty that a potential solution can become a success. This threshold must be reached before investments for the realization of the final prototypes, MVPs, or any other relevant investments can be released to the design thinking and innovation team.

EXPLORATION MAP WITH PROBABILITIES

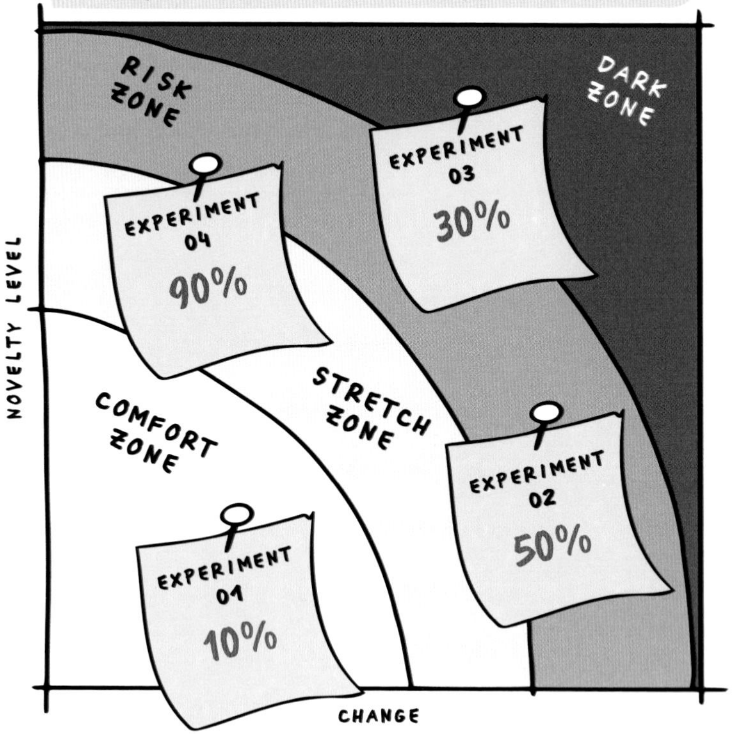
RISK ZONE
DARK ZONE
EXPERIMENT 03
30%
EXPERIMENT 04
90%
NOVELTY LEVEL
STRETCH ZONE
COMFORT ZONE
EXPERIMENT 02
50%
EXPERIMENT 01
10%
CHANGE

Which Prototypes Have the Potential for Breakthrough Innovation

Example

The input variables of the probability of success with respect to different competing solutions or a portfolio of potential innovations can also be visualized on an exploration grid for market opportunities. Typical axes reflect the expected return and the degree of evidence. It is recommended not to include any potential solution with no evidence, because such visualizations of potential market opportunities are usually the basis for applying portfolios for EXPLORE and EXPLOIT, which have the goal of pursuing a balanced strategy of exploiting existing business and exploring new market opportunities.

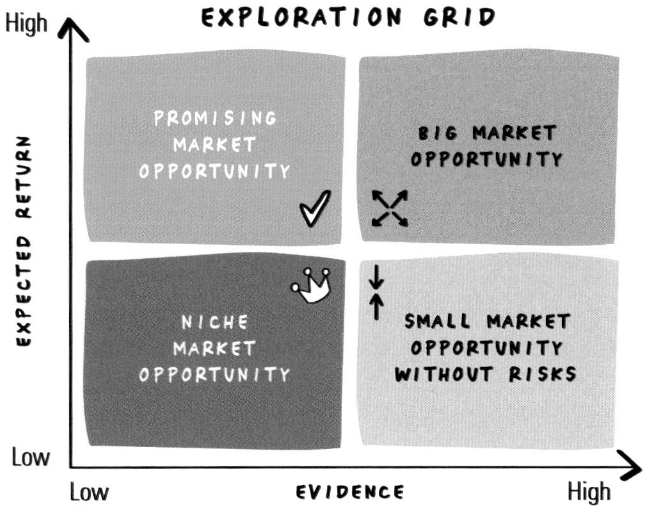
EXPLORATION GRID
High
Low
EXPECTED RETURN
PROMISING MARKET OPPORTUNITY
BIG MARKET OPPORTUNITY
NICHE MARKET OPPORTUNITY
SMALL MARKET OPPORTUNITY WITHOUT RISKS
Low
EVIDENCE
High

PROMISING MARKET OPPORTUNITY
- High financial potential
- Weak evidence of success

NICHE MARKET OPPORTUNITY
- Low financial potential
- Weak evidence of success

SMALL MARKET OPPORTUNITY WITHOUT RISKS
- Low financial potential
- Strong evidence for success

BIG MARKET OPPORTUNITY
- High financial potential
- Strong evidence of success

Examples of Indicators Related to the Exploration Grid:

- **Ratio of big, promising, niche, and small market opportunities in the exploration portfolio:** The metrics indicate the innovation risk on the exploration portfolio.
- **Budgets for teams gathering evidence over the entire design thinking cycle for potential market opportunities:** The quantification is an input for the expected return of investments in teams to provide evidence on the desirability, feasibility, and viability of potential market opportunities.

Alternatively potential market opportunities refer to core, adjacent, and transformational. Mostly used and applied in combination with the "where to play" and "how to win" framework.

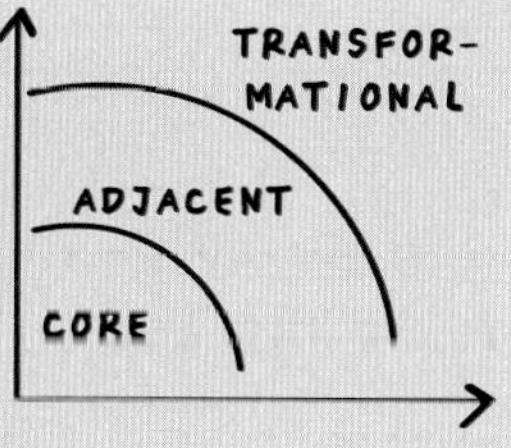
TRANSFOR-
MATIONAL
ADJACENT
CORE

Prototype Question Checklist

Another simple way of dealing with near-final prototypes, implementation, and defining measurements and metrics is to use appropriate question checklists to help teams determine all contingencies in the run-up to the key question: **how to bring it home**. The design thinking team creates a set of questions that need to be answered in order to proceed with a prototype/final prototype. Several questions and checklists can be considered, but it is most important to focus on what questions need to be answered to take something from a final prototype to a true market opportunity with the potential to thrive.

The Questions and Checklist Help the Design Thinking Team:

- to check important aspects beyond the desirability, feasibility, and viability.
- to actively prepare for potential roadblocks that could jeopardize realization during implementation.
- to estimate how long the implementation will take and what dependencies there are on other initiatives or the operational teams.
- to prepare good planning and moderation of EXPLORE to EXPLOIT.
- to consider new requirements that have gained importance, for example, regarding the environmental, social, and governance (ESG) criteria framework as shown in the example on page 127.

Design thinking includes exploring the problem space, prototyping an idea, and implementing the solution to achieve innovation success. But without impact, there is no innovation.

The questions vary from design challenge to design challenge, but examples help to formulate customized checklists:

- What immediate or short-range gains or results can be anticipated?
- How simple or complex will the idea's execution or implementation be?
- What might be potential roadblocks from key stakeholders in the organization?
- How soon could the solution be put into operation?

Additional questions around the market, legal environment, competition, or impact on ESG might be relevant as well, depending on the prototype and degree of innovation:

- Is the solution legal?
- What is its environmental impact?
- Will it have any negative effect?
- Who might be a current or potential competitor?
- Who will buy it first (early adopters)?
- How much will it cost to get the solution to market?
- How can ESG and impact be balanced in the best possible way?

How Can ESG and Impact Be Balanced in the Best Possible Way

Example

One of the evolving key questions in the realization of new market opportunities for design thinking and innovation teams is how organizations that prioritize both ESG and impact are best positioned to create sustainable value. Organizations that focus on both ESG and impact typically strive for the outcomes listed in the top-right quadrant of the matrix depicted below. However, even with the best intentions, this is not always possible due to a general lack of resources and time to conduct and build comprehensive ESG and impact assessments and reporting. For this reason, as part of the Prototype Questions Checklist, it is important for design thinking and innovation teams to have already focused and prioritized their efforts by not only considering the metrics recommended by leading frameworks (ESG and impact), but further narrowing down this list through a double materiality assessment. This includes soliciting feedback on the issues and metrics that have both the greatest business value and the greatest relevance to stakeholders, and then taking action.

Combine the business and innovation metrics with the ESG measurement frameworks to generate benefits for all stakeholders.

ESG PRIORITIZATION RISK & OPPORTUNITY MATRIX

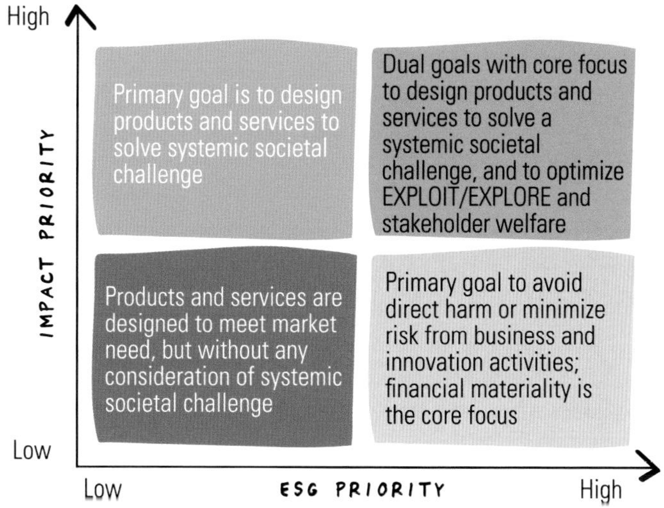
High
IMPACT PRIORITY
Low
Primary goal is to design products and services to solve systemic societal challenge
Dual goals with core focus to design products and services to solve a systemic societal challenge, and to optimize EXPLOIT/EXPLORE and stakeholder welfare
Products and services are designed to meet market need, but without any consideration of systemic societal challenge
Primary goal to avoid direct harm or minimize risk from business and innovation activities; financial materiality is the core focus
Low
ESG PRIORITY
High

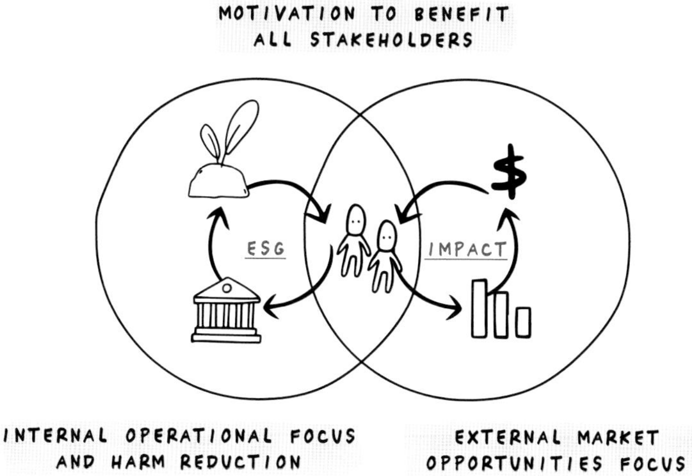
MOTIVATION TO BENEFIT ALL STAKEHOLDERS
ESG
IMPACT
$
INTERNAL OPERATIONAL FOCUS AND HARM REDUCTION
EXTERNAL MARKET OPPORTUNITIES FOCUS

Measuring at the Tool Level and Beyond

Examples Input Variables:

- Actual collaborating culture
- Design thinking activities
- Reflecting on activities on various levels
- Effectiveness of experiments
- Reaching defined success

Main Purposes of Measurements:

- Improving collaboration
- Building new capabilities
- Transformation
- Change of culture and mindset
- Impact of design thinking process steps and applied tools/activities

Examples of Measurements / Metrics:

- Confidence level of teams
- Design thinking maturity levels
- Future success based on mindshift changes
- Hard financial facts and outcomes and impact on brand value, customer satisfaction
- Reduced cost from start to implementation
- Final prototype quality (right level of functionality and experience; provides solutions to pain points, repeatable applicable; initiates excitement at the user to spread the word about solution; intuitive way of using the service, product, or experience)

Documentation of Design Thinking Activities

Within the EXPLORE phase the design thinking mindset and its associated methods and tools have the potential to impact on many different dimensions (collaboration, capabilities, transformation, culture, and innovation success). The applied tools in design thinking have the objective of reducing the customer and solution risk. Additional quantification, measurements, and metrics help to generate additional evidence and build the basis for minimum viable exploration metrics and the development of an innovation measurement system (IMS). Common metrics are related to how to collaborate, how ideas are implemented, and what is driving the business forward. Ultimately, all metrics help balance purpose with profit maximization, and more importantly, hold the key to a culture where a balanced portfolio of EXPLORE and EXPLOIT can exist.

The Design Thinking Canvas helps to document the respective findings over the entire design cycle and to create a good basis for the later validation of the product/market fit. Especially when new team members join or a handover to another team takes place, the most important elements, including the requirements for an MVP, are well documented.

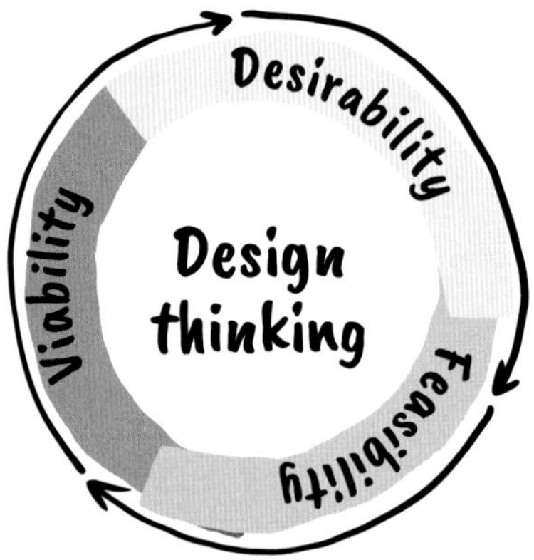

Desirability
Design thinking
Feasibility
Viability

Design thinking has the aim of minimizing the risks of market flops by engaging the customer/user through an iterative approach of collecting insights and building different prototypes to learn, test, and refine the potential solution.

Template download

DESIGN THINKING CANVAS

HOW TO DEFINE SUCCESS
- What is success for the customer?
- What is success for us as a company/team?

VISION (PROTOTYPE)
- What is the contribution to the North Star?
- What needs are met?
- What are the benefits for the customers?

REQUIREMENTS FOR MVP
- Which prototypes should be implemented as MVPs?
- What experiences/functions should be tested and validated first?
- What is the backlog of functions?

PROBLEM SPACE

DIVERGE – CONVERGE

PROBLEM STATEMENT
- Where is the problem?

EMPATHY MAP
- What are the deeper needs of the customers?
- What drives them?
- What assumptions about the customers are confirmed?
- Which ones should be discarded?

JOBS-TO-BE-DONE
- What is the underlying motivation?
- What are the jobs to be done?
- Which jobs have the highest priority?

CRITICAL ITEMS
- Which critical experiences and functions are of great importance to the customer?
- What should be a part of a potential solution in the future?

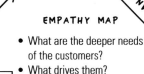

HMW QUESTION
- What is the point of view that can be derived from the findings of the previous phases?

SOLUTION SPACE

DIVERGE – CONVERGE

IDEATION
- How could the problem be solved?
- How do others solve it?
- What if...?

PROTOTYPE & TEST
- What critical functions and experiences can be built and tested?
- How can the solution be implemented and tested in a lo-fi prototype?
- How can the prototype be improved with the lessons learned?

FINAL PROTOTYPE
- Is there a problem/solution fit?
- Does the solution correspond to the customer needs?
- Is the solution implementable?
- Will the solution be economically successful?

UNDERSTAND – OBSERVE – DEFINE POINT OF VIEW – IDEATE – PROTOTYPE – TEST

DESCRIPTION OF THE PERSONA
- Who has the problem?
- What are the pains/gains for the customers, jobs to be done, and use cases?
- What are the current and future customer needs?

DOCUMENTATION OF THE PROTOTYPES WITH AN EXPLORATION MAP
- What worked? What didn't work?
- Was the respective prototype able to meet and satisfy individual experiences and functions of the customer needs?

LEAN CANVAS
- How can the short concept be described?
- How do we prepare for the next design lens, including initial ideas on key figures, customer segments, early adopters?

DOWNLOAD TOOL
www.design-metrics.com/en/design-thinking

Lean Start-Up: Product/Market Fit

After the comprehensive development of the problem/solution fit, the development of the product/market fit is a potential next step in reducing risks. The transition from design thinking toward lean start-up is overlapping and part of the extended design thinking toolbox. The main objective is to iterate the minimum viable product (MVP) through the build-measure-learn loop. The core elements of lean start-up are customer development and business model design based on some principles derived from the lean manufacturing paradigm. With regard to measuring, it can be observed that innovation teams apply and integrate very often the principles of innovation accounting for business model validation based on MVPs. However, in day-to-day practice, it is often the observed case that teams try to take a shortcut and omit the important design thinking activities before the lean start-up phase. In addition, attempts are often made in the lean start-up phase to deal with potential revenue results in absolute numbers, which are based on predictions that cannot be verified at the start of the product/market fit. Consequently, innovation teams should apply first the design thinking tools and second the validation of the product/market fit, indicating a range of possible revenues. The initial critical and unknown assumptions (see page 109) might need some updates based on the final prototypes and first validations in design thinking. Based on the distribution curve, kill metrics (see pages 182–183) based on Monte Carlo simulation might be applied to stop innovation initiatives with confidence, if the desired objectives are not reachable.

The Lean Canvas (see page 137) helps to document the comprehensive market validation. The measurement extends to customer value, retention pricing, and elements associated with recurring, repeat, up-sell, adaptation, or consumption rates. The Lean Canvas thus gives a more comprehensive picture of the business and monetization model. Further, it is recommended to define variables which can be tested with well-known statistical and research methods from the measuring toolkit (see page 237). The proper application of the lean start-up philosophy helps to reduce the risks further from the problem definition to the desired product/market fit.

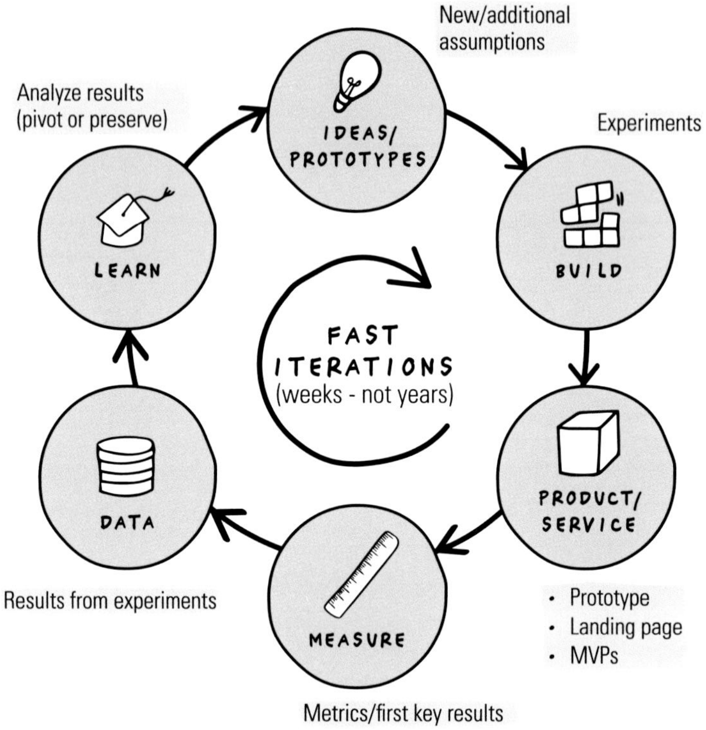
New/additional assumptions
IDEAS/ PROTOTYPES
Analyze results (pivot or preserve)
Experiments
LEARN
BUILD
FAST ITERATIONS
(weeks - not years)
DATA
PRODUCT/ SERVICE
Results from experiments
MEASURE
• Prototype
• Landing page
• MVPs
Metrics/first key results

Examples of Indicators Related to Lean Start-Up:

- Determining if new product, service, or experience is creating value for the customer/user
- Measuring if customers/users are returning
- Quantifying how much a customer/user is willing to pay for a new product, service, or experience
- Potential range of revenues of a new offering
- Total investment from problem identification to product/market fit

Innovation Accounting

Measuring innovation success is an essential part of the lean start-up methodology. It enables design thinking and innovation teams to objectively demonstrate that the right conclusions are drawn from extended experimentation with MVPs, which are then used to improve the MVP and a potential business model. A particular MVP or multiple MVPs can be the subsequent basis for designing a business ecosystem in which multiple actors generate a shared value proposition. Early simulations of revenue figures can be performed and measured with so-called driver trees (see page 132).

Innovation Accounting Helps the Innovation Team:

- to determine the starting point as to which MVP with its associated functions and experience should run through the build-measure-learn cycle.
- to carry out fine-tuning by testing the relevant hypotheses (e.g., with regard to viability).
- to create scenarios based on different hypotheses (e.g., with testing different acquisition and retention rates).
- to decide whether something should be discarded (pivot) or retained (preserve).

Sample Questions:

1. **Acquisition:** How do customers/users find us?
2. **Activation:** Do the customers have a first positive experience?
3. **Retention:** Are the customers/users coming back?
4. **Revenue:** How does the offer or function make money?
5. **Referral:** Do the customers/users recommend the offer to other people?

For many design thinking and innovation teams the lean start-up methodology is applied to verify the most risky business model assumptions by way of empirical testing with customers/users.

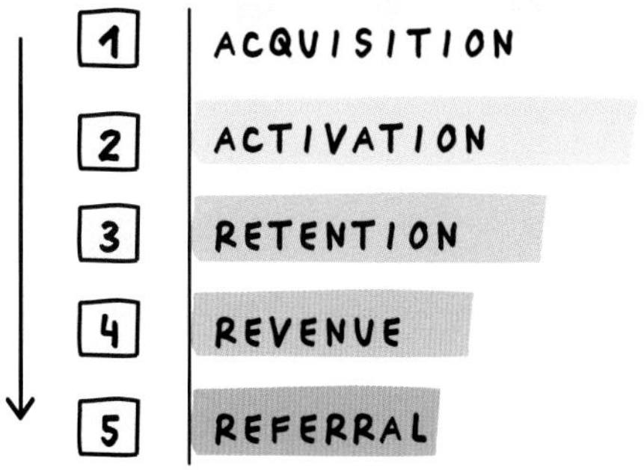

1
ACQUISITION
2
ACTIVATION
3
RETENTION
4
REVENUE
5
REFERRAL

Examples of Typical AARRR* Measurements:

- Number of unique visitors per channel
- Time to value; customer aha moments; conversion rates
- Retention versus churn rate; customer churn; retention length
- Referred customers; viral coefficient; viral cycle time
- Revenue churn, lifetime value; acquisition costs; average order value; conversion from free to paying customers

* AARRR metrics depend on business model, industry, and business ooocyctom configuration

Example

DRIVER TREE

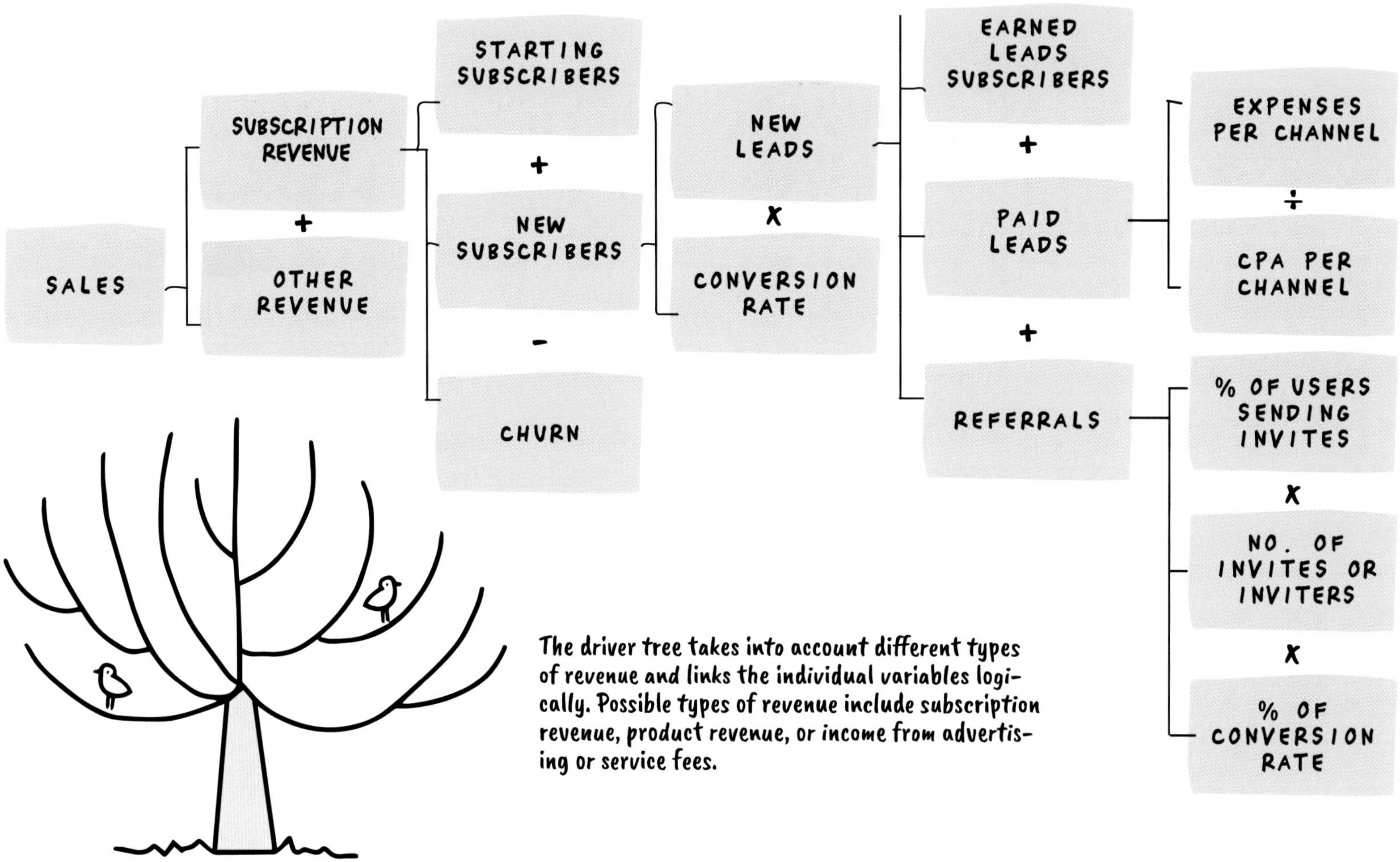

SALES
SUBSCRIPTION REVENUE
+
OTHER REVENUE
STARTING SUBSCRIBERS
+
NEW SUBSCRIBERS
-
CHURN
NEW LEADS
X
CONVERSION RATE
EARNED LEADS SUBSCRIBERS
+
PAID LEADS
+
REFERRALS
EXPENSES PER CHANNEL
÷
CPA PER CHANNEL
% OF USERS SENDING INVITES
X
NO. OF INVITES OR INVITERS
X
% OF CONVERSION RATE

The driver tree takes into account different types of revenue and links the individual variables logically. Possible types of revenue include subscription revenue, product revenue, or income from advertising or service fees.

Willingness-to-Pay (WTP) Analysis

For certain offers, pricing can be tested in an early phase and be validated as part of an MVP. Well-known models for pricing are cost-plus pricing, comparison pricing, or the willingness-to-pay analysis described here. There are different ways to determine the willingness to pay. One option is the "discreet choice analysis," either based on actual purchase data or on asking the customers for their preference among alternatives with different sets of attributes. In most cases, the potential customers are not able to articulate the value they place on a feature directly; what they can do is compare two alternatives in an A/B test and express their preference. Given a sufficiently large number of votes, this method (conjoint analysis) can be an implicit evaluation of each feature by the customers.

The Willingness-to-Pay Analysis Helps Innovation Teams:

- to obtain an initial idea about the willingness to pay for products and services on the part of certain customers and segments.
- to establish an estimate of demand that can serve as a basis for predicting the overall size of the market or segment for the nascent ecosystem.
- generally develop an assessment on whether the costs are in a healthy relationship with the validation from the willingness-to-pay analysis.
- to begin with tactical thoughts on pricing, business model, and ecosystem design and evaluate individual functions and experiences.

Procedure and Template:

1. An open question in the interaction with the potential customer and the MVP may mark the starting point: What is the attribute/feature worth to you?
2. Is there already an indication of what it may cost, or what the initial situation for the willingness-to-pay analysis is?
3. As part of A/B tests, questions about various pricing aspects can be asked.
4. The analysis may relate to the overall offer, individual features, or specific services (e.g., neutral advice).

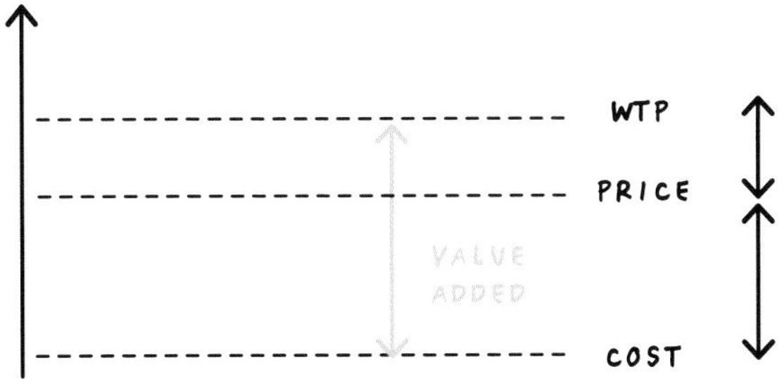
WTP
PRICE
VALUE ADDED
COST

It is unlikely that each customer has the same willingness to pay for a particular product or service, and it is recommended to create a market demand curve that shows how many customers will buy at a given price; from this, the market price elasticity can be calculated.

WHAT IS THE OFFER WORTH TO YOU?
OR WHAT IS THE FEATURE WORTH TO YOU?

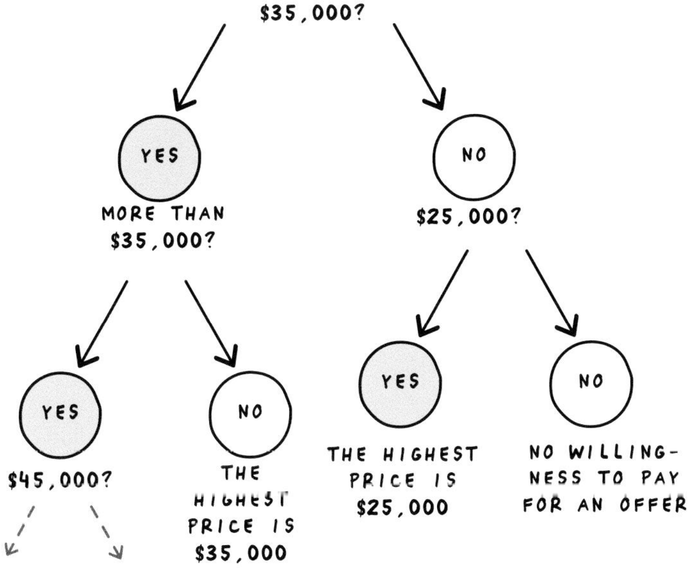
$35,000?
YES
MORE THAN $35,000?
NO
$25,000?
YES
$45,000?
NO
THE HIGHEST PRICE IS $35,000
YES
THE HIGHEST PRICE IS $25,000
NO
NO WILLINGNESS TO PAY FOR AN OFFER

MVP Portfolio and MVP Portfolio Planning

Successful design thinking and innovation teams think in opportunities and understand that, beyond the initial MVP, they need to reach new target customers continually with their offers. Target portfolios, defined and implemented incrementally, are good for this. At the beginning, an Ansoff matrix can be used. The original matrix shows four portfolio fields and contains the dimensions of market penetration, product development, market development, and diversification. This matrix can be adjusted to the development of MVPs, where it helps to evaluate the projects. Products in the form of existing offers and existing markets normally don't require lean start-up considerations. The focus in design thinking and lean start-up is on innovative customer development (new markets/new offers).

MVP
DISCOVERY
DELIVERY

Corporate practice shows that MVP is one of the most referenced and least understood concepts in modern product development. A deep dive into the work of Steve Blank and Ash Maurya's methodology is highly recommended to understand the concepts.

ANSOFF MATRIX

	EXISTING OFFERS	NEW OFFERS
EXISTING MARKETS	USUALLY NO MVP NEEDED FOR THE REALIZATION OF A UNIQUE VALUE PROPOSITION	INNOVATIVE OFFER DEVELOPMENT (DIVERSIFICATION)
NEW MARKETS	NEW CUSTOMER DEVELOPMENT (FIRST-MOVER ADVANTAGE)	INNOVATIVE CUSTOMER DEVELOPMENT

In the mindset of integrated and rather traditionally managed companies, innovative customer development is seen as risky because there is often little room for the exploitation of existing know-how or for achieving economies of scale.

MVP Portfolio Planning

Business growth initiatives need a comprehensive combination of functions, experiences, and features that are explored, developed, and provided by the actors. There are various options for portfolio planning and portfolio presentation. The individual MVPs alternate between discovery and delivery. Individual functions should be validated and introduced to the market as quickly as possible, for instance. Projects with a high risk in terms of time to market usually require additional resources, implementation know-how, or dependency on new technologies.

Portfolio Planning Helps the Design Thinking for Business Growth Team:

- to prioritize individual MVPs and align them with the strategic goals.
- to align future functions and experiences with the business goals, e.g., "run, grow or change the business."
- to get an overview of the status of the individual portfolio elements.
- to coordinate the respective project phases, e.g., planning, implementation, project management.
- to carry out adequate resource planning and allocate the respective roles and resources accordingly.

Procedure and Template:

- The individual projects are entered on the portfolio grid; their progress is reflected upon every four weeks or at shorter intervals.
- The size of the bubble per initiative indicates the value added. Depending on the project, you can work with NPV, IRR, revenue, or other figures.
- The four phases show where the initiatives are in terms of implementation; a rough classification: discovery and delivery.
- The arrows indicate whether there are any shifts regarding the next phase or whether some initiatives go back to the discovery phase since the first usability test yielded new insights, for instance.

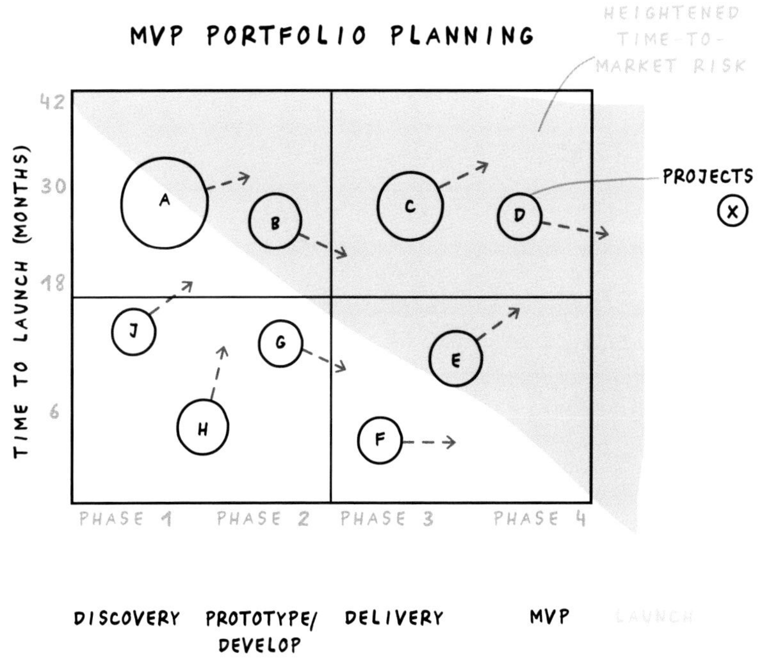

MVP PORTFOLIO PLANNING
HEIGHTENED TIME-TO-MARKET RISK
TIME TO LAUNCH (MONTHS)
42
30
18
6
A
B
C
D
J
G
E
H
F
PROJECTS
X
PHASE 1
PHASE 2
PHASE 3
PHASE 4
DISCOVERY
PROTOTYPE/ DEVELOP
DELIVERY
MVP
LAUNCH

Evaluating the data of an MVP is just as important as building the product part. Designing and testing metrics is part of value creation.

Innovation Accounting and Beyond

Examples of Input Variables:

- Acquisition channels
- From signing-up to happy customer/user
- Retention of customer/user
- Retention management activities
- Degree of virality (type, cycle time, coefficient)

Main Purposes of Measurements:

- Risk reduction
- Understanding cause and effect
- Defining next steps
- Tracking of individual customers/users
- Predicting success of new business

Examples of Measurements / Metrics:

- Confidence in the go-to-market concept
- Customer success
- Probability of scalability
- Market demand
- Probability of traction
- MVP - product and service quality (performance, features, reliability, conformance, durability, serviceability, aesthetics, perceived quality)

Documentation of Lean Start-Up Activities

Innovation accounting and the Lean Canvas supports the innovation team with structuring and visualizing the innovation project. In the practical use it is very common to apply the completed Lean Canvas to review the problem/solution fit and adjust if necessary. In this case, the collected data is compared to the best solution that fits the behavior and challenges of customers. The activities related to the Lean Canvas documents the first efforts of implementing the solution or business model. The main objective is to identify potential market risks entailed in the implementation. It should be noted that the respective metrics applied are different depending on the industry and type of business and should be used in different degrees of complexity. The AARRR framework by Dave McClure (see page 131) and similar frameworks like the RARRA or ATAR framework are applying a very simplified set of possible metrics that should be extended with minimum viable exploration metrics depending on the business environment and the business model.

The Lean Canvas helps to include and document the respective findings from design thinking and lean start-up. It is recommended to add with the documentation the customer/user profiles and experiment reports from earlier activities over the entire design cycle.

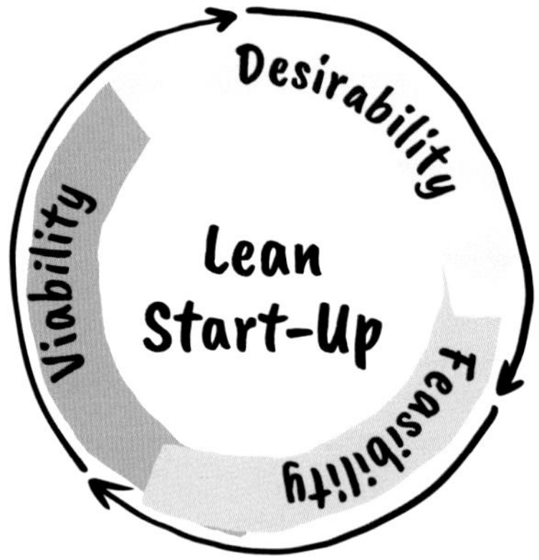

Desirability
Feasibility
Viability
Lean Start-Up

Activities related to lean start-up are about developing products, services, and experiences that have already proven to be highly desirable for customers/users, so that a market already exists once the market launch is complete.

Template download

LEAN CANVAS

WHO .... **WANTS** .... **FOR** .... **BECAUSE** ....

(Customers) (Product, service,...what) (Satisfaction of needs) (Motivation)

PROBLEM

- What are the main problems that the business idea has to solve?

Describe the biggest 1 to 3 problems of your customers.

SOLUTION

Describe a solution for each problems.

UNIQUE VALUE PROPOSITION

- What value should be provided to the customers?

A simple, clear message that explains why the solution is different and noteworthy.

-Gain creators and problem solvers.
-How does the solution support customers in fulfilling their job in specific use cases that are important to them?

UNFAIR ADVANTAGE

Something that makes it difficult for others to copy the solution.

CUSTOMER SEGMENT

List the target and user groups.

- For whom does the product or service add value?
- Who are the most important customers?
- Whose needs still have to be considered (e.g., partners, suppliers, external influencers, decision makers?)
- Who are the most important stakeholders and interest groups?

Draw a stakeholder map or business ecosystem.

KEY METRICS

- Which measurable figures show whether the solutions work?

CHANNELS

- Through which channels do your customer segments want to be reached?

EXISTING ALTERNATIVES

- How have these problems been solved so far?

HIGH-LEVEL CONCEPT

X for Y analogy:
Is there a simple analogy?
(e.g., YouTube as Flickr for video)

EARLY ADAPTERS

Describe the characteristics of the early adopters (those customers who will use the product first). The ideal customers may not necessarily be the first customers.

COST STRUCTURE

List the main fixed and variable costs.

REVENUE STREAM

List the sources of income.

DOWNLOAD TOOL
www.design-metrics.com/en/lean-canvas

Business Ecosystem Design: Systems / Actors Fit

The design and implementation of business ecosystems play an increasingly important role in the context of new products, services, and experiences. This section briefly discusses the systems/actors fit. The entire framework and discipline of business ecosystem design can be found in the book *Design Thinking for Business Growth*.

Design thinking and innovation teams that act as initiators of business ecosystems usually have few reference points from the past for defining how to measure such initiatives. Moreover, business ecosystems are dynamic entities in which actors can be added over time, for example to extend the core value proposition. However, capabilities of individual actors can also become obsolete, so they are replaced by other actors. In addition, most ecosystem initiatives have an exploratory nature, based on new and unique value propositions for which there is no data and evidence, for example from benchmarks or market studies. In an early phase of business ecosystem initiatives, many of the methods, tools, and measurement instruments from design thinking and the lean start-up phase take hold. For the design of the actual business ecosystem, the measurement points mainly relate to the configuration of the business ecosystem that is critical for success. Traditional metrics such as return on investment, profitability, cash burn rate, and revenue are of little help in terms of initialization, as these metrics are all lagging indicators. Other quantifications and measurements, in most cases, only measure the current trend and thus at most provide information about the number of users, subscribers, time spent from customers/users on the offer, retention, clicks, or social media exposure. The decisive factor, however, will be whether the business ecosystem is successful in the long term at the levels of value capture and value delivery, and whether the initiative creates ecosystem capital. Here, measurement should take place on several levels and the measurement criteria should fit the respective design, test, and implementation phase of the business ecosystem. In addition, the governance model often changes in the early phase and the operating model changes later as part of scaling. Consequently, metrics are needed that indicate performance and potential at the system level and at the level of the individual companies or actors participating in the ecosystem, as well as the ecosystem initiator or orchestrator. They need to be able to measure growth not only in terms of ecosystem participation, but also in terms of the supporting governance and underlying operating model. And most importantly, ecosystem initiatives need metrics that reflect the success factors unique to the development of the MVE and scaling as well as activities of enhancing the core value proposition over time.

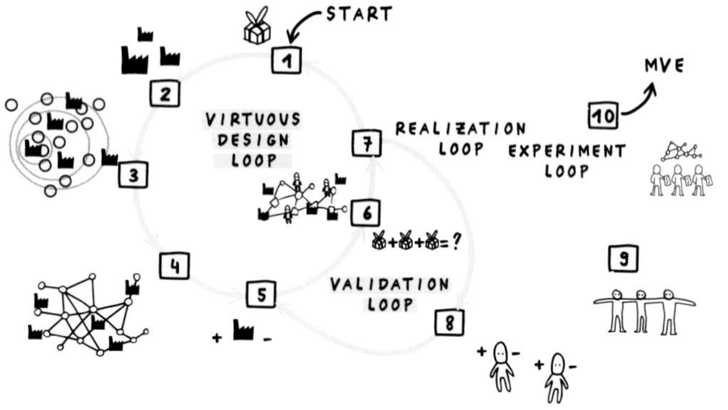
START
1
2
3
4
5
6
7
8
9
10
MVE
VIRTUOUS DESIGN LOOP
REALIZATION LOOP
EXPERIMENT LOOP
VALIDATION LOOP

The appropriate metrics help initiators/orchestrators to avoid the most common pitfalls in the design, implementation, and scale of business ecosystems.

Examples of Indicators Related to Business Ecosystem Design:

- Volume of real options in market opportunities versus relevant opportunities
- Number and engagement level of marquee business ecosystem actors contributing to the core value proposition
- Actual number realization of ecosystem capital in X-time
- Satisfaction of ecosystem partners/actors with expansion of the core value proposition over time
- Customer satisfaction with the value proposition delivered jointly by the actors in the system

Ecosystem Capital

An important metric for business ecosystem initiators and orchestrators is ecosystem capital, which is used to quantify the value of network connections, configuration, and value streams. On the one hand, traditional companies often lack this component because their business models and strategies have not allowed room for business ecosystem development. They have a high proportion of physical and human capital, which is sufficient for linear growth but limits the achievement of greater margins and the exploitation of new growth opportunities. Design thinking and innovation teams, on the other hand, that have understood the importance of ecosystem capital for their organizations, which consists of the interaction of all actors in the system and create win-win situations through a multidimensional view of business models, are well equipped to realize corresponding opportunities to leverage exponential growth.

To date, various elements have been included in the measurement of ecosystem capital. Mostly it includes a combination of knowledge, value streams, and number of connections that different organizations have to each other to deliver a core value proposition to the customer. On the one hand, the measurement is challenging because business ecosystems are dynamic, and it is a measurement that mainly consists in the external relationship of several actors in the system. Moreover, well-configured business ecosystems allow additional innovation at the edge of the business ecosystem as well as access to the customer, which is helpful for realizing exponential growth but does not make quantification and measurement easier.

In an early phase (MVE), however, the calculation is still manageable, since the value of the fomented and consolidated relationships is based on the initial considerations of the business ecosystem and the number of actors is typically limited to the minimum. The origin of ecosystem capital can therefore refer to the business relationship of each company in the system to the other companies and to the orchestrator, and be constituted, for example, in the context of an ecosystem/community experience in the metaverse, by the totality of the mostly intangible goods that are made available to the customer/user collectively.

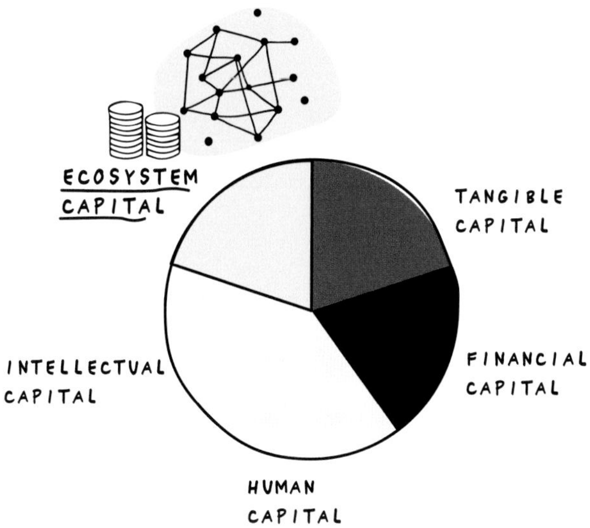

ECOSYSTEM CAPITAL
TANGIBLE CAPITAL
FINANCIAL CAPITAL
HUMAN CAPITAL
INTELLECTUAL CAPITAL

Typical Elements to Consider in Measuring Ecosystem Capital:

- Formal established and orchestrated relationships between actors in the system
- Established value streams and benefits for all actors in the system
- Exploitation of network effects and scale effects of actors in the system
- Adaptability of actors in the system to cope with the dynamic environment
- Alignment of the frequency of customer interaction, share of wallet, and retention to the system
- Configuration of the ecosystem with regard to openness, transparency, and market roles in the system
- Application of new and game-changing technology and the use of big data analytics for agile customer development and innovations

Business Ecosystem Design

Examples of Input Variables:

- Effectiveness of minimum viable ecosystem (MVE)
- Access to customers/users via ecosystem actors
- High-value data points generated by the ecosystem itself and ecosystem actors
- Customer feedback about the core value proposition delivered

Main Purposes of Measurements:

- Prove viability of ecosystem to initiator/orchestrator, customers, and actors
- Impact of ecosystem configuration on customers/users, actors, value streams, and growth
- Engagement level of actors and customers/users
- Market acceptance of the core value proposition

Examples of Measurements/Metrics:

- Activities in conjunction with the ecosystem (e.g., active users, customers, actors)
- Successful transactions, unit/product/service cost
- Satisfaction of actors participating and customer engaging with the ecosystem offerings
- Share of revenues and distribution of value streams across the ecosystem
- MVE – configuration quality (use of channels, data quality, alignment of actors, quality of value delivery, speed of adaption to new market dynamics, and changing customer needs)

Documentation of Ecosystem Design Activities

Business ecosystem design focuses on adaptability (the needs of the actors and their ability to jointly carry a value proposition), value enhanceability (design of sustainable value streams and benefits for all actors in the system), as well as feasibility (concretization of technology components and interfaces for the realization of an ecosystem approach). Through the respective design loops, the minimum viable ecosystem (MVE) is developed, which, in turn, is the basis for scaling the system. Depending on the market and configuration, the ecosystem is limited in upward growth (i.e., growth slows down). Usually, similar ecosystems are initialized by different ecosystem initiators, which compete with their own initiative. Thus, focusing on increasing the frequency of existing customers as well as strengthening the bond with ecosystem partners becomes a key element for strengthening the market position. In addition to the already ongoing customer development for the value proposition, the ecosystem will again explore the new and changing customer needs in order to expand the value proposition step by step.

The Business Ecosystem Design Canvas documents the activities over the virtuous, validation, realization, and experiment loop.

Adaptability
Feasibility
Value enhanceability
Ecosystem Design

Activities related to ecosystem design aim to reduce the ecosystem risk by designing and testing MVEs with a selected (minimum) number of actors to test the value streams, collaboration, and provision of the core value proposition.

ECOSYSTEM DESIGN CANVAS

Template download

DETERMINE THE NEEDS OF THE USERS OR CUSTOMERS

- Who is the customer or user?
- Describe the customer/user profile (pains, gains, jobs to be done, and use case).
- What problem is to be solved?

CORE VALUE PROPOSITION

- What is the core value proposition for the user/customer?

DEFINITION OF THE VALUE STREAMS

- What are the current and future (positive and negative) value streams?
- What are the product/service streams, money/credit streams, data, and information flow?
- What are the digital and digitizing value streams/assets?

DESCRIBE THE ACTORS

- Who are the actors in the business ecosystem?
- What is their function and role in the system?
- How high is their motivation to participate in the business ecosystem?

DESIGN / REDESIGN

EXPLORE

BUILD/TEST

DESIGN

- Which actors are pivotal for the provision of the core value proposition in the business ecosystem? (For the placement, go from the inside to the outside.)
- Place the actors with advanced and complementary offerings and enabling functions, and who are directly or indirectly part of the system.

REDESIGN

- Do various scenarios exist with different actors?
- Which actors can be eliminated?
- Are there actors who scale multidimensional value streams or better value streams?
- Is the business ecosystem robust and able to survive in the new scenario?

PROTOTYPE, TEST, AND IMPROVE THE BUSINESS ECOSYSTEM

- What are the MMFs/MVPs the ecosystem starts with?
- How can the first MVE be tested?
- What interactions/testing and measurement methods help us to improve the value streams, business models, and the role of actors in the ecosystem iteratively?
- What is the minimum version of the first MVE?

ANALYSIS OF THE ADVANTAGES AND DISADVANTAGES OF EACH ACTOR

\+ -

- What are the advantages and disadvantages for each actor?
- What are his strengths/weaknesses and opportunities/risks in the system?

MULTIDIMENSIONAL VIEW OF THE BUSINESS MODELS

- What does the resultant business model and value proposition for each actor look like?
- How does the respective business model contribute to the core value proposition?
- Is the defined core value proposition the result of the sum of the value propositions of all actors?

DOWNLOAD TOOL
www.design-metrics.com/en/ecosystem

Scale: Manage Transition

Depending on the focus, scaling/exploiting comes after the product/market fit or the MVE (systems/actors fit). Scaling focuses mainly on realizing growth and, as a result, maturity and evolution, and the metrics change. These range from metrics that capture information about the initial scaling, to maturity and evolution, and should be chosen to allow the right conclusions to be drawn. In addition to the chosen business model, the possibility of scaling depends very much on the respective product, service, and experience. In the case of highly specialized services and products with a very small addressable market, other levers are relevant than in the case of solutions for the mass market and rapid adaptation by customers due to new or changed customer needs. In both cases, however, there is a need for different measurements and the application of the appropriate mix of relevant exploitation metrics. To scale successfully, decision makers and growth teams must have the courage to think exponentially (i.e., to deal with levers greater than 5, 8, and 10 of the current value). The term *10x* is often used for exponential growth. In EXPLOIT many of the traditional metrics for gaining efficiency and IT scalability are applied. The focus shifts gradually from EXPLORE toward launching robust value-creating products or services, with compelling value propositions. Growth teams focus on designing virality, scalability, and customer loyalty; building brands with purpose, while gathering customer feedback for core value enhancement activities.

In addition to feasibility (expansion, professionalization, and leverage of the technology components), scaling has two focal points: rhythmicability (network effects, repetitions) and captivateabilty (number, intensity) of interactions in the business ecosystem. These both must increase and at the same time reduce the unit cost of each interaction/transaction in the system.

Since the focus of this book is on design thinking and innovation metrics, this toolkit will conclude with a brief discussion of the transition from EXPLORE to EXPLOIT, leaving the space open for applying tools and methods beyond the scope of this book. As in EXPLORE, there are different layers to consider in the growth and EXPLOIT phases (see layers on page 36) and guidance for creating an appropriate set of exploitation metrics (see page 187).

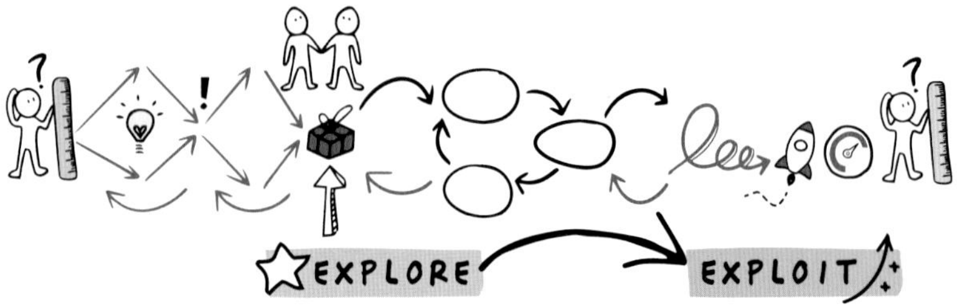
EXPLORE
EXPLOIT

PLAN HOW TO BRING IT HOME!

In transition, two mindset worlds collide, and good planning, facilitation, and an understanding of both mindsets and associated management stylers are key for bringing it home.

Examples of Indicators Related to Growth, Scale, and EXPLOIT:

- Number of customer adoption in X-time versus market size
- Customer stickiness to an offering versus all customers in X-time
- Solutions sold to new customer segments versus all segments
- Unit cost per transaction versus IT investments to scale

Growth/Scale/Exploit

Examples of Input Variables:

- Cost per transaction
- Usage of resources
- Throughput (e.g., transactions or interactions processed)
- Time per interaction engagement

Main Purposes of Measurements:

- Understand business/innovation performance, business health (health metrics), and prospects
- Retrain product, service, and experience quality throughout expansion and exploitation
- Maintain positive relationship with customers/users

Examples of Measurements/Metrics:

- Sales revenue with new services/products
- Customer loyalty and retention
- Market share
- Process efficiency
- Time between visits and/or contribution
- Quality of product and service scalability (social referral rate, engagement metrics, customer open-ended comments, degree of mass-customization, and pleasure of usage)

More portfolio measurements beyond EXPLOIT are listed on pages 190–191.

Documentation of Scale Activities

Most initiatives have the ultimate objective of scaling, which means that the responsible teams and decision makers are looking for ways to grow more efficiently, so that the gains exceed the previous and current investments. Growth is often used synonymously but is not always the result of innovation work. Companies also grow, for example, through mergers and acquisitions, or the simple hiring of more employees, and growth often leads to losses and gains. Scale tools and methods are varied and some tools for scaling can be found in the book *Design Thinking for Business Growth*. In terms of measuring scaling activities, it is necessary to distinguish between the type of business to be scaled. For appropriate measurement, the processes and interactions that drive scaling must be well understood. Metrics are often chosen that are critical to how these processes and interactions can be made repeatable and frequency can be raised. Many initiatives today are about determinants of engagement and repeat usage. Traditional scaling efforts tend to focus on per-unit efficiencies.

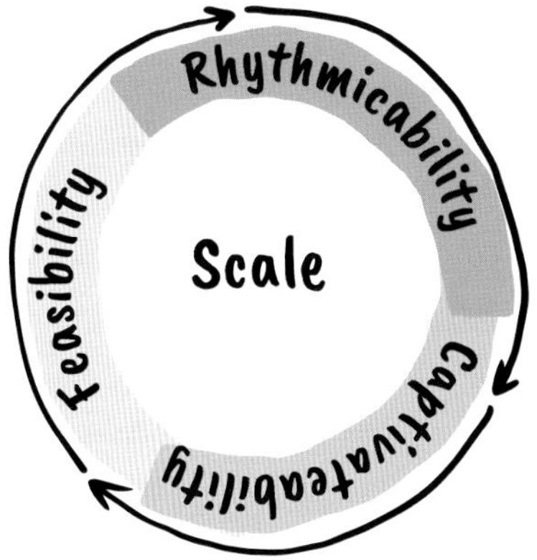
Rhythmicability
Captivateability
Feasibility
Scale

The Scale & Growth Canvas (see page 144) helps to document the scaling activities, which deal with the definition of the appropriate channels, network effects, and culture. The canvas is based on the considerations of a potential business ecosystem and should be adapted accordingly if the focus is, for example, on a single product and without the participation of partners/stakeholders in the system.

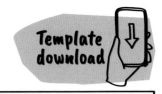

Template download

SCALE & GROWTH CANVAS

LEVERAGE CAPABILITIES OF PARTNERS OR ECOSYSTEM ACTORS

Which partners help to drive efficiency, innovate, provide relevant data, or accelerate the market opportunity?

SCALABLE PROCESSES, IT, DATA ANALYTICS

What activities are necessary to align the processes, the IT, and analytics with the requirements and growth?
Which algorithms help in customer interaction?

ECOSYSTEM CULTURE AND NETWORK EFFECTS

How can cross-company collaboration be realized with the team-of-teams idea? How can network effects be used for the growth of the ecosystem?

EXPANSION OF THE VALUE PROPOSITION

What other needs and customer problems can be addressed/ solved?
How is the value proposition expanded?
What is the experience, and which functions are offered?
Which new offers can be derived from data points and algorithms?

BUILDING THE CUSTOMER BASE AND COMMUNITY

Which mechanisms and methods are used to increase the number of customers, interactions, and ties to the system?

LEVERAGE OF DIGITAL, PHYSICAL, AND HYBRID TOUCH POINTS

Which channels are needed?
How can an opti-channel strategy be developed on the basis of data?

SOLVING PROBLEMS OF MANY

Who are the customers?
Have the segment considerations changed?
How are new needs and target groups dealt with?

OPTIMIZED COST STRUCTURE

How can the cost of user acquisition be kept lower than that of the lifetime value generated, for example?

EXPANDED VALUE STREAMS

What new value streams are rewarded by the customer?
Where does willingness to pay exist? Where are there options for bundles, cross-selling, or up-selling?

DOWNLOAD TOOL
www.design-metrics.com/en/scale

MEASURING 101

From Observations to Real Metrics

Almost Anything Can Be Measured

In the current discussion of iterative development and application of innovation measurement systems, it is believed that the work of design thinking and innovation teams is subject to constant change. For this reason the full range of statistical tools should be used to keep pace with this change. Measurements should be made on two levels: a carefully described object, on the one hand, and, on the other hand, productivity as a measure of innovation. Thus, besides the qualitative indicators, the quantitative output indicators are an important element in understanding whether individual measures from building new capabilities to implementing a strategy are effective and successful.

While the perception of innovation measurement systems evolved over the years and will evolve even more quickly in the future, some basic concepts of measuring, creating metrics and statistical methods remain. The well-known statement "anything can be measured" is not surprising, but it is crucial to define what to measure, to apply the appropriate methods, and to understand the concept of measurements and metrics in order to avoid surprises.

It is also of paramount importance not to confuse the notion that everything that can be measured should be measured.

One of the biggest challenges in terms of measurements and metrics is to understand the **WHY, HOW,** and **WHAT**, which will be addressed with this introduction. In addition, the wording is crucial because in many organizations' pure observations, measures and metrics are often lumped together and used synonymously. This often results in a colorful mix of miracle numbers (both vanity metrics and meaningful measurements) on visually appealing dashboards that are regularly distributed throughout organizations.

However, not all efforts provide good indication for steering and taking actions. Knowing how to map validated observations into numbers or using a specific amount in relation to an already defined measurement system or condition helps to manage the discipline of measuring.

SOMETHING IS GOING ON!

METRICS MIGHT NOT BE A BAD IDEA!

Ultimately, the goal of measuring should be to make better decisions or anticipate trends based on leading indicators.

In this 101, metrics are a derivative of measures which have a defined meaning in the innovation and business context used (e.g., innovation metrics based on a calculation between two or more measures). Other measurement systems, like measuring performance via OKRs, are usually an agreed metric value of achieving an individual, team, or organization's goals based on distinct confidence levels. More details on OKRs can be found in Performance Measurements 201.

Definition of Terms and Applications

[The terms around measurements and metrics must be clear, and the concept of measurements understood.]

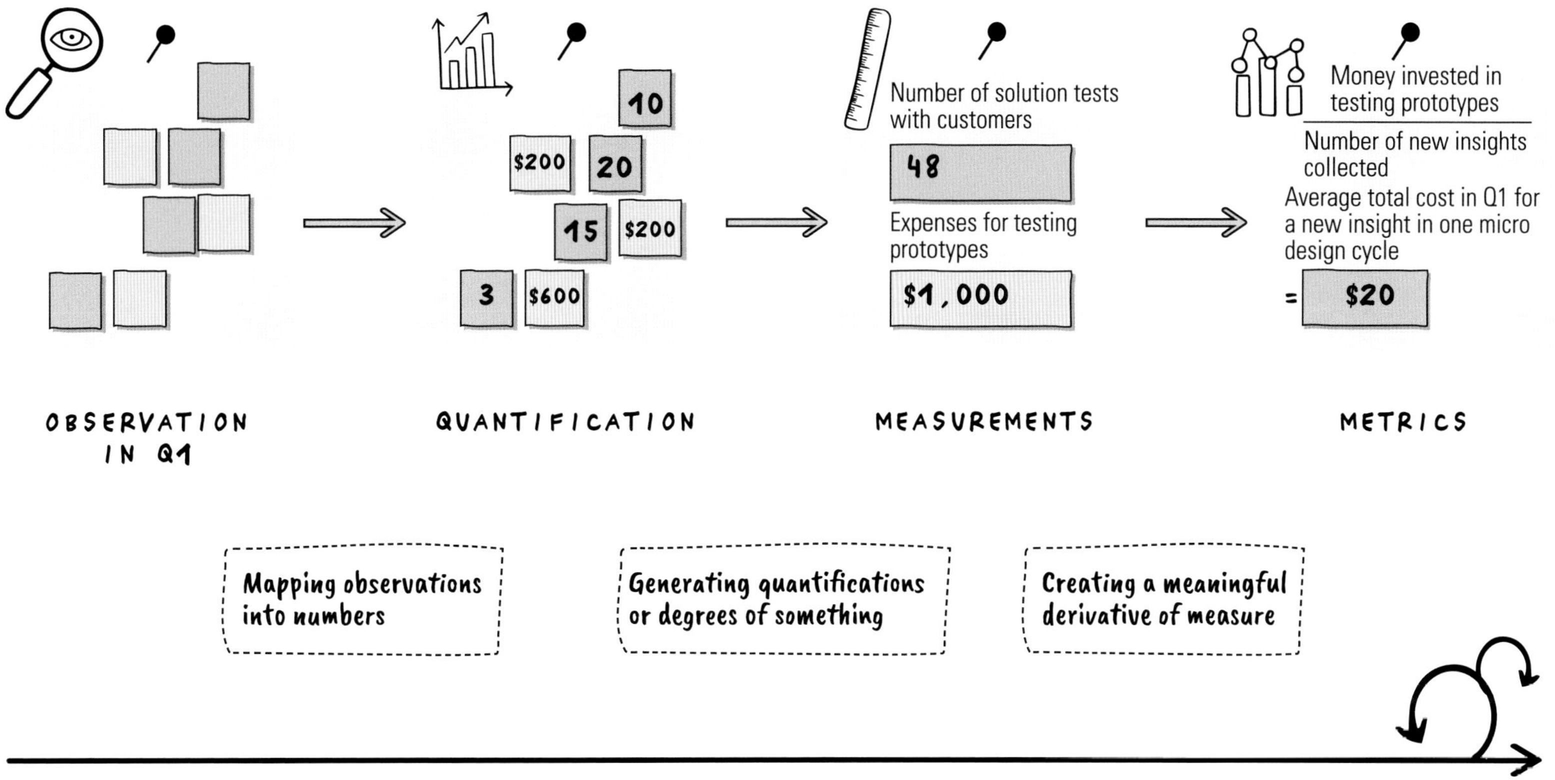

10
$200
20
15
$200
3
$600
Number of solution tests with customers
48
Expenses for testing prototypes
$1,000
Money invested in testing prototypes
Number of new insights collected
Average total cost in Q1 for a new insight in one micro design cycle
= $20
OBSERVATION IN Q1
QUANTIFICATION
MEASUREMENTS
METRICS
Mapping observations into numbers
Generating quantifications or degrees of something
Creating a meaningful derivative of measure

An Iterative Approach to Creating Metrics and Measurement Systems

After the introduction to design thinking and the initial definition of the terms of measuring, the question inevitably arises as to how the iterative approach of design thinking might be applied to the development of an innovation measurement system. This starting point for applying design thinking in developing and evolving a measurement system is excellent because such a system is a wicked problem, and there is a need for appropriate metrics in the organization and respective teams. There are many solutions for measuring systems in EXPLORE, and it seems that an iterative approach is suitable to develop one step by step. For these and many other reasons, innovative companies are using the design thinking mindset to build and evolve measurement systems, ask the appropriate questions for data analytics (see page 224), and lay the foundation for an AI-driven/supported measurement system in the future (see page 355).

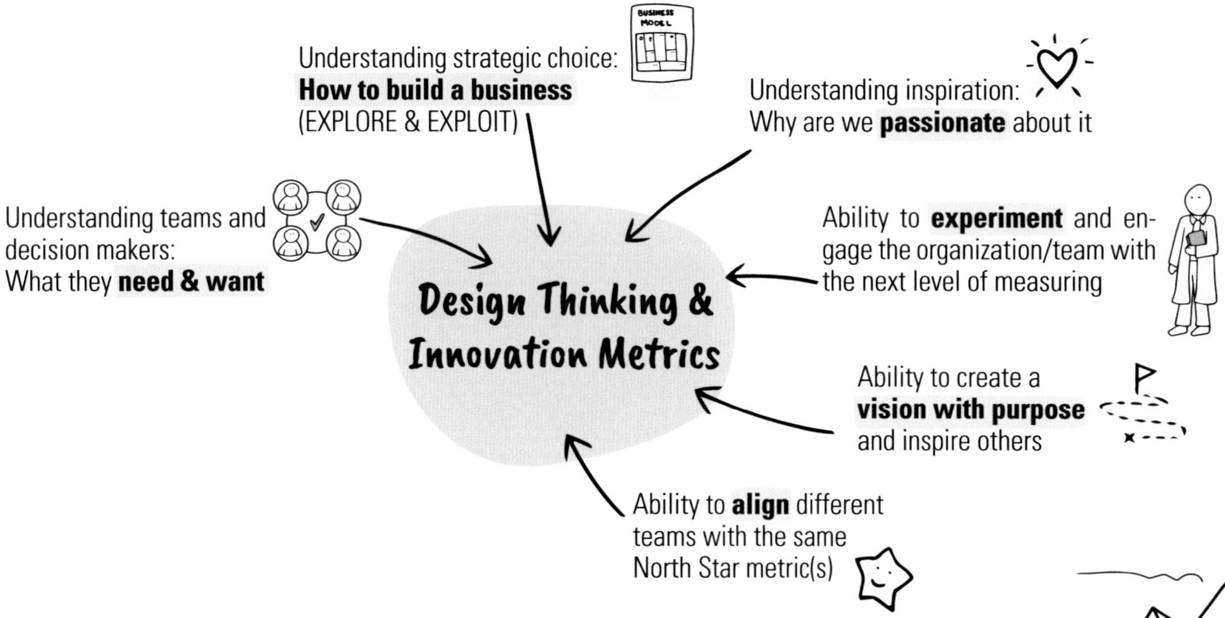

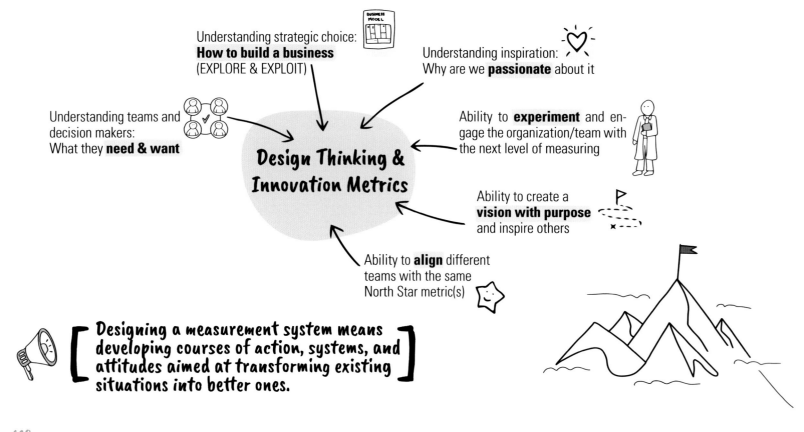

Understanding strategic choice: **How to build a business** (EXPLORE & EXPLOIT)
BUSINESS MODEL
Understanding inspiration: Why are we **passionate** about it
Understanding teams and decision makers: What they **need & want**
Design Thinking & Innovation Metrics
Ability to **experiment** and engage the organization/team with the next level of measuring
Ability to create a **vision with purpose** and inspire others
Ability to **align** different teams with the same North Star metric(s)

[Designing a measurement system means developing courses of action, systems, and attitudes aimed at transforming existing situations into better ones.]

Iterating the Innovation Measurement System (IMS)

The development and professionalization of an innovation measurement system is best done step by step and in iterations. The explorative approach with the Minimum Viable Exploration Metrics presented in this section (see pages 176–181) as well as the utilization of North Star metrics (see pages 168–174) can quickly achieve initial measurement results. In addition to the project/business dimension, it is also important to include the team/culture dimension. There is a need for both, the awareness for value creation and a new business perspective that goes beyond value delivery. For example, increasing design thinking maturity, as presented on page 93, helps measure change based on evolving design thinking capabilities. Other existing measurement systems can also be leveraged to measure change at the detailed levels of each dimension to create the appropriate actions. In addition, redefined performance measurement systems support the transformation if the purpose of the implementation includes the aforementioned ambitions.

Innovation Measurement System

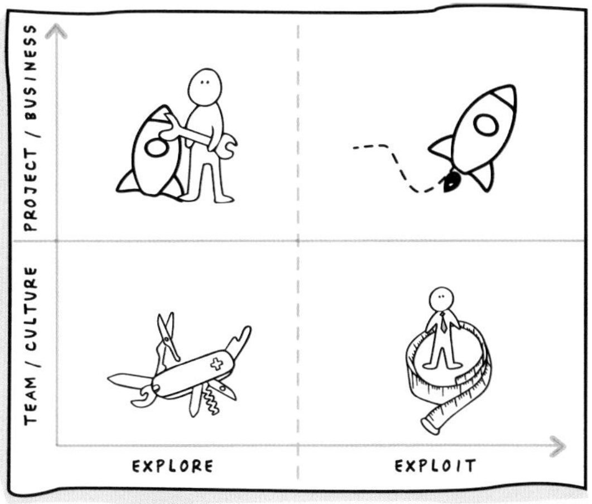

PROJECT / BUSINESS
TEAM / CULTURE
EXPLORE
EXPLOIT

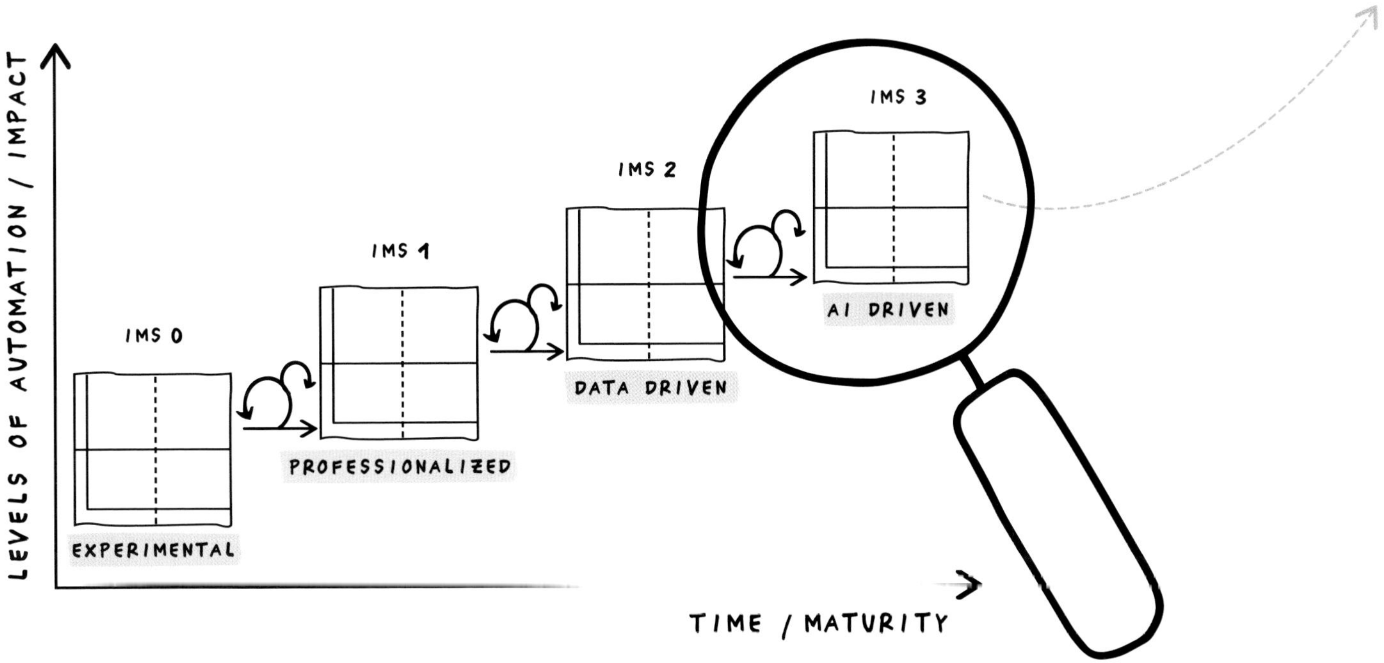

LEVELS OF AUTOMATION / IMPACT
IMS 0
EXPERIMENTAL
IMS 1
PROFESSIONALIZED
IMS 2
DATA DRIVEN
IMS 3
AI DRIVEN
TIME / MATURITY

Applying the IMS Maturity Model

Corporate practice shows that many companies start at IMS level 0 or 1 with their existing measurement systems in the definition of metrics for EXPLORE. Occasionally, global innovation leading companies with a strong focus on utilizing new enabler technologies have innovation measurement systems at level 2. Often, this also includes traditional companies that have managed to expand their capabilities in the area of EXPLORE over the last few years and have recognized the importance of data and measurement as part of the digital transformation. In addition, it should be noted that not every company is aiming to reach IMS 3. For many organizations and teams, the first steps in EXPLORE are already a major milestone, involving organizational and cultural change, customer centricity, and a new perception of innovation and design thinking work as an important component of future business success. Consequently, the IMS must fit the aspiration level of the company, its culture, and its business alignment, and should evolve iteratively and dynamically according to these parameters.

	IMS 0 (Experimental)	IMS 1 (Professionalized)	IMS 2 (Data Driven)	IMS 3 (AI Driven)
PROJECT/BUSINESS	• Creating awareness for a balanced EXPLORE and EXPLOIT portfolio • Distinguishing between metrics for EXPLORE and EXPLOIT • Understanding the impact of value creation for long-term business success	• Managing the transformation from EXPLORE to EXPLOIT • Balanced leadership attention for EXPLORE and EXPLOIT • Established and accepted measures for EXPLORE	• Application of advanced forecasting techniques • Dedicated leadership for EXPLORE and data analytics • Application of hybrid model (customer led and data led)	• Apply AI to drive/support EXPLORE and EXPLOIT • Leadership supported by higher machine autonomy • Quick pattern recognition and definition of sophisticated measures
TEAMS/CULTURE	• Learning about new mindsets, behavior, and culture • Review of existing performance measurement systems • Building capabilities for EXPLORE and awareness and skills for measuring	• Metrics accepted and understood by the team • Alignment between North Star and company and team objectives established • Innovation metrics are part of the company/team culture	• Data/measuring competencies in all functions and teams • Coordinated measurement • Culture of learning, exploration, and continuous innovation • Data driven for innovation and efficiency	• AI and humans as a team • Addressing new or unknown problems • Advanced measurements • Improvements based on predictions and advanced prescriptive analytics
AUTOMATION	• Limited, with ad hoc analysis • Manual data collection • Static information for decisions	• Key metrics are operationalized • Automated (single mode) • Dynamic information for decisions	• High trust in insights from data • Automated (multimodal) • Real-time information for decisions	• AI part of objective setting • Automated (dynamic multimodal) • AI-driven/supported decision

START → FUTURE — TIME

The Essentials of Measuring

For a basic understanding of measurements and metrics, the central questions about measuring will be answered before going into the types of metrics, how they are defined, and, finally, how an innovation measurement system can be established through the concept of North Star metrics, minimum viable set of exploration metrics, and the selection of appropriate exploitation metrics. However, it is also important to reiterate that metrics and measurement activities should not limit creativity or obscure the goal of completing design challenges and projects in the appropriate depth, with the necessary number of iterations and with appropriate latitude. At the end of both 101s, it should be understood that not only creative confidence can be learned, but also the skills that help to lose one's insecurity about measurements.

As a Starting Point toward the Essentials of Measuring, Three Central Questions Are Relevant

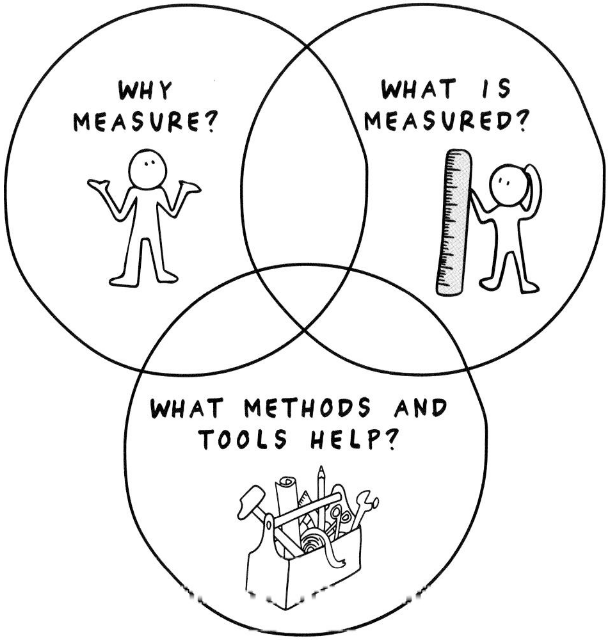
WHY MEASURE?
WHAT IS MEASURED?
WHAT METHODS AND TOOLS HELP?

Why Measure?

The spontaneous answer is that we want to compare an unknown quantity with a known fixed quantity of the same kind. But actually, it is about something much more important: achieving a reduction of quantitative uncertainty based on observations. Here it is important to understand that even a marginal reduction in uncertainty can be extremely valuable.

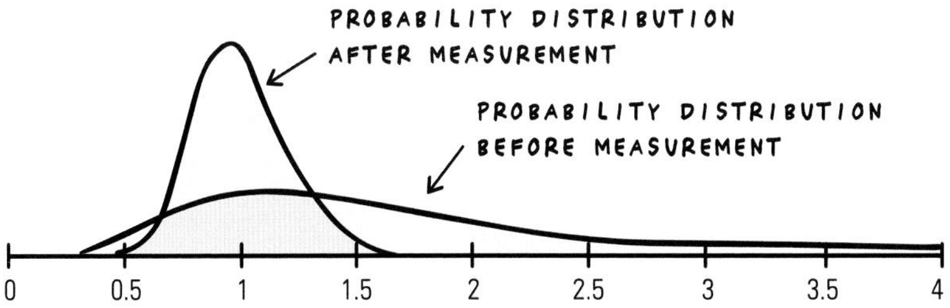
PROBABILITY DISTRIBUTION AFTER MEASUREMENT
PROBABILITY DISTRIBUTION BEFORE MEASUREMENT
0
0.5
1
1.5
2
2.5
3
3.5
4

With regard to innovation outcomes, measurements help to understand better the return on investments (ROI), establishing the starting point for building new capabilities or knowing better about the drivers for creativity, growth, or cost saving.

Anything in the innovation space that can be observed should lend itself to some type of measurement methods.

What Is Measured?

Similar to any problem-solving process, what is to be measured can be well defined or ill defined. For measuring EXPLOIT, many well-defined measurements exist (see pages 189 and 195). The well-defined ones are typically metrics that have a clear and precise designation and a universally valid definition. By contrast, for ill-defined metrics, it is essential, on the one hand, to make the problem tangible, and, on the other hand, to describe the object to be measured. As soon as the innovation team finds possibilities for defining what is meant by a certain phenomenon, ways are found to measure it. Typical examples of ill-defined measurements are purpose, creativity, employee experience, mindset, future skills adoption, or brand image.

Before measuring anything, five key questions should be answered to help in analyzing, applying temporally, or operationalizing a metric:

- **Why** is it important to make a measurement, and what decision could be derived from it?
- **What** are the consequences that can be observed when the decisions are implemented?
- **How much** knowledge about the problem to be solved exists and is known to the team?
- **When and where** does a value or metric make a difference?
- **How much** is the additional information worth and what impact will it create?

Often the measurement process starts with a predefined word that habitually already has certain assumptions associated with it. But instead of falling in love with a word, the development of a metric should begin by thinking thoroughly with the team about what real quantity is to be measured. As an example, in the context of innovation, people often talk about the design thinking mindset, and that this contributes to the success of innovations. What is it about this mindset that contributes to better business success? And why? What real-world behavior is it associated with? What does this behavior look like and when does it happen? Who is involved? Is the subject to investigate the co-creation with the potential user/customer, or rather the willingness-to-pay analysis that is central to MVP development? To avoid this trap, instead of *mindset* it is recommended to simply call it X as placeholder, which should stand for something very specific without room for interpretation. This way discussions about already existing metrics, misinterpretation, and any assumptions can be avoided.

What Methods and Tools Help?

For measuring, all statistical concepts reaching from descriptive statistics to simulation are helpful (see toolkit from pages 238–255). Within this introduction of metrics and measurements, selected methods are presented to define North Star metrics (page 172), exploration metrics (page 179), exploitation metrics (page 194), and creativity metrics (page 209). As uncertainty often exists with regard to measurements and metrics, the respective measurement systems are visualized using simple examples. Corresponding templates and canvas models can be downloaded for daily use to facilitate the handling and development of the metrics for every innovation team and beyond.

Linkage or connection between metrics is particularly important in managing innovation projects, where there is a variety of metrics that define the innovation success.

Putting Measurements and Metrics into a Broader Spectrum of Impact

In the context of measuring innovation and activities related to EXPLORE, it is particularly important to understand the concept of leading and lagging indicators, as not every metric is suitable to reflect the information that is currently needed. Various innovation measurement and performance measurement systems measure progress toward a specific goal and use financial and non-financial indicators.

Leading versus Lagging Indicators:

- **Leading indicators** measure the activities the team needs to complete to reach the goal. They can be tracked on a more ongoing basis.
- **Lagging indicators** usually take a long time to change and confirm the later-stage results of the innovation efforts and activities.

LAGGING INDICATORS	LEADING INDICATORS
Metrics used to measure **past** performance	Metrics used as a **predictive** measure of **future** performance

North Star, exploration, and exploitation metrics might be leading or lagging, and very often work best when combined.

Example

How to Make Them Work Together

The key for meaningful measurements is very often to use a combination of both leading and lagging indicators. The adaptation of the leading indicators, for example, via exploration metrics, helps to make changes that increase the chances of success in the innovation project. In this way, the leading indicators complement the lagging indicators and help, for example, to measure progress more quickly or to make improvements at the process level. The following is a simple example of how the different scenarios can be combined.

Example of Combining Indicators

Measuring Design Thinking Macro Cycle Length (Lagging) + Average Design Thinking Process Step Length (Leading)

If the time from problem identification to final prototype is too long and the aim is to reduce the time it takes to come up with a validated final prototype, the lagging indicator might be macro cycle length. Without going through the entire design thinking macro cycle, teams decision makers might not understand how already performed capability-building programs, up-skilling initiatives, enlargement of teams or other improvements are effective on the macro cycle length. To track the progress of the design challenge, it might be worth also measuring the average design thinking process step length as well. This splits up the macro cycle into activities of teams working on a specific process step or even with a specific tool. For example, the average length for preparing, engaging with the customer, and collecting insights might be a leading indicator for spending time. Looking into all activities and micro-process steps individually enables teams to identify bottlenecks that extend the overall time from finding the appropriate problem to solving it.

Working with Qualitative and Quantitative Data

The distinction between quantitative and qualitative metrics does not require much explanation. However, it is often the case that the best data is qualitative data that can be quantified. Increasingly, there is also the possibility of obtaining quantifiable data on insights that were previously only available through qualitative surveys. The last section of this book explains, for example, how neuroscience quantifies and makes measurable what was previously only qualitative data on behavior and human motivation. It is more emotional in nature, and therefore more difficult to quantify with traditional methods. Quantitative data, on the other hand, is anything that can be tracked on a numerical basis. It is more commonly used to understand what is going on and to visualize broader trends.

Both data sets serve a specific purpose, and both have their pros and cons, as compared below. Qualitative data is more expensive to obtain, but it is richer and more directional. Quantitative data lacks depth, but it is easier to collect and, of course, more statistically robust. As a result, quantitative metrics are good to explain **WHAT** and **HOW MANY** of a given hypothesis. The **WHY** is usually better captured through qualitative research methods.

When selecting measures and metrics for innovation to track, it has proven beneficial to focus on metrics that provide a balance of qualitative and quantitative evidence.

Qualitative

PROS

- Rich data
- Personal
- Directional information

CONS

- Expensive
- Difficult to interpret
- Difficult to make comparisons

Quantitative

PROS

- Less costly
- Easier to validate
- Easier to collect
- Statistically robust

CONS

- Data is limited/missing
- Lacks depth and can be misunderstood
- Needs > Sample size

The Influence of External Factors

In the world measurements and metrics, external influences cannot be avoided. For this reason, metrics that predict the future should always be compared with current macroeconomics studies, outlooks, and scenarios. Measurement experts include the external world in the internal measurement system. For example, at the time this book was written, many national and global factors indicated that a global recession was on the horizon, along with new geo-political challenges that made it necessary to rethink certain parameters from the past. These parameters and causes and effects must be understood in order to adapt the measurement system and make the accurate measurements with the appropriate metrics. It is hard to predict, for example, the price elasticity of supply and demand without knowing that a potential recession will occur. Companies that have maintained successful measurement systems over years know their leading and lagging indicators very well and can reduce the uncertainty of forecasts based on these data and measurement points.

For example, for companies in the consumer goods sector, leading indicators related to consumer price sentiment will be relevant (in the described scenario of a recession and high inflation). On this basis, prices can be raised or varied with package sizes that are less dramatically perceived by consumers, such as a price increase of 5 or even 10 percent. Thus, for specific metrics with regard to business and product success, basic knowledge of the economic theories is very helpful in understanding if supply and demand are in equilibrium. Another example related to the design of global ecosystem initiatives for companies in future scenarios, which are important for assessing potential network effects or the split of ecosystem initiatives into more national and smaller subsystems. The illustration below depicts potential scenarios of globalization from WEF (2022). Each scenario of the future might have massive impact on the configurations of a value proposition or the business ecosystem itself.

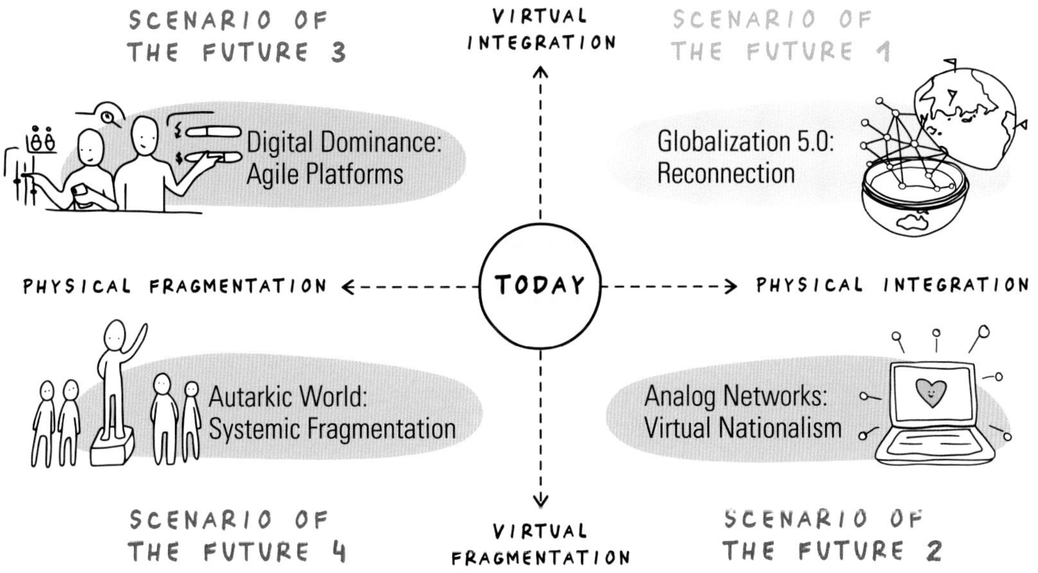

SCENARIO OF THE FUTURE 3
Digital Dominance: Agile Platforms
VIRTUAL INTEGRATION
SCENARIO OF THE FUTURE 1
Globalization 5.0: Reconnection
PHYSICAL FRAGMENTATION
TODAY
PHYSICAL INTEGRATION
Autarkic World: Systemic Fragmentation
Analog Networks: Virtual Nationalism
SCENARIO OF THE FUTURE 4
VIRTUAL FRAGMENTATION
SCENARIO OF THE FUTURE 2

Example

What if the envisioned Globalization 5.0 will not take place as predicted?

Dimensions in Observing, Measuring, and Predicting the Future

In the context of measuring and metrics, the applied dimensions and views influence the measurement setting and metrics system. Some dimensions, like the time horizon of analyses and forecasts, are well defined (near versus distant, lagging versus leading) and need no further reflection. Other dimensions, like the influence of external factors and related imagination of the future, are more crucial, because their character might be based on dystopias or utopias.

Core beliefs and the logic of actions might be based, for example, on causation or effectuation, which influence decision making and resource allocation. Another scenario is exploration with the purpose of delivering real value creation, or focusing on identifying opportunities at the right time. The following four pairs of distinct dimensions help to reflect worldviews, mindsets, and the cultural environment in which a potential measurement system will operate.

Logic of action rationality for metrics and measurements: causation versus effectuation

CAUSATION		EFFECTUATION
The future is predictable, and the course of action is goal oriented. Decision making is based on selecting from alternatives with maximum return and procuring the necessary resources. Contingencies are avoided by planning from stakeholders performing competitive analysis.	VS	The future is controllable, and the course of action is means oriented. Decision making is based on selecting a satisfying alternative where a possible loss of resources is affordable. Contingencies are opportunities to be exploited and stakeholders aim to establish partnerships.

Transported imagination of the future and potential opportunities: dystopia versus utopia

DYSTOPIA		UTOPIA
The future perspectives are worst-case scenarios based on criticism of current trends, society, norms, and systems. Innovation teams have a strong inner conflict with the system (rules, laws, and society); turning points trigger tendencies to overthink the system and values.	VS	The future perspectives are ideal-case scenarios based on the perfect place, state, or condition of the system, society, and other factors. Innovation teams actively question the current system with the aim of positive change; this supports the team and organization in recognizing the positive aspects.

Activity level of measurements and foresight management: reactive versus proactive

REACTIVE		PROACTIVE
The future perspective is based on waiting until a problem becomes obvious. Activities are triggered to solve the problem and to go back to the comfort zone. Decision makers and innovation teams feel stressed, lost, and reluctant to act. Teams have a weak understanding of their purpose.	VS	The future is based on embracing new situations. Innovation and design thinking teams see them as a challenge and learning opportunity. Decision makers and teams feel confident, strong, and in control of the design cycle. Teams are purpose driven.

Form and actions of exploration: opportunity recognition versus value creation

OPPORTUNITY RECOGNITION		VALUE CREATION
Decision makers and innovation teams have a low uncertainty at the beginning: Outcomes and probabilities for success are well known by decision makers and teams. The applied mindset and cognition include classical market analysis tools and instruments, which lead to incremental improvements.	VS	Decision makers and innovation teams have a high uncertainty at the beginning. Outcomes and probabilities are unknown in the beginning by the decision makers and the teams. The applied mindset and cognition are embedded in a design thinking and growth mindset, which leads to breakthrough innovation.

DEFINING AND APPLYING METRICS

Define and Review Metrics

After this brief recap of the essentials of measuring, the question of how to define effective metrics inevitably arises.

How to Define Effective Metrics for a Measurement System

A metric is an already introduced measure that, in and of itself, gives meaning to a condition. It is a measurement of a particular problem, issue, process, or initiative. For this reason, the 10 most important principles that must generally be met for already established metrics to be meaningful are briefly summarized on the right. The principles are generally valid and do not only refer to the complexity of measuring innovation.

Often, innovation teams and decision makers encounter metrics or an entire measurement system that has grown historically, and the effect of the measurement needs to be questioned in relation to current and future activities. In these cases, it is a matter of fine-tuning, changing, deleting, and adding to the current metrics. For an initial simple validation of the applied metrics, it should be reflected whether the current measurements are really evidence based.

In General, Existing Metrics or Measures Can Be Divided into Four Categories:

1. **No evidence:** Common sense shows that potential employees will join a specific company if it will pay them more than their direct competitors.
2. **Irrelevant evidence:** We compare three randomly selected world-class companies that have nothing to do with our business model and analyze their revenue multiples.
3. **Unreliable evidence:** Employees assume they are more innovative when they work by themselves in their home offices.
4. **Evidence-based:** No interaction with potential customers reduces the likelihood of innovation success.

The 10 Most Important Principles for Assessing Existing Metrics & Measurement Systems

- ✓ Good metrics have significance and importance (e.g., in terms of strategy, ambition, or objectives).
- ✓ Meaningful metrics describe a whole phenomenon (e.g., the summary of different indicators of innovation success or behavioral change).
- ✓ Key parameters for the metric are accessible and manageable (e.g., through recurring surveys or access to data points).
- ✓ The effort of collection and the value of measurement is in a healthy relationship, otherwise the effort is not worthwhile.
- ✓ The metric is understandable and comprehensible (e.g., the recipients of it immediately know which activities to develop).
- ✓ A good system of metrics is precise, and the resulting measures are clear.
- ✓ Ideally, the metric is created in real time or in a timely manner, so that, for example, rapid action or improvements are initiated.
- ✓ Metrics results are communicated transparently and made available to the respective teams/organization.
- ✓ All applied metrics are meaningful and credible from different perspectives so that they are widely accepted.
- ✓ Finally, the metrics applied are as factual and evidence based as possible, which is best achieved within the framework of the aforementioned principles.

! **The easiest way of evaluating metrics is to perform a sanity check. Ask what the team did differently if figures went up, went down, or stayed the same. If there is no answer, it is recommended to put those metrics aside and look for better ones.**

Examples of General Application Cases

Example

With the measurement and the appropriate metrics, meaningful options for action can be derived. The purposes of metrics or a well-designed measurement system are diverse and vary depending on the application and organization. In companies, organizations, and teams, the following application cases are the most common derived from observations and re-designing of measurement systems and programs. The metrics might be defined as input, output, outcome, or process metrics, or a well-designed combination of input, output, and outcome (see example page 160) .

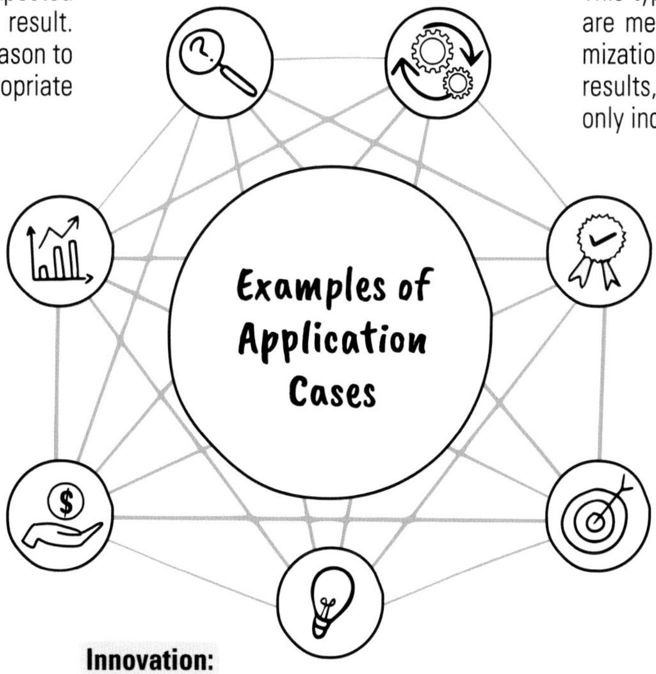
Examples of Application Cases

Problem analysis:
This is typically a gap between the expected result of a measurement and the actual result. The measurement result provides the reason to determine the cause and initiate appropriate corrective actions.

Trend development:
Measurement over time provides information about the dynamics of organizational measures and processes. Often, only the passage of time indicates which trends are significant and where one-time effects occur.

Budgeting:
Measurements from the past are often used for budget decisions in the future. Especially if significantly better results can be achieved through higher budgets.

Process improvements:
This typically involves quality, costs, or time, which are measured and provide information about optimization potentials. Depending on the measurement results, the adjustments can be very extensive or only incremental changes.

Recognition:
These are metrics that are used to reward individuals for performance. In the context of a performance dialogue, 360-degree feedbacks serve to initialize a holistic assessment about employees and to derive targeted coaching measures from it.

Objectives & key results (OKRs):
Modern performance measurements typically allow the teams to define their own metrics that provide information on whether defined objectives are on track. The metrics also serve to verify or validate indicators of the actual metric.

Innovation:
This is typically a whole set of metrics which are applied to build new capabilities, change behavior, and measure the effectiveness and the impact of the innovation work.

Differentiation Between Input, Process, Output, & Outcome Metrics

In applying metrics, it is important to define clearly the meaning of input, process, output, and outcome metrics. In the consideration of innovation and business success, a distinction is usually made between the immediate-term result (=output) and the results that arise immediately after the end of a defined project activity (=outcome).

Typical examples of output metrics are:
- Financial and nonfinancial results of innovation activity
- Number of in-depth interviews with customers/users
- Increased use of agile toolbox by teams

Typical examples of outcome metrics are:
- Number of new customers won with a new service
- Employee satisfaction rate
- Degree of solution fit with MVP trails of paying customers

Equally important are input metrics, which are associated with measuring assets and resources, as well as process metrics, which measure the efficiency or productivity of the innovation or business process.

Typical examples of input metrics are:
- Budgets for design thinking trainings programs
- Quality of insights generated by data and observations

Typical examples of process metrics are:
- Time to complete validation of product/market fit
- Time-to-market for expansion of core value proposition

The impact based on certain inputs and activities leads, for example on the team/culture level, to concrete impact in conjunction with behavioral change (see example below for better collaboration across teams).

Example of input, process, output, and outcome for design thinking training

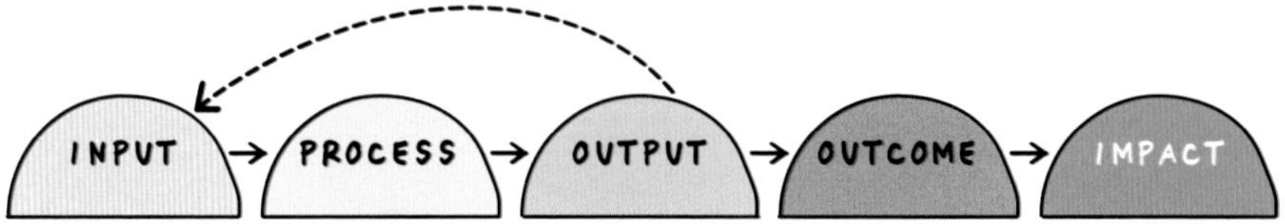
INPUT
PROCESS
OUTPUT
OUTCOME
IMPACT

- INPUT: $ budget for training; # design thinking coaches/trainers
- PROCESS: # training hours/coaching hours; # trainings performed
- OUTPUT: # employees trained; # employee satisfaction
- OUTCOME: # maturity level design thinking; # individual skill levels
- IMPACT: products with outstanding customer experience

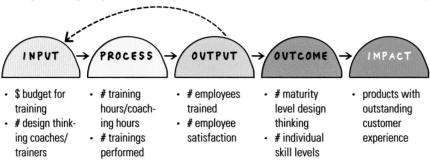

In the application of performance measurement systems, the input of teams and capacity is leading to output which is monitored, for examples, on quarterly outcomes set by objectives and key results (see example below).

Example of input, output, and outcome in conjunction with OKRs

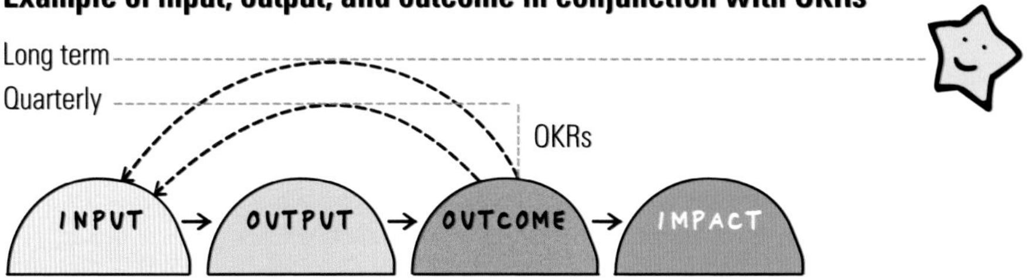
Long term
Quarterly
OKRs
INPUT
OUTPUT
OUTCOME
IMPACT

- INPUT: planned initiative/mission; # engineering & design resources
- OUTPUT: # of new features; # customer tests
- OUTCOME: degree of customer problem solved with new features
- IMPACT: driving-up customer satisfaction with new features?

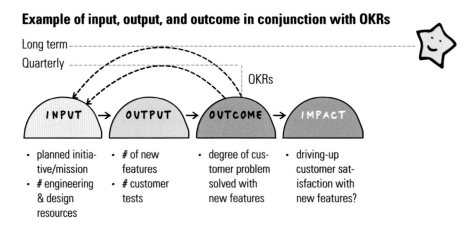

Measurements related to the innovation performance of the process are usually quantifications related to time, quantity, development costs, quality of innovation realized, as well as risk management indicators related to the developed product, service, or experiences.

Connecting the Dots of Measuring

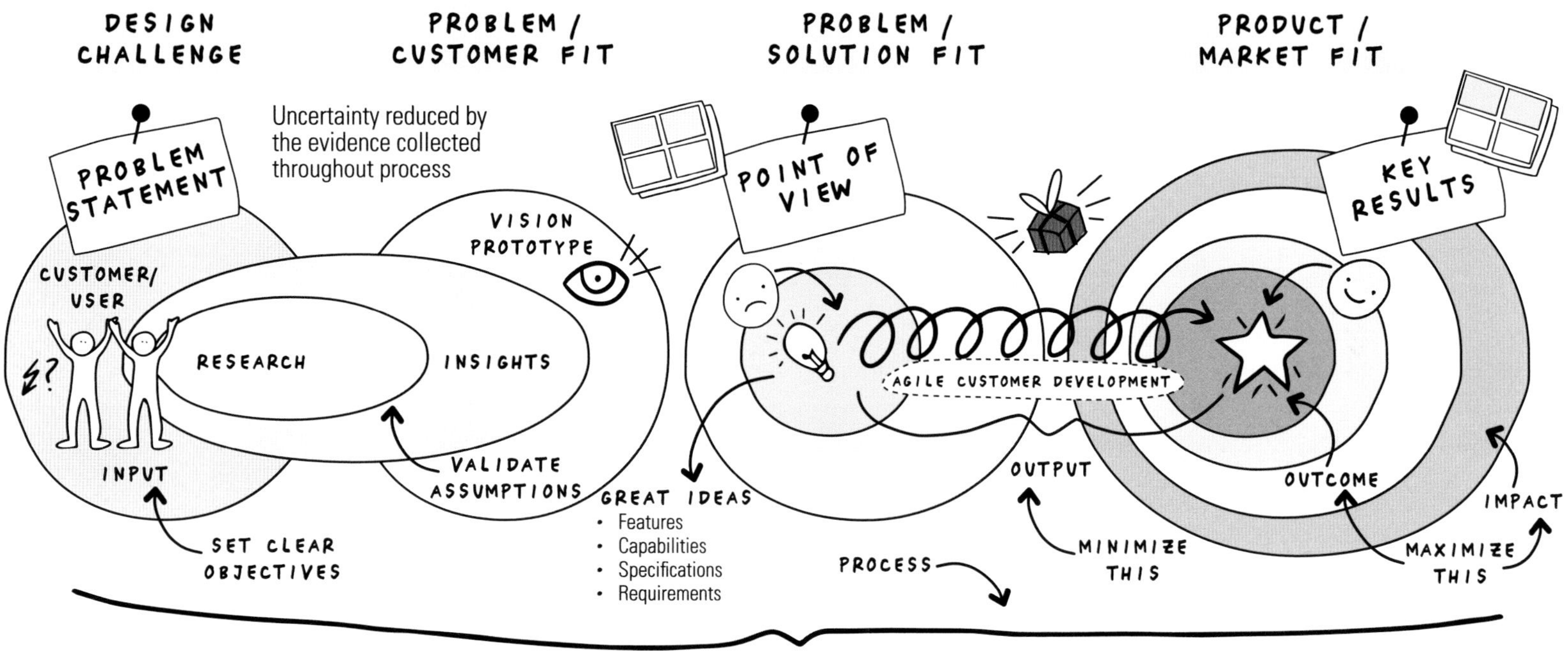

DESIGN CHALLENGE
PROBLEM / CUSTOMER FIT
PROBLEM / SOLUTION FIT
PRODUCT / MARKET FIT
PROBLEM STATEMENT
Uncertainty reduced by the evidence collected throughout process
POINT OF VIEW
KEY RESULTS
VISION PROTOTYPE
CUSTOMER/ USER
RESEARCH
INSIGHTS
AGILE CUSTOMER DEVELOPMENT
INPUT
VALIDATE ASSUMPTIONS
GREAT IDEAS
Features
Capabilities
Specifications
Requirements
OUTPUT
OUTCOME
IMPACT
SET CLEAR OBJECTIVES
PROCESS
MINIMIZE THIS
MAXIMIZE THIS

Input

NORTH STAR **PROJECT**

- Mandate to innovate
- Team size
- Budgets
- Foresight

Process

SPEED

EFFECTIVENESS **EFFICIENCY**

- Progress of activities
- Time-to-market across EXPLORE/EXPLOIT portfolio
- Return on investments on various levels

Output

ACTIVITIES **PARTICIPATION**

- Additional valuable ideas
- New functions required from users and potential customers
- Value proposition validated via MVP testing

Outcome

SHORT TERM

- Learnings
- Awareness
- Attitudes
- Mindset
- First steps of cultural change

MID TERM

- Increase in revenue from new customer/ product category as % of total sales

Impact

LONG TERM

- Total added value of the project to all value the company generates from EXPLORE activities

AMBIGUITY → SUCCESS

Measuring over the Entire Design Cycle

The previous example showed the cause-effect relationships of measuring at a high level across the entire design cycle. However, these can be broken down even further to the respective activities from design thinking to lean start-up and the scaling of the solution. No matter what the view of the potential measurement system looks like, it is important the metrics should primarily pursue the goal of supporting the appropriate next steps and activities. Some of the defined metrics will correlate (similarity between two metrics), and other metrics will have the ability to show causation. Causation is very important because it provides valuable insights to trigger change before things might go into the wrong direction. However, it is not always direct, and a specific event might have multiple causes. In best-case scenarios, the knowledge about the causality of two metrics supports any future related decisions and activities.

Purposes of Innovation Measurements and Metrics:

- Provide strategic direction by indicating shifts in priorities
- Direct the (re)allocation of resources
- Assess the effectiveness of innovation spending
- Make teams and leaders accountable and incentivize them to achieve objectives (see also 201, pages 258–284)
- Analyze and enhance innovation performance and culture
- Change behavior and processes

There are no blueprints for the measurement of ill-defined metrics in EXPLORE. For example, the respective minimum viable measurements for EXPLORE are to be defined depending on the situation, the culture, and a top-line metric. To show the range and magnitude of measurements and metrics, examples of input, output, process, or outcome metrics, which might find application over the entire design cycle, are displayed in the table on page 163. For EXPLOIT, more traditional and well-defined metrics come into play to help improve mainly efficiency and to validate predictions about outcomes. However, the danger remains that outdated measurement systems and approaches don't match the new ways of working and innovation style, and are thus measured incorrectly or with little significance.

For the interactive design of measurement systems applying design thinking (see page 148), it has proven useful to roughly divide the step-by-step procedure into the dimensions of learning, action, and impact zone.

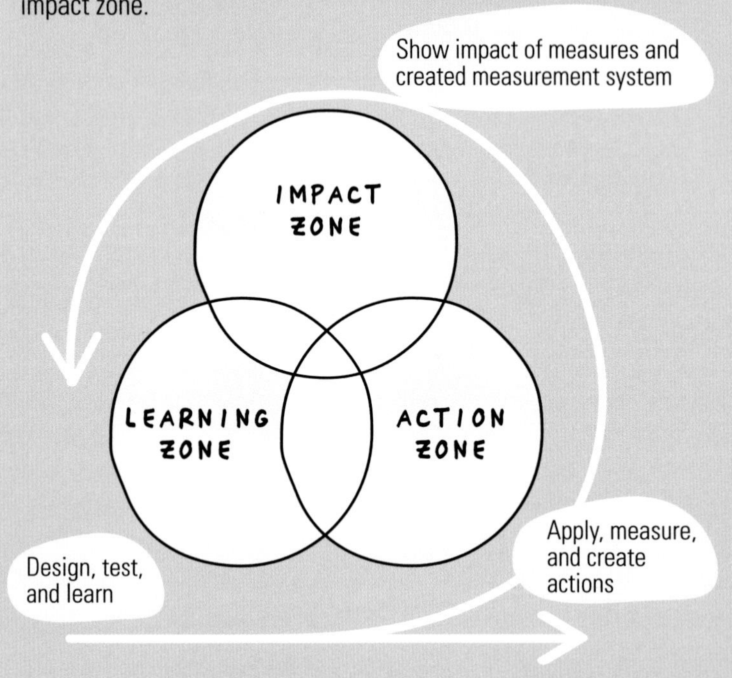
Show impact of measures and created measurement system
IMPACT ZONE
LEARNING ZONE
ACTION ZONE
Design, test, and learn
Apply, measure, and create actions

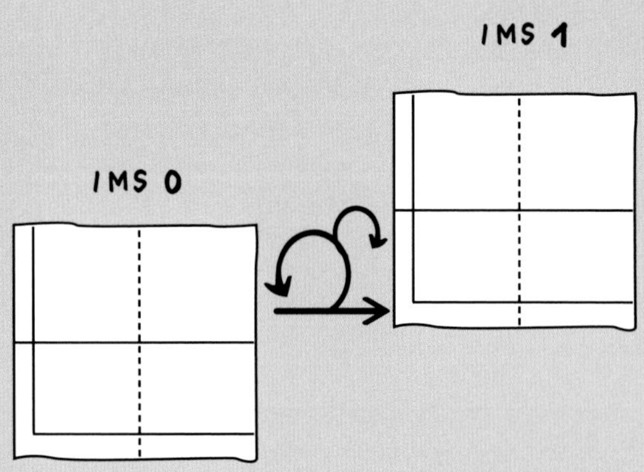
IMS 1
IMS 0

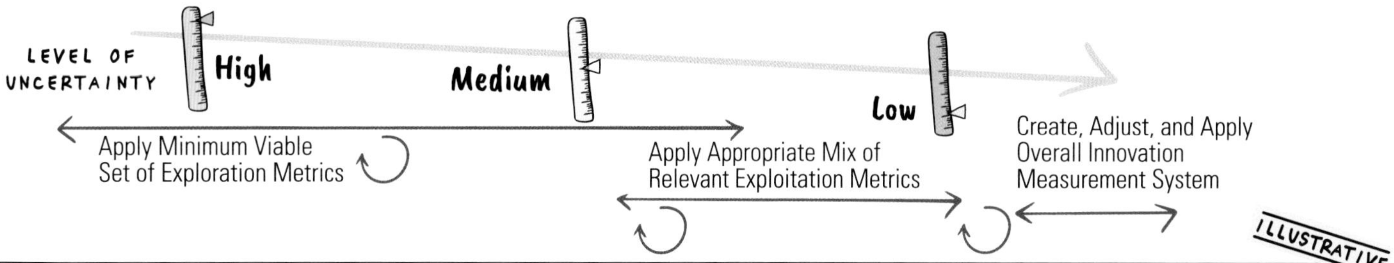
LEVEL OF UNCERTAINTY
High
Medium
Low
Apply Minimum Viable Set of Exploration Metrics
Apply Appropriate Mix of Relevant Exploitation Metrics
Create, Adjust, and Apply Overall Innovation Measurement System
ILLUSTRATIVE

MEASUREMENTS DEMONSTRATE	• Design thinking mindset applied • Exploration leads to learning	• Knowledge enhancement • New solutions are developed, tested, and re-designed or killed	• Validation of viability • Validation of value proposition • Agile customer/MVP development	• Innovation work (design thinking, lean start-up, and ecosystem design) creates the desired impact • Configuration and validated scalable unit economics in different dimensions
EXAMPLES OF INPUT METRICS	• Freedom given by leadership team to EXPLORE beyond core business • Management commitment to defined culture of exploration and renewal	• Cost to final prototype for extended design thinking team from problem identification to potential solution	• Number of lead customers/users participating in MVP testing with a heterogeneity of profiles representing all potential person a groups. • Cost from problem definition to product/market fit related to number of sprints to solve the customer problem	• Organizational growth and # of employees engaged in team-of-teams activities to create new market opportunities • Scalability of the operational tasks related to a new product, service, or experience to engage with more customers/users
EXAMPLES OF OUTPUT METRICS	• Improvement of design thinking maturity based on # of project-based learning interactions • # of critical and unknown assumptions identified	• # of experiments which contribute significantly to defined learning outcomes	• Additional valuable ideas for add-on functions from users/customers collected in MVP testing • Increased use of agile toolbox of teams which attended project-based training and capability build program.	• Revenue growths from exploration activities • ROI on innovation activities and design thinking challenges • Customer engagement with new products, services, or experiences • Increase of application of new mindset/culture through project-based learning activities
EXAMPLES OF PROCESS METRICS	• Time to insight • Take-up time from the team to adjust to new processes	• Innovation portfolio balance indicates the # or volume of EXPLORE versus all projects	• Time to market • Intensity of co-creation with customers	• Time to adoption (measured in time or number of customer adoptions in X-time versus the market size)
EXAMPLES OF OUTCOME METRICS	• Degree of fit of persona developed with strategic mass-market requirements defined in corporate strategy	• Degree of compatibility of solution with existing technology, applications, and IT (feasibility)	• Degree of solution fit with MVP trails of paying customers/users • Degree of market fit with conversations (e.g., CAC, ARR, NPS)	• Ratio of (expected) profits from users/customers from EXPLORE in comparison to adjustments of target segments in EXPLORE • (Expected) increase in revenue from new customer/new product category as a percentage of total sales

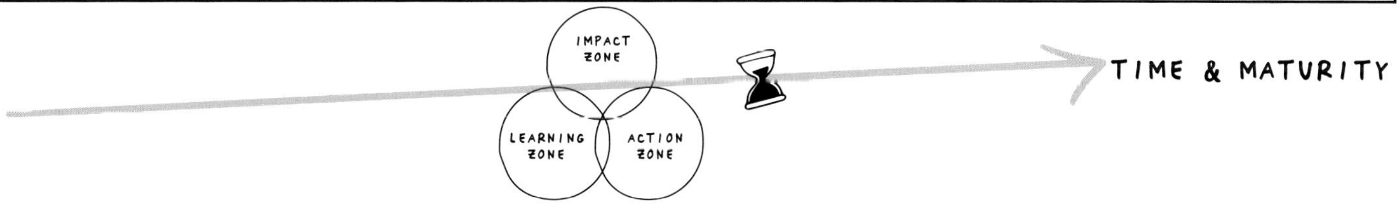
IMPACT ZONE
LEARNING ZONE
ACTION ZONE
TIME & MATURITY

Metrics Can Also Become a Target for Manipulation and Subversion

Especially in large organizations with top-down driven measurement systems and efforts, it can be observed that metrics are either gamed over time or motivate teams to operational hustle and bustle that doesn't have any real effect on value creation, delivery, and capture. For this reason, the achievement of OKRs, for example, is usually not directly linked to bonuses for the team or single team members. It can be also observed that companies with a good measurement culture avoid pushing for standalone vanity metrics that encourage, for example, articulating and documenting a high number of ideas. In theory, the number of ideas provides the basis for upstream innovation; but without putting the ideas into a broader spectrum of impact, this kind of measurement might be impractical or lead to poor behavior, as illustrated below.

! Some metrics might be gamed over time, and it is no surprise that one team might get more resources than another.

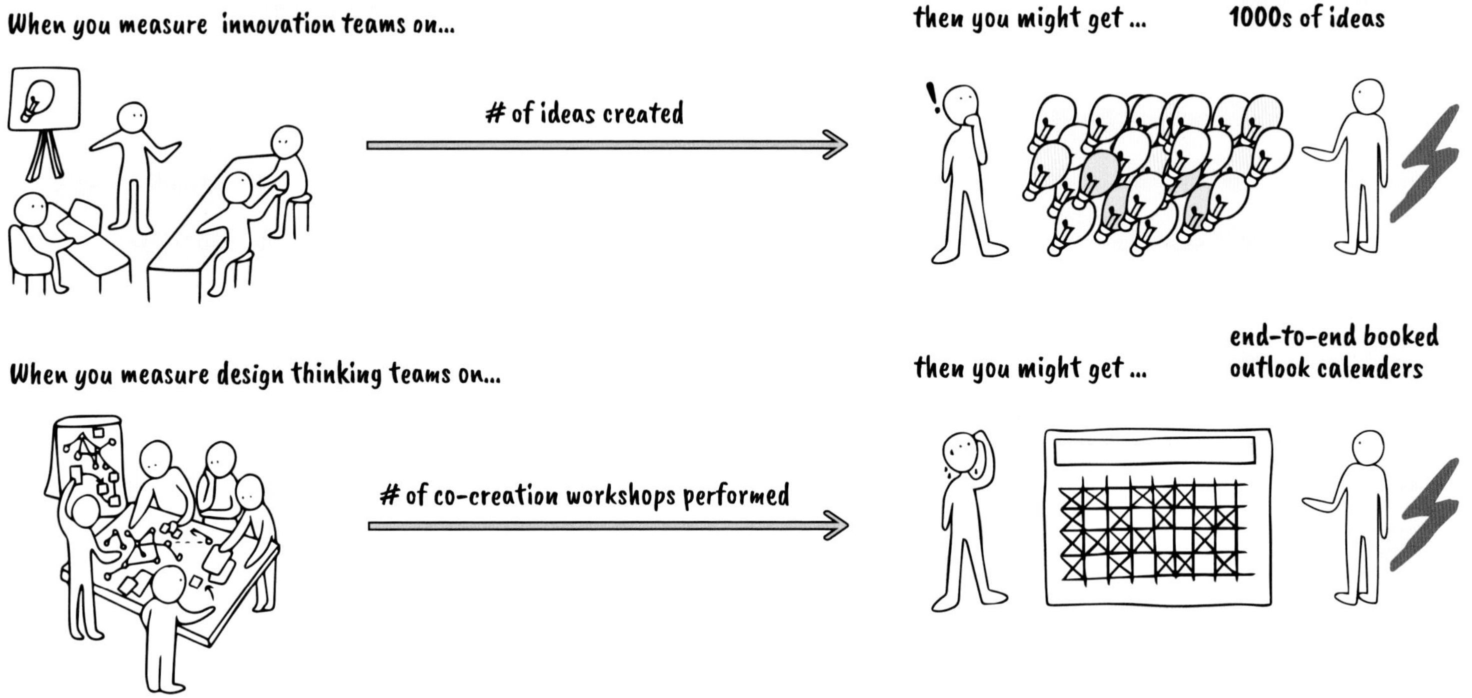
When you measure innovation teams on...
# of ideas created
then you might get ...
1000s of ideas
When you measure design thinking teams on...
# of co-creation workshops performed
then you might get ...
end-to-end booked outlook calenders

North Star, EXPLORE, and EXPLOIT Metrics

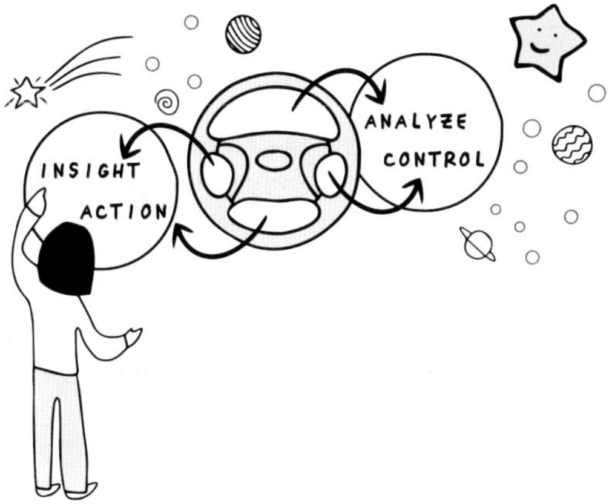
INSIGHT
ACTION
ANALYZE
CONTROL

It has already been pointed out that the respective metrics should be actionable and refer to ratios and unit economies rather than to gross quantities (see also general recommendation on page 166). This is particularly important in the context of innovation, as actionable metrics offer many advantages in the context of EXPLORE, based, for example, on thoughtfully designed North Star metrics. All of them help to make business decisions by comparing means to ends. Exploration metrics are usually applied up to the product/market fit on different levels. Afterwards the concept of an appropriate mix of relevant exploitation metrics is increasingly applied, as these measurements and metrics are common indicators for business growth, quantity, and the basis for scaling and exponential growth. The cause and effects outlined in the table below explain the differentiation.

	North Star Metric(s)	Minimum Viable Set of Exploration Metrics	Appropriate Mix of Relevant Exploitation Metrics
DESCRIPTION	Focus on growth ambitions and directions to manage product, innovation, and business success	Inform business decisions by measuring means relative to ends	Quantifies business performance to improve scaling, efficiency, and cost savings
BEST TIME TO BE APPLIED	Definition of vision, strategy, and strategic choice	EXPLORE (before product/market fit)	EXPLOIT (after product/market fit or system/actors fit)
CAUSE & EFFECT	Focus on impact and actions	Focus on learning, improving, and effectiveness	Focus on effectiveness and efficiency
REPRESENTS MOSTLY	Company, organization (support for understanding better the why, how, and what)	Company, organization, team/individual behavior (support evidence for innovation work)	Performance of company and activities (support for monitoring the health of the company)
TYPES OF METRICS	Behavior change, ratios, rates	Ratios/rate, new capabilities, new behavior, unit economies	Gross quantities and unit economies
EXAMPLES	Attentions, engagements, transactions, frequencies, productivity, and reach	Time-to-insight, budget-to-development, time-to-adaption, maturity level of design thinking	Profitability ROI Growth of users/customers

The following five sections discuss the practical approach to defining North Star metrics, working with minimum viable exploration metrics, and selecting the appropriate exploitation metrics. Toward the end of this section, the exciting and often mysterious measurement of creativity with different ways of measuring is discussed with common measurements on all levels of EXPLORE reflected. The respective suggestions for designing the measurement system work best when the entire range of possibilities is applied. However, all of them can also be applied individually depending on the situation as part of a step-by-step implementation or focus on specific metrics or measurement systems.

At this point it is important to point out that well-defined North Star metrics are also indispensable for working with objectives and key results (see 201, pages 258–311) within a modern performance measurement system, besides guiding teams in all potential activities related to EXPLORE and EXPLOIT.

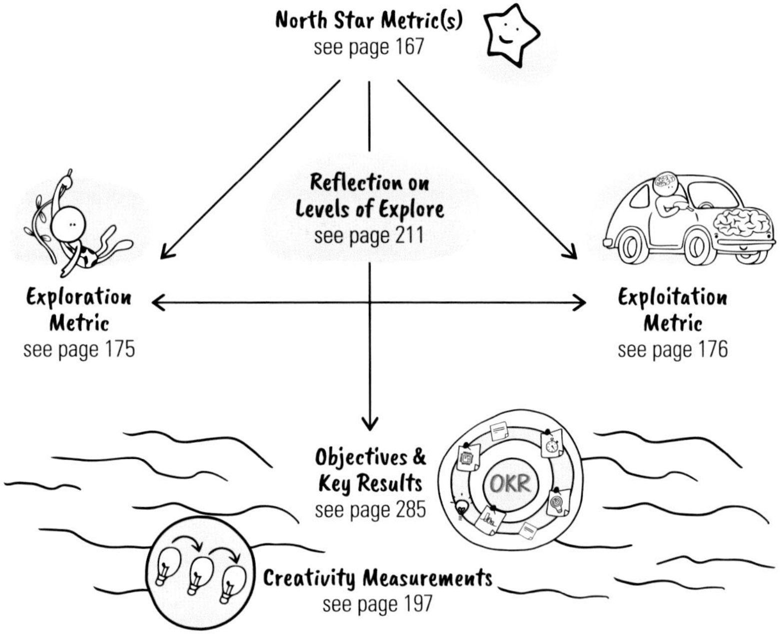
North Star Metric(s)
see page 167
Reflection on
Levels of Explore
see page 211
Exploration
Metric
see page 175
Exploitation
Metric
see page 176
Objectives &
Key Results
see page 285
OKR
Creativity Measurements
see page 197

General Recommendation to Manage Metrics

It has proven helpful to consider the following four recommendations in the design, application, and management of metrics related to innovation and business success.

1. Actionable and/or Behavior Changing
- Trigger motivation for change
- Define actions related to measurements

2. Rate or Ratio
- Analyze movement, numbers, interaction, or other relevant elements over time

3. Easy to Understand
- Mind-map metrics, correlations, and causality
- Test whether the team understands the reason for the measurement

4. Comparisons and/or Directions
- Showcase the direction the project, product, team, or individuals are moving
- Create timelines and find clues for comparisons

DEFINING THE NORTH STAR METRIC(S)

Top-Line Metric(s)

Prerequisites of Building a Solid Measurement and Metrics System

Whether developing new products or services with the design thinking mindset, measuring innovation, or defining and applying objectives and key results, the guiding principles and defined North Star help to provide direction for many activities. Well-crafted top-line metrics can ensure teams are on track to achieve results, engage leaders and employees on collaborative objectives, and promote the value of innovation activities.

The design of the related metrics and an effective measurement system is highly dependent on the ambition and strategy of the organization, and the problems for which metrics are to be developed vary depending on the organization's culture, processes, capabilities, and, of course, the overall direction which focuses on EXPLOIT, EXPLORE, or, ideally, a balanced portfolio of both of them. Consequently, the North Star includes financial, operational, and customer-related metrics.

For example, there are innovation teams that measure the impact of design thinking with the broader organization by the design thinking tools used, including how frequently teams engage with potential customers/users in collecting insights, prototyping, and customer feedback, while developing new, high-impact market opportunities. This might be triggered by a supplement of the North Star metrics which reflects the ambition of creating aha moments of customer value.

Like other metrics proposed in this book, the top-line metrics must be measurable and progressive. The metrics change over time based on the activities created from the measurements, moving from leading to lagging. For example, the revenue of new products, services, and experiences in the first quarter after market launch (lagging) is directly tied to all inputs of efforts put into the development and market launch activities (leading).

Well-defined North Star metrics are also indispensable for working with objectives and key results (see 201, page 258) within a modern performance measurement system, besides guiding teams in all potential activities related to EXPLORE and EXPLOIT.

An outstanding North Star metric is best described by properties such as comparable, understandable, a rate/ratio, and, in most cases, behavior change.

Proactively Managing the S-Curve from Problem to Growth & Scale

- The work of design thinking and innovation teams has the primary focus of realizing new market opportunities that will ensure future growth.
- Business growth triggers and encounters strategic inflection points over the sigmoid curve (S-curve).
- The North Star metrics definition allows us to set a well-defined focus for growth at each stage of the S-curve of growth.
- In many cases the inflection points are those points in time where the organization is asked to make fundamental shifts to maintain and/or capitalize on the forward momentum.
- Without a proper alignment, communication, and awareness of the North Star metrics, organizations tend to take the wrong turn, or persist in stagnation or limited linear growth.

From Start to Scale

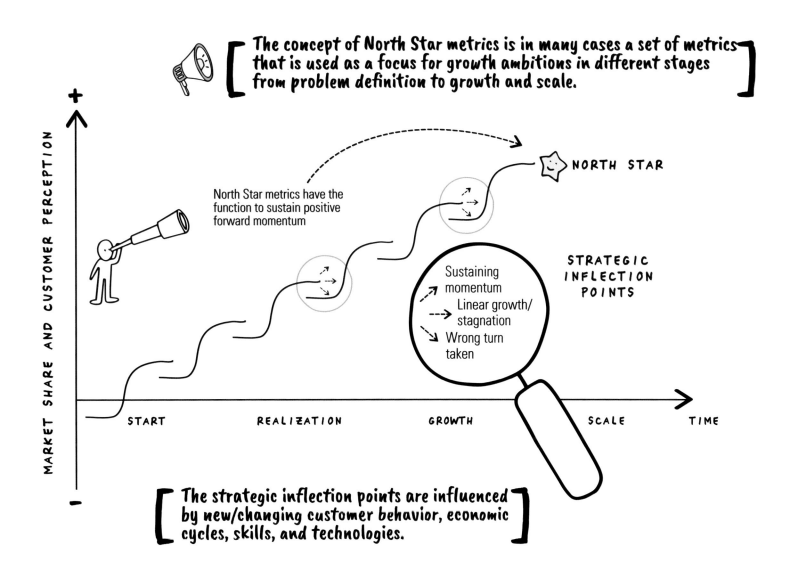

[The concept of North Star metrics is in many cases a set of metrics that is used as a focus for growth ambitions in different stages from problem definition to growth and scale.]

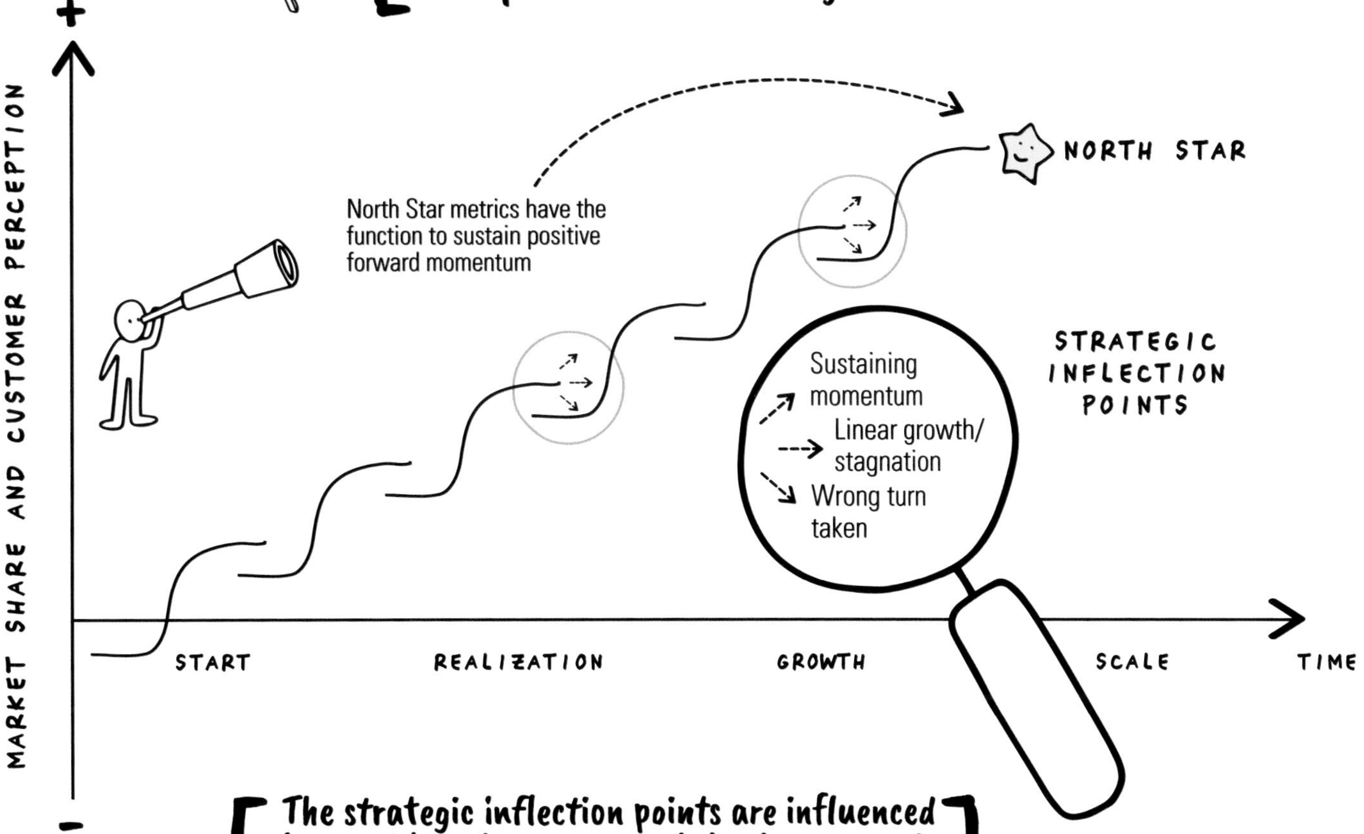

+
MARKET SHARE AND CUSTOMER PERCEPTION
-
North Star metrics have the function to sustain positive forward momentum
NORTH STAR
STRATEGIC INFLECTION POINTS
Sustaining momentum
Linear growth/ stagnation
Wrong turn taken
START
REALIZATION
GROWTH
SCALE
TIME

[The strategic inflection points are influenced by new/changing customer behavior, economic cycles, skills, and technologies.]

How to Define the North Star Metric(s)

Defining one or more North Star metrics is often not such an easy task. However, the importance for teams is very powerful, as exploration metrics are formulated, activities are managed, and EXPLORE and EXPLOIT portfolios are aligned with their guiding principles based on the North Star. For these reasons, the North Star also has an impact on the entire innovation measurement system.

Corporate practice also shows that the North Star definition is particularly challenging in business ecosystem design and orchestration tasks, as ecosystems are made up of different elements, and actors' distinctive priorities need to be aligned with them. In the end, it is important that the elements in the system work well together so that coevolution as well as symbiosis emerges.

However, if done right, the North Star can perform miracles in the achievement and alignment of organizations. Often it is seen that companies and decision makers focus too long on a North Star metric that does not fit.

Therefore, it is important to know the signals of the S-curve and to act accordingly (see page 251). If the signals are not taken into account, organizations tend to focus on short-term optimization and miss new market opportunities. As a result, innovation teams and other parts of the organization work aimlessly on initiatives to look operationally busy. A powerful North Star shows where the journey is going next and where the energy and effort of the teams should be directed. Also, in the formulation of OKRs and FAST Goals (see page 258–263), the importance of the North Star is emphasized again and again.

The North Star Metric(s) Canvas (see page 172) supports the definition of different sets of top-line metrics, which usually represent an output metric with defined results. The input metrics are in most cases drivers and actions of work that result in innovation success and/or business growth. Understanding cause-and-effect, enriched with data analytics and statistic concepts (see toolkit pages 224–255), supports initially in asking the appropriate questions and to tap into the power of predictive scenarios beyond the well-known rearview perspective. A data-driven approach brings additional competitive advantages leading to business growth.

Why North Star Metrics Matter

- **Growth Strategy:** Ambitions of new growth opportunities in business ecosystems or with radical innovations
- **Customer Growth Strategy:** Increase in number of paying customers/users
- **Growth Efficiency Strategy:** Massive gains in efficiency by applying scale effects and automation
- **User Experience Strategy:** Improved customer experience based on product and service experience of the entire customer journey
- **Consumption Growth Strategy:** Higher-intensity usage of services, experiences, and products
- **Business Development Strategy / Go-to-Market Strategies:** Higher revenues for existing portfolios of products and services
- **Engagement Growth Strategy:** Increased engagement of users/customers with services, products, and experiences

Working with the North Star Metric(s) Canvas

The North Star Metric(s) Canvas supports the design of cause and effect and documents the leading and lagging indicators. In the initial step, the canvas reminds everyone about the core principles of designing powerful metrics:

- ✓ Metrics are quantifiable.
- ✓ They represent vision.
- ✓ They are ligned to strategy.
- ✓ They are leading indicators.
- ✓ They are actionable.
- ✓ They are easy to understand.
- ✓ Vanity metrics are omitted.

The North Star might focus on the company, business unit, ecosystem, or product level, depending on the business and the governance model applied. The results might be discussed mid- or long term, depending on the dynamics in the industry (see potential turning points and indicators of the S-curve). The directly related input metrics need also alignment to the type of business. For example, a common model about the design of new products, services, and experiences in the digital age refers to breadth, depth, or frequency of customer/user engagement. The last step is to align the activities and work to be done with the North Star.

The North Star Metric(s) Canvas is also an excellent communication tool, which supports, for example, the design thinking and innovation and growth-hacking teams in better understanding the WHY, HOW, and WHAT. When formulating OKRs, the North Star Metric(s) Canvas provides orientation and facilitates focused innovation work in given bandwidths. Of course, there sometimes might be also innovation initiatives that are launched without North Star guidance, and innovation teams are asked to think of previously unimagined growth possibilities. But these initiatives also need a clear mandate to the teams to disregard the North Star and existing guardrails.

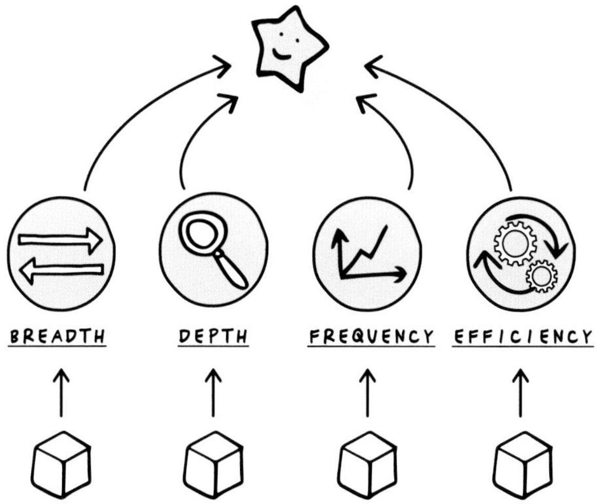

BREADTH
DEPTH
FREQUENCY
EFFICIENCY

Experience in recent years shows that well-defined top-line metrics have a positive impact on targeted resource allocation. In addition, team collaboration is improved, and the frustration and disorientation that are unfortunately still felt in many organizations are reduced. It could also be observed very often in companies without a clear North Star that almost all teams are struggling with the WHY, as it is difficult to identify the purpose and the role of each individual and the team in the larger context. Such adverse circumstances do not help in the purposeful development of new market opportunities (EXPLORE), nor in the optimizing and scaling of the core business (EXPLOIT).

It is recommended to regularly revisit the North Star metrics to ensure they reflect the strategy and predict results.

Template download

NORTH STAR METRIC(S) CANVAS

THE WORK TO BE DONE « METRICS »
Apply for example minimum viable exploration metrics for new product, service, or experience initiatives to contribute to the North Star

4

LEADING

DIRECTLY RELATED INPUT METRICS

Create leading indicators based on type of business (e.g., breadth, depth, or frequency of customer/user engagement)

Input 1: Input 2: Input 3:

3

Input 4: Input 5: Input 6:

NORTH STAR METRICS
What is the company, business unit, and ecosystem initiative's North Star?

Define metrics by considering core principles:

- ✓ Quantifiable
- ✓ Represent vision
- ✓ Aligned to strategy
- ✓ Leading indicators
- ✓ Actionable
- ✓ Easy to understand
- ✓ No vanity metrics

1

MID-TERM BUSINESS RESULTS AND CUSTOMER VALUE
What is the mid-term impact for the business?

2

LONG-TERM BUSINESS RESULTS AND CUSTOMER VALUE
What is the long-term impact for the business?

LAGGING

DOWNLOAD TOOL
www.design-metrics.com/en/north-star

Example of Defining Input Metrics Based on a North Star Metrics

A typical North Star metric is related, for example, to increasing customer interactions, engagement, or time spent on a platform or with an ecosystem. Let's imagine a company is growing into a market of streaming content, like music, podcasts, or other content relevant to a broad audience with a subscription or freemium business model. The main objective might be for the company to grow the time customers/users spend on the platform.

Based on this example, the flow of input metrics **(B)** and output metrics **(A)** seems to be a good start because it directly indicates how likely a customer/user is to stay with the provided services or leave for alternative offerings. However, the setting of the metric is far from optimal, as the North Star metric (e.g., time spent engaging with content, in the chosen example) is a lagging indicator, meaning that this data point is adjusted far too late for tactical/strategic changes or the reprioritization of innovation projects to be made.

Other input metrics, based on exploitation and exploration metrics, will most likely be needed **(C)**, which can be optimized and analyzed accordingly to predict the impact on the most important metrics. Assumptions can be made, such as that features are necessary for time spent on the platform. For example, initializing a higher frequency of visits or measures that increase time spent per visit. To bring users/customers back more frequently, process ideas can be developed that encourage their engagement, such as highlighting new content creators and/or offering AI-based recommendations and/or peer-to-peer recommendations. If these measures work and have been validated, by applying iterations and the concept of minimum viable exploration metrics, the next level is to test and measure mechanisms that keep users/customers on the platform longer. Potential ideas might emerge in terms of playlists that are data-generated and thus custom fit to the needs, current moods, or schedule of the active customers.

Example question in identifying the output metrics (A):
- What drives growth and revenues on our platform?

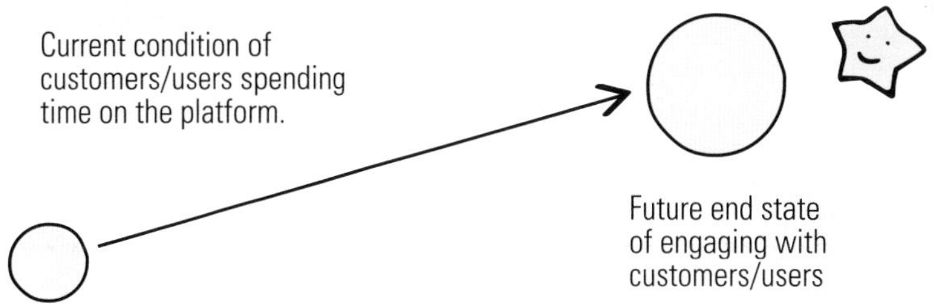
Current condition of customers/users spending time on the platform.
Future end state of engaging with customers/users

Example questions to get started in identifying input metrics (B):
- What is the current condition?
- What is needed now?
- How can the situation be improved?

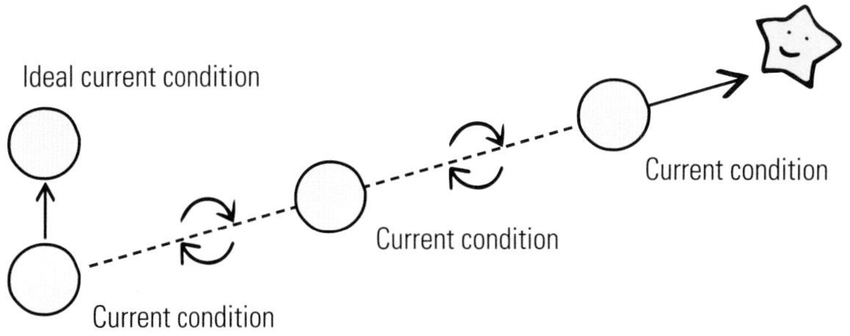
Ideal current condition
Current condition
Current condition
Current condition

Example questions in the application of input metrics (C):
- How to align the current efforts more closely with the chosen inputs?
- What decisions should be taken now and in the near future?
- Are the causes and effects defined supporting those decisions?
- Is it possible to trace the work to be done?
- What is the confidence level for the chosen inputs and North Star metrics?
- What are the lessons learned after the first implementation of activities?

The visualization of the causes and effects provides confidence about the defined action fields for innovation teams by focusing on the ideate state of the current conditions based on the North Star.

Example

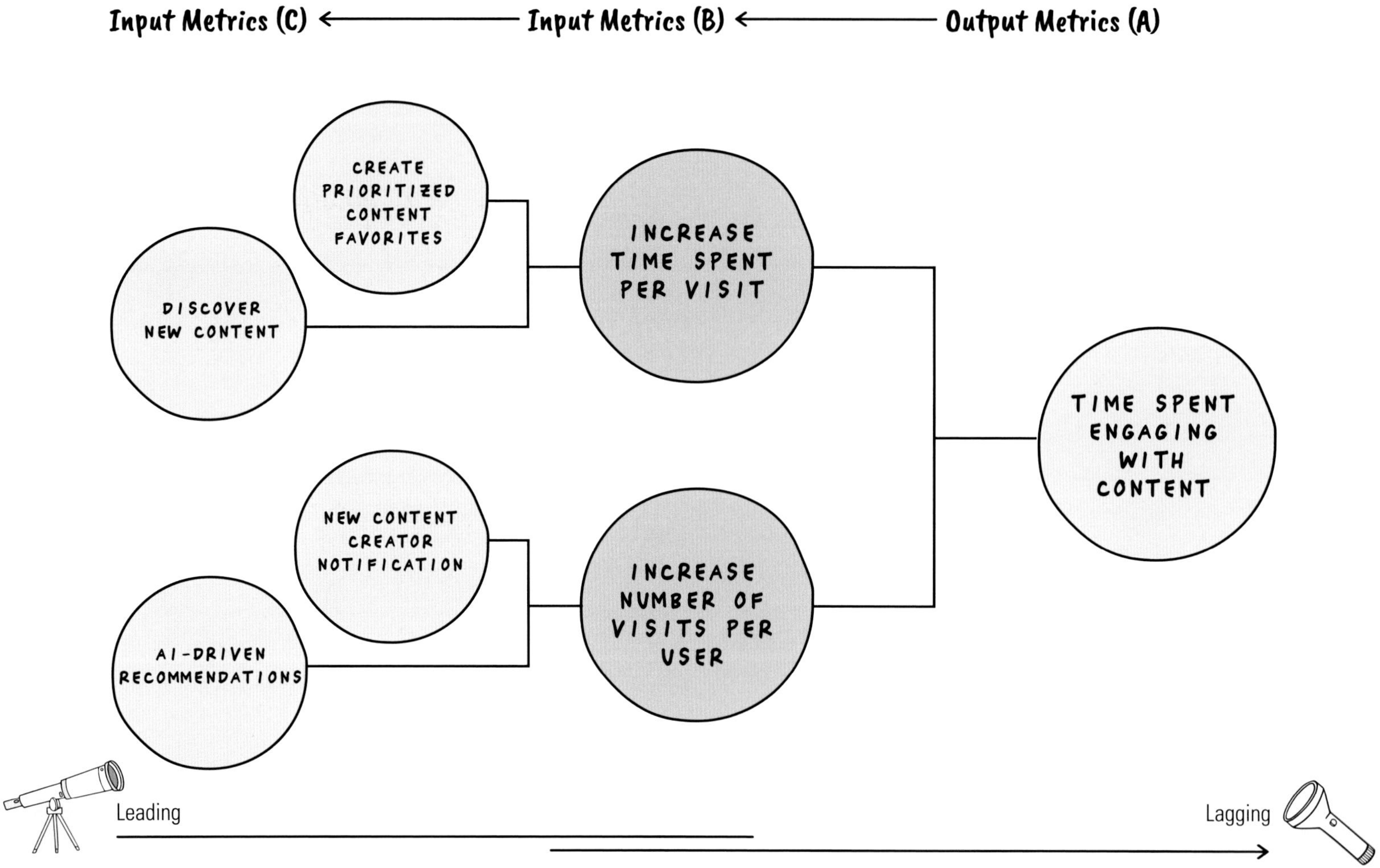
Input Metrics (C)
Input Metrics (B)
Output Metrics (A)
CREATE PRIORITIZED CONTENT FAVORITES
DISCOVER NEW CONTENT
INCREASE TIME SPENT PER VISIT
TIME SPENT ENGAGING WITH CONTENT
NEW CONTENT CREATOR NOTIFICATION
AI-DRIVEN RECOMMENDATIONS
INCREASE NUMBER OF VISITS PER USER
Leading
Lagging

WORKING WITH EXPLORATION METRICS

MMV

A Minimum Viable Set of Exploration Metrics

The term *Minimum Viable Exploration Metrics*, which has been mentioned frequently in this book, has been deliberately chosen in order to distinguish this type of metric from all the other efforts and existing key performance indicators typically circulating in organizations. For all activities related to EXPLORE, working with these metrics has proven practical because it avoids overwhelming the patience of decision makers, while providing the necessary space, time, and appropriate framework for transformation and creating radical new ideas and solutions. At the same time, minimum viable exploration metrics aim to fulfill all principles of state-of-the-art metrics and measurements. The approach is appropriate because it fits well into the current belief that experimentation is part of creating value in a dynamic environment.

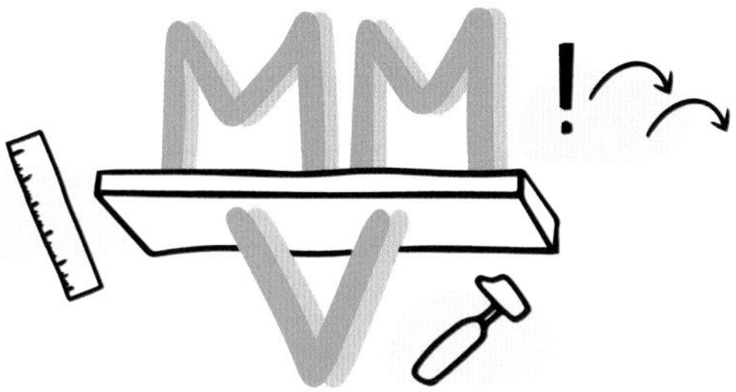
MMV

However, the intimate set of metrics is based on the organization's maturity and prevailing and changing culture, and how the company and the teams are managed and organizationally anchored. Within the definition of these metrics, there are usually leading, lagging, nonfinancial, and financial indicators for working in the problem and solution space, as well as for building capabilities and improving the maturity of a targeted mindset.

Once the design thinking and innovation teams as well as the leadership team embrace the concept of measuring minimum viable exploration as valuable, something amazing happens. The more the design thinking and innovation teams can minimize the risk in their actions and throughout the design cycle, the more trust is created, and the support and sponsorship activities increase. This fact is due to the basic principles of resource allocation and investment decisions in companies that need insights, data, and a gut feeling for their decisions that reduces the investment risk.

The decision-making and support process can be compared to that of a venture capital firm looking for evidence of value creation, delivery, and capture in different rounds of investing. This starts with the evidence and confidence of the team in working on a real customer problem and continues to the product/market fit to the team/product fit for the later implementation of the solution. Within companies, organization, and teams, there are additional dimensions with regard to new capabilities, cultural change, and organizational design to be measured.

As a result, minimum viable exploration metrics find application in measuring progress and effectiveness for both projects/business as well as teams/culture. In any effort to change culture and mindset, minimum exploration metrics provide evidence and are functioning as a supporting element that reinforces, for example, the value of thinking from the customer's perspective, listening, questioning, and relearning. Progress might be measured initially, for example, in the form of insights, customer feedback, or impact of activities on the transformation of teams or the entire organization. Some of these metrics will have a short lifespan as the initial measurement goal is achieved after a while. Other metrics will turn out to be very useful and will be operationalized later. Yet other metrics will not meaningfully improve anything, provide insights for concrete actions, or be useful in measuring progress.

The application of exploration metrics usually combines already established, powerful, approved metrics with the experimental minimum viable exploration metrics.

The supporting data-analytics process, with all steps from pivot to operationalization of measurement and metrics, is described in the measuring toolkit (pages 228–334).

From Preparation to Adaption

MINIMUM VIABLE METRICS

[Minimum viable metrics (MVMs) inform business decisions by measuring means relative to ends. They are best applied before product/market fit, with a strong focus on learning and improving.]

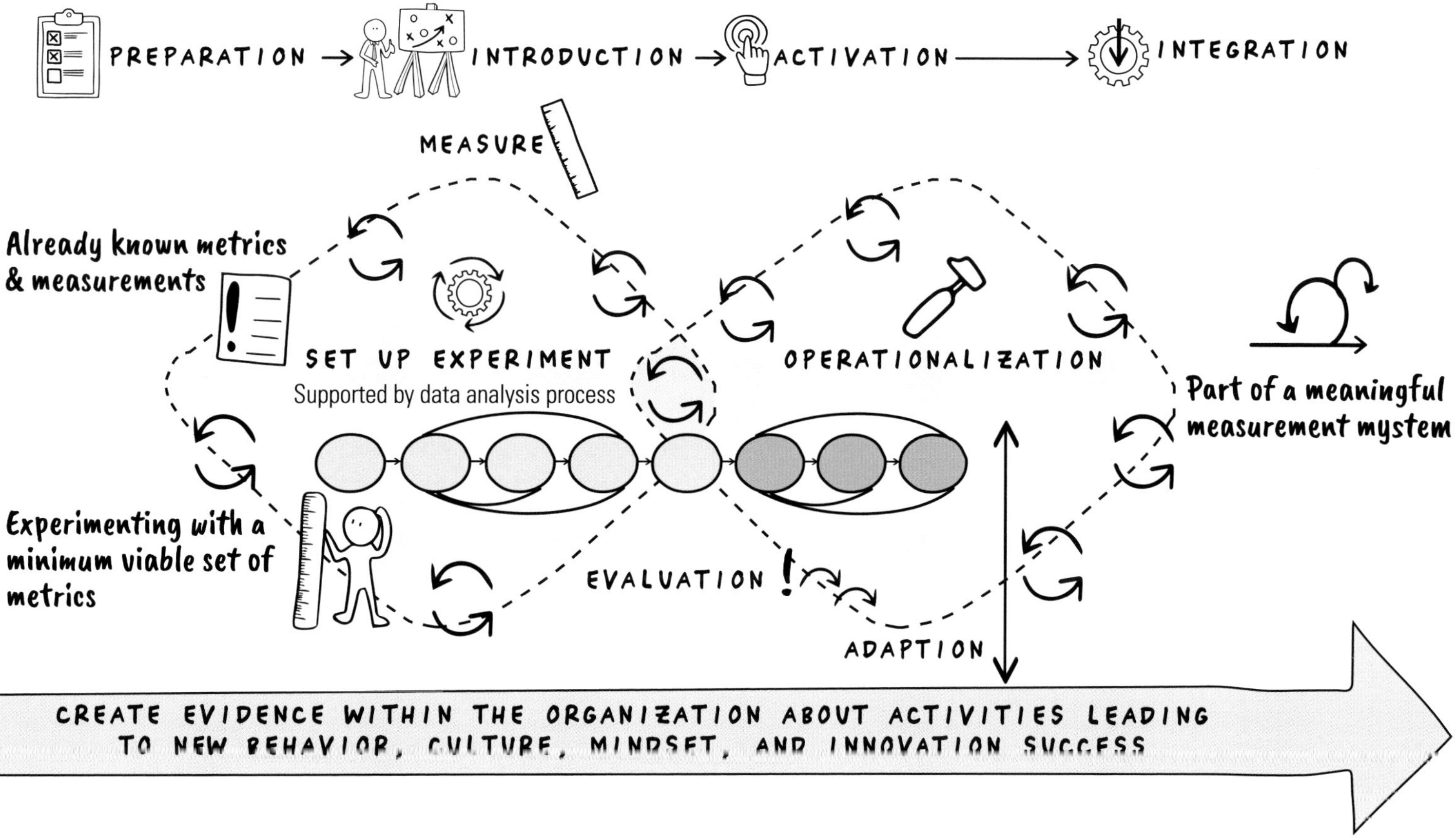

PREPARATION → INTRODUCTION → ACTIVATION → INTEGRATION
MEASURE
Already known metrics & measurements
SET UP EXPERIMENT
Supported by data analysis process
OPERATIONALIZATION
Part of a meaningful measurement mystem
Experimenting with a minimum viable set of metrics
EVALUATION
ADAPTION
CREATE EVIDENCE WITHIN THE ORGANIZATION ABOUT ACTIVITIES LEADING TO NEW BEHAVIOR, CULTURE, MINDSET, AND INNOVATION SUCCESS

Working with the Minimum Viable Metrics (MVM) Canvas

The objective of the application of minimum viable metrics is to help create a culture in which innovation thrives. Moreover, the teams working on tomorrow's market opportunities will always come to the point where they are asked if their work can be measured or if they can prove that the potential solutions have worked before. Especially in times of limited budgets and competing initiatives, it is understandable that more and more evidence is required.

The Minimum Viable Metrics (MVM) Canvas supports exploration of initial measurements at various levels in Explore and documents leading and lagging indicators. As a first step, the canvas captures the reasons for the measurement, the intended audience, and potential activities that could be derived from it. Afterwards, the metrics are introduced, activated, adapted, and, if practical, operationalized. For those who have focused their work only on novelty, creativity, and invention, the notion of an evidence-based practice will be difficult at first. However, a step-by-step approach to measurements and the establishment of a meaningful measurement system does not limit experimentation, learning, and the breaking of new ground.

However, caution is warranted, as an overly evidence-based approach can also inhibit creativity, out-of-box thinking, and particularly radical innovation, thus killing great opportunities for market opportunity in the nucleus. For true innovation, which by definition means something novel or new, it is often difficult to bring in evidence or rigorous research to support this. For many reasons, it makes sense to choose a path outlined in the Minimum Viable Metrics Canvas, which opens up the possibility to establish a measurement system that measures value and success in conjunction with processes, outcomes, and inputs. Practical experience shows that initially four to five metrics are sufficient to see if better decisions can be made.

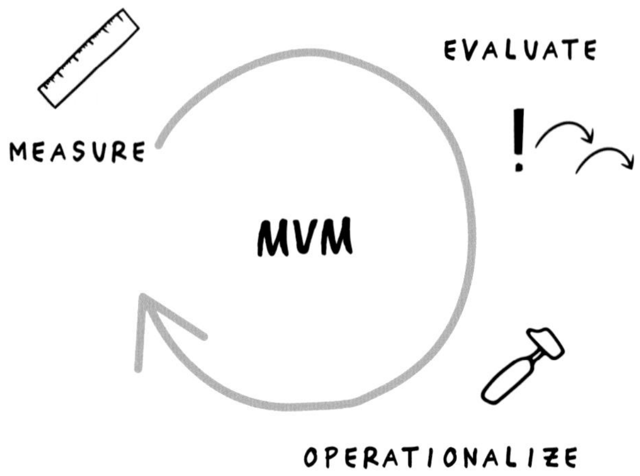

EVALUATE
MEASURE
MVM
OPERATIONALIZE

A Minimum Viable Set of Exploration Metrics Helps the Team:

- to define meaningful leading indicators and build step-by-step a measurement system.
- to identify metrics that allow the later operationalization of real-time analysis for actions on the ground.
- to iteratively identify the most pressing problems and challenges to solve with regard to culture, collaboration, processes, and so on.
- to overcome the dilemma of using only one kind of measurement for improving for example process efficiency.

The primary purpose of experimenting with minimum viable metrics is to positively change behavior and innovation work, and not to further optimize it or limit its degree of freedom.

Template download

MVM – MINIMUM VIABLE METRICS CANVAS

NAME OF MINIMUM VIABLE METRICS
Give the metrics a name or call them X as a starting point

PREPARATION

PROBLEM TO TACKLE 1
What is the focus of the exploration metrics?
Which problems or challenges are addressed?
How has it been measured so far?

DECISION MAKER / TEAM
Who does the metric help?
At what level does the measurement have the most effect?

POTENTIAL ACTIVITIES
Which activities might lead to new success in behavior, culture, mindset, and innovation derived from the metrics?

INTRODUCTION / ACTIVATION

VISION & INNOVATION STRATEGY 2
How to envision the future?
What are the North Star metrics?
What is the innovation strategy?
What are important capabilities to be developed?

DESIGN OF ACTIONABLE METRICS 3
What is the best way to describe the metrics?
For example: Formulate for the potential metrics in simple and understandable questions for the team/decision maker

MEASURE 4
Run measurement from leading and/or lagging indicators and measure at the appropriate level (e.g., process, input, or outcome).

COSTS, SCHEDULE, & FREQUENCY
What are the costs of measuring and how frequent/long should measuring be?

ADOPTION / INTEGRATION

CONCLUSIONS / NEXT STEPS 7
Derive questions, indicators, and actions from the results of measurement.
What are the most important findings from measuring?
Can the measurement scenario be adapted?

OPERATIONALIZE
How often are measurements taken and does the series of measurements have an expiration date?
How are the results visualized, shared, and communicated?
Is there a possibility to automate the measurement?

SUMMARY OF THE IMPACT MEASURING 6
What improvements have the metrics led to?

EVALUATE 5
What should be learned from the initial measurement?
How can the metrics be gradually improved?
Have the metrics enough impact to be operationalized?
Compare for example the outcomes between teams or similar projects.

DOWNLOAD TOOL
www.design-metrics.com/en/mvm

Example of Designing a Minimum Viable Exploration Metric

Similar to the work in design thinking, the exploration of viable metrics needs a problem, initial solution ideas for a potential metrics, and experiments to determine the impact. As a rule, leading indicators should be used to determine which activities to apply. Often, for good metrics, behavioral changes need to be found that lead to business impact as a result of the innovation work.

For the example of creating minimum viable metrics, a common question with regard to design thinking activities might be in focus:

The team intends to prove that needfinding efforts have a positive impact on the success of a new product.

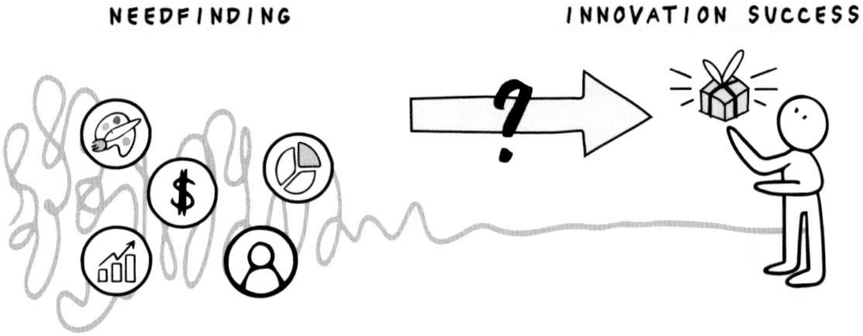
NEEDFINDING
INNOVATION SUCCESS
?

The team applies the appropriate parts of the MVM canvas to design the measurement experiment:

Step 1: Assumptions about a potential minimum viable exploration metric in conjunction with needfinding and product success
Step 2: What is the vision/North Star and principles of the innovation strategy?
Step 3: Design of potential metrics leading to actions
Step 4: Start measuring
Step 5: Evaluate actionable metrics (over the entire design thinking cycle)
Step 6: Discuss impact of operationalizing the metrics
Step 7: Derive questions, indicators, and actions from the results

Step 1: Assumptions about a potential minimum viable exploration metric in conjunction with needfinding and product success

- Design thinking teams with more insights about the customer/user have a better chance of success.
- To get more insights about the customer/user, design thinking teams need to conduct customer research and in-depth interviews.

→ As a result, the more research and needfinding interviews, the better the likelihoods of success.

Step 2: What is the vision/North Star and principles of the innovation strategy?

The core principles of the innovation strategy are:
- Innovation comes from everywhere
- Focus on the customer/user
- De-risk through early experiments and prototypes

✓ Needfinding is a core element which contributes to two out of three principles.

Step 3: Design of potential metrics leading to actions

Metric 1: Measuring velocity of needfinding interviews (customer interview velocity)
Metric 2: Measuring velocity of conducting insights (insight velocity)

→ Both might be leading indicators of success.
Formulate for the two metrics simple and understandable questions for the team:

→ Did the design thinking team conduct at least five needfinding interviews this week? Yes or no? How many?
→ Did the customer interviews generate any insights this week? Yes or no? How useful are the insights on a scale from 1 to 10?

Step 4: Start measuring

Example

- Compare, for example, four teams working on the same design challenge globally for local markets.

→ Design Thinking **Team A** generates useful insights **40%** of the time engaging with customers/users.

→ Design Thinking **Team B** generates useful insights **30%** of the time engaging with customers/users.

→ Design Thinking **Team C** generates useful insights **70%** of the time engaging with customers/users.

→ Design Thinking **Team D** generates useful insights **55%** of the time engaging with customers/users.

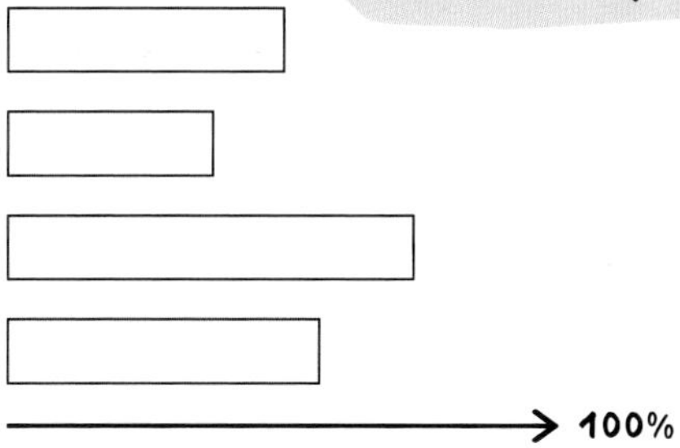
100%

Step 5: Evaluate actionable metrics (over the entire design thinking cycle)

→ **The measurement shows a wide range in terms of the conversion of engagements with customers and valuable insights.**

Step 6: Discuss the impact of operationalizing the metrics

→ **The team has noted that the week-over-week data should be aggregated into a two-week-sprint rolling metric (based on changing process steps and different focus in each Double Diamond).**
→ **The metrics provide an early warning sign if the team is not leaving the comfort zone by engaging with potential customers/users.**

Step 7: Derive questions, indicators, and actions from the results

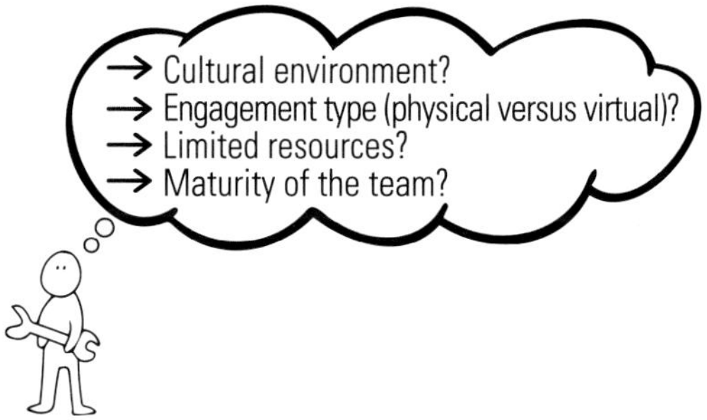
→ Cultural environment?
→ Engagement type (physical versus virtual)?
→ Limited resources?
→ Maturity of the team?

The team reflects and questions the potential causes and defines actions:

→ **What is the cause of Team D's better insights compared to Team A and B's lower scores?**

Example Actions:
→ **Hands-on project coaching to upskill teams with more needfinding techniques and professional ethnographic research methods**

Continue to apply the metrics based on a defined setting from step 6 to see if the number of useful insights increases, or test other potential actions for a defined time.

Applying Kill Metrics

Leading indicators related to ROI based on early experiments conducted.

Many experienced chief innovation managers and decision makers are aware that part of the impact comes from stopping innovation projects, initiatives, and design challenges. Innovation work is also defined by the saying "failure is success."

Metrics like insight velocity support decision makers by separating the success of each idea (ROI) from the success of the team. This means innovation and design thinking teams can be successful by showing that an innovation initiative will be unsuccessful.

Starting something isn't the hard part—killing an idea is. Defining success and evaluating the confidence levels makes it easier to make the tough decisions before it's too late.

Most innovation projects and design thinking challenges use tools like "Define Success" (see page 107) to start the project. However, they usually lack measurability and thus remain without effect throughout the entire design cycle. Minimum viable exploration metrics can also act as very project-specific kill metrics, if used correctly.

Google, for example, uses consequent kill metrics in selected design challenges with moonshot ambitions. For the start of such a design challenge, the design thinking team is encouraged to define several kill metrics as measurable criteria that determine whether an otherwise promising project should be terminated. Kill metrics also help bring teams back to the facts, as the riskiest parts of a design challenge are identified from the beginning, before the team becomes too emotionally engaged.

In addition, kill metrics help direct the available budget from the initiatives that generate real value, while saving many hours, resources, and money that would otherwise be wasted over a long cycle of agile development without impact.

However, the work with kill metrics needs a healthy balance between the optimistic and visionary attitude of design thinking and an enthusiastic skepticism that critically questions the visionary ideas with the possibilities of reality. It has also proven to be good when teams finish their ideas themselves as soon as there is evidence for it. A culture and reward system that supports such behavior does not put the team in the position of having to finish something by hook or by crook.

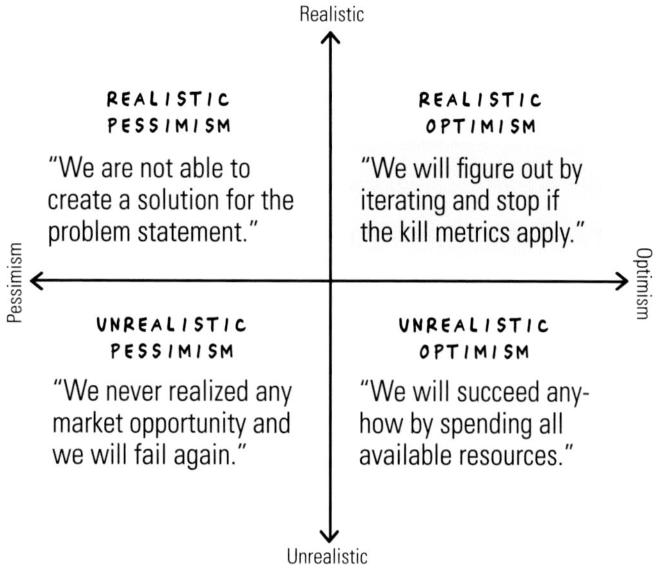
Realistic
REALISTIC PESSIMISM
"We are not able to create a solution for the problem statement."
REALISTIC OPTIMISM
"We will figure out by iterating and stop if the kill metrics apply."
Pessimism
Optimism
UNREALISTIC PESSIMISM
"We never realized any market opportunity and we will fail again."
UNREALISTIC OPTIMISM
"We will succeed anyhow by spending all available resources."
Unrealistic

Realizing and informing that an innovation initiative will not be successful generates immediate ROI simply by stopping the project.

For example, a typical business ecosystem design project takes at least a year to go from problem definition to product/market fit to the realization of a first minimum viable ecosystem (MVE). The assumption is that the cost is at least $2.2 million in resources. By applying heavily ethnographic research, collecting insights, and creating many point of views as a basis for rapid prototyping and testing, the team might find out that the problem to solve does not even exist, or potential solutions are not fitting customer needs.

"If X hasn't happened until Y, we'll shut it down."

By defining a powerful and appropriate kill metric in an early stage of the design cycle, millions of USD are saved, because the initial exploration might have cost only $200,000, instead of spending $2.2 million over an entire year. The design thinking team has saved the money by not spending it on a potential solution that they have proven isn't going anywhere. That's actually real and tangible ROI. The work of the design thinking team is returning $2 million to the business that would have otherwise been burnt without any business impact.

Typical Kill Metrics:

- Early customer feedback on various aspects of the vision prototype, which might influence replacing it with existing alternatives.
- Customers' willingness to pay (with different monetarization options) for an experience, service, or product has no evidence for significant revenues.
- Very high costs for creating a first functional prototype or expenditures exceeding the maximum available budget for making it happen.
- Critical functions and experiences do not meet quality and quantity expectations for realizing the solution.
- Technology is not ready for realizing the prototype or other constraints and dependencies that cannot to be changed at present.

More and more companies are recognizing the value of saving money by applying proper design thinking to project selection and increasing profit from better customer/solution fit. Revisit the examples on quantifying the financial impact of design thinking projects on page 68.

Example

How Might We Design the Mega Metaverse Wellness Oasis Ecosystem

META WELLNESS OASIS

Early customer/user feedback and first prototypes to test with neurotech revealed that virtual experiences related to wellness have limited acceptance and the impact on well-being is less intensive than physical or hybrid experiences.

→ Based on three kill metrics:
1) Improvements of mental and physical health by 30%
2) Customer/user acceptance to visit every week
3) Willingness to pay for treatment (based on immersion of light, sound, travel, social interaction from different communities) > $45

Beginning at the beginning, with the design thinking mindset, tools, and methods as well the set of metrics for the minimum viable exploration, saves money by aligning the design challenge to the kill metrics and not taking a full year to validate market acceptance. In the example outlined, based on easy-to-understand metrics, only $200,000 is spent and $2 million in savings are realized.

HOW TO ADAPT AND OPERATIONALIZE THE INNOVATION MEASUREMENT SYSTEM

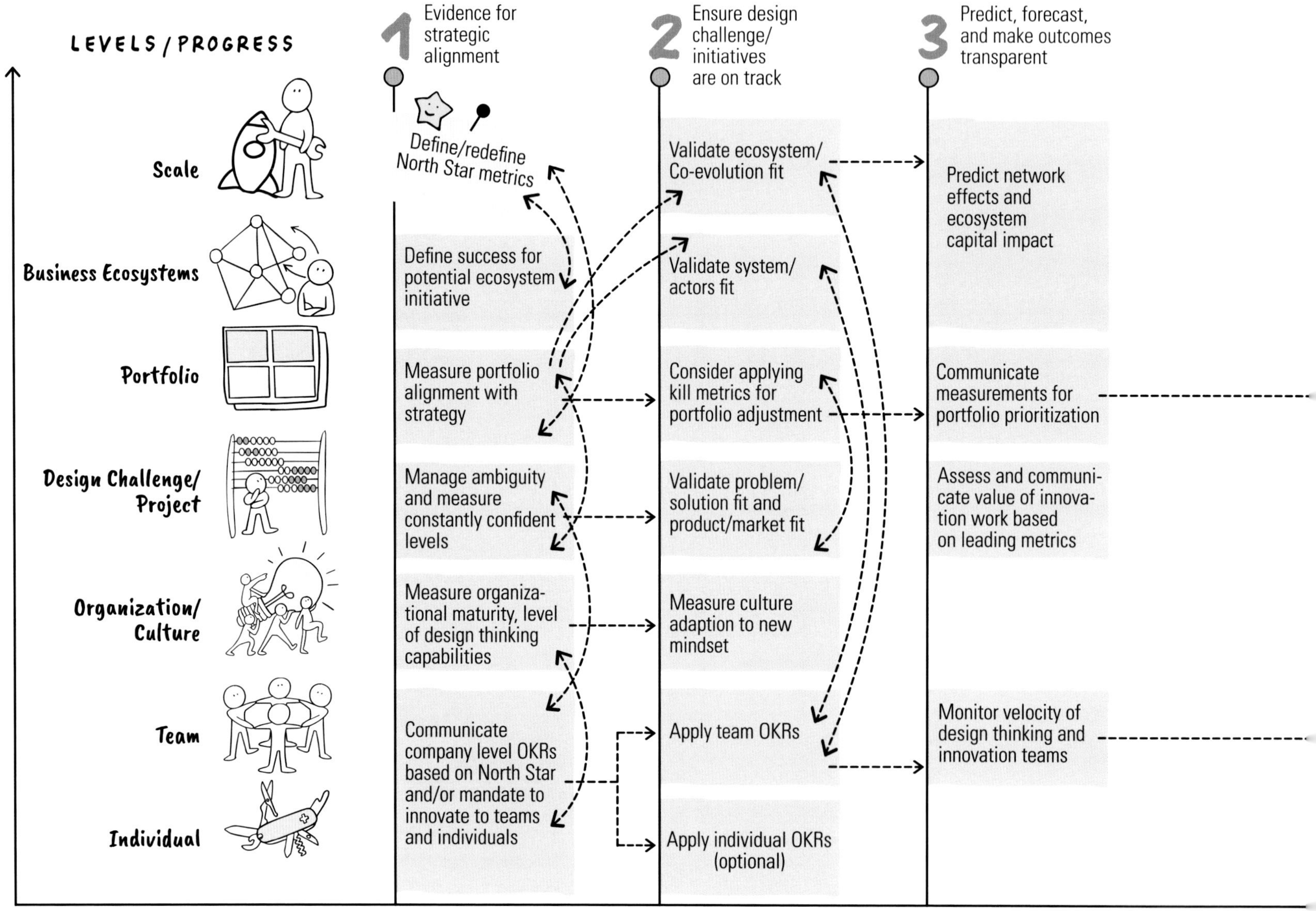

LEVELS/PROGRESS
Scale
Business Ecosystems
Portfolio
Design Challenge/ Project
Organization/ Culture
Team
Individual
1 Evidence for strategic alignment
2 Ensure design challenge/ initiatives are on track
3 Predict, forecast, and make outcomes transparent
Define/redefine North Star metrics
Define success for potential ecosystem initiative
Measure portfolio alignment with strategy
Manage ambiguity and measure constantly confident levels
Measure organizational maturity, level of design thinking capabilities
Communicate company level OKRs based on North Star and/or mandate to innovate to teams and individuals
Validate ecosystem/ Co-evolution fit
Validate system/ actors fit
Consider applying kill metrics for portfolio adjustment
Validate problem/ solution fit and product/market fit
Measure culture adaption to new mindset
Apply team OKRs
Apply individual OKRs (optional)
Predict network effects and ecosystem capital impact
Communicate measurements for portfolio prioritization
Assess and communicate value of innovation work based on leading metrics
Monitor velocity of design thinking and innovation teams

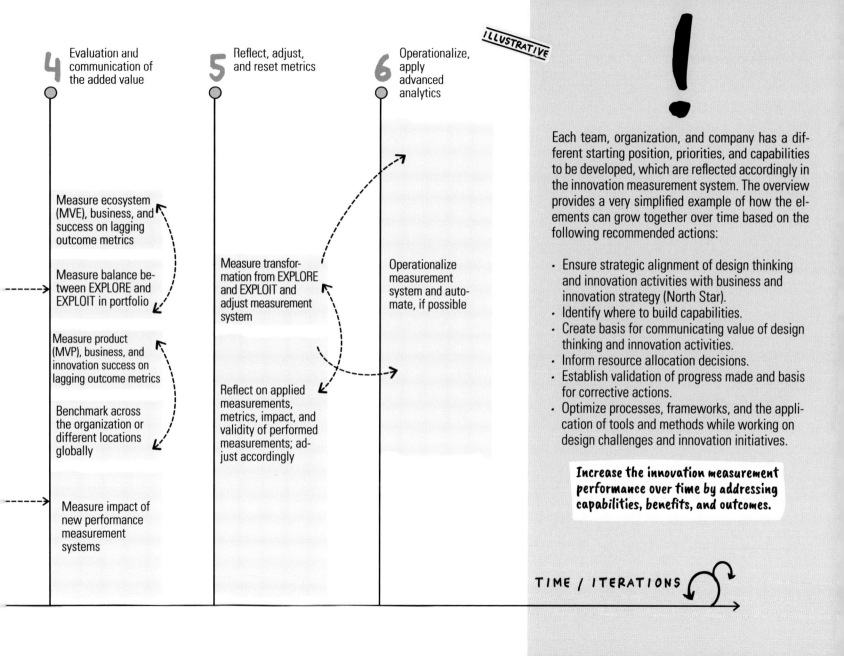

ILLUSTRATIVE

4 Evaluation and communication of the added value

- Measure ecosystem (MVE), business, and success on lagging outcome metrics
- Measure balance between EXPLORE and EXPLOIT in portfolio
- Measure product (MVP), business, and innovation success on lagging outcome metrics
- Benchmark across the organization or different locations globally
- Measure impact of new performance measurement systems

5 Reflect, adjust, and reset metrics

- Measure transformation from EXPLORE and EXPLOIT and adjust measurement system
- Reflect on applied measurements, metrics, impact, and validity of performed measurements; adjust accordingly

6 Operationalize, apply advanced analytics

- Operationalize measurement system and automate, if possible

TIME / ITERATIONS

!

Each team, organization, and company has a different starting position, priorities, and capabilities to be developed, which are reflected accordingly in the innovation measurement system. The overview provides a very simplified example of how the elements can grow together over time based on the following recommended actions:

- Ensure strategic alignment of design thinking and innovation activities with business and innovation strategy (North Star).
- Identify where to build capabilities.
- Create basis for communicating value of design thinking and innovation activities.
- Inform resource allocation decisions.
- Establish validation of progress made and basis for corrective actions.
- Optimize processes, frameworks, and the application of tools and methods while working on design challenges and innovation initiatives.

Increase the innovation measurement performance over time by addressing capabilities, benefits, and outcomes.

In the context of applying minimum viable metrics, the measurements might occur on the methods and tools level as well as on the project, design challenge, and collaboration level before an actionable and evidence-based measurement system is created. The very simplified example on the previous double page depicted how the individual measurements help to initiate the change and how an innovation measurement system can emerge and evolve over time.

	Methods and Tools Level	Project, Design Challenge, Collaboration Level	Metrics and Measurement System Level
Measure: PUTTING MINIMUM VIABLE METRICS TO WORK	**Start small and operationalize later** • Some experiments of measurements can easily be implemented and tested on a tool basis. • Encourage teams to perform metrics on different iterations over the design cycle. • Build step-by-step acceptance for a system of measurements and show how even single measurements help to become better, faster, and more successful.	**Start with what is already known** • Set the basis for a starting point (e.g., maturity in design thinking, definition of success on a project and team level). • Integration of existing meaningful individual, team, organizational, and cultural metrics to enhance effectiveness, impact and purpose.	**Share what is known in value creation** • Attract resources by demonstrating collective results. • Show how to allocate scarce resources in a better way. • Be transparent and share costs of information and data collection for measuring. • Focus on actions based on measurements to create immediate impact.
Evaluate: LEARN WHAT IS NEEDED	**Get used to measurements** • Apply them and see if the team will game the metrics or accept it. • Reflect the outcomes and actions derived from the measurement activities. • Become better, faster, and more successful.	**Learn what is needed to know** Evaluation that serves the purpose: • Test assumptions about what to measure. • Target specific unknowns in innovation failure or success. • Demonstrate impact of measurement.	**Build a learning agenda and system** Create an actionable and evidence-based measurement system • Enable individuals, teams, and decision makers to access and to repropose what is known. • Prioritize evaluation efforts.
Operationalize: CREATING REFERENCE POINTS	• Make metrics a routine in different phases over the design thinking micro and macro cycles.	• Fill gaps with new variants, new or adjusted measurements and metrics.	• Operationalize meaningful metrics and include in measurement system.

SELECTING EXPLOITATION METRICS

How to Select the Appropriate Mix of Exploitation Metrics

Most of the standard metrics for EXPLOIT are well known, and a quick look into the existing reporting documents reveals many known KPIs related to profit, loss, and the full rearview of spending for projects, functions, and activities. Many existing approaches to managing portfolios in connection with EXPLOIT and EXPLORE refer to the known KPIs for EXPLOIT without making any further distinction. This is the reason why in this book, on the one hand, a distinction has been made in the measurement of EXPLORE and EXPLOIT and, on the other hand, in 201, an in-depth explanation of OKRs for innovation teams is provided. Practical experience shows that a suitable mix of metrics is also required for the management of exploitation. The selection and associated measurement tools alone would be material for another book. At this point it is important for the entire company to understand that the exploration and exploitation metrics should work together as visualized with the yin and yang. This section aims to create awareness of the different measurement options in EXPLOIT, and to present hints and approaches for reorganizing the existing system of mostly lagging indicators. The biggest challenge of the existing classic KPIs is that they are mostly not aligned to the new ways of working and are not very informative about the levers of possible change and transition in terms of behavior and processes. For this reason, in Measuring 101, a major focus has already been placed on the pioneering North Star metrics for growth and minimum viable exploration metrics, which enables teams to decipher the layers of a potential customer in today's and tomorrow's interaction with the innovation. The application of OKRs contributes also to measuring exploit and provides a strong link to business results, goals, and outcomes. The respective performance implications tend to be short term, and from the results of the measurement refinement, efficiency and execution take place. The most important aspect in Exploit is to optimize the existing portfolio, to increase efficiency and to measure whether the predictions come true and whether the company remains viable.

In the context of this book, the mix of exploitation metrics is usually applied after the product/market fit to measure growth, scale, maturity, and decline over the portfolio cycle.

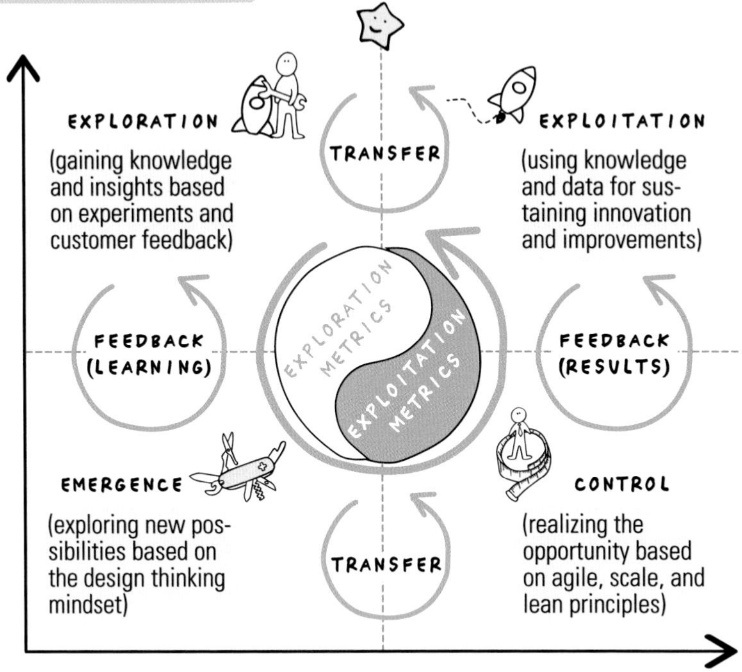

EXPLORATION
(gaining knowledge and insights based on experiments and customer feedback)
TRANSFER
EXPLOITATION
(using knowledge and data for sustaining innovation and improvements)
FEEDBACK (LEARNING)
EXPLORATION METRICS
EXPLOITATION METRICS
FEEDBACK (RESULTS)
EMERGENCE
(exploring new possibilities based on the design thinking mindset)
TRANSFER
CONTROL
(realizing the opportunity based on agile, scale, and lean principles)

In Many Cases the Mix of Exploitation Metrics Includes:

- Indicators for mapping and managing portfolios
- Well-defined growth metrics based on a North Star
- Classic KPIs that are well known, impactful, and accepted in the organization
- About five to seven health metrics that describe a business's vital signs and are measured more frequently than classic KPIs
- Key metrics dealing with potential worst-cases scenarios to respond with suitable actions
- Common financial reporting indicators which support the measurement of the KPIs and health metrics

Classic Key Performance Indicators Overview

Many of the classic KPIs for EXPLOIT that are used in companies are not true metrics as described in the introduction in Measuring 101. They neither help to identify if the company is vital nor do they support in making decisions. Correcting this is relatively easy, and it could be considered a learning experience on the way to better managing EXPLORE and EXPLOIT. But even for those with years of expertise, cognitive biases can sometimes lead to poor KPI selection. The selection and definition of the respective measurements should be based on the activities over the next periods, the chosen business model and the larger vision, which in the best case is defined with top-line metrics (North Star). The focus of KPIs can thus be aligned with the acquisition of new customers to the measurement of marketing campaigns within EXPLOIT. Most KPIs in EXPLOIT have the goal to increase efficiency and to monitor the operational business. A KPI is by definition a measurable value that demonstrates how effectively a company, organization, or team is achieving business objectives. The key questions and examples of KPIs are a good starting point for reviewing existing KPIs, and changing them leads to better results on the different levels for already realized design thinking challenges, innovation projects, and business growth initiatives.

Examples of common key questions for managing exploit:

Efficiency Metrics:
Does the company, organization, or team use resources efficiently?

Liquidity Metrics:
Does the company, organization, or team meet immediate spending needs?

Profitability Metrics:
Does the company or product/service earn acceptable margins?

Valuation Metrics:
What are the company's prospects for future earnings?

Growth Metrics:
Are revenues, profits, and market share growing in total or portfolio elements?

KPI Classics

- Revenue growth
- Revenue per client
- Net profit
- Customer satisfaction

Alternative KPIs

Revenues
- Conversion rates
- Costs per acquisition
- Average sales price

Profits
- Cross profit percentage
- Overhead recovery rate

Quality KPIs
- Net promoter scores (NPS)

Efficiency KPIs
- Process efficiency

EXPLORE/EXPLOIT portfolio KPIs
- Ratio of EXPLOIT/EXPLORE versus all projects
- Return on investment (e.g., comparison of two to three innovation projects in defined periods after market launch)

Effectiveness KPIs
- Resource profitability (average cost per hour/billing rates)
- Number of change requests after market launch (e.g., impact on quality, timelines, resources, and budgets)

Budget KPIs
- Planned value (e.g., planned versus actual costs for market launch and operations)

Training KPIs
- Training/research needed for implementing new products, services, and experiences

Operating with Explore and Exploit Portfolios over the Entire Cycle

A portfolio life cycle view shows that after EXPLOIT, activities transition to sustain and retire. Managing EXPLORE and EXPLOIT is already challenging, and the other phases also need governance and measurements. Both are important to prioritize between multiple offerings in the portfolios, ensure that resources are allocated to the most appropriate initiatives, and that a pipeline of new market opportunities is realized.

The great danger is that EXPLOIT portfolios are overweighted because they contain the current cash cows. Particularly in large companies, practice shows that attempts are often made to keep products, services, and experiences alive at great financial expense because they have formed the core of commercial activity over many years. It is therefore important to look at each offering in the respective portfolios objectively and to use an up-to-date and evidence-based assessment. Moreover, portfolio management is not a linear process. Not every market opportunity will make the full journey through the life cycle. They can fail and drop out at any point in the journey. The granularity of the respective measurement points over the entire cycle depends on the industry, business models, and the market environment.

In terms of governance of the entire cycle, it is important that all relevant stakeholder views are taken into account, that the appropriate set of metrics is used, and, above all, that supposed assumptions of the opinion leaders and bias are removed from the continuous monitoring and team coaching.

Consequently, portfolio governance should be perceived as coaching the team to ask the right questions and to stick to the agile customer development approach through the phases of explore, exploit, sustain, and retire.

A portfolio management approach based on a defined cycle and aligned with customer development, companies, and teams can look at all the offerings in their portfolio in a holistic way. This means that the EXPLORE portfolio can focus on learning more about the customers/users (Operating in the jungle) and the EXPLOIT portfolio can monitor the superhighway with all contingencies. This allows a balance to be achieved between exploration and exploitation, and resources to be deployed where they are most needed.

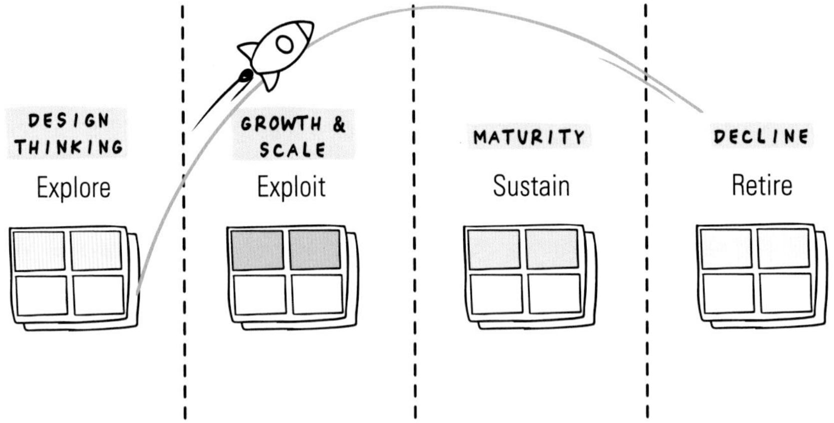
DESIGN THINKING
Explore
GROWTH & SCALE
Exploit
MATURITY
Sustain
DECLINE
Retire

Explore
- Articulate the customer problem, market opportunity, and business need.
- Obtain evidence that there is a real customer need that is strong enough to invest in a solution.
- Build the prototype and minimum solution that meets the customer needs to validate the market demand.

Exploit
- Scale the validated business model or ecosystem to generate business outcomes (revenue, reach, and return on investments).

Sustain
- Seek to maintain business outcomes while minimizing costs and optimizing efficiencies.

Retire
- Move the offering out of the portfolio or support the existing base with no additional investments.
- Apply an evidence-based approach with updated data to prove that growth potential, strategic benefit, or technological viability do not exist.

Why Measuring and Managing the Portfolio over the Entire Cycle Provides Value

The mandate to the teams for exploration and the associated mindset is the basis for the realization of future growth opportunities. The validation via the problem/customer fit up to the product/market fit opens the possibility for resources to be used in the best possible way. This approach via the problem to scale and growth framework ensures that only validated market opportunities (based on real problems) progress from exploration to product development, and that only validated products progress to the growth and marketing stages. The application of minimum viable exploration metrics supports learning and gaining knowledge. Measuring at the tool level supports the effort of mastering design thinking capabilities and at the same time gives confidence to the teams that are applying a new mindset. Documenting failures and learning successes helps all teams move to the next level of learning. While the organization and teams profit from what is learned, the investments are reasonable, because in EXPLORE the upfront investment is limited to validating the next identified assumptions based on desirability, viability, and feasibility. Having two portfolios allows for mapping all design thinking challenges and innovation initiatives against the entire portfolio. Mapping should be performed on the product, service, experience, business model, and business ecosystem to assess all relevant elements. In most cases it reveals that not enough time, effort, and management attention are dedicated to the EXPLORE portfolio or other important elements of diversification.

Operating with EXPLORE and EXPLOIT Portfolios Helps to:

- encourage teams to be curious about new or changing customer needs.
- focus on learning and capability build in the explore phases.
- free up resources through lo-fi prototypes and MVPs, as developers and marketing teams are not needed to validate the customer's problem.
- to minimize the risks across all phases, as the respective financial resources are only allocated for the next activities that are needed to validate critical assumptions.
- run a sanity check on the business/innovation strategy and ensure that the execution is matching the ambition and objectives.

Examples of Elements to Map and Actively Manage Portfolio Activities

- Percentage between core, adjacent, and transformative offerings
- Ratio between retire and explore
- Market size of opportunities in MVP/MVE status (large, medium, small) versus portfolio elements in retire (large, medium, small)
- Mapping of new market opportunities in development phase with North Star, defined strategic priorities, or defined mandate to innovate
- Addressable market size per phase (growth and scale, maturity, decline) based on defined categories for large, medium, small
- Portfolio fade time in exploitation based on the time after product/market fit and/or not reaching all phases of the cycle
- Lagging indicators* (Ratio of innovation spending to revenues with new offerings; comparison of existing versus new profitability in exploit and sustaining, all time bound measurements between phases of the entire cycle)

*Lagging indicators provide insights for reflecting management decisions but provide limited information to actively manage the portfolio.

Applying Health Metrics

For an appropriate mix of exploitation metrics, it is in many cases valuable to question classic KPIs to see if health metrics are not the better option of measurement for deriving the desired action. As the name implies, health metrics describe a business's vital signs, much like physical vital signs. Especially in turbulent times, health metrics can be the better choice, as they help to find out how resources and time are used. This makes in many cases the difference between surviving and thriving. Typical health metrics are those that are the real five to seven critical measurements in terms of viability for the business: the frequency is weekly rather than monthly or yearly; the decision maker or team responsible can generate direct impact on the numbers; there is a link to the strategic priorities/North Star; and, finally, the measurement is focused on thresholds rather than targets.

A good starting point for establishing health metrics is to review the current KPIs to see which ones have the potential for health metrics. In addition, new health metrics can of course be designed and applied based on the current business situation and growth initiatives being implemented. Again, it is important to note how the measurement is performed and the metrics are calculated. In terms of threshold, for example, the performance of the last 12 months can be taken as a baseline. For the first measurements it is recommended, similar to the minimum viable exploration metrics, to first evaluate the measurement and to see if concrete actions can be derived from it. The health metrics are excellent for weekly team reviews and prioritization sessions. If the health metrics are really critical for the viability of the company, organization, or team, they should be operationalized and automated if possible.

Selected Shifts: From Classic KPIs to Health Metrics

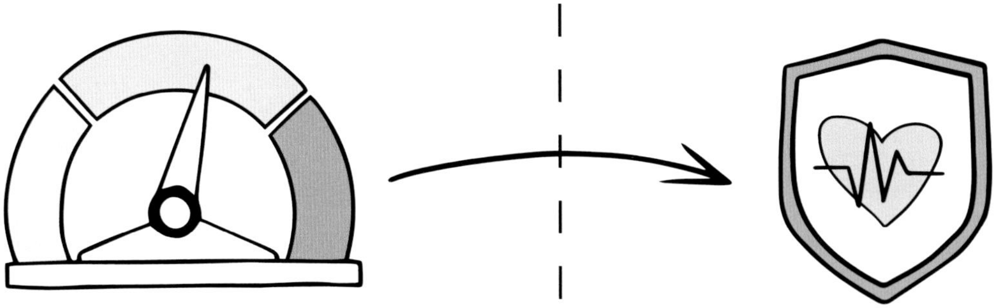

A critical review and reflection of the current KPIs in the company help in the adjustment and a shift toward health indicators that support the monitoring.

Typical Health Metrics

Example

The outlined building blocks below present a small sample of health metrics ranging from financial viability to the measurement of implemented performance measurement systems that perform a temperature check on the team's health. Especially for metrics in the area of operational efficiency, traditional KPIs tend to monitor status rather than directly leading indicators of team and resource efficiency. The outlined examples on new customer acquisition and customer retention, by contrast, are best known from the lean start-up approach and are often already used as health metrics in the context of the application and later implementation of new products, services, and experiences. The Health Metrics Canvas helps with a simple structure and provides the most important questions for the selection of the most important metrics that feel the pulse of the organization and beyond.

Financial Viability
- Cash reserves
- Operating surplus/gap
- New sales revenues to target or previous month

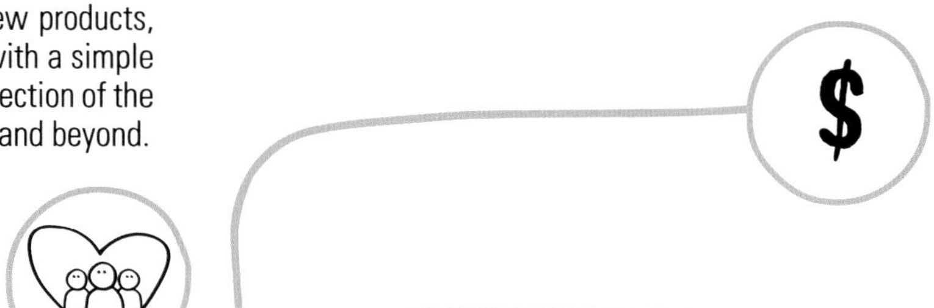

$

New Customer Acquisition
- Sales growth year-to-date or month-over-month
- Qualified leads (high likelihood to close in period)
- Lead conversion or close rate (number of leads/number of new customers)

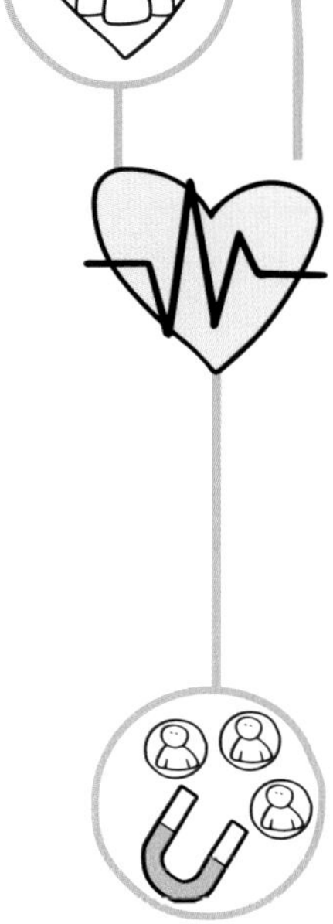

Operational efficiency
- Delivery times (committed time to delivery)
- Errors rate (% of defective product to total)
- (Prioritized) commitments for improving existing products, services, or experiences (% of tasks delivered to those committed)

Current Customer Retention
- Committed customers (number of active customers)
- Retention rate (number of customers at end of month - # of new customers/# of customers at beginning of the month)
- Backlog (budgets contracted but not used)

Team & Individual
- # of 1:1 to number of employees
- Regular sentiment check (weekly one-question survey)
- OKRs on track/achieved (% of items on track or achieved across all team members)

Template download

HEALTH METRICS CANVAS

HEALTH CHECK PERIOD: Indicate reporting cycle and level of measuring		**NAME OF ORGANIZATION / TEAM INITIATIVE:**		
1 IDENTIFICATION OF 5-7 METRICS	**2 IMPACT**	**3 PARAMETERS**		
DESCRIBE METRICS Why is the metric a health metric? How was it measured before? 1. 2. 3. 4. 5. ...	**LEVEL OF CONCERN** Describe the threshold of concern	**SOURCE OF DATA** What is the data source? How will you calculate the metrics?	**PREVIOUS AVERAGE** What is the previous average for a defined period of time (e.g., 12 months)?	**METRIC OWNER** Who is the owner for the metric?

DOWNLOAD TOOL
www.design-metrics.com/en/health-metrics

Applying Financial Reporting

The purpose of financial indicators is to monitor the business and make intelligent business decisions. There are many financial indicators and the selected examples can only reflect a small selection of them. However, decision makers and teams with P&L responsibility need to know how to use the metrics to create accurate, actionable, and predictive financial analysis. The ultimate goal should be to close knowledge gaps at all levels of an organization. Transparency between business groups on key exploitation metrics is one of the most widely recognized ways to effectively monitor the health of the business and achieve short- and long-term business goals. Typical examples of financial indicators are related to profit and loss, working capital, or operational metrics related to payrolls.

Working Capital Metrics

Working capital is an excellent indicator to understand the immediately available capital. The working capital metrics create a picture of your business's financial health by evaluating available assets that meet short-term financial liabilities. The metrics are calculated by subtracting current liabilities from current assets, and it includes assets such as on-hand cash, short-term investments, and accounts receivable and liabilities such as loans, accounts payable, and accrued expenses.

Payroll Metrics

Metrics related to payroll costs are an important way of understanding spending in labor-intensive industries. An accurate value provides insight into whether increasing or eliminating staff should be considered and makes financial sense.

Typical Metrics Include:

- Cycle time to process payroll
- % of untimely payroll payments
- % of manual payroll payments
- Payroll recalculation
- # of Payroll errors per month
- # Payroll corrections and changes
- Average hourly pay

Example

Profit and Loss Measurements

Knowing where the breakeven point is and regularly checking (monthly) whether a growth initiative is operating at a profit or loss is a key exploitation metric for decision makers. Trying to estimate this important metric or waiting until the end of the year for an accurate report may result in no action or the wrong measures being implemented.

P & L Measurements Include (Depending on the Industry and Kind of Business):

- Revenue mix
- Revenue by customer
- Outstanding revenue
- Revenue per employee
- Operating income
- Profit increase/decrease year over year
- Cost of goods sold
- Profit per product
- Net profit
- Net profit margin
- Current revenue versus expected revenue

From business practice, for example, the comparison of current and expected revenues and expenses has proven to be very helpful, as these metrics provide information on whether a growth initiative is meeting expectations or not. It also serves to identify areas for improvement early. This allows the respective decision makers to change direction before the gap between expectations and actual values becomes insurmountably large. Present performance figures in conjunction with future business plans should also provide an impression of anticipated revenues and expenses to enable appropriate long-term planning.

How to Prepare Exploitation for Economic and Market Downturns

As mentioned earlier, external factors can influence performance and should be considered in the definition of measurements and metrics. One possibility is to ask the appropriate questions with different frequency, for example, in the case of specific events or in the worst-case scenario of observing revenues that are declining close to operating expenses. From practical experience the following three scenarios most commonly lead to concrete actions in dynamic business environments.

1. Proactive Moves

- What is this year overall forecast and monthly average?

Revenue target: ________________
versus monthly average revenue: ________

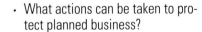

- What actions can be taken to protect planned business?

Examples of actions:

- Acquisition of new customers
- Retain customers
- Reduce expenses without harm on operations

Impact

- Estimated annual savings: ________
- Additional annual revenues: _______

2. Defensive Moves

- What is the cash runaway?
- How long is the organization to operate if revenues takes a hit?

Cash reserve: ________________
Monthly operating expenses: ________
\# of months

- What actions can be taken if revenues fall close to operating expenses?

Examples of actions:

- Change of acquisition strategy
- New marketing strategy
- Value proposition adjustments
- Reduce headcount

Impact

- Estimated annual savings: ________
- Additional annual revenues: _______

3. What if… Moves

- What already known trigger events exist (e.g., if revenue is within X% of OPEX)

Trigger #1: ________________
Trigger #2: ________________
Trigger #3: ________________

- What actions can be taken if a specific trigger event occurs?

Examples of actions:

- Resetting of strategy
- New orientation of sales
- Creation of incentives
- Cutting holding capital expenses

Impact

- Estimated annual savings: ________
- Additional annual revenues: _______

MEASURING CREATIVITY

Measuring Creativity of Individuals and Teams

The measurement of creativity is probably one of the most difficult metrics, and research in this area has its limits. Creativity is not available at the push of a button, and it is even less evident when we demand it. However, creativity is one of the most important human qualities in innovation work, and it is a central element in design thinking (divergent thinking). From observations of teams in various settings, it becomes evident that team members who are flexible and creative often achieve the most ingenious solutions. However, depending on the measurement method, the definition of creativity may vary. In the context of design thinking and innovation work, for example, creativity is an act of turning new and imaginative ideas into reality. Creativity is characterized by perceiving the world in new ways, finding hidden patterns, making connections between seemingly unrelated phenomena, and finding radically new solutions with an interdisciplinary team for a customer/user problem. In design thinking, creativity consists of thinking and doing. Since the 1950s two main streams of measuring creativity have evolved: divergent thinking and imagination. Test-based research on creativity originated with Guilford's (1950) theory of divergent thinking. Based on Guilford's tasks, measuring the characteristics of divergent thinking gave rise to numerous tests in the 1970s and 1980s, for example the Torrance Tests of Creative Thinking. The second stream focuses on everything related to the activity of visual imagination, which encompasses creating, interpreting, and transforming vivid mental representations. Over the years, the conviction has matured that the creative imagination is one of the most important abilities that contributes to the effective use of the creative potential of individuals and teams. As of today, there is no measurement tool that can feasibly measure the full breadth and complexity of creativity, which leads to the fact that only different characteristics can be measured (see examples of common measures). Combined tests are limited as well because, for example, a weak relation between divergent thinking and imagination exists. As a result, multiple measures are used to try and see a convergence of the results or an interconnection of different elements. Through the application of neuroscience technology, there will be better ways to understand and measure creativity in the future (see page 360ff).

Knowing the reason for measuring creativity or the kind of creative work one wants to evaluate is the first step in demystifying the creative space.

Four Common Ways to Measure Creativity

Depending on the measurement method, the definition of creativity may vary. In some measurements, creativity refers to the deviant production (in what quantity creativity occurs or how diverse it is). In other measurements, creativity refers to the novelty of the design or prototype.

The four models presented in this section provide different ways of measuring creativity or creative work. Thus, it is possible to look at output, influence, program criteria, or sociocultural acceptance. The choice of model depends on the situation.

For example, if individual team members are to be measured or compared to each other, Guilford's approach is a good choice. When measuring the origin or influences of a work, taxonomy is often considered as a model. For objective measurement of creative tasks, a coherent program should be included, and for questions about the cultural value of a work, it is best to follow Csikszentmihalyi's model to study the response of the field.

From Single Metrics to Hierarchical Frameworks

CREATIVITY

[**Creativity might be humans' highest cognitive function. It`s everyone`s superpower and it can be developed and refined, albeit with effort and practice.**]

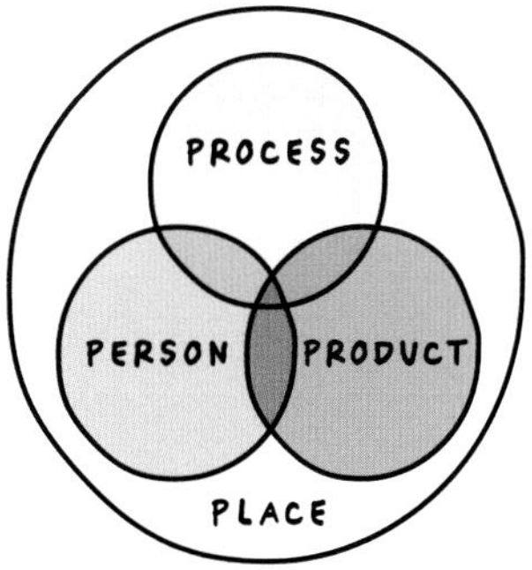
PROCESS
PERSON
PRODUCT
PLACE

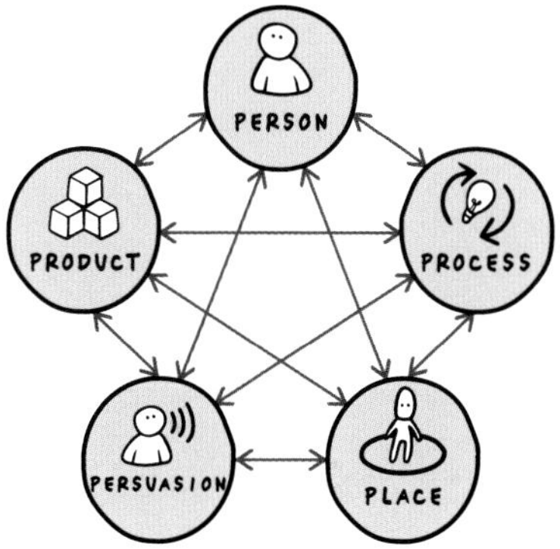
PERSON
PRODUCT
PROCESS
PERSUASION
PLACE

Measuring the creative person (individuals and teams) based on skills, talents, knowledge, and experience involved in creative work.

Measuring creativity in a combination of cognitive, conative, and emotional factors which interact with the environment dynamically.

Measuring creativity based on the interconnection of person, place, process, product, and persuasion.

SINGLE METRIC (1P) → 4Ps → FRAMEWORK 5Ps+

Divergent Thinking Is an Essential Capacity for Creativity

> **Design thinking combines empathy for the context of a problem, creativity in the generation of insights, and rationality to analyze and then fit solutions to the context.**

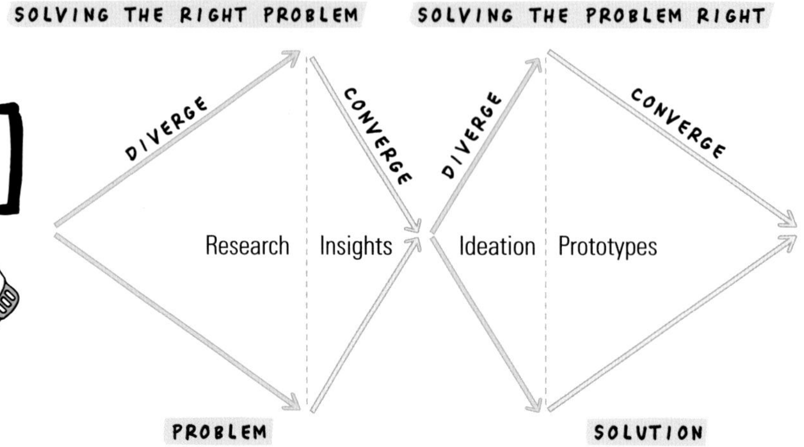
SOLVING THE RIGHT PROBLEM
SOLVING THE PROBLEM RIGHT
DIVERGE
CONVERGE
DIVERGE
CONVERGE
Research
Insights
Ideation
Prototypes
PROBLEM
SOLUTION

The basic elements of creativity are the ability to copy, transform, and combine.

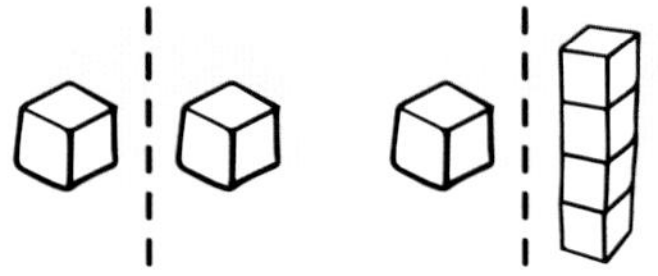

Value creation starts with the problem for a significant time before even trying to solve it.

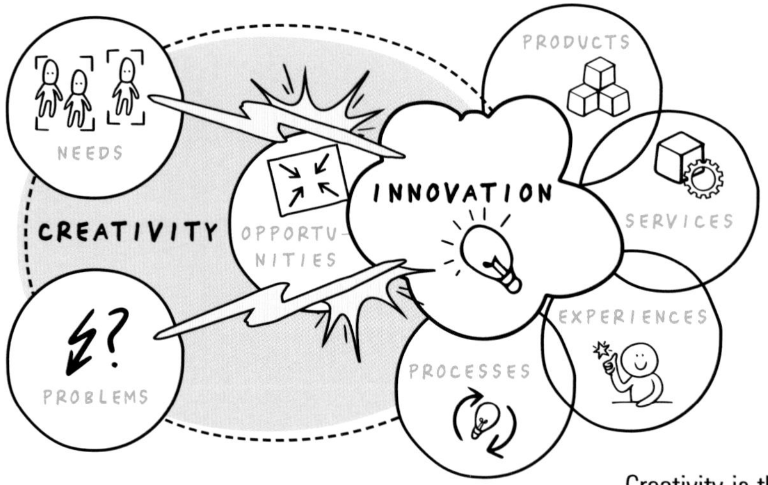
NEEDS
CREATIVITY
OPPORTUNITIES
PROBLEMS
INNOVATION
PRODUCTS
SERVICES
EXPERIENCES
PROCESSES

The most creative teams have the ability to operate in a childlike way of play and to explore ideas.

Creativity is the process of having original ideas that have value.

Taxonomy of Creative Thinking

In practical application Nilsson's Taxonomy of Creative Design (2011) is often applied. The model has a proximity to the design thinking mindset in the environment of daily innovation work and is thus in line with the creative problem-solving methods presented in the Design Thinking 101. The taxonomy offers a progression from imitation to original creation, which allows creative work to be measured along a spectrum of form and content. In this way, it places creative work within a comprehensive, unifying canvas that, in the context of this book, serves not only as an analytical lens through which to measure creative work, but also as a methodological approach for developing creative skills within capability build programs for design thinking.

- Imitation: the development of a similar or almost similar product as one in existence.
- Variation: the making of slight alterations to the existing object in a way that makes it look different despite being able to maintain the original object's identity.
- Combination: the mixing of two or more products to an extent that they can be described as either all or both.
- Transformation: the recreation of a product in a novel milieu to an extent that it acquires a number of the original product's attributes. Nevertheless, it cannot be referred to the original product.
- Innovative creation: the creation of a product that seems to have no single discernible quality of any existing idea or object.

The measurement of a work is done in the context of its previous history, seeking solutions to questions such as "How far does the creative work depart from previous works?" and "How big a leap has the creator made in terms of content and form?" Among the strengths of this measurement tool is its ability to evaluate creative works relative to other existing works by measuring their originality and influence. A weakness in the measurement is that it does not say anything about the value, relevance, or effectiveness of the work. To overcome limitations, the increasingly complex creative thinking processes as well as different knowledge domains can be used or constructed to varying degrees at each level.

The illustrated combined approach is the basis for clarifying, for example, the requirements for the different levels of thinking so that teams can identify where new capabilities are needed and how best to assemble the team to achieve mission-critical tasks. The assessment might be based on an open dialog with the team around needs for creativity, individual development needs, and data on the perceptions and barriers to embracing creative thinking, as well by observing design thinking and innovation teams. Based on the measurements, training and capability build programs are designed and scaled in the wider organization.

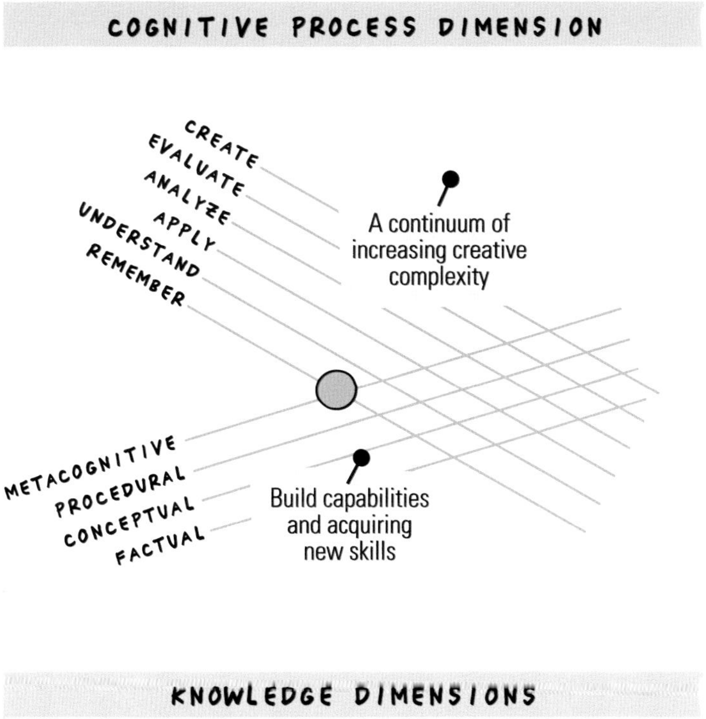
COGNITIVE PROCESS DIMENSION
CREATE
EVALUATE
ANALYZE
APPLY
UNDERSTAND
REMEMBER
A continuum of increasing creative complexity
METACOGNITIVE
PROCEDURAL
CONCEPTUAL
FACTUAL
Build capabilities and acquiring new skills
KNOWLEDGE DIMENSIONS

Example Measuring Team Creativity of Work Performed

Often, we see the creativity of teams only in the retrospective, which usually came to full effect through the application of the appropriate creativity methods. Based on the taxonomy of creative design (see page 201), the design thinking activities related to creative techniques like working with analogies or creating dark horse or funky prototypes can be measured. In retrospect, the respective ideas and prototypes can be classified, for example, as an imitation of an already existing product or service, as a variation of an existing solution, as a combination of two or more ideas, or simply as a radical transformation of an idea in the opposite direction that was not previously imaginable. The following very simple example of a design challenge, from *The Design Thinking Toolbox* (pages 187–195), can be used to assess team performance in terms of creativity based on five key questions derived from the taxonomy:

- Is the solution idea the same or nearly the same as something that already exists?
- Is it a slight modification of an existing product or service so that it is different but still retains the identity of the original solution?
- Is it a mix of two or more functions and experiences so that it can be said to be both or all?
- Is it a re-creation of something in a new context, so that it has some characteristics of the original solution, but can no longer be said to be the same?
- Does it have no recognizable characteristics of pre-existing solutions or ideas?

These questions are an analytical tool for evaluating the originality of teams' initial ideas, prototypes, and solutions. However, it does not evaluate the level of difficulty, how far a potential solution goes beyond what was there before, and some of the mechanisms for how it achieves these gains. It is reasonable to say that this measurement of the creative work a team has performed is not applicable in measuring the relevance, value, or effectiveness of the solution.

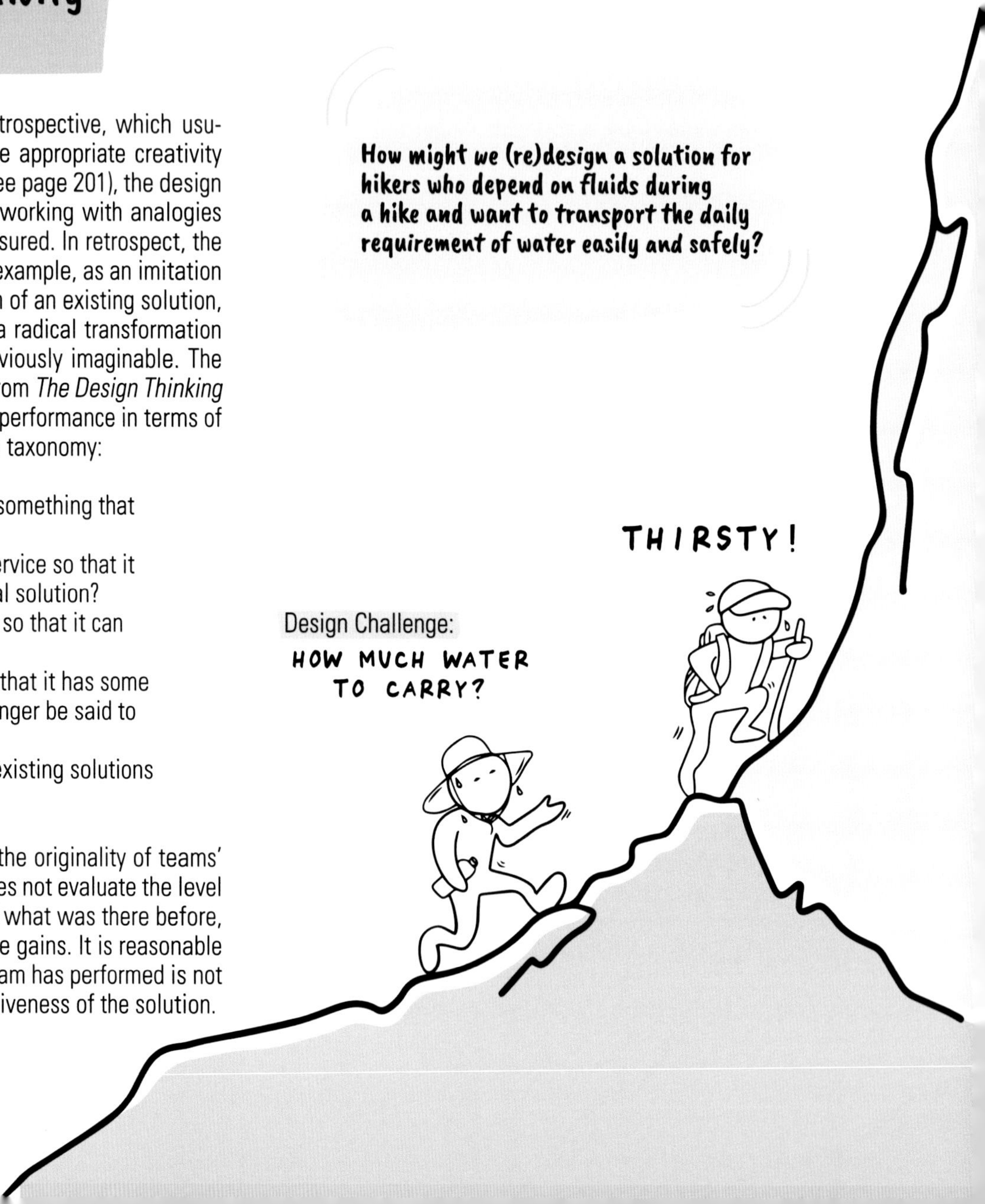
How might we (re)design a solution for hikers who depend on fluids during a hike and want to transport the daily requirement of water easily and safely?
THIRSTY!
Design Challenge:
HOW MUCH WATER TO CARRY?

Example

1 **The team wants to initially learn more about the user and their problem through experimentation.**

Critical Function F2 "Open" **Critical Experience E4 "Design"**

Performance: Imitation of something the team has seen before.

2 **The team creates a first solution based on insights, learning, and existing experience.**

Solution: Better grip and new design

Performance: Staying with the core idea and tweaks it a little bit, which adds some novelty, but it very much remains of the solution of the existing alternatives

3 **The team creates a prototype from two ideas.**

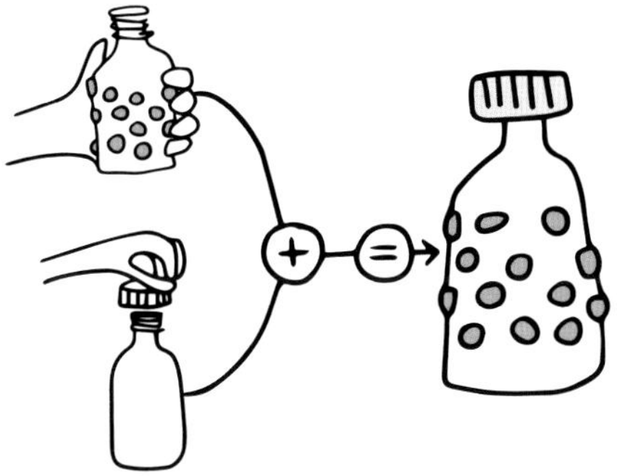

Performance: Combination of two ideas to improve the overall experience and functionality

4 **The team leaves the existing frame to build something new to the market and new to the customer.**

Performance: The solution appears to be an original creation. From all ideas, the flying water has the greatest novelty and could be classified as the most creative.

Guilford Model

The Guilford model is a commonly used creativity measurement system. The model primarily measures the creativity of individuals, based on four key measures that relate to the various properties of an individual team member. Each measure can be expanded and exercised, and all focus on creative performance in terms of instant contextualization in a set of responses (see example on page 205). The original model measures fluency, elaboration, flexibility, and originality. In terms of fluency the model measures the volume of responses. It measures the type of responses to assess flexibility. Originality is measured by the unusualness of the answers. Finally, elaboration evaluates the details of the responses. Similar to the previously presented models, the Guilford model is not all-inclusive in assessing creativity; it performs a specific type of individual psychometric assessment by measuring the type of productivity quotient, if that is what a person wants to achieve. One advantage of the Guilford model is that it assesses productivity output in a descriptive and quantifiable way. Its main disadvantage is that it does not indicate anything about the relevance or value of work in terms of its creative output. Over the years, the measurement criteria have been expanded vastly, and a full range of tests for creativity has been developed that includes even finer categories. For example, psychologist Ellis Torrance has developed a series of tests based on Guilford's work called the Torrance Tests of Creative Thinking. The tests try to measure divergent thinking and other problem-solving skills, which can be assigned to each category either visually or verbally. Today, Guilford model measurements can be conducted and validated by examining the brain using imaging techniques such as magnetic resonance imaging (MRI) or electrical activity monitoring methods like electroencephalography (EEG) that show how neurons and circuits in the brain are firing. Other techniques, like fractional anisotropy (FA), are applied to measure the structure of white matter in the brain. White matter is mainly made up of myelinated axons, long connections between neurons that carry brain signals. The combination of these techniques makes it possible to get an approximation of the makeup and activity of neurons during the creative process. Read more about applying neuroscience in assessing creativity, teamwork, and the application of design thinking tools on pages 366–367.

Example of Measuring Creative Productivity

A very quick and easy way to capture the creative productivity of team members can be done in the form of a familiar warm-up (related to the Guilford model measurement). Each team member is provided with a sheet of paper with 12 circles. Everyone has three minutes to paint as many circles as possible and let their creativity flow. After the three minutes of warm-up, which is mainly aimed at the output of the divergent thinking of each team member, different metrics can be applied. This approach might not be comprehensive in measuring creativity, but it achieves a kind of psychometric evaluation of team members. However, the output is clear and quantifiable.

Potential Measures after This Warm-up Exercise Are:

Fluency: How many responses are there?
Flexibility: How many types of responses are there?
Originality: How unusual are the responses?
Elaboration: What is the level of detail of the responses?
Connections: What connected symbols have been created?

The Warm-up Exercise Helps the Team and Facilitator:

- to obtain a more psychometric evaluation of the team member.
- to mainly measure the productivity quotient.
- to quickly generate quantifiable ways to evaluate the output.
- to measure output in a clear and quantifiable way, however without indications of comprehensively measuring real creativity.

Procedure:

1. Create a sheet with 12 circles and hand it out to the team.
2. Ask the team to create as many ideas as possible within the circles provided.
3. Share all the sketches on a large wall and start assessing the creativity based on fluency, flexibility, originality, elaboration, and connections.

In Many Cases the Comparison of the Drawings in the Circles Provides a Great Indication of Already Divergent Thinking.

Example

These responses might be evaluated in the following ways:

- Patrick drew the most drawings, but his drawings were all faces.
He has the highest fluency.

- Peter drew the most types of responses, but he has fewer total responses than Patrick.
He has the highest flexibility.

- Marc drew two wheels and a ball, which have nice geometry!
No prize, alas.

- Lily drew only two responses, but no one else drew a balloon or a bomb.
She has the highest originality.

- Rukaiya drew only three faces, but with more detail than the others.
She has the highest elaboration.

- Michael connected the elements and drew mainly physical products.
He has the ability to combine systems thinking and design thinking.

FILL ALL CIRCLES IN 3 MINUTES

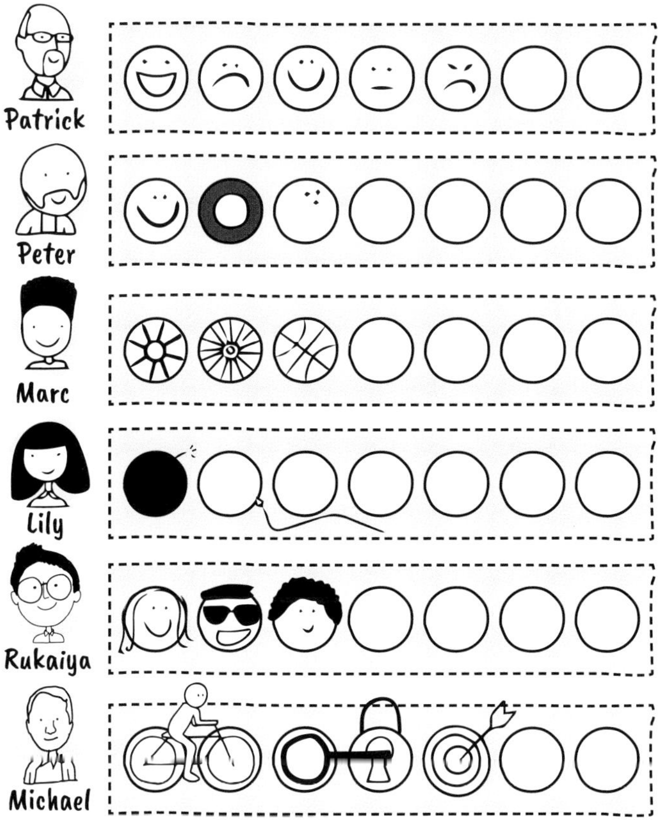
Patrick
Peter
Marc
Lily
Rukaiya
Michael

Requirements Model

The requirements model is another measurement tool that evaluates creative work by using criteria that are defined before the product is built. In design thinking, for example, critical functions and experiences for a potential solution are defined during the design challenge.

The final prototype provides the basis for a series of measurements, (e.g., the appropriate number of functions, the multi-user capability of the solution, or the time the customer/user is engaged with the solution). Other evaluations are a bit more complicated, such as emotional engagement or flow moments. Although they seem to be subjective measures, they can actually be measured concretely with targeted questions and increasingly in combination with new neurotechnologies. In addition, the requirement model helps to evaluate a solution directly if one has set precise requirements.

Among the advantages of the requirement model is that it is able to measure relevance, efficiency, and value against clearly defined requirements. The main disadvantage of the model is that it works best when a solution is compared to itself and not to other solutions. For measurement, it is also recommended that there are clear program requirements or that it includes the taxonomy of creative design (see page 201). The higher the task is in the taxonomy, the greater the need to apply external requirements to ensure that measurement is easy to perform.

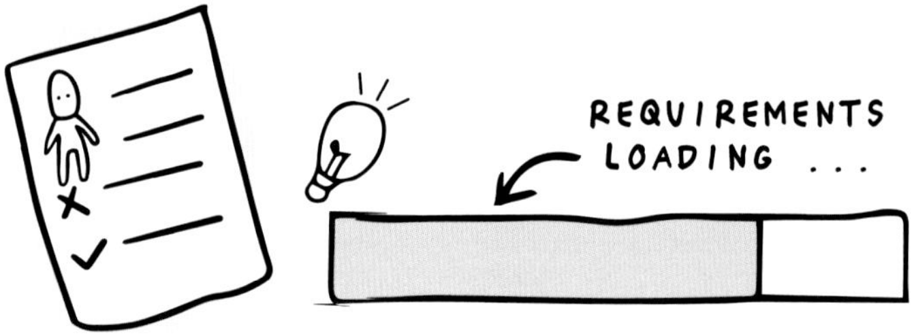
REQUIREMENTS
LOADING ...

Csikszentmihalyi's Model

Csikszentmihalyi's (1999) system model is another preferred tool for measuring creativity because it examines the social value of work. The basis of measurement is assumed to be a system that examines and measures the value of a work in the relationships between three important parties. The parties fall broadly under three major categories: the person, the domain, and the field.

The evaluation of creativity depends on both cultural and social responses to a work. Creativity is seen in this measurement as more than the mere provision of any old and dissimilar production. The model forbids calling a work creative, per se. Much more important in the measurement is the approval of the field, and in the creation of a work, the field is used as the main determinant of whether the product can be sustained within the domain or whether it will fade into oblivion.

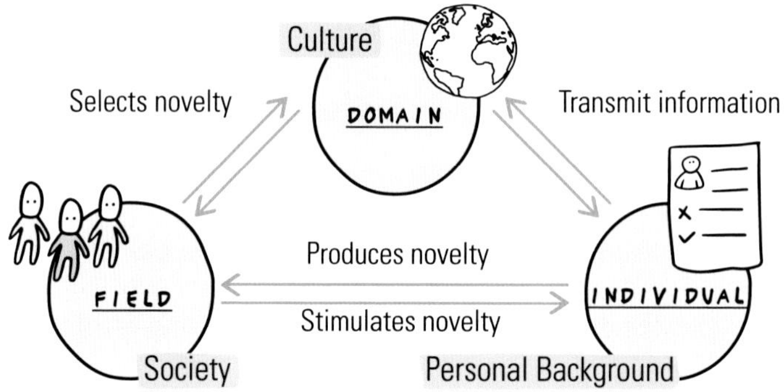
Culture
DOMAIN
Selects novelty
Transmit information
FIELD
Society
Produces novelty
Stimulates novelty
INDIVIDUAL
Personal Background

The main advantage of this model in measuring creativity is that it assesses the relevance and value of a work by considering the community context. A significant disadvantage is that the model can be subjective. As a cultural and social tool, the model is based on constantly changing parameters; therefore, the field tends to change over time, even as works emerge and disappear into anonymity. Thus, for example, in measuring the creativity of a product, no absolute value can be represented, but an awareness at a particular time and in a defined context can be reelected.

How to Apply Extended Frameworks for Measuring Creativity

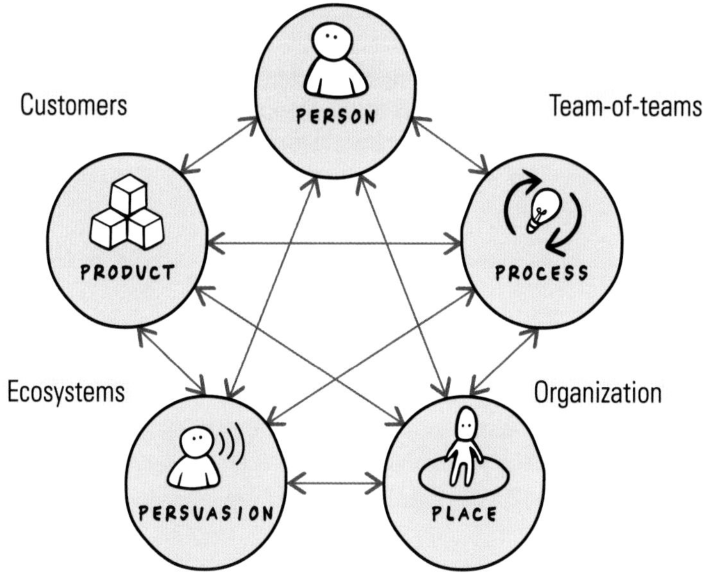
Customers
PERSON
Team-of-teams
PRODUCT
PROCESS
Ecosystems
Organization
PERSUASION
PLACE

The four common ways of measuring creativity have been related to different setting, for example, measuring the creativity of individuals and teams with their skills, talents, knowledge, and experience involved in the creative work. Having people with the right skills in design thinking and innovation teams is important for innovation success.

Besides these person skills, three other factors become very important in influencing creative outcomes: process, place, and product. Process is the development path of a creative team, the steps and actions it takes to get the job done. Place refers to the environment in which the work is done and includes everything from access to material resources to high-level organizational policies and leadership to the culture and social norms of an organization. Product is the creative work itself—the new object, idea, or behavior that the organization hopes to commercialize. Together, person, process, place, and product form four of the five Ps.

Design thinking goes beyond these four Ps to realize new market opportunities. Cross-team coordination and working in team-of-teams mode is another key factor. An interdisciplinary team that wants to implement a creative idea must convince other teams, managers, executives, and even external stakeholders that what they want to do is novel and useful enough to merit support.

For this reason, another factor, persuasion (the fifth P) is added in the measurement. Thus, in an expanded model, first, there is creative potential, which includes people, process, and place (i.e., the elements that influence how creative the new product, idea, or behavior might be). Second, there is the creative output, which is composed of product and persuasion, that is, the creative outcome, which is composed of the novelty and usefulness of the product, idea, or behavior. Both are driven by the effectiveness of the persuasion needed to convince others that the effort of creative work is worth doing.

The five Ps help in defining activities and measures around the important topics of people, processes, and places that are conducive to the development and implementation of creative ideas. Capability builds programs related to creative performance aim to remove creative barriers, (re)define creative products, and persuade people to create space for ideas to flourish. Companies that are in transformation and want to establish a new mindset need to initiate measures related to creative potential as well as creative performance if they want to foster creativity. When the focus is only on creative potential, the scope for implementing creative ideas is usually limited—opportunities are set aside due to lack of time and resources or the company's inability to deviate from standard EXPLOIT procedures. Similarly, measures that address only creative performance—activities aimed at creating the organizational space for creativity—without at the same time initiating measures into the creative potential needed to generate new and useful ideas, EXPLORE, are unlikely to succeed.

Assessing Creativity for Innovation Teams

A first attempt to present the different models and approaches holistically is the 5Ps Creativity Canvas. Based on the five Ps, the most important aspects for measuring creativity are summarized and the appropriate questions for reflection and the definition of activities analogous to the diverging and converging design activities are presented. The goal of the canvas is to document all elements at a glance for measuring and shaping the creative innovation engine.

The canvas focuses on the WHY in relation to the person and the team, with explorations of curiosity, purpose, knowledge, and reflection, and then goes into process, place, persuasion and product. The different dimensions allow organizations to address creativity from the system perspective to the individual/team perspective. They also provide a basis for measurements of the creative potential to the creative performance of new products, services, and experiences.

It is recommended to use the appropriate methodology and models for the respective measurement of single elements of the canvas. The results of the respective measurements can be used as the basis for the development of training and capability-building programs, as shown in the previous example, up to the mirroring of activities in divergent thinking with innovation objectives.

The canvas should be adapted to and supplemented for the respective cultural requirements, since, for example, creative frustration or the handling of ambiguities differs depending on the culture, mindset, and team spirit.

Start to bridge the gap between scientifically oriented theories and metaphorically oriented applications of measuring creativity.

WHO ELSE?
WHAT?
WHY?
HOW?
WHERE?
$

Template download

5Ps CREATIVITY CANVAS

PERSUASION

NETWORKS & COMMUNITIES
Who are the people who can support the creative pursuit?
How do you create and maintain a fair and healthy relationship with these people and groups?

SOCIAL ENVIRONMENT
What social environment is needed for creative work?
To what extent can the environment be actively shaped?

COLLABORATION
What does the collaboration look like?
Which stakeholders are to be involved in the process and when?
How is collaboration with others during creative activity?

PROCESS

PROBLEM
Is there a problem worth solving, and if so, what exactly is the problem?
Why should it be solved?

INSIGHTS
How is the emergence of insights supported and shared?
How are interruptions and re-framing applied?

IDEATION
Which thinking tools are applied to develop new approaches?
Which techniques are appropriate, and how can they be trained?

EXPLORATION
How radical should a potential solution be, and what influence should it have on a domain?
To what extent do questions, thoughts, and design activities influence the domain?

PERSON / TEAM

CURIOSITY
What is the basis of curiosity and why?
How can teams be curious more often and/or increase the chances of unforeseen situations?

PURPOSE
What are the real drivers and motivators, and why?
What is the inner core that is being expressed?

KNOWLEDGE
What knowledge and skills are needed to achieve a particular creative goal?
How can certain skills and abilities be built up?

REFLECTION
How can creative action be reflected upon and learned from?
How can parts of the creative process be planned?

PRODUCT

SUCCESS
What does success look like at the project, team, and behavioral-change level?
What outcome is good enough to achieve?

QUALITY / QUANTITY
How is the quality of creative performance evaluated?
How many functions and experiences from the requirements have been implemented?

USEFULNESS
How useful is the service, product, or experience from the perspective of the customer and other reviewers?
How close is the emotional bond of the customer/user?

NOVELTY
How new is the product to the customer and the market?
How radically different is the solution from existing solutions?

PLACE

MINDSET
What are the characteristics and perspectives that promote or hinder creativity?
How can attitudes and behavior change in the short, medium, and long term?

RESOURCES
Which resources are needed, and which are available?
Is it possible to replace lacking resources, perhaps with external support?

ROOM
What does the physical, hybrid, or virtual work environment look like to accomplish the tasks?
What additional elements are needed to support the creative process?

DOWNLOAD TOOL
www.design-metrics.com/en/5p-creativity

Example of Measurements and Metrics

Example

Creativity is a skill that can be developed and a process that can be managed. Creativity begins with a foundation of knowledge, learning a discipline, and mastering a way of thinking. Individuals and teams learn to be creative by experimenting, exploring, questioning assumptions, using imagination, and synthesizing information.

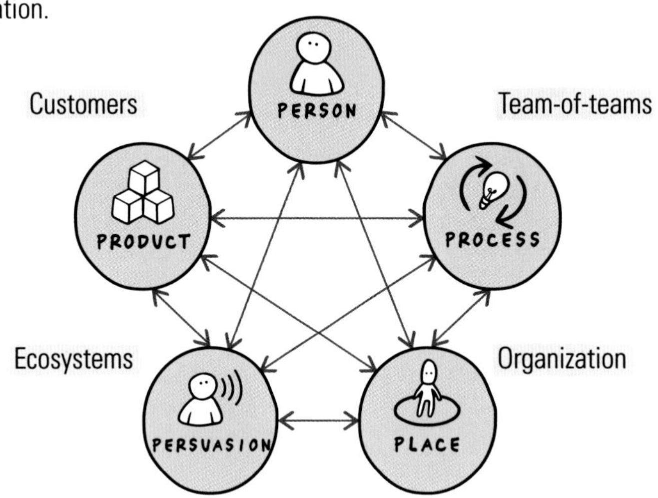
PERSON
PRODUCT
PROCESS
PERSUASION
PLACE
Customers
Team-of-teams
Ecosystems
Organization

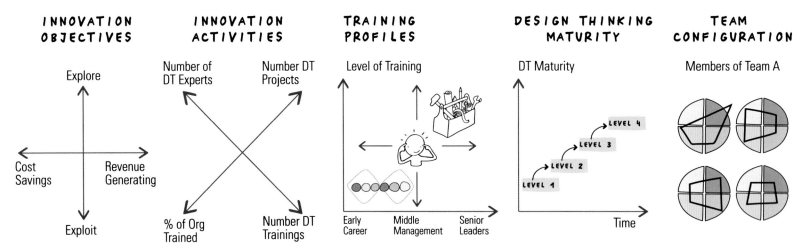

INNOVATION OBJECTIVES

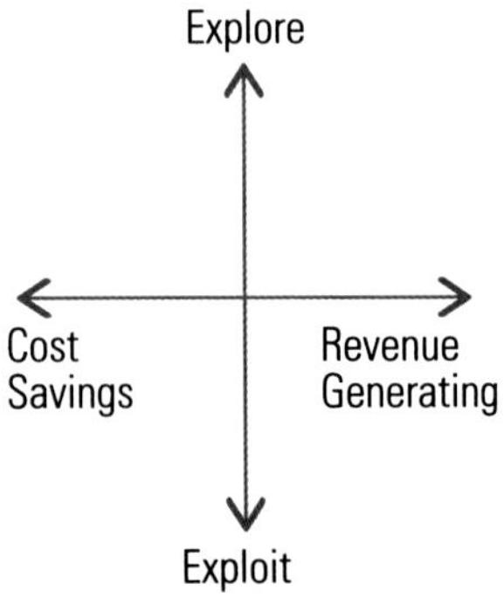
Explore
Cost Savings
Revenue Generating
Exploit

INNOVATION ACTIVITIES

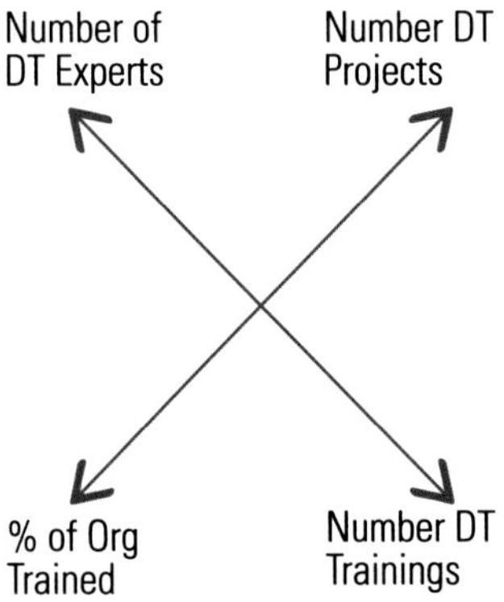
Number of DT Experts
Number DT Projects
% of Org Trained
Number DT Trainings

TRAINING PROFILES

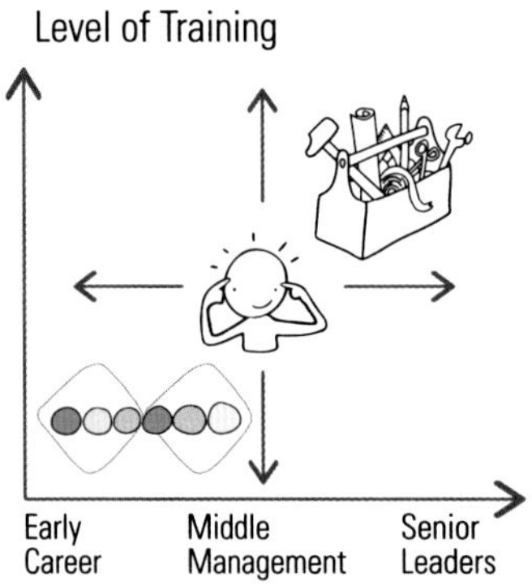
Level of Training
Early Career
Middle Management
Senior Leaders

DESIGN THINKING MATURITY

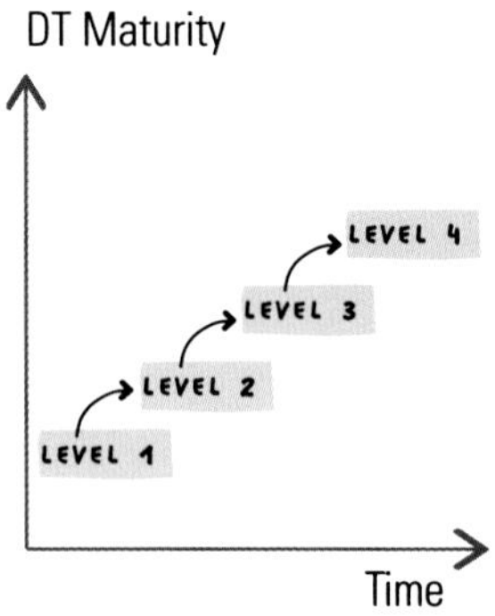
DT Maturity
LEVEL 1
LEVEL 2
LEVEL 3
LEVEL 4
Time

TEAM CONFIGURATION

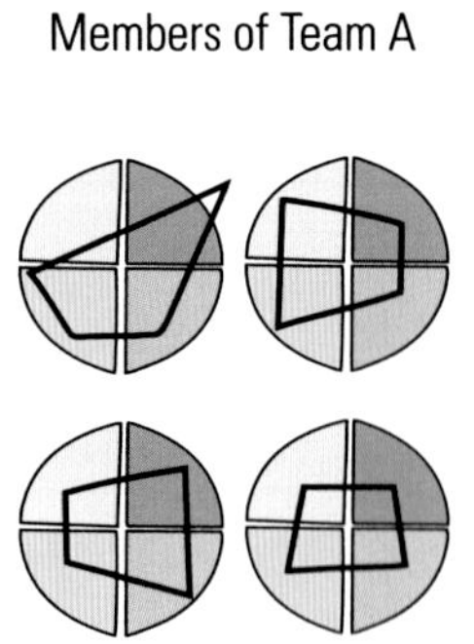
Members of Team A

REFLECTING MEASUREMENTS

Measurement Options for Each Level of EXPLORE

Now that the basic considerations for building a measurement system are understood and the organization is ready to perform initial measurements, this section presents examples of measurements for each level with regard to EXPLORE. The measurements presented here should serve as inspiration for creating new exploration metrics, but they should by no means simply be adopted for the organization or the team without question. As described in the introduction of this 101, the central question always is, why should something be measured and what decisions can be derived from it. Also, how much is the additional information worth and what impact will it create?

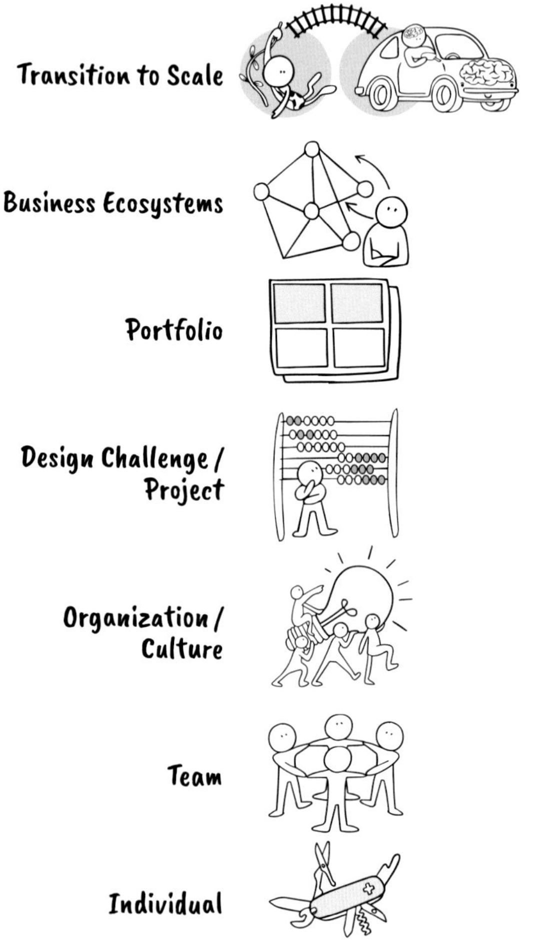

Transition to Scale
Business Ecosystems
Portfolio
Design Challenge / Project
Organization / Culture
Team
Individual

The structure of the examples presented in this section follows the logic of the model initially presented in the introduction on pages 35–36 (from bottom to top). In the context of this book, the focus of the chosen examples for all topics is related to increasing innovation performance. With regard to the prioritization of the levels for measurements, there is also no universally valid conclusion. Depending on the tasks to be mastered, from restructuring to cultural change or major changes in strategic orientation to building new capabilities, different priorities are to be set, for example, as part of the North Star discussion. The concept of minimal viable exploration metrics supports the initial measurements, helps to build up the measurement system, and provides a flexible framework for keeping the measures in view even in turbulent and dynamic times. A matrix and a meticulous breakdown of measurements according to stakeholder and function in the company is not recommended, because with premises such as transparency, openness, and a moderated transition from EXPLORE and EXPLOIT, it is more necessary today than ever before to understand the different measurement worlds.

> **Unless we determine what shall be measured and what the yardstick of measurement in an area will be, the area itself will not be seen.**
> **- Peter Drucker**

In presenting the examples, initially measurements related to the team and culture (individual, team, organizational culture) and then measurements relevant to the project and business (design challenge/project, portfolio, business, and innovation ecosystem) are presented. Finally, the transition to EXPLOIT and the preparation to scale is reflected. As already highlighted before, for individuals, teams, and culture it is mainly about the measurement of human behavior, mindsets, and capabilities to be developed. The measurements on the design challenge/project as well as on the portfolio level range from measurements on the tool level to the adjustment of the balance between EXPLORE and EXPLOIT in the portfolio of a company.

Navigating through the Measuring Spheres

MEASUREMENT OPTIONS

[The measurements to apply depend on the overall objectives, purpose, and desired outcomes. The map of exploration metrics and measurements will look different for every organization and team.]

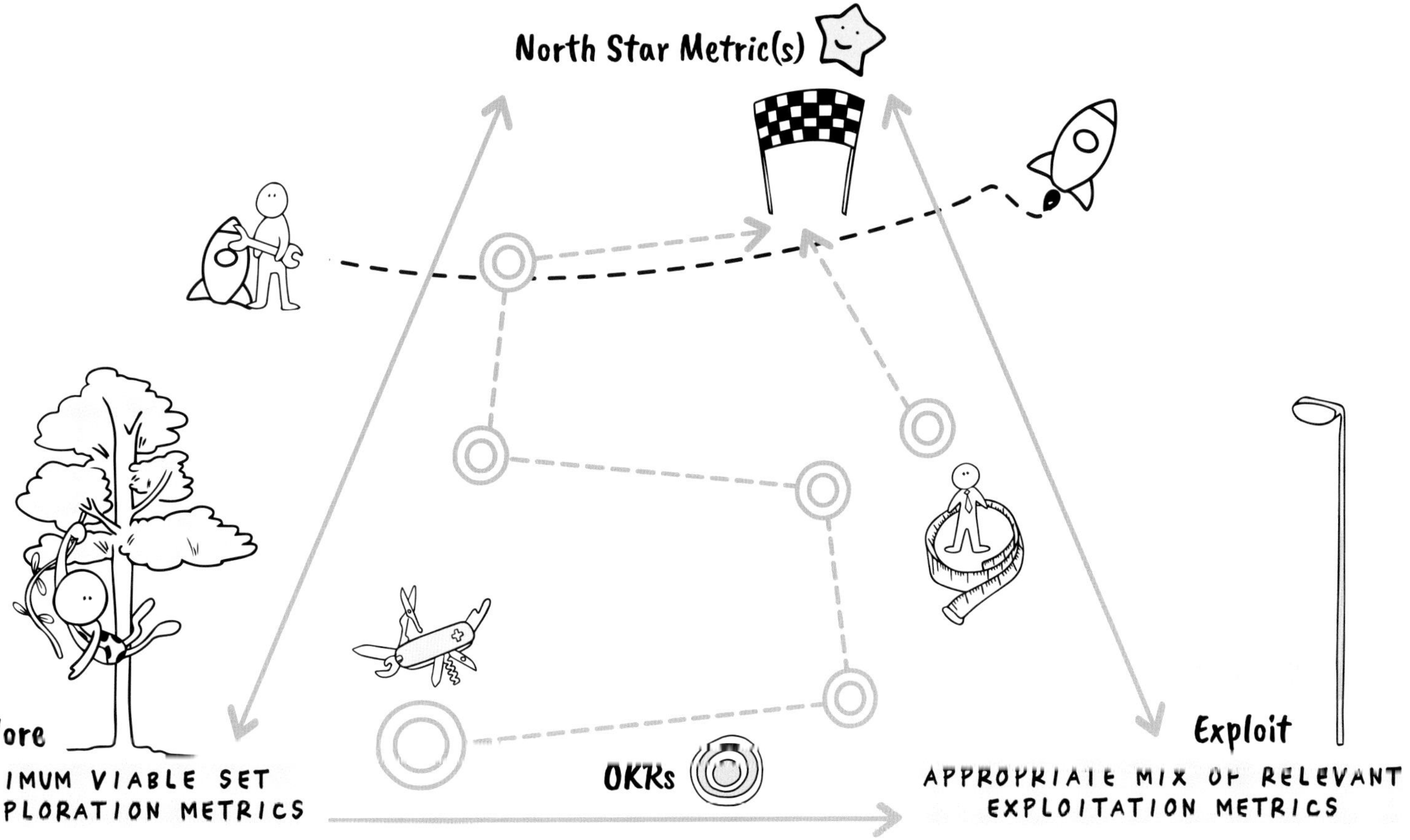
North Star Metric(s)
Explore
MINIMUM VIABLE SET OF EXPLORATION METRICS
OKRs
Exploit
APPROPRIATE MIX OF RELEVANT EXPLOITATION METRICS

How to Measure Individuals

Measuring on the individual level can be linked to the performance measurement systems (see 201) or to HR measurement systems. Both kinds of measurements have their challenges, and for concrete measures in EXPLORE, the impact of the work of individuals might be even more puzzling to measure. The individual work in EXPLORE is influenced and reinforced by many factors. For example, for design thinking it depends very much on how the interdisciplinary team is formed, access to customers/users is provided, the overall mandate to innovate is articulated, and of course the measurement system is defined. Therefore, it is recommended to focus more on the behavior and mindsets of individuals, which are related to what team members do and say. The previously outlined measurements related to creativity (see page 205) might also provide relevant measures in examining, for the example, the individual divergent thinking ability or indication of flexibility versus persistence. In the case of OKRs, it is recommended not to agree on individual OKRs in EXPLORE, as different skills and collaborative efforts are usually required to be successful.

Measuring Individuals Based on Behavior and Mindset Helps:

- to obtain an understanding of whether individuals are team players.
- to identify entrepreneurial behavior and attitudes.
- to extend the view of skills and capabilities based on T-shaped profiles.
- to learn about individual thinking preferences in an effort to create a holistic thinking pattern in a team.
- to detect tendencies toward a fixed or growth mindset.

Leading indicators related to individuals provide confidence to the team and organization that the project has the appropriate skills and capabilities to make progress and achieve the desired outcomes. The measurements of transformation with regard to behavior and mindset are supporting outcome metrics, which relate to the total value the company generates from all forms of EXPLORE activities. Research also shows that linear organizations have a linear function of employees and profit, while exponential organizations have an exponential function of employees and profit. Individual measurements based on forced rankings or vitality curves do not fit the purpose of creating high-performing innovation teams and a culture of U-shaped teams. Consequently, it is not recommended to rank employees by asking managers to just create a list of the best to the worst employee, as a way of measuring individual performance.

Examples of Measuring Individuals

- Tracking activities performed by a team member to reach the goal
- Tracking of individuals, timelines, progress, and confidence level contributing to the design challenge
- Tracking revenues per employee (new products versus existing products)
- Soliciting 360-degree feedback (preferred for individuals with customer interaction) or simplified 180-degree feedback
- Using human capital innovation ROI to assess the value of human capital (e.g., knowledge, habits, and social and personal attributes) related to the work in EXPLORE
- Measuring individuals based on the three-item growth mindset scale (Dweck, 2006, 2016)
- Learning and development of new skills based on qualitative feedback about behavioral change
- Increasing knowledge and skill of the individual based on learning, training programs, or project-based learning activities

The individual contribution and a growth mindset are important, but this will only thrive if it contributes to the achievement of set objectives by the team.

How to Measure Teams

The creation of interdisciplinary teams based on thinking preferences and T-shaped profiles is a viable procedure for setting the basis for high-performing teams. Different measurement methods are available to measure, for example, the innovation self-efficacy score developed by Berny Roth from Stanford University. The approach measures the growth mindset in relation to design thinking. However, in many cases the practical application shows that team members working based on their own motivational effort in EXPLORE already have a growth mindset applying characteristics associated with the design thinking mindset (see page 64). Therefore, it is recommended to add, for example, an internally judged measure of outcome value and novelty, based on an average of anonymous ratings provided by the team members. In the context of capability programs for design thinking, it is common to measure employee satisfaction scores with the aim to analyze and illustrate how the positive reinforcement of design thinking had affected employees.

Measuring Teams Based on Innovation Work Performed and Collaboration Helps:

- to create understanding about team composition and dynamics.
- to assess how the team delivers on agreed objectives and key results.
- to observe alignment of mindset principles related to design thinking/agile.
- to explore the team's safety for interpersonal risk taking.
- to explore how single process steps of the design thinking process are declared and accepted as the appropriate behavior by the team (e.g., failing smart via testing of assumptions and solution prototypes).
- to obtain indication of the design thinking maturity levels of the teams/organization.

More sophisticated measurements aim to analyze the impact of design thinking on the working culture (see page 216) through employee motivation, engagement, team collaboration, and effectiveness. These measurements show over time the extent to which different parts of the organization are developing creative confidence. They also show the positive effect on individual and team innovation capacities and creative work that highly motivates, for example, the extended team engagement in work related to EXPLORE. On a team level other powerful measurements try to find out how teams are collaborating to analyze the important questions about why individuals and teams do something.

Examples of Measuring Teams

- Team diversity and degree of interdisciplinary capability within the team
- Ratio of people with design, ethnographic, and user research skills compared to other team members
- Increase in speed of innovation based on new mindset and behavior
- Time needed for the team to make decisions to achieve the right impact, at the right time, and with the right intent
- Degree of trust between teams and top management
- Number of conflicts within the team that were resolved constructively
- Measuring the authenticity of team members in workshops, meetings, and bilateral conversations
- Measure learning agility determined by applying reflective thinking and tracking of the changes instigated within the team as a result of this kind of learning
- Propensity to initiate conversations in which the status quo is challenged based on regular rounds of feedback
- Whether all team members are present and connected during workshops and meetings (indicators includes phones are turned off, interruptions and repetitions are kept to a minimum, and details are memorable for everyone)

How to Measure Organization / Culture

Design thinking has become increasingly important in the context of organizational design and culture. In addition to the actual innovation work, it is also being used successfully to accompany the organizational and cultural transformation. For example, the design thinking mindset and tools are used to introduce OKRs (see pages 286–311) and at the same time establish a team-of-teams organizational culture, if this has been one of the objectives to work with a redefined performance measurement system. In terms of measuring culture, it is important to understand that the assumptions and beliefs of each team member drives behavior, and the collective behavior of all teams determines the results. For example, in the context of OKRs set to change behavior, the results measure performance and indicate, at the same time, if the North Star metric and company objectives have been achieved. The culture of an organization has a significant impact on innovation performance and often reflects the core values of the organization. Culture can also be seen as a direct reflection of organizational leadership.

Measuring the Culture and Organizational Dynamics Help:

- to pro-actively design the organization and culture for both EXPLORE and EXPLOIT.
- to build a strong organizational culture that helps to achieve objectives.
- to improve the engagement and retention of team members and teams.
- to manage proactively diversity, equity, inclusion, and belonging.
- to distinguish between opinions/rumors and real behavior in the organization.

The measurement of culture is also challenging because a lot of factors are relevant in determining the appropriate set of metrics. More sophisticated measurement systems apply typologies of organizations ranging, for example, from a culture of adhocracy, clan-thinking, and hierarchy to market culture. The competing values are mapped against organizational dimensions referring to internal-external and stability-flexibility, leading to the four organizational types that indicate whether a culture fits, for example, EXPLORE with attributes of dynamic, entrepreneurial, and people orientation. For the assessment of single teams, very often behavioral observations based on a pre-defined scale are sufficient. The organizational values based on desirable behaviors that represent them are the basis of the culture.

Examples of Organization / Culture Metrics

- C-levels/decision makers have available calendar time for EXPLORE
- Number of employees performing extra and on top activities to explore new market opportunities
- Number of employees attending capability build programs in design thinking and applying learning in project work
- Amount of improved internal collaboration and knowledge sharing based on team-of-teams setting
- Talent benefits based on the impact on recruiting and retaining talent, triggered, for example, by externalized showcasing of innovation success
- Satisfaction and purpose score of teams and individuals
- Measuring work performed between exploring, exploiting, and sustaining the business
- Openness and co-creation activities with potential customers and users in problem space and solution space
- Ratio between number of design challenges and breakthrough innovations realized

Measuring helps to ensure the right teams collaborate on the most ambitious design challenges in the most effective way to accomplish innovation success and strategic development.

How to Measure Design Challenge / Project

The design thinking toolkit (see pages 104–144) already provided various examples of measuring design thinking activities related to achieving the problem/customer fit, problem/solution fit, and product/market fit. In addition, the measurement of activities related to design thinking is related to measurements on the tool level, creating clarity with regard to customer needs and critical assumptions about velocity in various dimensions. In addition, measurements with regard to confidence find application from problem definition to scale. Examples of measuring confidence are outlined on page 109.

Measuring on the Design Challenge / Project Level Helps:

- to assess the quality of design work by different teams.
- to measure and quantify the additional value of the design work.
- to assess how knowledge is gained and developed over time.
- to obtain a feeling about the confidence level of the team over time.
- to understand the importance of process steps in creating new products, services, and experiences.

In the context of design thinking and measuring on a tool level, the responsibility for choosing a quantification, measurement, or metric is mostly with the design thinking team. The metrics applied must be aligned with the activities performed and occur while hunting or transporting, based on the declared behavior (see pages 90–91). Therefore, many design thinking teams and facilitators apply the core principles of minimum viable exploration metrics to measure, for example, activities and learning. In addition, design thinking is applied for various problem statements that take the team to the dimensions of product, service, to process innovations. The mindset is also applied to build measurement systems (as described in this book) to set the initial validation in business ecosystem design within the paradigm of design thinking for business growth. Sub-disciplines and derivates of design thinking (e.g. service design) apply; for example, consider self-assessments of the designed service based on a five-level scoring system (0–4) ranging from good service to bad service. The scoring is applied to validate existing and (re)designed services based on pre-defined user criteria (e.g., easy to find) and performance criteria (e.g., respond rate to change).

Example of Measuring Design Challenge / Project

Metrics Related to the Work over the Design Cycle:

- **Understand:** Number and quality of assumptions tested
  Number and quality of hypotheses developed
  Participation rate of relevant stakeholders in co-creation
- **Observe:** Number of intentional versus unintentional learnings and depth of interactions with customers/users
- **PoV:** Number and depth of new insights derived from understanding, observing, and testing
- **Ideate:** Quality of ideas generated
  Diversity of applied creativity techniques
- **Prototype:** Number of prototypes built and distributed over the exploration grid
- **Test:** Number of intentional versus unintentional learnings
  Overall efficacy of prototypes built to prove assumptions/hypotheses
  Number and depth of customer interviews and tests per cycle

- Team confidence about problem to solve or prototype to build
- Time to problem/solution fit or product/market fit
- Time-to-market
- Improved assumptions to knowledge ratio
- Usability and customer effort scores
- Adoption rate of features contributing to value proposition

How to Measure the Portfolio (EXPLORE versus EXPLOIT)

Most companies and organizations with the ambition to innovate run two portfolios: one portfolio for EXPLORE, and a second portfolio managing exiting products, services, and offerings (EXPLOIT). The exploitation portfolio refers to efficiency innovations or incremental product or service innovations to sustain current business. The focus for design thinking and innovation teams might be partly on exploitation portfolio in activities to re-design processes or the previously mentioned (re)design of business models, but the core activities and leadership attention should be on the exploration portfolio for creating future business success. Explore deals with transformational or radical innovations, which are often new for the customer/user and new for the market. The mandate for the teams must be clear with regard to how radical the innovation efforts might be or which guiding principles from the North Star apply.

Measuring the Portfolio Helps:

- to understand the amount of resources invested and outcomes generated.
- to obtain insights about the strategic alignment and relative performance.
- to establish a forecast about future value and possibilities to scale.
- to de-risk over the entire design cycle based on learning and knowledge gains.

Companies that take innovation work seriously and go beyond mere lip service create structures that allow the EXPLORE portfolio and the associated measurements to be managed just as professionally as EXPLOIT. Responsible leaders reserve between 50-100 percent of the available time for the support, alignment, financing, and measuring of these important activities. Some companies even go so far as to divide the CEO position between two people who manage EXPLORE and EXPLOIT, respectively. As a closely aligned leadership team, they might have an aligned North Star or two different North Star (depending on the situation of the business and urgency to reinvent the business) and implement different cultures based on the activities of EXPLORE and EXPLOIT (see pages 190–191). However, in the transition from EXPLORE to EXPLOIT, the two leaders are aligned to moderate and lead the process together (see also page 221). The best results in EXPLORE are achieved when consistently applying the problem to growth and scale framework. That starts with the problem statement, validates assumptions through design thinking, applies lean start-up, and prepares the organization for scaling. Every iteration and every failure contributes to learning and developing solutions over time that allow creation of the desired impact, which can be measured.

CEO
CEO

Examples of Measuring the Portfolio

Measuring Portfolio, Funnel Health, and Strategic Alignment:

- Number of innovation initiatives or design challenges per playing field
- Money invested per design challenges over the entire design cycle
- Number of design challenges killed based on kill metrics
- Number of pivot or preserve decisions made per design challenge and/or over the entire portfolio
- Number of portfolio/funnel decisions made by the design thinking team versus top-down decisions
- Ratio between initiatives based on current customer needs versus future customer needs
- Percent of original total of customer problems (today and tomorrow) observed in the market
- Size comparison of initiatives based on potential innovation success and impact
- Percent of portfolio elements moving along the problem to scale and growth framework
- Percent of portfolio elements aligned with North Star
- Portfolio vitality index (percent of revenues from realized market opportunities over the last three years)
- Expected commercial value of the portfolio (risk-adjusted NPV)

Problem to Scale and Growth Innovation Funnel

Organizations with large and diverse portfolios apply different metrics at different stages. The core business of mature companies focuses on profitability and revenue growth (EXPLOIT). Activities related to sustaining the business focus on market share, while design thinking and ecosystem design projects ignore internal numbers and focus on the growth of untapped market opportunities to determine if the transformation is worth it (EXPLORE). In contrast to a stage-gate process for innovation, the Problem to Scale and Growth Innovation Funnel (see more details in The Design Thinking Toolbox, pages 263–266) has the purpose of tracking the developments of different design challenges and initiatives and documenting the respective validation steps.

Once the innovation teams have been given the freedom to innovate radically or the mandate to explore within the North Star (i.e., strategic context), the only confidence needed is in the team and its capabilities throughout the design cycle. This approach allows that the decisions are made by the team itself, which has the most information about the customer, market, and willingness to pay. This approach works best when decision makers and boards allocate funds according to their risk tolerance, leaving the persevere, pivot, or kill decision to the team.

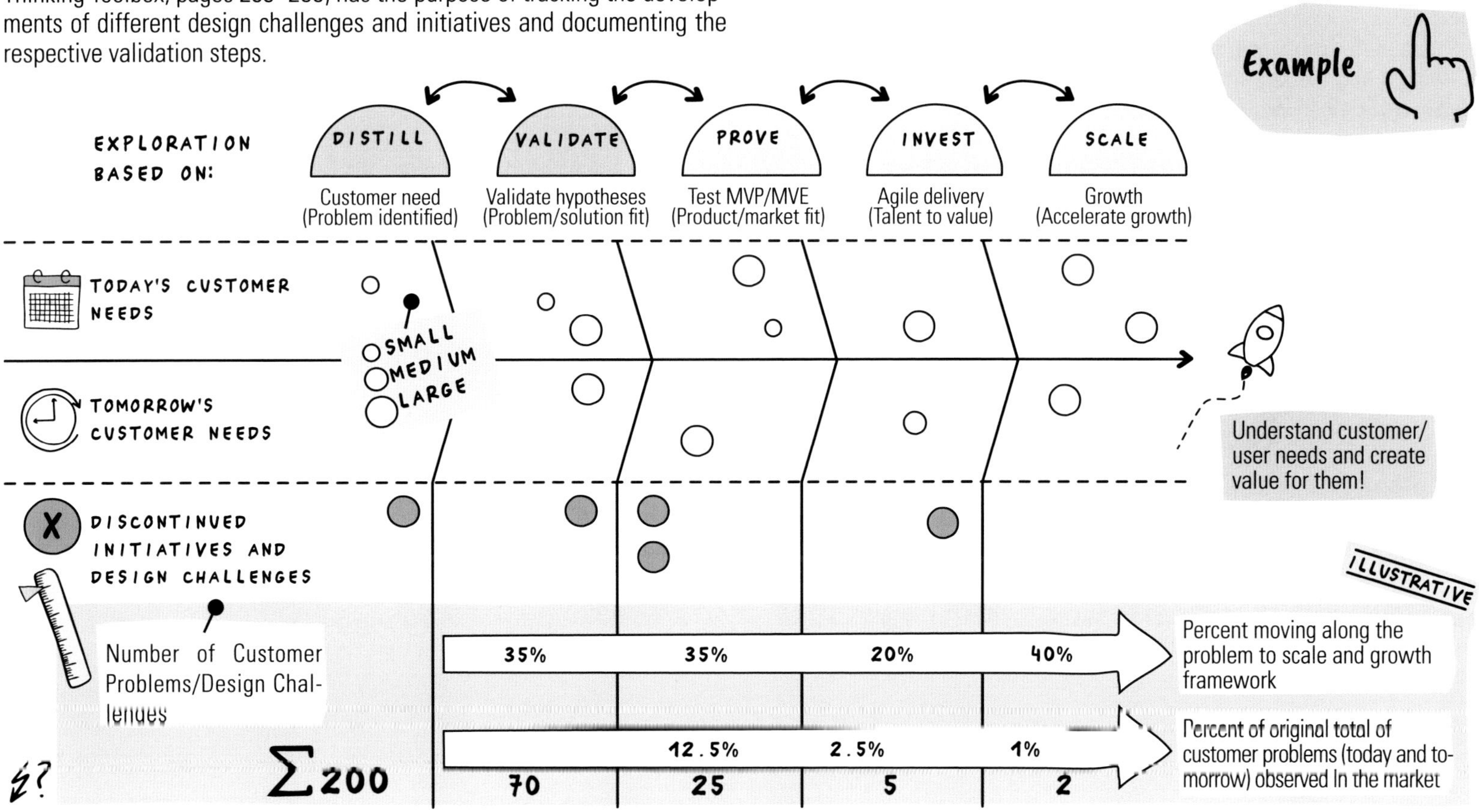
Example
EXPLORATION BASED ON:
DISTILL
VALIDATE
PROVE
INVEST
SCALE
Customer need (Problem identified)
Validate hypotheses (Problem/solution fit)
Test MVP/MVE (Product/market fit)
Agile delivery (Talent to value)
Growth (Accelerate growth)
TODAY'S CUSTOMER NEEDS
TOMORROW'S CUSTOMER NEEDS
SMALL
MEDIUM
LARGE
Understand customer/user needs and create value for them!
DISCONTINUED INITIATIVES AND DESIGN CHALLENGES
ILLUSTRATIVE
Number of Customer Problems/Design Challenges
35%
35%
20%
40%
Percent moving along the problem to scale and growth framework
12.5%
2.5%
1%
Percent of original total of customer problems (today and tomorrow) observed in the market
Σ200
70
25
5
2

How to Measure the Innovation and Business Ecosystem

With regard to ecosystem terminology, it should be distinguished that there are innovation ecosystems and business ecosystems. The innovation ecosystem expresses, in the context of value creation, the phenomenon of a loosely connected network of actors collaborating with the goal of mostly creating innovations related to a specific technology. The purpose of the innovation ecosystem is mainly to new solutions through co-creation and a shared set of knowledge and skills between the participating actors. Such initiatives often gather in clusters, for example, the FinTech cluster in Singapore, or the IT security cluster in Tel Aviv. In business ecosystems, companies apply the business ecosystem design methodology to collaboratively create new value propositions for customers/users orchestrated by one or two actors in the system. Very often, the business ecosystem builds on existing innovation, transaction, or knowledge ecosystems to realize a new offering for customers. As a result, two ecosystems can be measured: the effectiveness of the innovation ecosystem and the performance of the business ecosystem, if a business ecosystem has been chosen as a path to realize a new value proposition to customers/users.

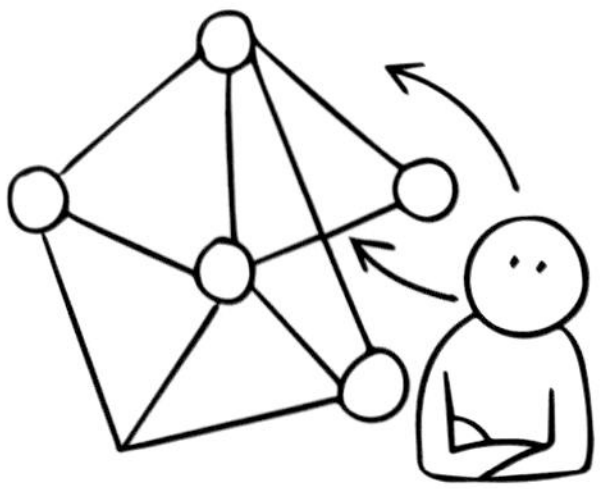

Examples of Measuring the Innovation Ecosystem

- Estimated savings of innovation ecosystem initiatives relative to the comparable costs for in-house problem-solving
- Increase in corporate-wide open innovation culture through the collaboration of innovation ecosystems
- Increase of reputation among other actors in the innovation ecosystem
- Degree of top management commitment to open innovation initiatives
- Degree of freedom given to innovation teams to establish search fields beyond the North Star
- Customer benefits from innovations created and/or provided by the actors of the innovation ecosystem
- Expected increase of revenues from new customers/users as percentage of total sales based on collaboration with actors of the innovation ecosystem

Examples of Measuring the Business Ecosystem

- System/actor fit (realization of MVE)
- Engagement level of core actors in the system to provide the value proposition to the customer/user
- Number of new active customers obtaining services, products, and experiences from the business ecosystem
- Created ecosystem capital based on, for example, number of actors, value streams, and reach to customers
- Satisfaction of business ecosystem actors
- Customer satisfaction with delivery of value proposition from actors in the system
- Contribution margin per transaction
- Revenue per customer/user
- Retention costs per customer and/or actor
- Acquisition costs per customer and/or actor
- Share of revenue from expansion of core value proposition with new products, services, and experiences

How to Manage the Transition from EXPLORE to EXPLOIT Preparation to Scale the Solution

Finally, the transition from EXPLORE to EXPLOIT is also central. It is important to be clear what the rules and culture are in EXPLORE and what the rules and culture are in EXPLOIT. This important distinction has already been described in the short introduction on pages 46–47 with a metaphor related to the jungle and the activities on the corporate superhighways. Some tips and examples can be derived from practical experience to facilitate the transition, mutual understanding, and evolution of metrics. One simple way is that the incentives of selected board of directors and the CEO are linked to exploration metrics. For example, up to 20 percent of compensation can be tied to measurements in the Explore portfolio and/or the impact of activities in EXPLORE. Other companies go a step further and nominate a co-CEO who is only concerned with EXPLORE and new growth opportunities. Both of these actions send a clear message about the importance of EXPLORE and its measurement systems. This is often accompanied by dedicating more budget and resources to EXPLORE and de-investing in the EXPLOIT portfolio. A good rule of thumb is that 10 percent of current revenues and 1 percent of profits should be allocated to EXPLORE activities (excluding R&D in technology) to stay on track for success in the medium to long term. These references vary depending on the industry and market dynamics. The money is also an investment in the learning, action, and impact zone as described on page 162. It is often linked to a mandate to leave previous principles and guiding principles of the core business and to "fail as much as possible." It also allows multiple design challenges and innovation initiatives to be launched in parallel, in many cases even as a competition between regions and teams for the same design challenge, which not only gives new opportunities for measurement in EXPLORE, but also increases the probability of innovation success. By consistently focusing on the customer and their needs as well as measuring and applying the appropriate methods and tools from problem/customer fit to product/market fit, few true innovations remain that have the potential of scalable market opportunities with significant contributions to new business. Another way of linking EXPLORE with EXPLOIT is based on the North Star metrics. If the North Star for example is based on transaction volumes related to an existing platform business, the EXPLORE activities relate to certain correlations and heuristics that in particularly trigger more transaction volume directly or indirectly.

Depending on the type of business, the minimum viable exploration metrics in EXPLORE will indicate test customer experience measurements or pre-defined test customer efforts that will be used later in EXPLOIT and provide predictions about the later impact of the platform's business. For more radical innovation initiatives without a direct link to the existing business, the EXPLOIT metrics should be defined as part of the overall design cycle to ensure that the appropriate set of exploitation metrics is used in exploiting and sustaining the business.

Design thinking and innovation teams of large organizations with a high maturity in terms of mindset, processes, and tools usually realize one or two breakthrough innovations over a given time, based on hundreds of design challenges. Teams with a low maturity and limited appropriate skills and measurement systems need even more effort, time, and failure to realize one radically new and successful market opportunity. In addition, knowing how to bring it home (see pages 90–91) is key to success, as any final prototype, MVP or MVE is only a multiplier of well-planned implementation and professional scaling activities. A well-known rule of thumb demonstrates what the impact of excellence in value creation, and implementation might have on growth and scale.

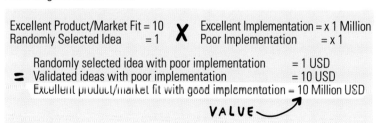

Excellent Product/Market Fit = 10 **X** Excellent Implementation = x 1 Million
Randomly Selected Idea = 1 Poor Implementation = x 1

=
Randomly selected idea with poor implementation = 1 USD
Validated ideas with poor implementation = 10 USD
Excellent product/market fit with good implementation = 10 Million USD

VALUE

To the Point!

Innovation can be measured at different levels and, regardless of how diffuse the measurement is, it is still a measurement if it provides the team or organization with more information about the phenomenon than previously known.

In any organization or team, the development of new market opportunities or business models requires investment, and at different stages throughout the design cycle, the effectiveness of the applied innovation approach might be evaluated.

All measurements, at their best, provide valuable information that can serve as an incentive for decision makers, teams, and individuals and hold them accountable.

When reliable North Star, exploration, exploitation, and creativity metrics are established, they have great influence in setting strategic direction, including deciding which problems to solve first, how to evaluate new ideas, and how to allocate resources and apply future thinking.

MEASURING TOOLKIT

1.
2.
3.

Essential Concepts in Analytics and Statistics

The measurements toolkit aims to provide a simple and easy-to-understand approach to statistics and the basic concepts of data analytics. An overview of the data analysis process is provided and basic concepts from statistics are described. Data analytics has also become more relevant with more sophisticated innovation measurement systems. The individual components are carefully introduced and illustrated with easy-to-understand examples. To keep things simple, mathematics and theorems are covered in as little depth as possible. With this introduction it should be possible to handle concepts such as outcome, event, random variable, expectation, mean, variance, and probability distribution. This toolkit introduces the world of statistics and data, which is of paramount importance for the future application of design thinking and data-driven innovation (see page 315).

Statistics and Data Analytics Can Be Broken into Five Key Types:

- Descriptive, which answers the question: **"What happened?"**
- Diagnostic, which answers the question: **"Why did this happen?"**
- Dashboard, which answers the question: **"What is happening now?"**
- Predictive, which answers the question: **"What will happen?"**
- Prescriptive, which answers the question: **"What should be done?"**

Basic concepts of statistics and data analytics are valuable tools for design thinking and innovation teams aiming to increase revenue, innovation, and product success, and retain customers/users. Each type of data analysis helps the teams reach specific goals, and they should be used in tandem to create a full picture of data that informs decision makers and innovation teams about actions to take for specific activities, capability-building programs, or process optimization. Descriptive analysis can be used on its own or as a foundation for dashboards and the other analysis types. While descriptive analysis initially provides simple and accessible ways to formulate initial minimum viable exploration metrics, the data analytics process aims to tap into more sophisticated tools for understanding the why, what, and how. Forecasting techniques in particular are becoming increasingly important for innovation and strategy work in these rapidly changing times.

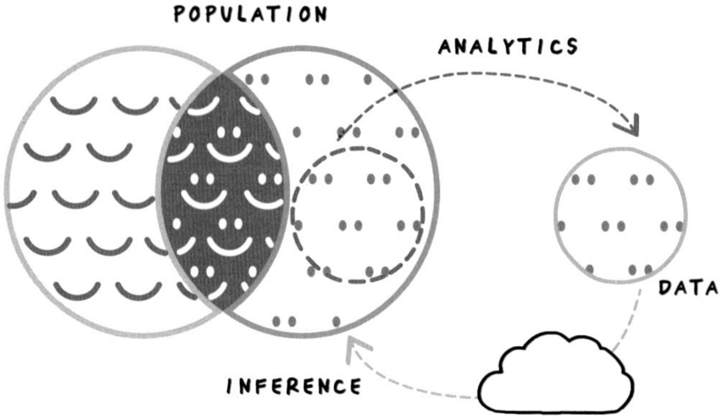
POPULATION
ANALYTICS
DATA
INFERENCE

Preview of Suitable Methods, Tools, and Processes:

Spectrum of Applied Statistics & Analytics

Toolkit

[A good understanding of statistics and data analytics concepts supports the measurement of innovation, but it also opens up new dimensions of applying the hybrid model, AI, and neurodesign tools.]

WHAT HAPPENED?

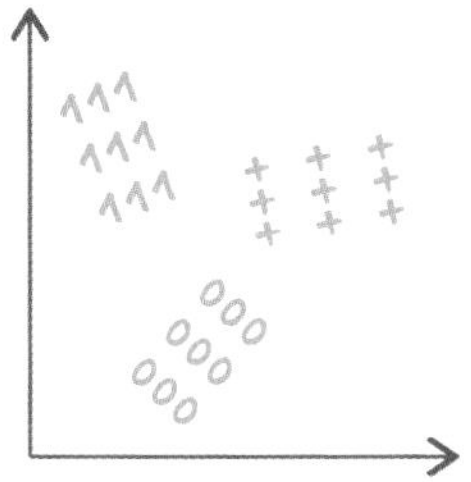

DESCRIPTIVE

- Descriptive Statistics
- Data Clustering
- Appropriate Exploitation Metrics (e.g., Financial Information, Business Performance, or Ratios)

WHY DID THIS HAPPEN?

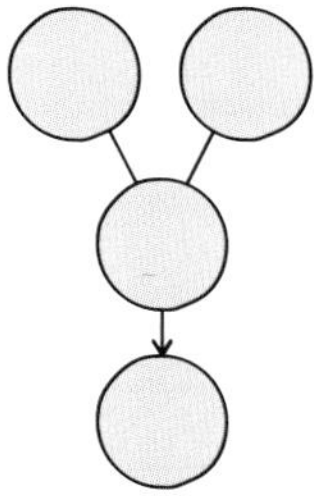

DIAGNOSTIC

- Sensitivity Analysis
- Design of Experiment

WHAT IS HAPPENING NOW?

DASHBOARD

- Strategic Dashboard
- Tactical Dashboard
- Analytical Dashboard
- Operational/Static Dashboard

WHAT WILL HAPPEN?

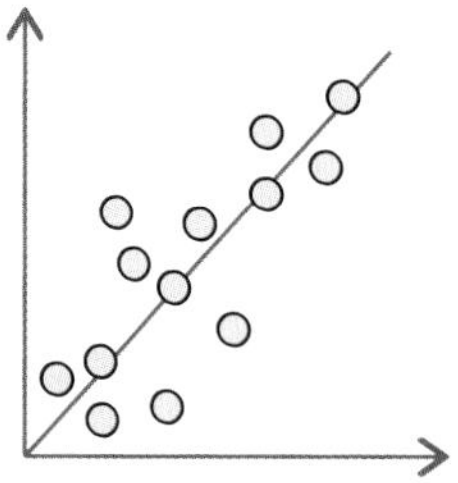

PREDICTIVE

- Linear and Logistic Regression
- Neural Networks
- Support Vector Machines

WHAT SHOULD BE DONE?

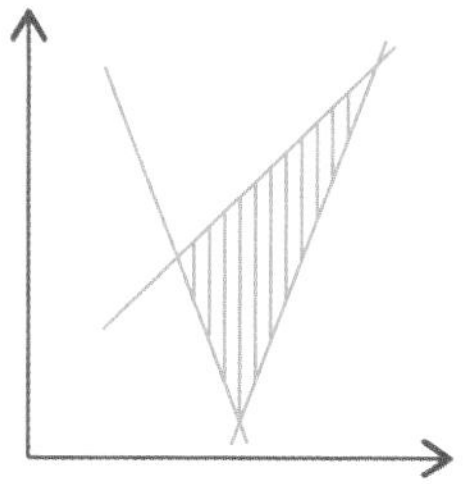

PRESCRIPTIVE

- Simulation (e.g., Monte Carlo)
- Optimization (e.g., Linear/Nonlinear Programming)

LOW — STATISTICAL & ANALYTICAL SOPHISTICATION — HIGH

The Data Ecosystem & Analytics Process

Since many design thinking and innovation teams will most likely be more concerned with data handling, data organization, and data processing than with statistical procedures, the activities in the data ecosystem are initially addressed, as are the activities of the data life cycle. The analytics process describes the path that data takes from initial collection to interpretation into actionable insights. The data ecosystem (external + internal) depends on the maturity and openness of the organization and is linked to the type of measurements and metrics performed.

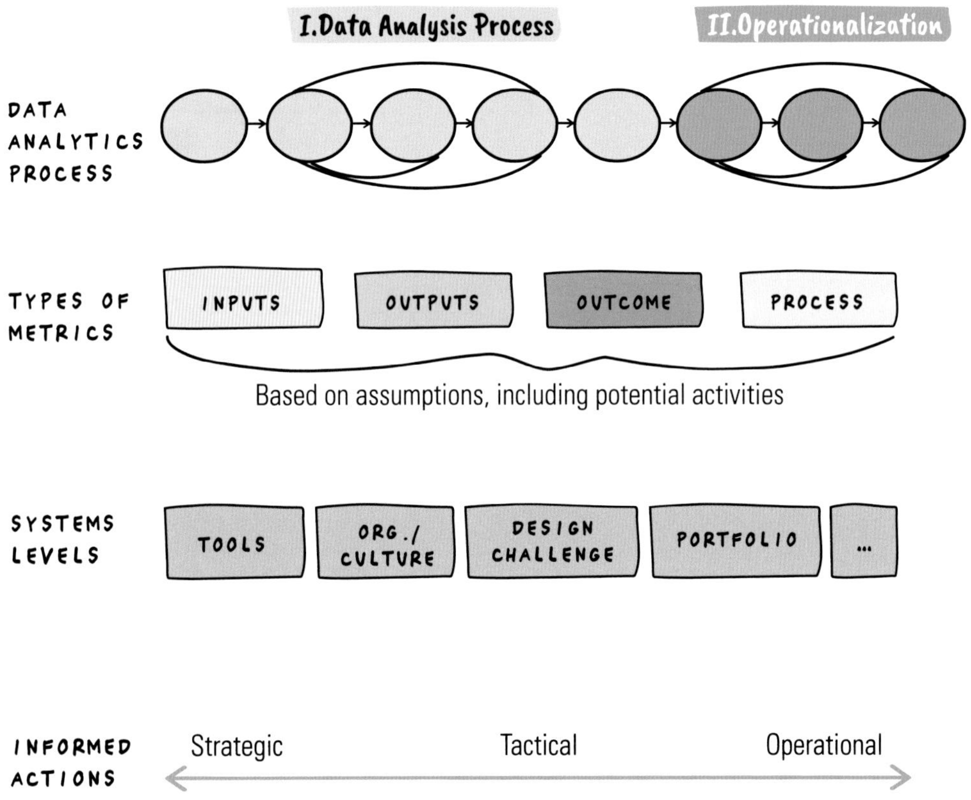
I. Data Analysis Process
II. Operationalization
DATA ANALYTICS PROCESS
TYPES OF METRICS
INPUTS
OUTPUTS
OUTCOME
PROCESS
Based on assumptions, including potential activities
SYSTEMS LEVELS
TOOLS
ORG./CULTURE
DESIGN CHALLENGE
PORTFOLIO
...
INFORMED ACTIONS
Strategic
Tactical
Operational

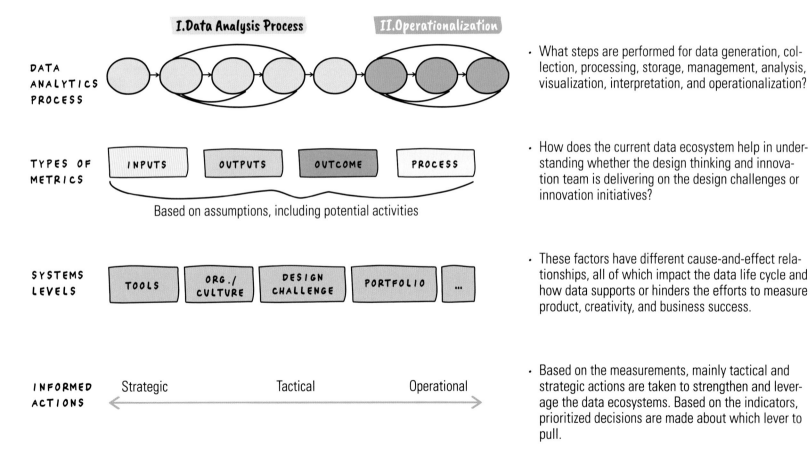

- What steps are performed for data generation, collection, processing, storage, management, analysis, visualization, interpretation, and operationalization?
- How does the current data ecosystem help in understanding whether the design thinking and innovation team is delivering on the design challenges or innovation initiatives?
- These factors have different cause-and-effect relationships, all of which impact the data life cycle and how data supports or hinders the efforts to measure product, creativity, and business success.
- Based on the measurements, mainly tactical and strategic actions are taken to strengthen and leverage the data ecosystems. Based on the indicators, prioritized decisions are made about which lever to pull.

To leverage the full potential of the data ecosystem and apply the appropriate metrics, it is recommended to form a team which is capable of asking the appropriate questions to get the right results from the measurement system. In many cases team members need to understand strategic direction, causes, and effects across the system's layers, and statistics and data analytics. The right questions will lead in most cases to the best results in applying minimum viable exploration metrics and finding the appropriate set of exploitation metrics. In addition, the design thinking and innovation team must be open to embracing and driving change for activities related to outcomes, processes, and outputs as well as changing the way they are measured. It is vital that decision makers and innovation teams are able to adapt and embed the outcomes from quantifications, measurements, and metrics to realize value over the entire design thinking cycle and beyond. The team and each of its members are key in successfully applying design thinking and innovation metrics. Within the definition of minimum viable exploration metrics and the selection of appropriate exploration metrics, it is clear that developing an innovation measurement system in a dynamic environment includes quickly delivering evidence and having a solid cycle of measurement. Similar to design sprints, the measure, evaluate, and operationalize cycle provides the tools and methods for iterating through powerful questions and first assumptions, starting simply and becoming more complex as the maturity of the team and organization evolve. Finally, the overall approach and some key measures are part of every design challenge and innovation initiative to be embedded in the IMS and innovation life cycle framework, respectively.

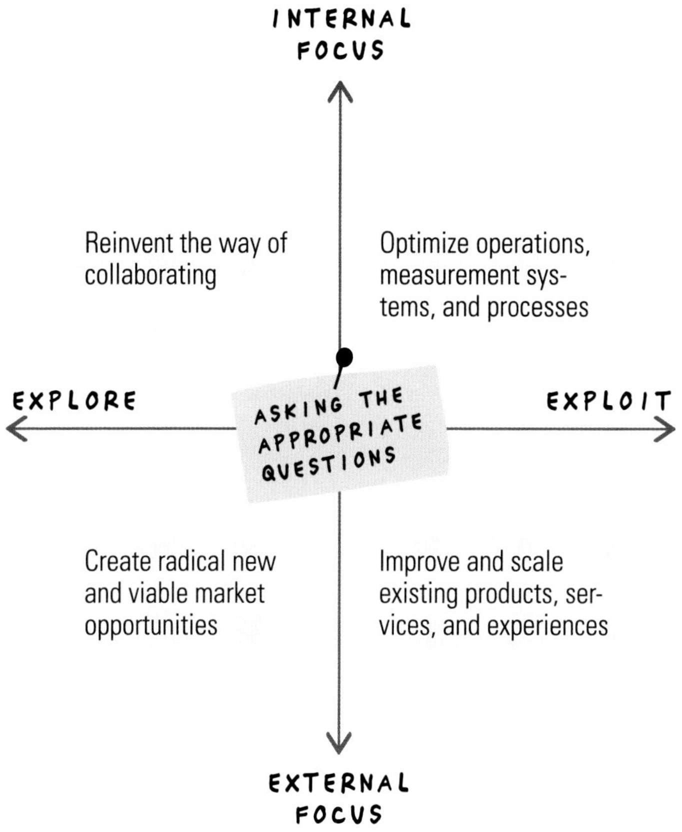
INTERNAL FOCUS
Reinvent the way of collaborating
Optimize operations, measurement systems, and processes
EXPLORE
ASKING THE APPROPRIATE QUESTIONS
EXPLOIT
Create radical new and viable market opportunities
Improve and scale existing products, services, and experiences
EXTERNAL FOCUS

Working with the Data Ecosystem and Different Metrics Types Help the Team:

- to tackle change for activities related to the innovation work and re-refinement of the measurement system.
- to understand how the appropriate questions can be embedded in the measurement system to find the right answers.
- to align activities to the overall strategic direction and to improve innovation work on various levels over the entire design cycle.

Having a powerful design thinking/innovation team in place is fundamental to asking and answering the right questions.

Overview of Data Analytics Process

From First Experiments to Business and Innovation Impact

Like any discipline, data analysis and subsequent operationalization can be a step-by-step process that helps provide guidance. Each stage requires different skills and knowledge. However, to gain meaningful insights, it is important to understand the process. An underlying framework is invaluable in producing results that stand up to scrutiny. However, working with data is also inherently messy, and the process anticipated is different for every project. Especially the iterations between designing, setting up, and conducting the experiment are not predictable. Sometimes data sets are not available, core analyses are misleading, or human error occurs in the process. The outlined process framework is a helpful reference while starting any data analyzing journey (for example, via minimum viable exploration metrics, as outlined below) or operationalizing any meaningful metrics to measure innovation and business success. In a future with predictive alignment over measurement systems and metrics, larger parts of the organization need access to sophisticated analytics tools, an understanding of the analytics process, and support in building sophisticated measurement systems step by step.

Design thinking and innovation teams need to know the analytics process to better understand and leverage the work of data scientists for innovation beyond the purpose of measuring.

I. Data Analysis Process | **II. Operationalization**

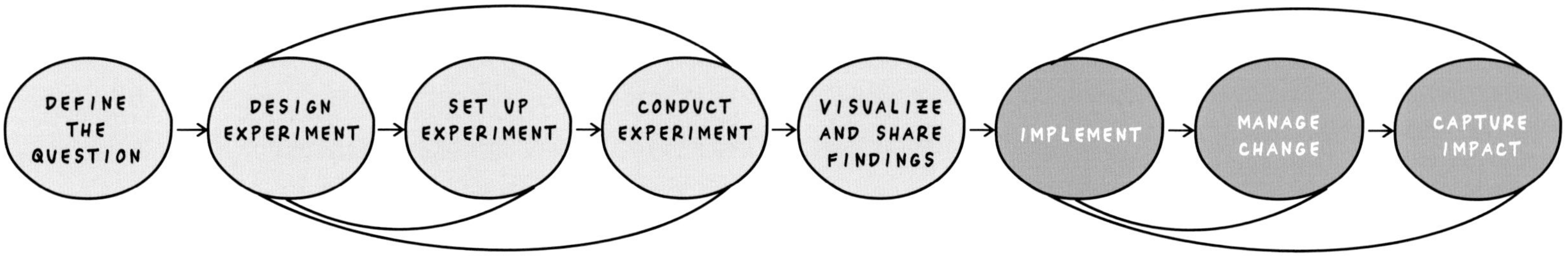
DEFINE THE QUESTION
DESIGN EXPERIMENT
SET UP EXPERIMENT
CONDUCT EXPERIMENT
VISUALIZE AND SHARE FINDINGS
IMPLEMENT
MANAGE CHANGE
CAPTURE IMPACT

Define the question
- Define and prioritize question
- Where is the problem?
- Estimate the size of opportunity

Design experiment
- Brainstorm hypotheses
- Interview user/experts/decision makers
- Observe and understand

Set up experiment
- Build, clean up, and merge data sets

Conduct experiment
- Test data sets
- Analyze, iterate, and re-define with the available sets of data

Visualize and share findings
- Present findings and align with decision makers and users of data
- Create knowledge transfer

Implement
- Enable real time analytics and establish meaningful metrics
- Incorporate metrics in decision and review processes

Manage change
- Define and implement change management procedures and communication plan

Capture impact
- Implement findings (new or adjusted processes, tools, and capabilities)
- Measure impact on innovation, business success, customer satisfaction, and so on

MVM CANVAS

MINIMUM VIABLE METRICS ! → EVALUATION → ADAPTION

I. Data Analysis Process

1. Define the Question

The first step in any data analysis process is to define the purpose of analyzing any data. As in design thinking, this is called the problem statement or as outlined in the Minimum Viable Metrics (MVM) Canvas, the problem to tackle (see page 179).

Defining the objective means coming up with a hypothesis and figuring out how to test it. Start by asking, "What business problem should be solved?" This may sound simple, but it can be harder than it seems. For example, the company leadership team might ask a question such as, "Why is there no market success with the new product?" The short and pragmatic answer would be that no one applied design thinking to understand the real customer problem. However, this may not get to the heart of the problem and would not be evidence based. The big challenge is to understand the company and its aspirations (North Star) well enough to find the appropriate problem, and also to formulate it correctly.

After defining the appropriate problem statement, the next three steps are interlinked and work best with iterations. For the phases in which you design, set up, and conduct the experiment, the business acumen of the teams is required.

2. Design Experiment

Perhaps it can be discovered through observation that the sales process works very well, and customers can be excited about the product at first, but later the customer experience leaves much to be wished for during installation. Could this be the reason that the word of this experience has spread and is keeping other customers from buying the product? From these findings, hypotheses can be derived, which are later tested with the appropriate data sources. The key question is which data sources can help answer the key questions and test the hypotheses described.

3. Set Up Experiment

In general, a large number of data sets are available. However, these are usually not directly accessible or must be collected in certain cases. Here it is essential to be aware of which data are needed for the experiment. This can be quantitative (numerical) data, such as sales figures, or qualitative (descriptive) data, such as customer feedback. All data falls into one of three categories: first party, second party, or third party data.

First-Party Data:
First-party data is collected by the company itself (e.g., from the customer). This is usually transactional data, information from the CRM system, or results from customer satisfaction surveys that are clearly structured and organized. Likewise, there may be evidence from focus groups, direct observations, and/or insights from exploration interviews.

Second-Party Data:
Second-party data enriches the data-analysis process. This is usually initial data from other companies or organizations that is either exchanged as part of a cooperation or can be acquired from a data marketplace. As a rule, such data is structured and reliable. Examples of data from secondary providers are websites, apps, or social media activities, such as online purchase histories or shipping data.

Third-Party Data:
Third-party data is collected and aggregated from numerous sources. Often (but not always) the third-party data contains a large number of unstructured data points (big data). Many companies and organizations collect big data to create points of view or to conduct market research (e.g., big 4 consulting companies, Gartner). In addition, many governments, countries, and cities have open data initiatives, such as Open Data Singapore, where data can be found on specific events.

Once the appropriate data sets are available, they must be cleaned. This is crucial to ensure that the data is of high quality. The cleaning is time consuming and usually represents the largest work package for the team. However, this preliminary work is extremely important as it will have an immense impact on the results later on.

Typical activities during data cleaning:

- Bringing structure to the data: Especially typos or layout problems should be corrected, which makes the assignment and later processing of the data easier.
- Fill in gaps in data series: If important data is missing, it should be filled and completed accordingly.
- Clean up major errors, duplicates, and outliers: Especially when merging data from numerous sources, appropriate attention must be paid here.
- Remove unwanted data points: Observations that are not relevant to the planned analysis should be extracted.

4. Conduct Experiment

Preliminary findings, however, can only be generated by conducting an exploratory analysis. Conducting data experiments helps to identify initial characteristics and trends, and to iteratively develop the hypotheses. Back to the example, if it is found that the customers are more likely to talk badly about the price instead of experience, there might be a correlation between price and customer satisfaction, and perhaps the initial hypothesis presents less of a problem than the market price for the newly launched product.

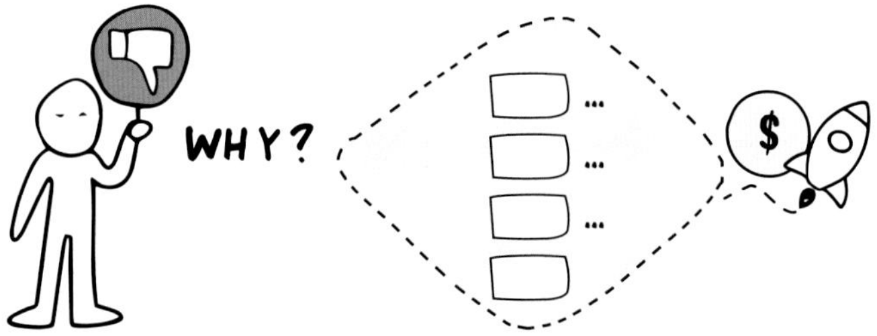

WHY?
$

5 Tips for Data Cleaning

- ✓ Cleaning Your Data Isn't Rocket Science.
- ✓ Bad Data = Bad Experiences.
- ✓ Good Data = Greater Efficiency.
- ✓ Duplicate Data Can Hurt the Bottom Line.
- ✓ Dirty Data Destroys Trust.

The type of data analysis to be performed depends largely on what the objectives are. As described in the second part of this toolkit on page 238, there are different techniques for this purpose. Univariate or bivariate analysis, time series analysis, and regression analysis are techniques that are relevant for processing the data. For most of the innovation and design thinking teams, it might be more important to understand how to apply them.

As a rule of thumb, all types of data analysis fall into one of the following four categories: A, B, C, or D

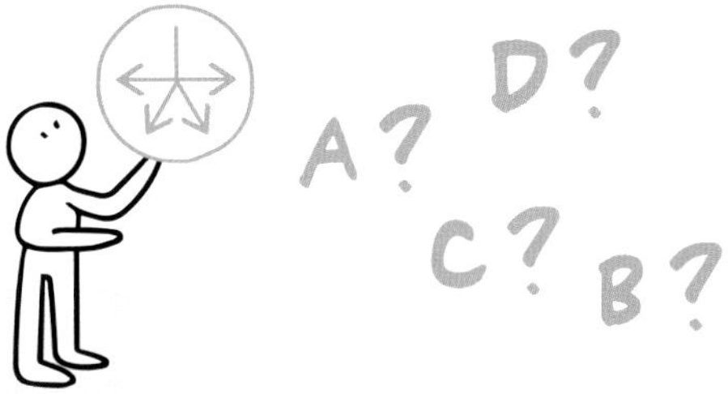
A?
D?
C?
B?

A) Descriptive Analysis:

The purpose here is to understand why something happened. It is literally diagnosing a problem, just as a physician uses a patient's symptoms to diagnose a disease. Back to the example: What factors are negatively impacting the customer experience? A diagnostic analysis would help answer that question. For example, it could help to establish correlations between the problem (difficulty in attracting repeat customers) and the factors that might be responsible (e.g., methodological approach for innovating, project costs, speed of delivery, and quality of personas).

B) Diagnostic Analysis:

Diagnostic analysis could be used, for example, to determine that retail customers are churning faster than other customers. This could indicate that the team or organization is losing customers because they lack expertise in that area. That would be a very useful finding!

C) Predictive Analysis:

The objective here is to detect future trends based on historical data. In business, predictive analysis is often used to forecast, for example, future growth. But that's not all. Predictive analytics has become increasingly sophisticated in recent years. The rapid development of machine learning allows companies to make surprisingly accurate predictions. Smartphone manufacturers typically use past data to predict when the optimal time is to launch a new smartphone. Similarly, online marketplaces use transaction data to predict future trends or determine the volume on a Black Friday sale to tailor their strategies. These simple examples show the often-untapped potential that predictive analytics can bring to innovation teams and the wider organization.

D) Prescriptive Analysis:

The value of prescriptive analysis lies in the fact that recommendations can be made for the future. This is the last step in the analytical part of the process and at the same time the most complex. This is because it involves aspects of all the other analyses described. A good example of prescriptive analytics are the algorithms that will drive the next generation of autonomous self-driving cars. In fractions of a second, these algorithms make myriad decisions based on past and current data as well as a defined value system to ensure a smooth and safe ride. Prescriptive analytics also helps design thinking and innovation teams decide on new products, services, business models, or the design of entire business ecosystems to be developed or to participate in.

5. Visualize and Share Data

After the analysis is complete and the findings are available, there is another important step that is often underestimated in its importance. The result needs to be shared with the extended teams and decision makers. However, this is more complex than just sharing the processed data; it's about interpreting the findings and presenting them in a way that is understandable to all relevant stakeholders. Since the information is often also shared with design thinking teams and decision makers, the findings that are presented must be 100 percent clear and unambiguous. For this reason, it is recommended to use dashboards and interactive visualizations that meaningfully convey the key messages (see dashboard examples on pages 254–255).

The way the results are interpreted and presented often influences the direction of decisions to make strategic thrusts. Depending on what is communicated, a whole company, organization, or innovation team might decide to restructure, launch a high-risk product, or even redesign the entire innovation process and measurement system. That's why it's very important that all the evidence collected is presented, not just selected data points. If everything is presented in a clear and concise way, it is very easy to see that the conclusions drawn are scientifically sound and based on facts. On the other hand, it is also important to highlight gaps in the data or to point out findings that may be open to interpretation. Communicating honestly is the most important part in this last step of the initiated data analytics process.

Beyond Analyzing Data

Some of the outcomes will have a short lifespan as the initial measurement goal is achieved. Other analytics/metrics will turn out to be very useful and will be operationalized. After evaluating the value, the next major part of the data analytics process is the operationalization of the data model. However, these days this is more than just creating and deploying analytics and data science models for specific use cases. It also involves managing, monitoring, and refining the models to keep them meaningful, relevant, and useful. Operationalizing analytics models will greatly increase the value of the business and innovation teams over time and should be part of any data analytics process—from first experiments to business and innovation impact.

Tools for Analyzing, Interpreting, and Sharing Findings

Tools to Define Objectives/KPI Dashboards:
- Dashbox
- DashThias
- Open-source software (e.g., Grafana, Freeborad, Dashbuilder)

Data Collection Tools:
- Data management platforms (DMPs, as those from Salesforce, SAS, Xplenty, and open-source platforms from Pimcor, D:Swarm)

Data Cleaning Tools:
- OpenRefinde (open-source tool)
- Python libraries/R Packages: e.g., Pandas
- Data Ladder (professional tool)

Data Visualization Tools:

Visualizations without coding:
- Goolge Charts
- Tableau
- Datawrapper
- Infogram
- Power BIQlik

Realted to Python and R:
- Plotly
- Seaborn
- Matplotlib

II: Operationalization

Operationalizing the data is the second part of the process of creating real value. From a distance, operationalization seems simple: bring artificial intelligence (AI) into production or operational environments, and the magic happens by itself.

In practice, however, operationalizing such models can be challenging, requiring further iterations and constant adjustments. Live data to rapidly growing data sets that change quickly need the appropriate attention. For example, if the model is to predict customer churn on changing trends, as in the example used, the operationalization process must be integrated with the CRM system to effectively predict churn.

1. Implement

Once the models and intimate measurements have been developed, refined, and visualized, another iterative process of operationalizing the models to account for business changes begins. Decision makers and teams need robust, automated feedback mechanisms for meaningful measurements and indicators, as well as tools and processes for model management on the fly.

The model pipeline requires code development similar to, but different from, that already described in the data analytics process. In operationalization, for example, SQL code is written to find the best model. This code must initially be based on a snapshot of the data. It is recommended to use so-called scoring engines in production so that they can evaluate the best models.

The data analysis and operationalization team for production monitors the performance of all models and evaluates population drift model degradation and user acceptance. It is also recommended that randomized control tests be planned, scheduled, and run to determine model lift so that accurate reports on model performance and return on investment can be generated.

Variability of Integration:
Integrating data and scoring pipelines with the multitude of systems requires a lot of integration work that is time consuming and very technical.

Model Monitoring:
The accuracy of the metrics and model predictions must be constantly monitored, and the models must be adjusted and retrained over time as the data changes, both of which are very time consuming.

Ability to Scale:
Initial models often rely on a smaller subset of the total available data set. For example, in a churn model, the models may be developed on less than 50 percent of the available data, but in production, the models must be scaled to handle 100 percent of the available customer data to predict, for example, churn.

Portability to Other Systems:
Often, tools used in the data analytics phase are significantly different from production environments. This means that adopting models and operationalizing them requires porting the code to platforms and systems that were not considered during model development.

Moving the code to production means that a fair amount of rework has to take place to recode the models (based on R/Python) using, for example, SQL code that is native to the production database.

2. Manage Change

Operationalization of processes and related communications includes rapid prototyping of models with dedicated teams, departments, or relevant stakeholders in the organization, integration of data with operational systems, and the ability to take prescriptive action with minimal human intervention.

Best practices include maintaining a strong security posture and applying it to new areas, establishing infrastructure as code and platform as a service where appropriate, and measuring results to make necessary improvements.

3. Capture Impact

The last capability involves creating metrics to measure impact on innovation, business success, and customer satisfaction. Meaningful metrics should relate to outcomes, and other measures of innovation and business success as described previously in this book and the respective levels.

Best practices include building and improving a program for affected teams and conducting field tests and experimentation to fail fast and to learn quickly.

> Maximize the value of data analytics and machine learning models in an effort to reach IMS level 2 and higher.

Finally, with the capabilities described, companies, organizations, and teams can increase their agility and, for example, implement DataOps, DevOps, and MLOps, create business cases that demonstrate value, enforce governance, create role definitions and assignments, and complete portfolio prioritization (EXPLOIT and EXPLORE). Data, analytics tools, and meaningful metrics form the foundation for advanced analytics applications to support business and innovation goals.

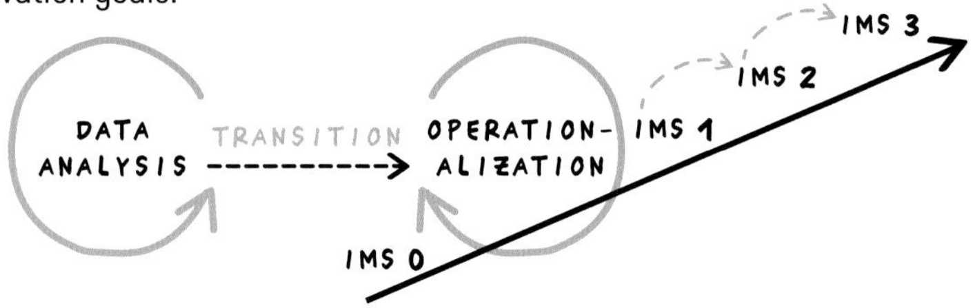
DATA ANALYSIS
TRANSITION
OPERATION-ALIZATION
IMS 0
IMS 1
IMS 2
IMS 3

Tools for Operationalizing Data Models

- ✔ Tools for business process integration
- ✔ Software automation tools
- ✔ Tools for predictive insights
- ✔ Anomaly detection tools
- ✔ Intelligent data fabric
- ✔ Flexible AI models
- ✔ Pre-built AI Services
  - Watson Studio (Build)
  - Watson ML (Run)
  - Watson OpenScale (Manage Metrics)

Dashboards

Dashboards are a great way to summarize the most important metrics for a team or the entire organization. Most of the off-the-shelf strategic dashboards provide a constantly available health check and status of company performance based on standard measurements and metrics. However, very often these dashboards do not reveal why the numbers are a certain way because this is not the purpose of such a dashboard. The dashboard design aims to constantly monitor mainly a set of exploitation metrics and to present key company metrics at a glance to investors, shareholders, and other stakeholders.

Minimum viable exploration metrics are mostly used on a deeper level. They provide a clear and precise overview of the current situation for the company, the innovation team, or a specific design challenge, or they are linked to activities on the tool level.

Over time, however, metrics develop from those that prove useful and are used recurrently. But again, less is more. From time to time, companies experience true dashboard fever. Often during this time, too many and sometimes countless metrics are designed, which often linger in the systems for a long time. Therefore, it is indispensable to question at regular intervals whether the dashboards still serve their purpose and, if necessary, to assign an expiration date to the metrics because at a certain point other things come into focus. This is especially the case when measuring activities relates to change or building design thinking capabilities based on the design thinking maturity model.

The visualization for simple measurements at the tool level and for metrics with an expiration date often just needs dashboards that are good enough to plan the next steps.

Out-of-the-box innovation measurement tools and strategic dashboards provide mainly standard metrics and very often vanity metrics to provide shiny readings of performance on innovation activities.

Off-the-Shelf Dashboards

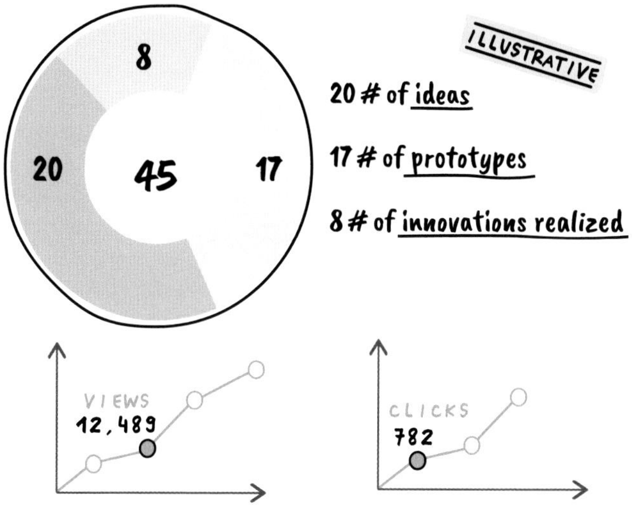

ILLUSTRATIVE
8
20
45
17
20 # of ideas
17 # of prototypes
8 # of innovations realized
VIEWS
12,489
CLICKS
782

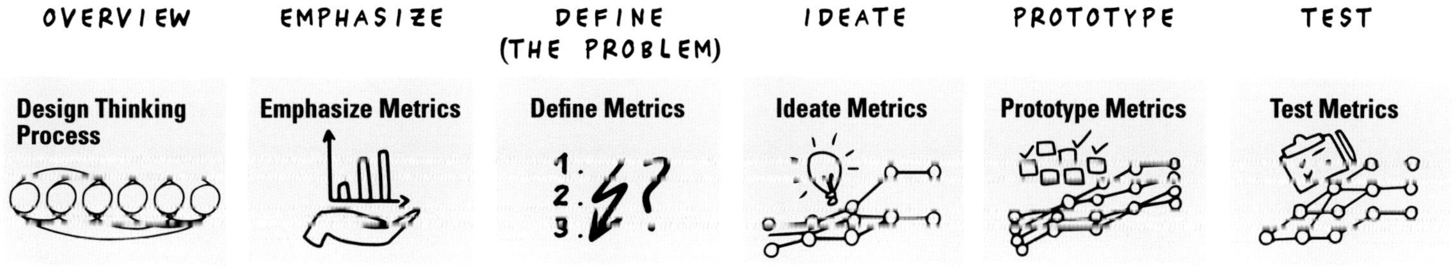

OVERVIEW
EMPHASIZE
DEFINE (THE PROBLEM)
IDEATE
PROTOTYPE
TEST
Design Thinking Process
Emphasize Metrics
Define Metrics
Ideate Metrics
Prototype Metrics
Test Metrics
1.
2.
3.

How to (Re)design Dashboards

If not depending on templates from off-the-shelf tools, the same considerations should be made when designing and arranging metrics on a dashboard as when placing digital content on a web page.

Key Questions for Dashboard Design are:

- What story should be told based on the data, measurements, and metrics, and based on the interests of the decision makers and the innovation teams using them?
- Does the design hierarchy of the dashboard reflect these intentions?

A good way to begin is to base the design of a dashboard on what the intended audience is paying attention to. If it is known that a particular metric should be particularly eye catching, and a columnar dashboard has been created, then that metric should not be placed on the right or bottom of the dashboard. Using eye-tracking techniques, it was possible to find out where users usually look first on a dashboard.

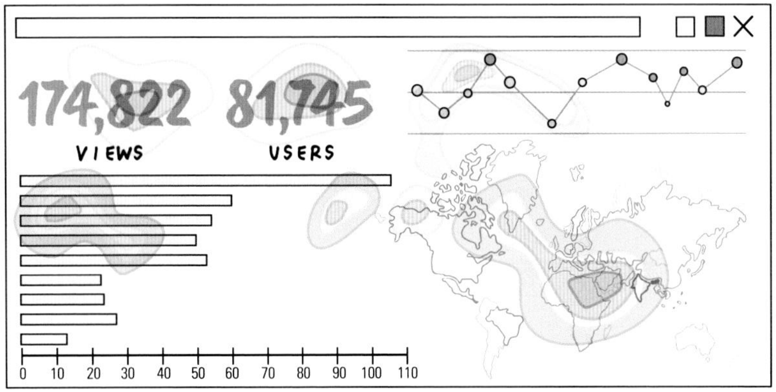
174,822
81,745
VIEWS
USERS
0 10 20 30 40 50 60 70 80 90 100 110

When big numbers are shown on a screen with another noticeable element, like a map, distributed attention is observed. Most of the team members and decision makers look at the big number and the map.

In general, the same patterns are observed for dashboards as for text-based web pages. For example, the F-pattern, which is defined as the propensity for humans to consume web pages starting in the upper-left and moving to the bottom-right, is also observed for dashboards.

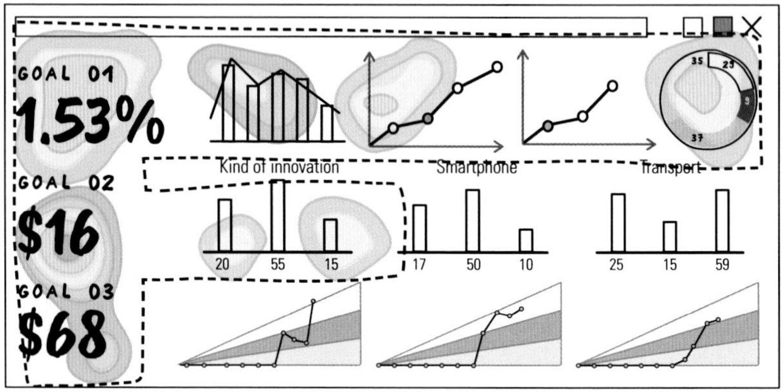
GOAL 01
1.53%
GOAL 02
$16
GOAL 03
$68
Kind of innovation
Smartphone
Transport
20 55 15
17 50 10
25 15 59

From various dashboard experiments conducted, it was possible to determine that innovation teams and decision makers tend to spend more time looking at the top portion of a dashboard. The application of the fixation duration metric allows, for example, counting of the total amount of time all team members and decision makers looked at a certain area.

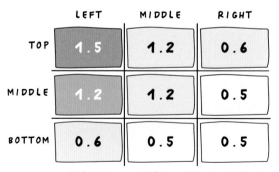

	LEFT	MIDDLE	RIGHT
TOP	1.5	1.2	0.6
MIDDLE	1.2	1.2	0.5
BOTTOM	0.6	0.5	0.5

All Participants: 10-Second Viewing Period

STATISTICS AND MEASUREMENTS TOOLS

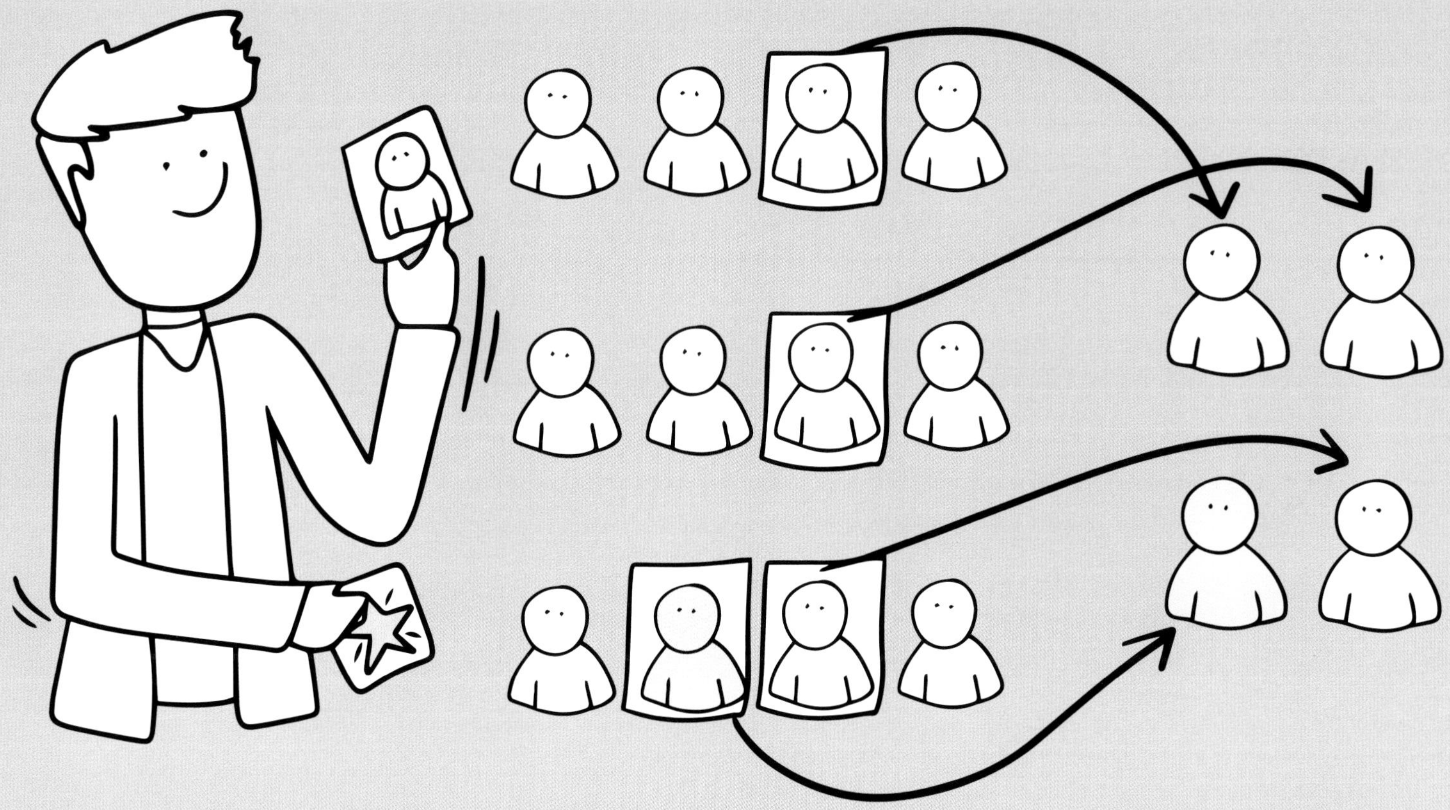

Levels of Measurement

In any type of data collection or analysis, it is important to understand the type of data being worked with. There are different variables in a data set, and these variables can be measured with different levels of precision. This is known as a level of measurement. There are four main levels of measurement: nominal, ordinal, interval, and ratio:

	NOMINAL	ORDINAL	INTERVAL	RATIO
CATEGORIES AND LABELS VARIABLES	✓	✓	✓	✓
RANK CATEGORIES IN ORDER		✓	✓	✓
HAS KNOWNS, EQUAL INTERVALS			✓	✓
HAS A TRUE OR MEANINGFUL ZERO				✓

The level of measurement refers to how precisely a variable has been measured.

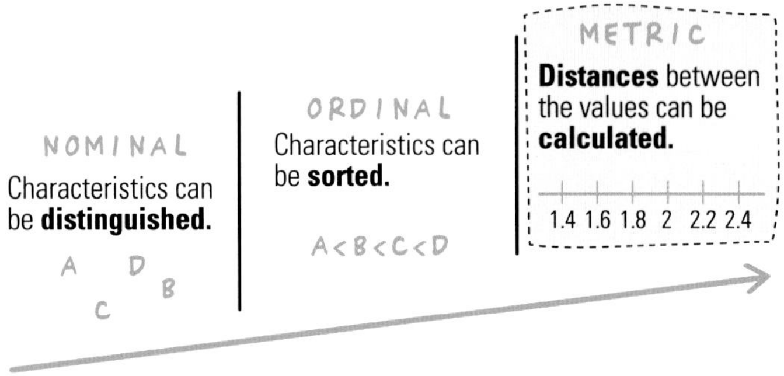
NOMINAL
Characteristics can be **distinguished.**
A D B C
ORDINAL
Characteristics can be **sorted.**
A<B<C<D
METRIC
Distances between the values can be **calculated.**
1.4 1.6 1.8 2 2.2 2.4

The level of measurement is important because it determines the type of statistical analysis that can be performed. Consequently, it affects both the type and depth of insight that can be gained from the data in question. Certain statistical tests can only be performed if more precise measures have been used. Therefore, it is important to plan in advance how the data will be collected and measured.

The Four Levels of Measurement:

1. The nominal scale simply categorizes variables according to qualitative labels (or names). These labels and groupings don't have any order or hierarchy to them, nor do they convey any numerical value. For example, the variable "Kind of Innovation" could be measured on a nominal scale according to the following categories: process innovation, service innovation, product innovation, and so on.

2. The ordinal scale also categorizes variables into labeled groups, and these categories have an order or hierarchy to them. For example, you could measure the variable "innovativeness" on an ordinal scale as follows: low innovativeness, medium innovativeness, or high innovativeness.

3. The interval scale is a numerical scale which labels and orders variables, with a known, evenly spaced interval between each of the values. An example of interval data is insights collected by the teams, where the difference between 10 and 20 insights is the same as the difference between, say, 50 and 60 insights.

4. The ratio scale is exactly the same as the interval scale, with one key difference: The ratio scale has what's known as a "true zero." A good example of ratio data is the absolute number of ideas. If there is no idea, it truly is 0 ideas—compared to the change of innovation culture (interval data), where a value of zero degrees doesn't mean there is no innovation culture; it simply means that it has not changed to a climate which might be better to realize innovation!

Example

Nominal Data

Nominal data divides variables into mutually exclusive, labeled categories.

EXAMPLE

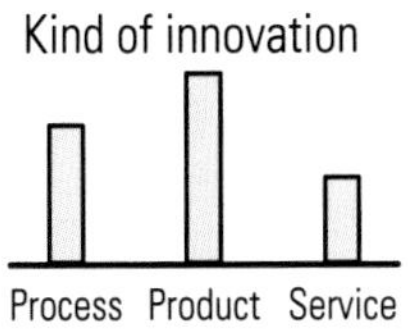
Kind of innovation
Process
Product
Service

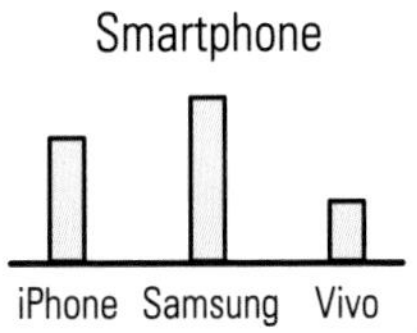
Smartphone
iPhone
Samsung
Vivo

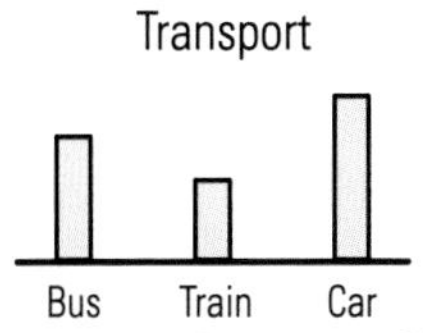
Transport
Bus
Train
Car

HOW IS NOMINAL DATA ANALYZED?

Descriptive statistic: Frequency distribution and mode

Non-parametric: Statistical tests

Ordinal Data

Ordinal data divides and classifies into categories that have a natural order or rank.

EXAMPLE

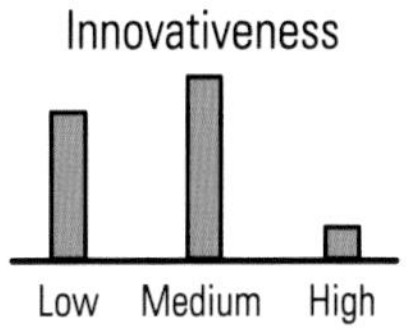
Innovativeness
Low
Medium
High

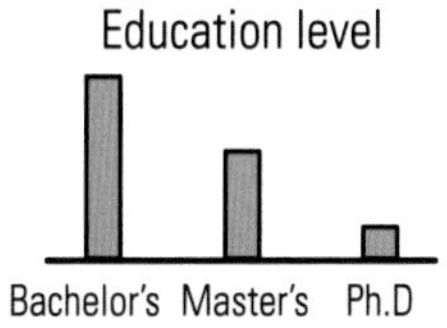
Education level
Bachelor's
Master's
Ph.D

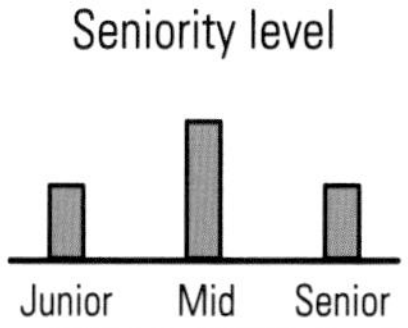
Seniority level
Junior
Mid
Senior

HOW IS ORDINAL DATA ANALYZED?

Descriptive statistic: Frequency distribution, mode, median, and range

Non-parametric: Statistical tests

Interval Data

Interval data is measured along a numerical scale that has equal intervals between adjacent values.

EXAMPLE

Team confidence
90%
80%
50%

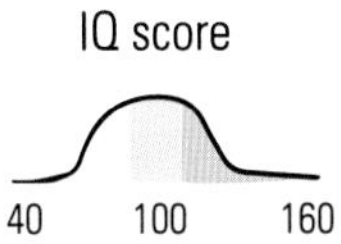
IQ score
40
100
160

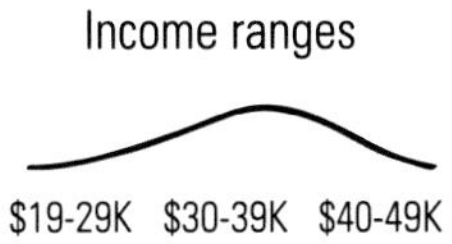
Income ranges
$19-29K
$30-39K
$40-49K

HOW IS INTERVAL DATA ANALYZED?

Descriptive statistic: Frequency distribution; mode, median, and mean; range, standard deviation, and variance

Parametric Statistical tests (e.g., t-test, linear regression)

Ratio Data

Ratio data is measured along a numerical scale that has distance between adjacent values and a true zero.

EXAMPLE

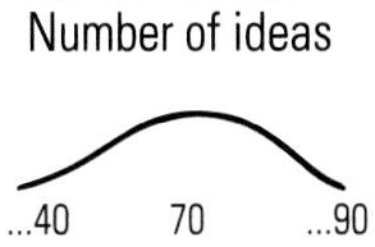
Number of ideas
...40
70
...90

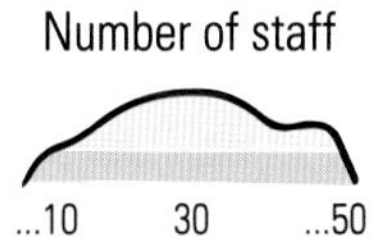
Number of staff
...10
30
...50

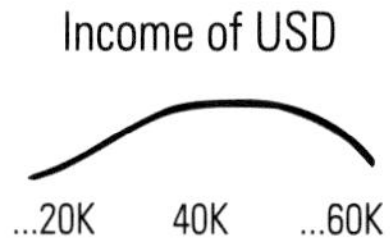
Income of USD
...20K
40K
...60K

HOW IS RATIO DATA ANALYZED?

Descriptive statistic: Frequency distribution; mode, median, and mean; range, standard deviation, variance, and coefficient of variance

Parametric Statistical tests (e.g. ANOVA, linear regression)

Data Collection

Without data there will be no measurement, and therefore it has to be considered and planned whether the necessary data for a measurement is available or must be collected. For the respective decision makers and teams, this means that data collection has to be carefully considered and planned. Main methods of quantitative data collection are based on first-, second-, and third-party data (see page 229). For measuring the complexity of innovation, it often is necessary to incorporate several data collection methods. Secondary data should not be ignored as it often is an accessible source of reliable and carefully collected and compiled data. In general, there are two approaches to sampling that can be considered: non-probability and probability sampling. If the intention is to make the samples representative of the population, probability sampling is recommended for the work with measurements and metrics because it allows inferences to be made from the sample to the population, and representativeness can be assessed using population measures.

A Well-Defined Data Collection Process Helps the Team:

- to identify a business, innovation, or process issue that needs to be addressed, and set goals for the design of a measurement activity or the development of metrics.
- to gather data requirements to answer the key question for the targeted outcomes.
- to identify the data sets that can provide the desired information.
- to set a plan for collecting the data, including the collection methods that will be used.
- to collect the available data and to start design and test a minimum of viable exploration metrics.

Data collection matters, because 80 percent of data cleaning problems are caused by poor data-gathering techniques.

Common Methods in Data Collection

- **Surveys and Questionnaires**
- **Interviews**
- **Direct Observation**
- **Focus Groups**
- **Existing Documents and Records**

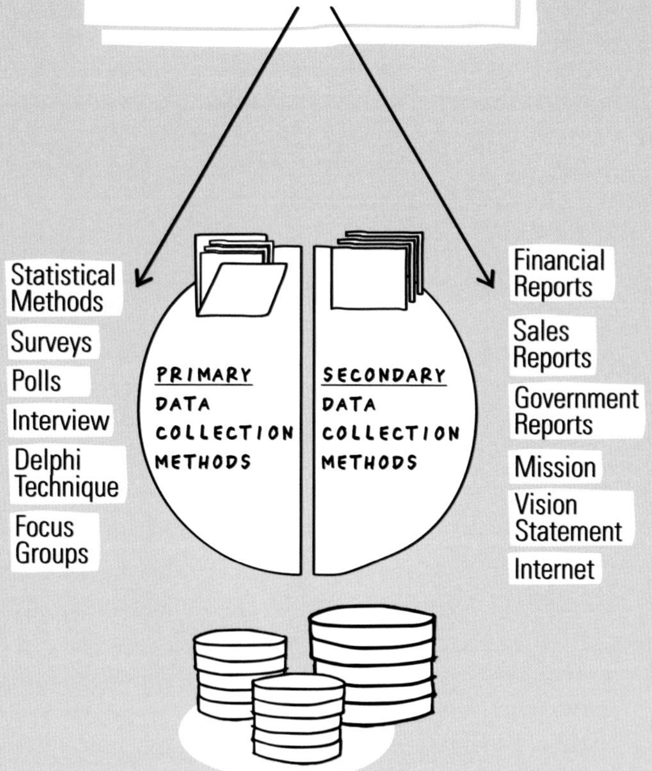
Statistical Methods
Surveys
Polls
Interview
Delphi Technique
Focus Groups
PRIMARY DATA COLLECTION METHODS
SECONDARY DATA COLLECTION METHODS
Financial Reports
Sales Reports
Government Reports
Mission
Vision Statement
Internet

Descriptive versus Inferential Statistics

Example

In the second part of the 101 toolkit, the basic principles of statistics are presented along with a simple measurement example with 286 datapoints indicating the type of innovations (e.g., process, product, service innovation, etc.). Especially the first examples show that the starting point with statistics is basically the collection, organization, analysis, interpretation, and presentation of data.

Distribution:

Data set: **286** innovations

PROCESS INNOVATIONS	130
PRODUCT INNOVATIONS	39
SERVICE INNOVATIONS	91
BUSINESS MODEL INNOVATIONS	13
BUSINESS ECOSYSTEM INNOVATIONS	13

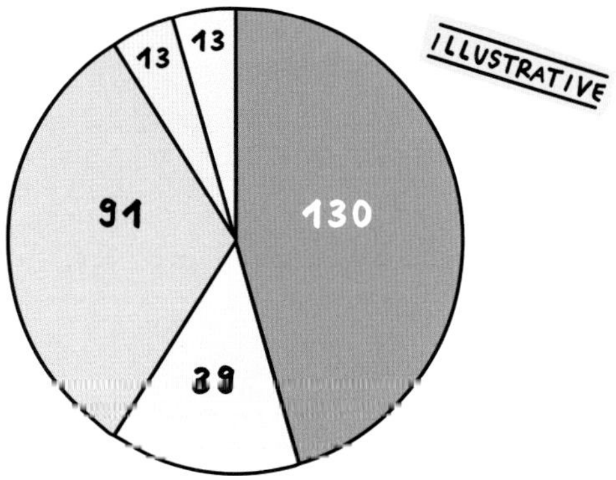

13
13
91
130
39
ILLUSTRATIVE

Descriptive statistics focus on describing the visible characteristics of a data set (a population or sample). Meanwhile, inferential statistics (see page 243) focus on making predictions or generalizations about a larger data set, based on a sample of those data. These are examples of applied descriptive statistics:

Central Tendency:

Common measures of central tendency include:

- **The mean:** The average value of all the data points.
- **The median:** The central or middle value in the data set.
- **The mode:** The value that appears most often in the data set.

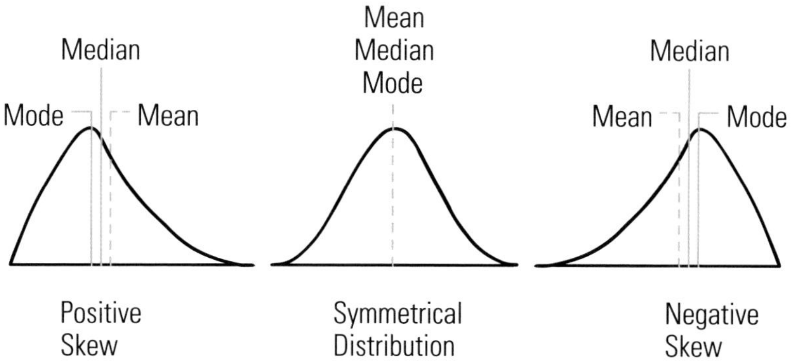

Median
Mode
Mean
Positive Skew
Mean
Median
Mode
Symmetrical Distribution
Median
Mean
Mode
Negative Skew

Once again, using our kind of innovation example, we can determine that the mean measurement is 57.2 (the total value of all the measurements, divided by the number of values), the median is 39 (the central value), and the mode is 13 (because it appears twice, which is more than any of the other data points).

Variability:

How values are distributed or spread out.

Standard Derivation:

For example, high standard deviation suggests that the values are more broadly spread out.

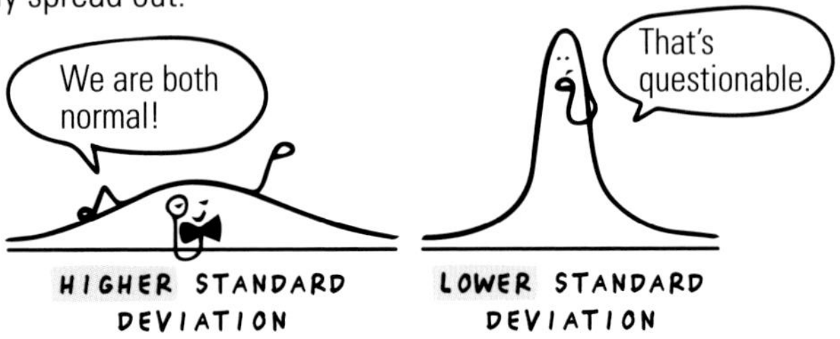
We are both normal!
That's questionable.
HIGHER STANDARD DEVIATION
LOWER STANDARD DEVIATION

Minimum and Maximum Values:

For example, in the case of innovations, the minimum and maximum values are 13 and 130, respectively.

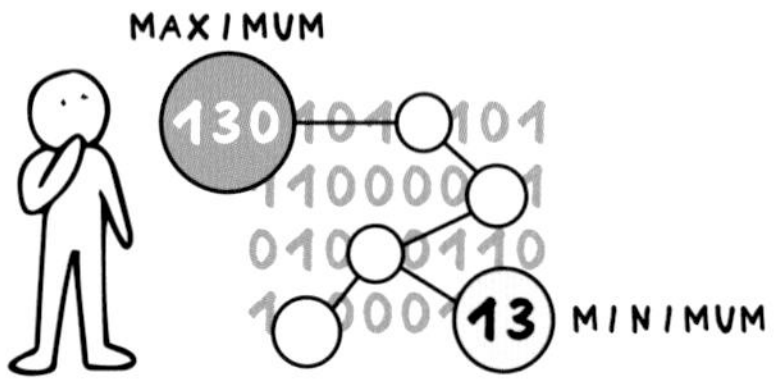
MAXIMUM
130
13
MINIMUM

Range:

For example, measures the size of distribution value which is in the case of the innovation example **130 – 13 = 117**.

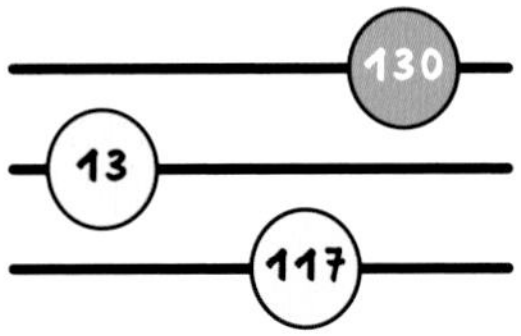
130
13
117

Kurtosis:

Measures the tails of a given distribution containing extreme values (many outliers are indicated as high kurtosis).

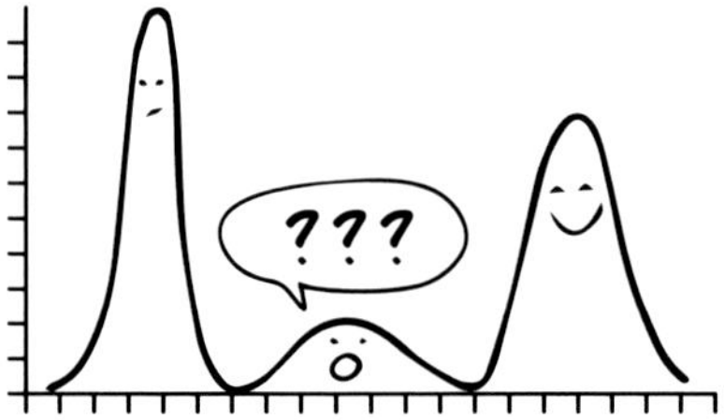
???

Skewness:

Measures the symmetry of data sets (long left-hand tail = negative skewness, and long right-hand tails = positive skewness).

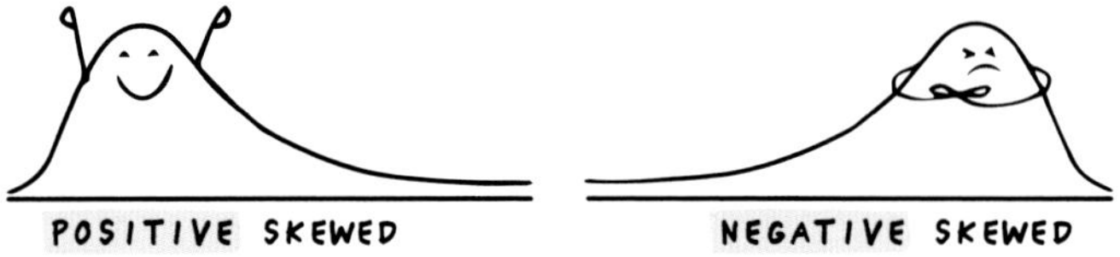
POSITIVE SKEWED
NEGATIVE SKEWED

Population & Samples in Statistics:

Two basic, but vital, concepts in statistics are those of population and sample:

- Population is the entire group to draw data from (and subsequently draw conclusions about). While in day-to-day life, the word is often used to describe groups of people (such as the population of a country), in statistics it can apply to any group from which information will be collected. This is often people in general, but it could also be customers, teams, objects, departments, and so on.

- A sample is a representative group of a larger population. Random sampling from representative groups allows organizations to draw broad conclusions about an overall population. This approach is commonly used in polling. Market research institutes, for example, ask a small group of decision makers about their views on certain topics. They can then use this information to make informed judgments about what the larger population thinks. This saves time, hassle, and the expense of extracting data from an entire population (which for all practical purposes is usually impossible).

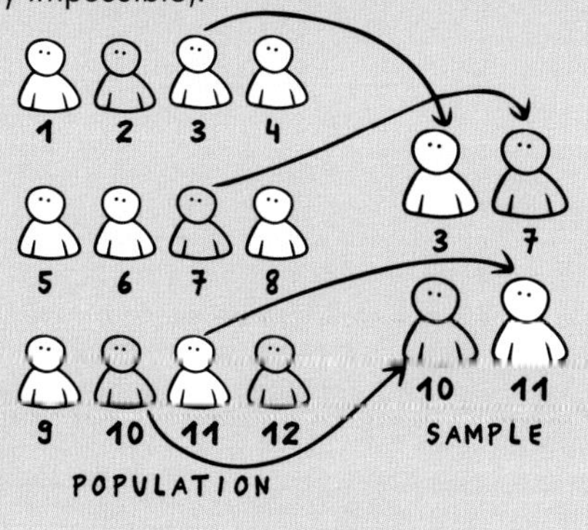
1 2 3 4
5 6 7 8
9 10 11 12
POPULATION
3 7
10 11
SAMPLE

Inference and Hypothesis Testing

While descriptive statistics basically summarize the most important characteristics of a data set, inferential statistics are concerned with making generalizations about a larger population based on a representative sample. Since inferential statistics focus on predictions (rather than stating facts), results are usually stated in terms of probabilities. However, descriptive and inferential statistics are often used in conjunction. Together, these powerful statistical techniques are the foundational basis on which data analytics is built.

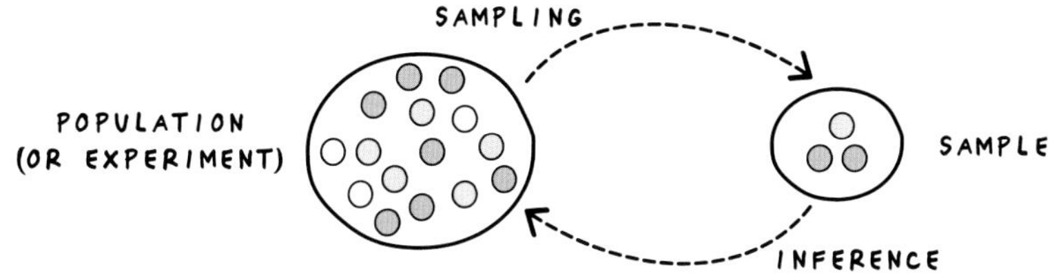
SAMPLING
POPULATION
(OR EXPERIMENT)
SAMPLE
INFERENCE

Starting with a Random Sample:

Random sampling can be a compound process and often depends on the particular characteristics of a population. However, the fundamental principles involve the definition of the population, decision of the sample size (big sample size = more representive results), randomly selected selection of a sample (e.g., applying random number generator or a defined algorithm), and analyzing the data sample. (Note: A random sample is representative of a population. It will never be 100 percent accurate and, therefore, it is important to incorporate a margin of error.)

Hypothesis | Null Hypothesis:

The Hypothesis Is a Test of Two Concepts:

1. the null hypothesis that there is no significant effect or relation, and
2. the alternative hypothesis that there is a significant effect or relation.

The null hypothesis is assumed unless evidence can be given with a high likelihood of being true that the evidence from the data supports the alternative hypothesis. Likelihood is assessed as the probability that the test statistic is different from zero with a preset degree of confidence.

Examples of Applied Inferential Techniques

Hypothesis Testing:

The formulation and test of a hypothesis shows if the sample repeats the results of the proposed explanation (the hypothesis). The general objective is to exclude the possibility that a particular result occurred by chance. An example would be the formulation of an alternative hypothesis development (H1, H2, H3, H4) on the effects of innovation on a firm's performance based on a conceptual framework.

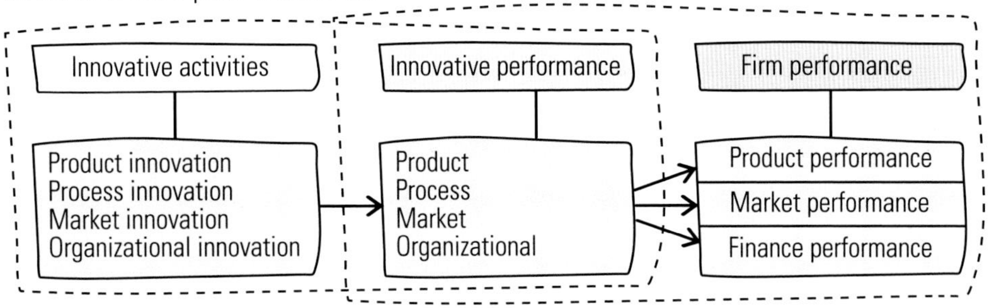

Innovative activities
Product innovation
Process innovation
Market innovation
Organizational innovation
Innovative performance
Product
Process
Market
Organizational
Firm performance
Product performance
Market performance
Finance performance

Hypothesis H1: The higher the level of innovation activities is, the greater the innovative performance is.

Hypothesis H2: The greater the innovative performance is, the greater the production performance is.

Hypothesis H3: The higher the level of innovative performance is, the greater the market performance improvement is.

Hypothesis H4: The higher the level of innovative performance is, the greater the finance performance improvement is.

Regression & Correlation Analytics:

Regression and correlation analysis are both techniques used for observing how two (or more) sets of variables relate to one another. Regression analysis is used to determine how a dependent (or output) variable is affected by one or more independent (or input) variables. It is often used for hypothesis testing and predictive analysis.

To predict future innovation success (an output variable), you might compare the recent capabilities at companies and innovativeness (both input variables) to determine how innovation success might happen in companies with a certain set of capabilities.

Confidence Intervals:

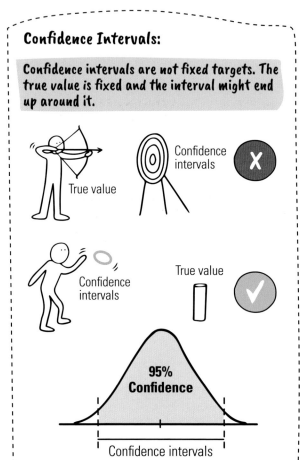

Confidence intervals are not fixed targets. The true value is fixed and the interval might end up around it.

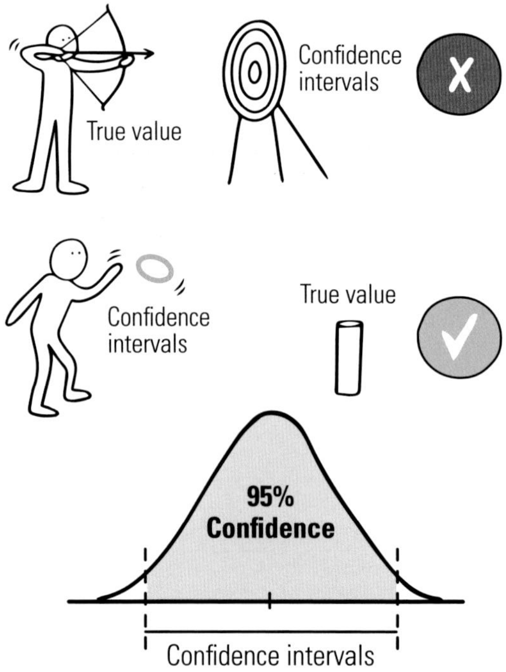

True value
Confidence intervals
X
Confidence intervals
True value
95%
Confidence
Confidence intervals

Confidence intervals are used to estimate certain parameters for a measurement of a population (e.g., the mean) based on sample data. The confidence interval does not provide a single mean value, but a range of values. This is often expressed as a percentage.

Measure of Association

The association measure is applied to different factors or coefficients to quantify a relationship between two or more variables. Association measures are used in various areas of innovation measurement, such as quantifying relationships between input and output metrics or innovation behaviors. A measure of association can be obtained through various analyses, including correlation analysis and regression analysis. (Note: While the terms "correlation" and "association" are often used interchangeably, correlation in the narrow sense refers to linear correlation and association refers to any relationship between variables.)

The method used to determine the strength of an association depends on the characteristics of the data for each variable. Data may be measured on an interval (=ratio), an ordinal (=rank), or a nominal (=categorical) scale. These three characteristics are usually continuous, integer/qualitative. Within this 101 only the most common methods are introduced (see overview on the right). Additional methods range from point-bi-serial correlations to performing Durbin-Watson tests for serial correlations. The only danger with all measured associations is the tendency to infer causality. Whenever one variable causes changes in another variable, there is an association. However, it does not always follow from the presence of an association that causality exists.

Measuring of Association Helps the Team:

- to fully understand the causes and effects of different measurements and metrics.
- to assess the direction (positive and negative) with the correlation coefficient as well as the magnitude and significance of the association.

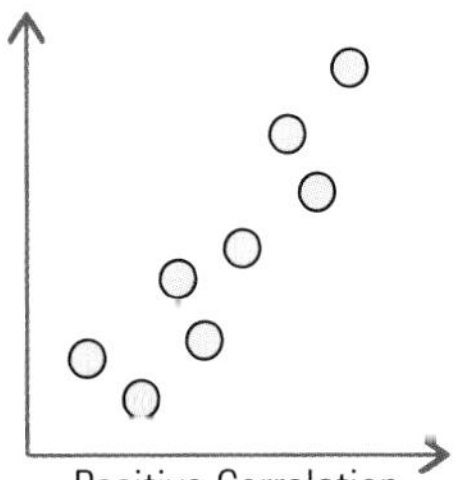

Positive Correlation

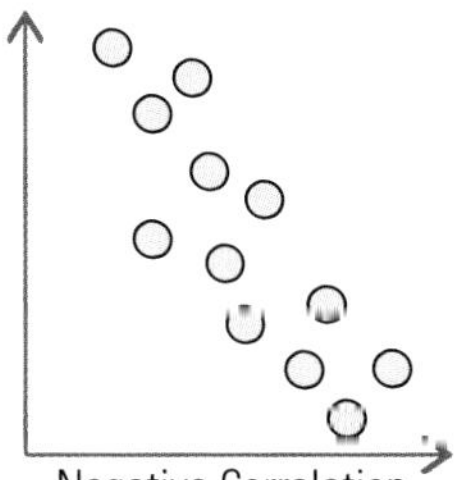

Negative Correlation

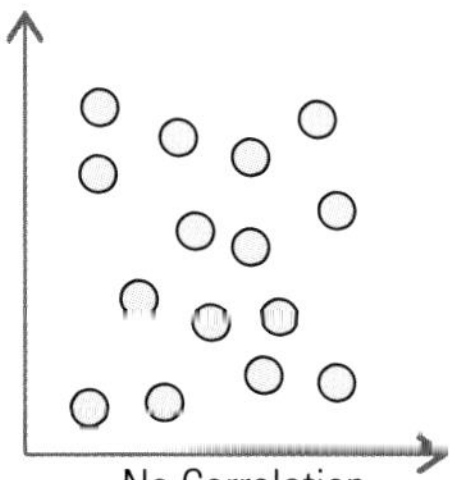

No Correlation

Most Common Methods

Pearson's Correlation Coefficient:

This coefficient provides information about the relationship between two variables measured on an interval/ratio scale. Both variables are measured on a continuous scale. The appropriate measure of association in this situation is Pearson's correlation coefficient, which measures the strength of the linear relationship between two variables on a continuous scale.

Spearman Rank-Order Correlation Coefficient:

This coefficient is used to measure the strength of a monotonic (in a constant direction) relationship between two variables measured on an ordinal scale or a rank scale. This approach is also used, for example, when one innovation variable of interest is measured on an interval scale, and the other on an ordinal scale.

Chi-Square Test:

The chi-square test for association (contingency) is a standard measure of the association between two variables. It can be used to assess the significance of the association between variables. However, the chi-square test, unlike the Pearson correlation coefficient or the Spearman Rho test, is a measure of the significance of the association and not a measure of the strength of the association.

Relative Risk and Odds Ratio:

Odds ratios are often used in research for the association between categorical variables. The relative risk is appropriately applied to categorical data, such as those from a cohort study. It measures the strength of an association by looking at the incidence of an event in an identifiable group (numerator) and comparing it with the incidence in a baseline group (denominator). For example, a relative risk of 1 means that there is no association, while a relative risk greater than 1 indicates an association.

Simple Bivariate and Multiple Linear Regression

Based on the previously introduced measures of association and the validated causality, the statistical modeling can be applied. The simplest variant of statistical modeling is simple linear regression (working with only two variables). Linear regression is a widely used method and is particularly useful when considering multivariate relationships, and the bivariate model can be easily extended. However, care should be taken to ensure that the independent variables are not collinear. The process of multiple linear regression modeling can be assisted and automated by the use of selection methods, but in the end, this should not replace human decision making.
The list of assumptions outlined on the right needs to be tested when fitting a regression model. If any of these assumptions are violated, the model might be biased and inefficient. In contrast to simple linear regression, the multilinear regression model is useful when there are multiple independent variables, as it estimates an individual slope of the regression line for each independent variable.

Simple Bivariate and Multiple Linear Regression Helps the Team:

- to explore and estimate the relationship between a dependent variable and one or more independent variables.
- to build on assumptions with regard to existing relationships between variables.
- to work with both categorical nominal and ordinal interdependent variables through the use of dummy or indicator variables.
- to visualize and plot the residuals for generating improved models.

Linear Regression Assumptions

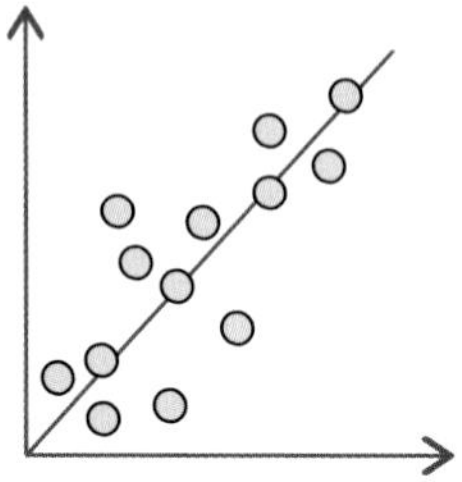

- Relationship between the variables is sensible.
- Relationship between the variables is linear.
- Independent variables are not highly associated (multicollinearity).*
- No presence of extreme outliers exists.
- Residuals are random and normally distributed.
- Residuals are independent of one another (time series data).
- Variance of residuals is fairly constant at different levels of the predicted variable (homoskedasticity).

*Note: This is only relevant for multiple regression models as simple regression only has one independent variable.

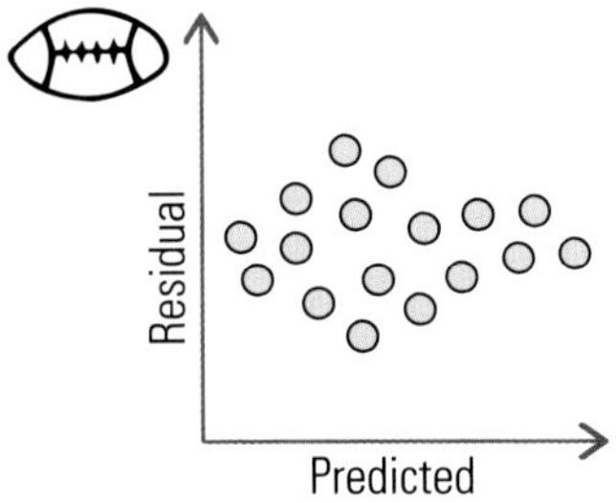
Residual
Predicted

Rugby ball shape suggests the need for more explanatory variables

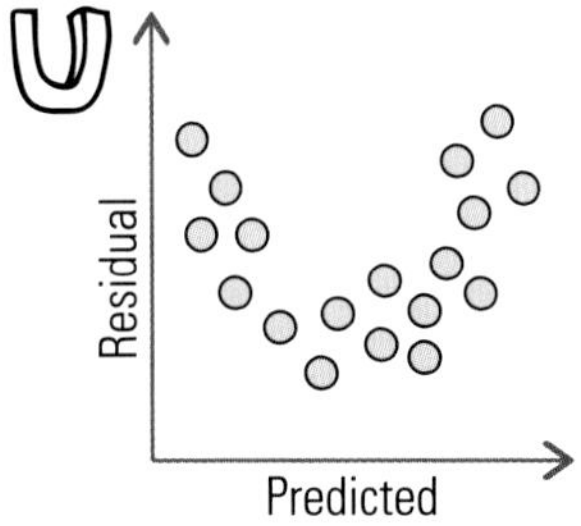
Residual
Predicted

U shape plot suggests non-linear relationship; try a term with a power (could also be n shape)

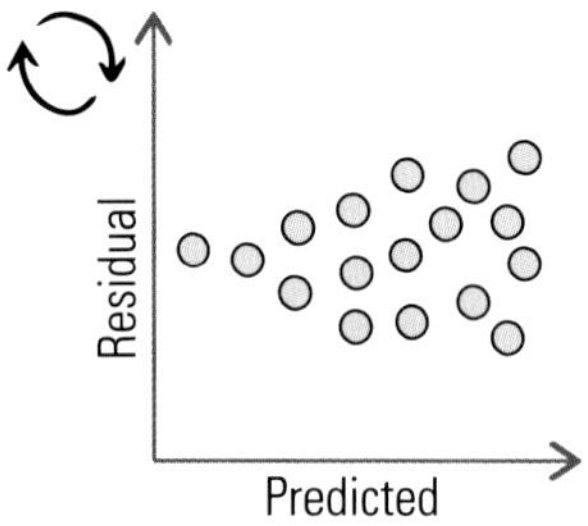
Residual
Predicted

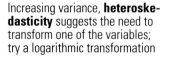

Increasing variance, **heteroskedasticity** suggests the need to transform one of the variables; try a logarithmic transformation

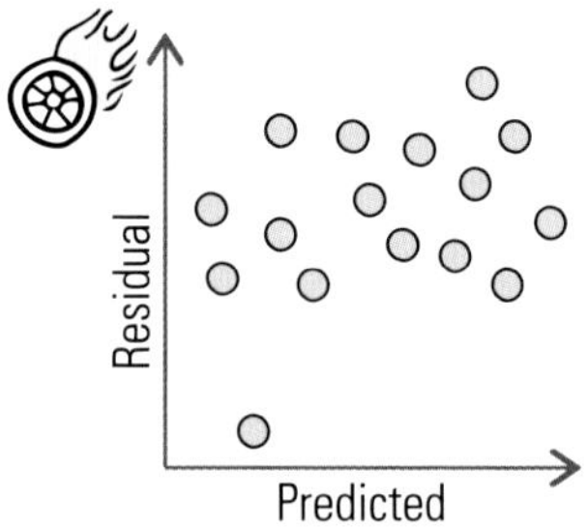
Residual
Predicted

Residual plot with an extreme value; try removing extreme value

Tools for Forecasting Trends

Forecasting is a very important technique in the context of innovation and business success. Forecasting tools aim to predict the future based on the results of previous data. It involves a detailed analysis of past and present trends or events to predict future events. Forecasting acts for decision makers and innovation teams as a planning tool that helps them to get ready for the uncertainty that can occur in the future. Forecasting is a subdiscipline of prediction in which time-series data is used to make forecasts about the future.

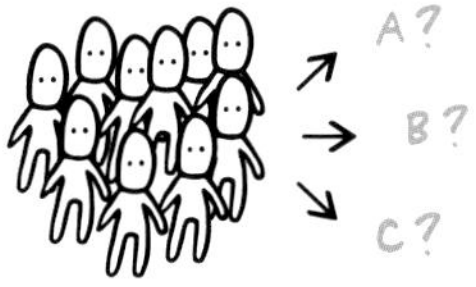
A?
B?
C?

Forecasting

"How many customers will visit the shop next month?"

Forecasting is anticipating, for example, the behavior of many customers/users on a longer timeline.

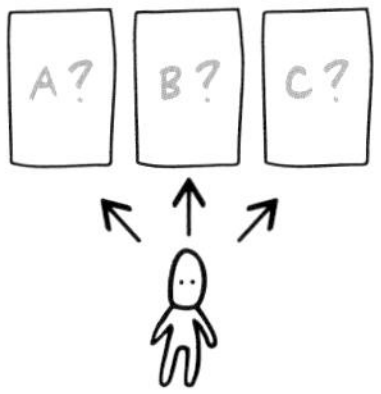
A?
B?
C?

Predictive

"Will the top customer buy at one of the competitors?"

Predictive analytics is anticipating the behavior of, for example, one customer/user on a shorter timeline.

Tools for Forecasting Trends Help the Team:

- to understand how things are currently.
- to predict in which directions things are likely shifting in the future.
- to recognize the big picture and forces behind the change.
- to optimize processes and resource allocation, and to reduce uncertainty about the future for all decisions related to the innovation work.
- to monitor even weak signals which are indications of change, emerging ideas, phenomena, and trends which will affect the company or its operational activities.

For a comprehensive analysis of the future, it is necessary to connect the dots at scale to spot and prepare for trends when it comes to technology, innovation, and business success. The recent developments in machine learning and natural language processing (NLP) technology allow organizations to understand trends in depth.

Most Common Methods

Qualitative Methods:

Used when historical data is not available and qualitative forecasting techniques are sufficient. They are subjective and based on the opinions and judgments of consumers and experts. They are usually used to make medium- or long-term decisions.

Quantitative Methods:

Predicting future data based on historical data is done through quantitative forecasting methods. If numerical data from the past is available and it is reasonable to conclude that one of the characteristics in the data will persist in the future, these methods are appropriate.

Average Method:

All future values are predicted to be equal to the average of the previous data.

Naïve Method:

A variation of the naïve method is allowing the previous month's actual values to be used as a projection for that period without making adjustments or attempting to identify causal factors. The naive method is used for economic, financial time-series, and health metrics.

Drift Method:

A variation of the naïve method is allowing the projections to increase or decrease over time, setting the amount of change over time (called drift) to the average change observed in the historical record.

Tips & Tricks for Generating Forecasts

Expert Tip: Exclude the Index in Certain Cases:

These days, an index or score is often used for forecasts. However, there are other possibilities for a suitable forecast model to predict a certain development. Typically, an index provides a number based on a set of indicators and a formula. When useful, the index should serve as the basis for specific decisions by predicting observable outcomes. However, if it only predicts predictable outcomes, it is better to omit it from the forecasting algorithm altogether. In addition, a combination of human-led judgement and AI might be another solution for generating forecasts. The Basic AI Canvas is a good starting point to ask the appropriate questions (see page 249).

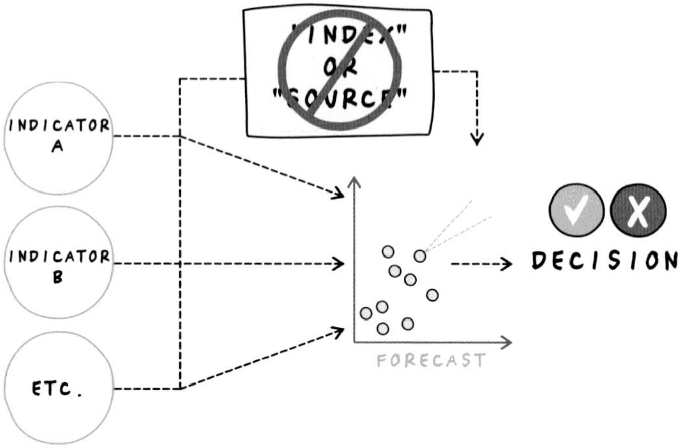
INDICATOR A
INDICATOR B
ETC.
FORECAST
DECISION

Expert Tip: Apply AI to Support Better Decisions

A new generation of artificial intelligence technologies have emerged that hold considerable promise in helping improve the forecasting process in such applications as product demand, employee turnover, cash flow, distribution requirements, manpower forecasting, and inventory. These AI-based systems are designed to bridge the gap between the two traditional forecasting approaches based on human-led team judgments and quantitative forecasting. For more on using AI for innovation and business success, see 301 (page 332).

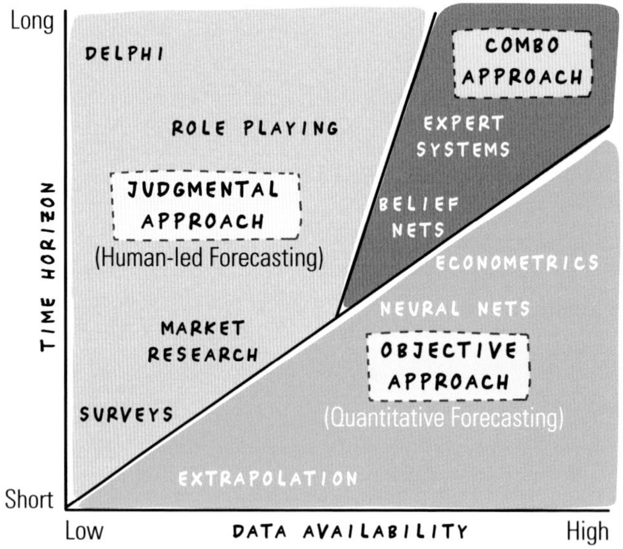
Long
Short
TIME HORIZON
Low
DATA AVAILABILITY
High
DELPHI
ROLE PLAYING
JUDGMENTAL APPROACH
(Human-led Forecasting)
MARKET RESEARCH
SURVEYS
COMBO APPROACH
EXPERT SYSTEMS
BELIEF NETS
ECONOMETRICS
NEURAL NETS
OBJECTIVE APPROACH
(Quantitative Forecasting)
EXTRAPOLATION

Key Questions to Ask for an AI-Supported Approach:

- What knowledge is needed to make the decision?
- How will different outcomes and errors be valued?
- What actions might be triggered from the outcome?
- What is the metric for a task to succeed?
- What data is needed for the predictive algorithm?
- What data is needed to train the predictive algorithm?
- How can outcomes be used to improve the algorithm?

Template download

BASIC AI CANVAS

DESCRIPTION OF BUSINESS PROBLEM OR FORECAST PROBLEM TO SOLVE
What is the problem the team wants to solve?

PREDICTION
What does the team need to know to make a decision?

JUDGMENT
How are different outcomes and errors valued?

OUTCOMES
What are the team's key metrics for task success?

ACTION
What does the team want to achieve?
What added value does it have for the customer?
Can activities be derived from it?

TRAINING
What data does the team need to train the predictive/prescriptive algorithm?

INPUT
What data must be available to run the predictive algorithm?

FEEDBACK
How can the team use the outcomes to improve the algorithm?

DOWNLOAD TOOL
www.design-metrics.com/en/basic-ai

Exponential Smoothing

One of the most widely used averaging forecasting methods in the innovation and business context is exponential smoothing. The technique allows organizations to identify historical patterns of trends or seasonality in the data and then extrapolate these patterns forward into the forecast period. In general, nine common data profiles can be predicted using common exponential smoothing techniques. They range in complexity from a constant data level to a more complex damped trend with a multiplicative seasonal influence. To ensure that the correct exponential smoothing method is chosen, a method with sufficient flexibility to fit the underlying data should be used. A good first step is to plot the data series to be forecast and then choose the exponential smoothing method that best fits the data.

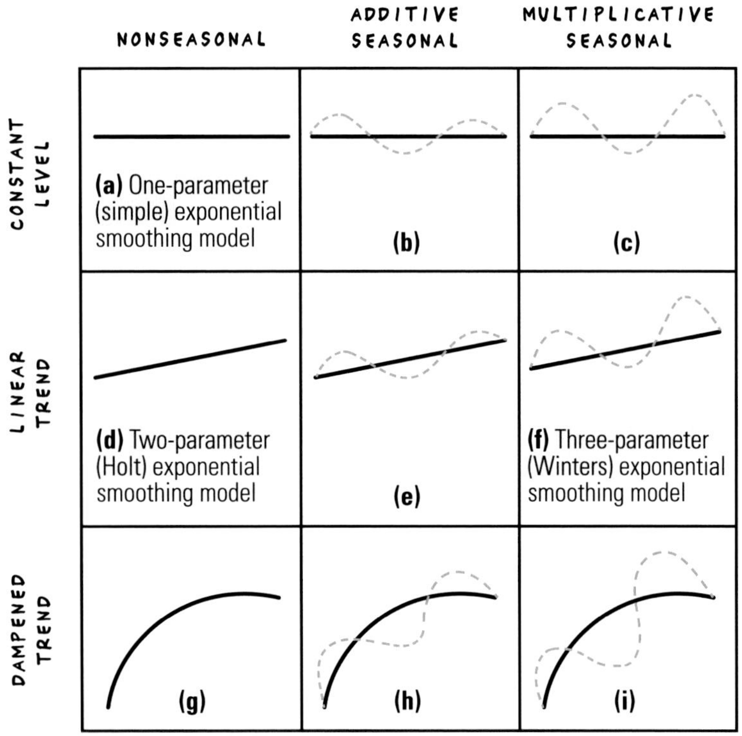
NONSEASONAL
ADDITIVE SEASONAL
MULTIPLICATIVE SEASONAL
CONSTANT LEVEL
LINEAR TREND
DAMPENED TREND
(a) One-parameter (simple) exponential smoothing model
(b)
(c)
(d) Two-parameter (Holt) exponential smoothing model
(e)
(f) Three-parameter (Winters) exponential smoothing model
(g)
(h)
(i)

The application of forecasting often refers to measurements and forecasts according to the product/market fit. Classic product life cycles often progress from introduction through rapid growth and market penetration to a mature phase of stable sales and periods of declining market share for the product (see pages 190–191). During this life cycle, various methods of sales forecasting may be appropriate.

Practical experience over the years has proven that it is good to conduct qualitative analyses and market experiments in the initial phase (i.e., before meaningful market data is available). Once the product is on the market and rapidly gaining acceptance, exponential three-parameter smoothing methods are often used, which include level, trend, and seasonal components. In a mature phase of the product in terms of sales stability, for example, two-parameter exponential smoothing models (or econometric models) that include level and seasonal components can be used. In the case of declining market shares, three-parameter exponential smoothing methods, which include level, trend, and seasonal components, become relevant again.

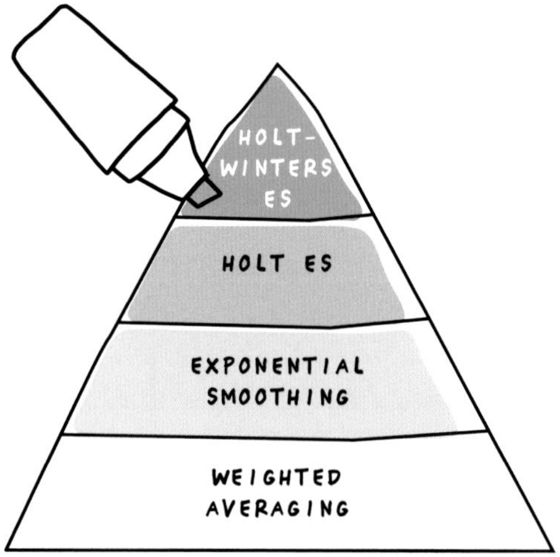
HOLT-WINTERS ES
HOLT ES
EXPONENTIAL SMOOTHING
WEIGHTED AVERAGING

Sigmoid Curve and Model

The S-curve as a metaphor for describing the dynamics of change is often used and a powerful tool (see pages 169–170 with regard to the definition of the North Star metrics). Without going too deep into the mathematical theorem, the Sigmoid curve shows how the cycle works. The circle that's marked is the inflection point, the place where decisions are made which will determine the future course of the cycle. In the context of innovation and business success, it has proven that the more quickly the team of decision makers identifies the approaching inflection point, the more prepared they will be to respond appropriately and put strategies in place to ensure the business follows the upward curve, and not the downward. The Sigmoid curve is usually applied to the business as a whole, a product, a service, an experience, or even how an offering is marketed or sold. The example illustrated with four types of innovation activities on the S-curve shows different patterns overserved in practice for turning points and a potential path of innovation success.

In the context of forecasting, the S-curve helps organizations, decision makers, and innovation teams to survive and grow. All of them should be more mindful of the elements of the EXPLOIT portfolio placed on the present life cycle on the curve and prepare for transformational change.

Especially in dynamic market environments, the speed of every S-curve seems to be increasing. To keep on growing, the successful growth and innovation teams must develop a second curve out of the first. Here, it is strongly recommended that the new curve starts before the first one peaks, in a time when all the evidence is that there is no need for change. In many organizations, the transition from one curve to another is confusing, because a culture of EXPLORE and radical innovation is facing a culture trimmed to efficiency and scale.

$$y = \frac{1}{1 + e^{-x}}$$

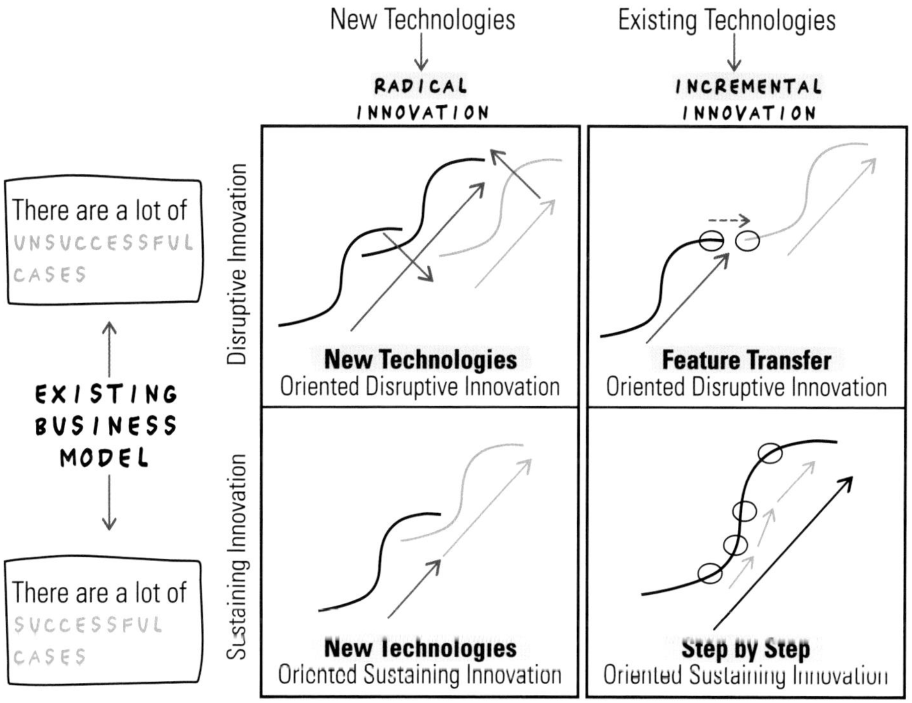
New Technologies
Existing Technologies
RADICAL INNOVATION
INCREMENTAL INNOVATION
There are a lot of UNSUCCESSFUL CASES
EXISTING BUSINESS MODEL
There are a lot of SUCCESSFUL CASES
Disruptive Innovation
Sustaining Innovation
New Technologies
Oriented Disruptive Innovation
Feature Transfer
Oriented Disruptive Innovation
New Technologies
Oriented Sustaining Innovation
Step by Step
Oriented Sustaining Innovation

Examples of Applied Statistics, Algorithms, and Use Cases

The possibilities of ML, AI, and deep learning are many and are explored in more depth in Data-Driven Innovation 301 (see page 336). The examples presented below show simple examples of application and use cases.

LINEAR REGRESSION

Understanding drivers for sales (prices of competitors, distribution, advertisement) to optimize price points and product price elasticities.

LINEAR QUADRATIC DISCRIMINANT ANALYSIS

Predict churn rate or predict sales lead's likelihood of closing.

NAÏVE BAYES

Analyze sentiment to assess product or service perception in the market or create any classifiers to filter relevant data.

REINFORCEMENT LEARNING

Optimize pricing in real time for an online auction of a product with limited supply.

LOGISTIC REGRESSION

Extension of linear regression with the aim to perform classification tasks for binary outcomes based on product size, shape, and color.

DECISION TREES

Provide a decision tree for hiring new talents for the design thinking and innovation teams or understand critical functions and experiences that make a product, service, or experience most likely to be purchased by customer/user.

SUPPORT VECTOR MACHINE

Predict how likely a potential customer/user is to click on an online ad.

ADABOOST

Adaptive boosting is a low-cost way to classify images with lower accuracy than deep learning.

Example

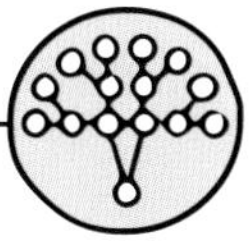

GRADIENT-BOOSTING TREES

Applying a combination of the results from all decision trees to generate product demand forecasts, or to predict pricing based on different features or characteristics.

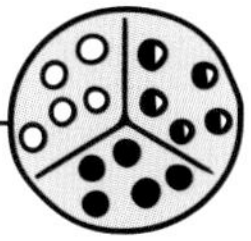

K-MEANS CLUSTERING

Segment customers/users into groups by distinct persona characteristics, like age segments, to prevent churn.

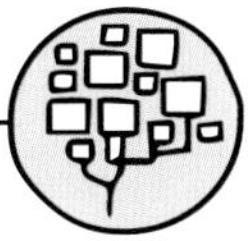

HIERARCHICAL CLUSTERING

Splits or aggregates clusters data to inform product/service usage or development by grouping customers/users based on mentioning keywords in social media activities.

DEEP LEARNING BASED ON CONVOLUTION NEURAL NETWORKS

Applied to understand brand perception and usage through images or create joint marketing opportunities based on logo detection on social media images and videos.

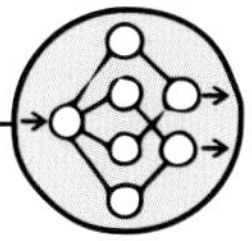

SIMPLE NEURAL NETWORK

Model with artificial neurons that is used to classify data or find relationships between variables in regression problems to predict whether registered users will pay or not pay a particular price for a product.

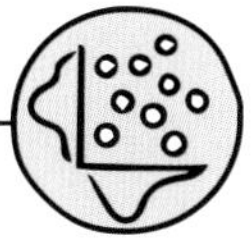

GAUSSIAN MIXTURE MODEL

Segment customers/users to better assign marketing campaign with less-distinct persona characteristics or product/service preferences.

RECOMMENDER SYSTEM

Cluster behavior prediction to recommend what kind of other services or products the customer/user may like based on preferences of other customers/users.

Graphing Distributions

Visualization of data and metrics helps to present a lot of information in a simple way. Neuroscience provided plenty of evidence that the human brain is not capable of processing hundreds or even millions of pieces of data to draw conclusions at a glance. For sharing information about quantifications, measurements, and metrics visual representations are often used to share the information with the innovation team or decision makers: Pie charts, line graphs, and histograms help identify problems and trends and make data comparisons that otherwise could not be communicated with the extended team in a focused way. Visualization tools such as Tableau, Power BI, Qlik, and DataStudio are an essential element of any story based on data to be shared purposefully today.

However, poor data visualizations do more harm than good. Either the team and decision makers are more confused than before, or, worse, they are misled.

Furthermore, poorly constructed data visualizations lead to lost time and effort, ultimately delaying the decision-making process and the derivation of appropriate activities and next steps.

Typical mistakes include using incorrect chart types, presenting data in a misleading way, not plotting enough data, and using different scaling or too much color. In addition, the core principles of dashboard design apply as introduced on page 236.

Beware Using an Incorrect Chart Type

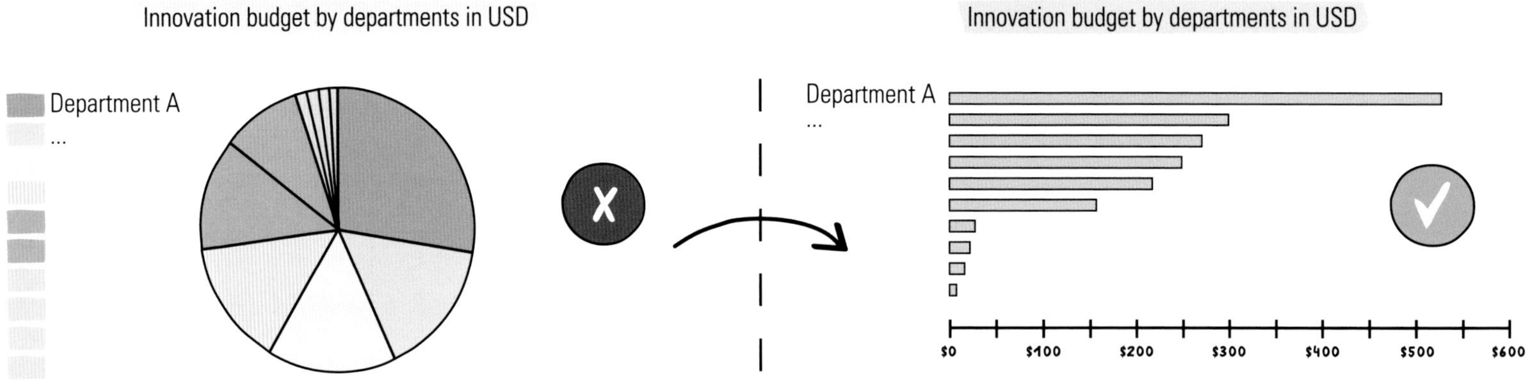
Innovation budget by departments in USD
Department A
...
Innovation budget by departments in USD
Department A
...
$0
$100
$200
$300
$400
$500
$600

The color and size combinations for each pie slice are unwelcome hurdles for the reader.

Sort the data in descending order to help users easily spot top spending categories.

Avoid Using 3D Charts

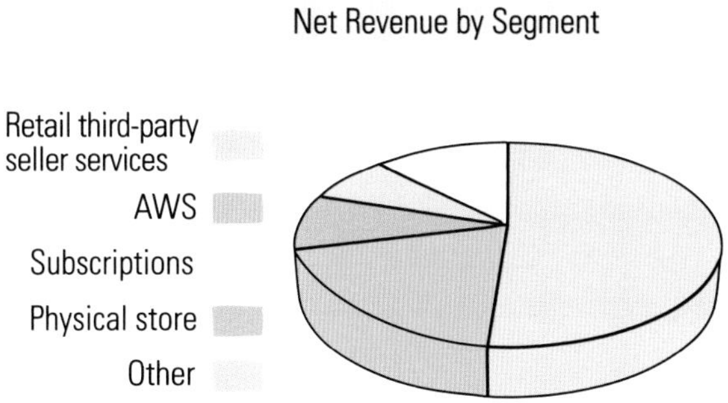
Net Revenue by Segment
Retail third-party seller services
AWS
Subscriptions
Physical store
Other

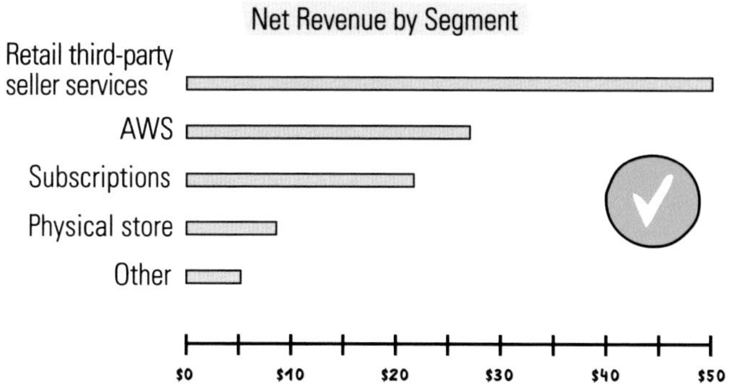
Net Revenue by Segment
Retail third-party seller services
AWS
Subscriptions
Physical store
Other
$0
$10
$20
$30
$40
$50

3D charts might create false representations since our eyes perceive objects as closer and larger than they truly are.

It's better to use a column chart to improve this graph's value and appearance.

Refrain from Presenting Misleading Data

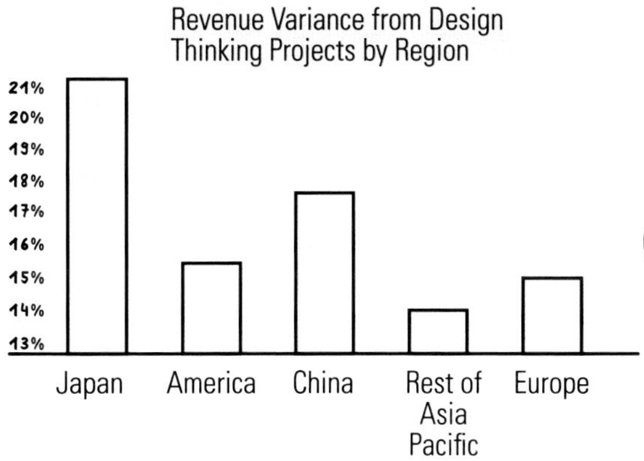
Revenue Variance from Design Thinking Projects by Region
21%
20%
19%
18%
17%
16%
15%
14%
13%
Japan
America
China
Rest of Asia Pacific
Europe

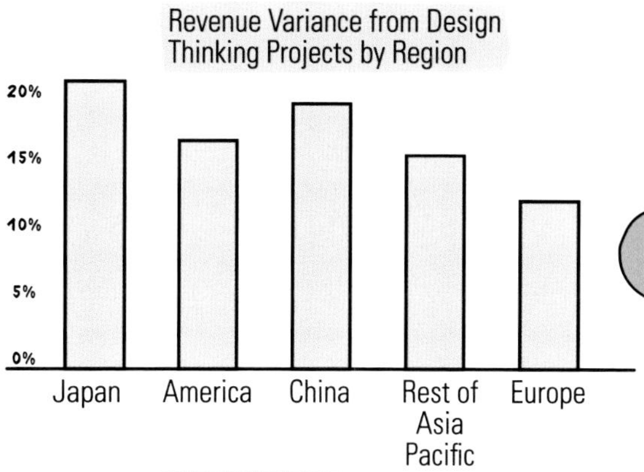
Revenue Variance from Design Thinking Projects by Region
20%
15%
10%
5%
0%
Japan
America
China
Rest of Asia Pacific
Europe

Shaky data comparisons not only erode credibility, but also lead to wrong measurements for future activities.

Correct data comparison helps readers to absorb the information faster and more efficiently. Setting the baseline to zero provides better ways to compare visually.

201

PERFORMANCE MEASUREMENTS

OKR

OKRs and Beyond

The two 101s already discussed the core concepts and tools to manage creativity, product, innovation, and business success. The 201 section is dedicated to the important aspects of aligning teams to objectives and key results. The overarching term of performance measurement systems was chosen for this purpose, as OKRs (objectives and key results) overlap with other management frameworks and fall somewhere between KPIs (key performance indicators), MBOs (management by objectives), and BSCs (balanced scorecards).

The main focus in this chapter is on the design of OKRs, which are significantly aligned with the goal of behavioral change, growth, and radical innovation in many companies. In particular, the *Objectives & Key Results Toolkit* (page 288) covers the methods and tools that help organizations, teams, and individuals to use the design thinking mindset to align, for example, moonshot thinking with the principles of modern performance management systems.

It is important to emphasize that the approaches and ways of thinking presented here do not aim to replace HR measurement systems. They can be used as a supplement in the context of the usual career discussions and team development but are primarily designed as an instrument for short- and mid-term prioritization and alignment with the North Star. Moreover, such an objective-setting system works best when the higher-level measurement system also meets these requirements. A step-by-step review and understanding the principles of North Star metrics, minimum viable exploration metrics, and appropriate EXPLOIT metrics, as described in Design Thinking 101 and Measurement 101, will help on this journey.

The basic principles, such as meaningful measurements and the formulation of criteria that can be acted upon with concrete actions, remain. Furthermore, the applied system for performance management must also fit the culture and the desired goals of the company. Without a concrete North Star, clearly defined ambitions, or the corresponding mandate to the teams to consciously radically rethink things, the respective practices are difficult to implement.

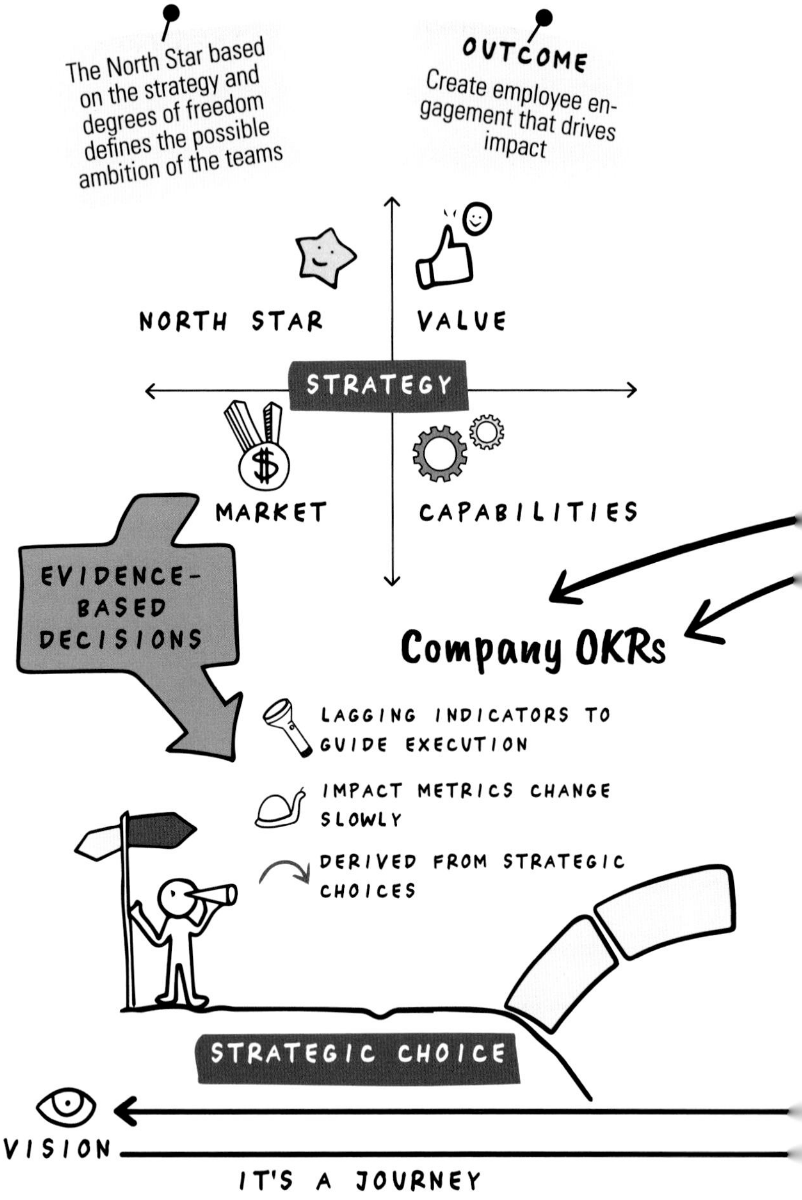

The North Star based on the strategy and degrees of freedom defines the possible ambition of the teams
OUTCOME
Create employee engagement that drives impact
NORTH STAR
VALUE
STRATEGY
MARKET
CAPABILITIES
EVIDENCE-BASED DECISIONS
Company OKRs
LAGGING INDICATORS TO GUIDE EXECUTION
IMPACT METRICS CHANGE SLOWLY
DERIVED FROM STRATEGIC CHOICES
STRATEGIC CHOICE
VISION
IT'S A JOURNEY

OKR Measurement Landscape

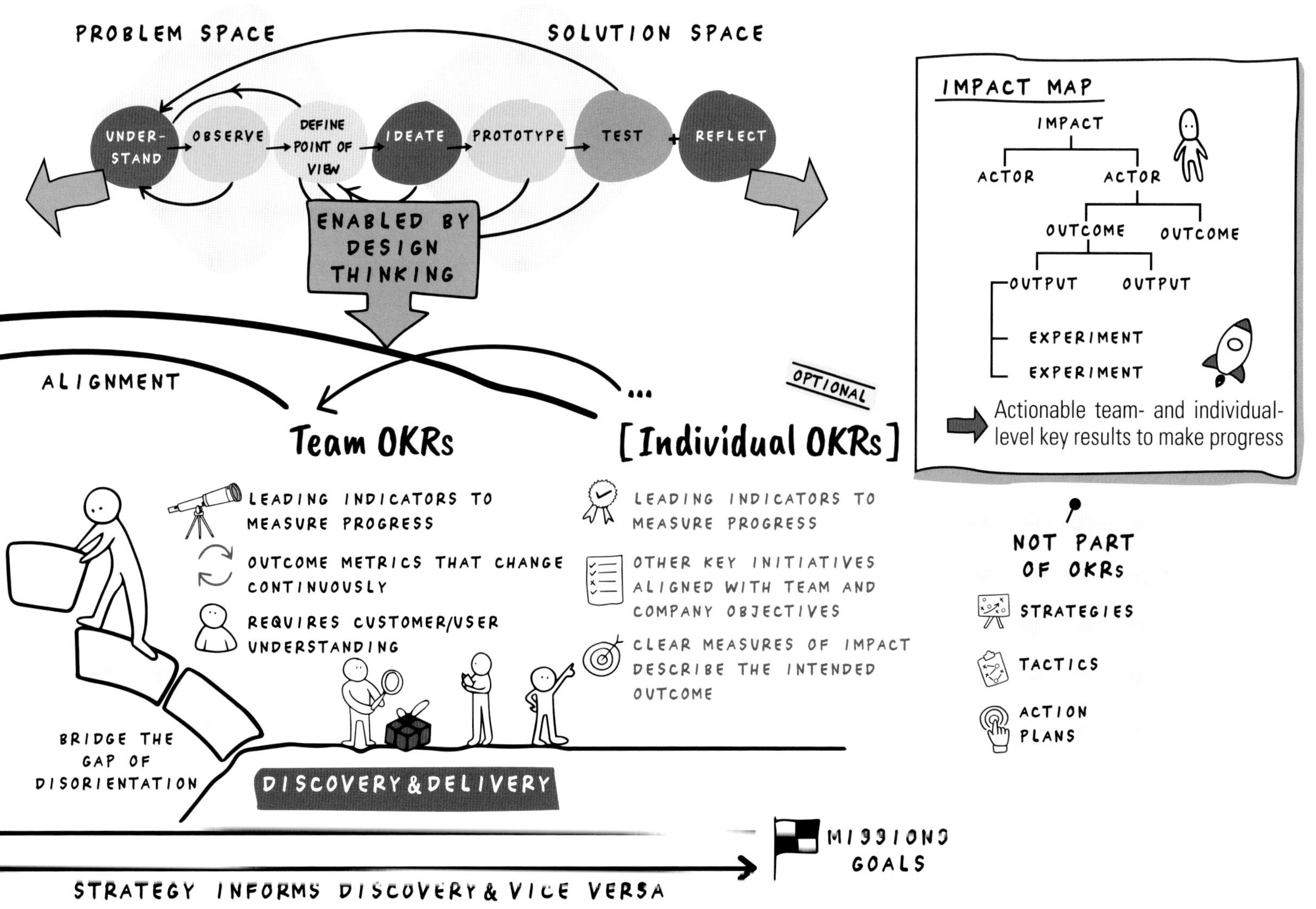
PROBLEM SPACE
SOLUTION SPACE
UNDER-STAND
OBSERVE
DEFINE POINT OF VIEW
IDEATE
PROTOTYPE
TEST
REFLECT
ENABLED BY DESIGN THINKING
ALIGNMENT
...
OPTIONAL
Team OKRs
[Individual OKRs]
LEADING INDICATORS TO MEASURE PROGRESS
OUTCOME METRICS THAT CHANGE CONTINUOUSLY
REQUIRES CUSTOMER/USER UNDERSTANDING
LEADING INDICATORS TO MEASURE PROGRESS
OTHER KEY INITIATIVES ALIGNED WITH TEAM AND COMPANY OBJECTIVES
CLEAR MEASURES OF IMPACT DESCRIBE THE INTENDED OUTCOME
IMPACT MAP
IMPACT
ACTOR
ACTOR
OUTCOME
OUTCOME
OUTPUT
OUTPUT
EXPERIMENT
EXPERIMENT
Actionable team- and individual-level key results to make progress
NOT PART OF OKRs
STRATEGIES
TACTICS
ACTION PLANS
BRIDGE THE GAP OF DISORIENTATION
DISCOVERY & DELIVERY
MISSIONS GOALS
STRATEGY INFORMS DISCOVERY & VICE VERSA

The Roots of Measuring Objectives

Every era has its performance measurement systems, and a quick look in the rearview mirror shows the concepts that were appropriate at a given time. Despite big changes, digitization, and a new generation of decision makers, concepts from the 1960s and 1980s still exist in our structures today. However, the purpose of performance measurements should be to measure and check the performance of companies, organizations, and teams that are necessary for the current and future challenges. Only in this way can learning take place and activities, business models, value propositions, and associated products and services be adjusted in a targeted and rapid manner. Traditional metrics determine how a business is performing against its goals. Modern frameworks and their corresponding metrics establish a mindset based on how teams are meeting their objectives. Based on OKRs, an adapted measurement system for innovation and design thinking teams is presented in 201, which introduces terms that are appropriate for innovation work. The chosen definitions of *Moonshot, Everest,* and *Roofshot* objectives in this section are descriptions of what a team wants to achieve (long and short term with distinct characteristics). The Moonshot objectives are, for example, meant to be bold and ambitious in order to challenge the team. The concept of key results and the respective scoring are the measurements set by the team to track the progress toward each objective in a defined time. The basic considerations of OKRs were invented by Andy Grove and made their way to Google with John Doerr in the 1990s. OKRs still serve their purpose in transformation efforts as first steps toward agile working and networked organizations. However, potential future measurement systems should respond to all the requirements that VUCA (volatility, uncertainty, chaos, and ambiguity) demands and also create the basis for providing scope for creativity and the exploration of new market opportunities in addition to ambitious objectives. In particular, the new growth opportunities that can be realized in business ecosystems beyond sector boundaries need elements that allow even more experimentation V and bring the innovation teams even closer to the interface with the customer. A future performance management system will allow teams and organizations to set flexible goals, apply different criteria for different decisions, and obtain data that matters and supports every team member to become successful based on thinking preferences, capabilities, and T-shaped profile developments.

DREAM BIG, SET OBJECTIVES, AND TAKE ACTION
OKR

Objectives should be bold, inspirational, and focused on an area of great impact at the individual, team, and organizational levels.

Andy Grove

The Evolution of Setting Objectives

PERFORMANCE MEASUREMENT SYSTEMS

[In times of agile leadership, companies implement strategy execution tools, which allow them to guide result-focused work and create alignment and engagement around measurable goals.]

EARLY DAYS

Objectives
Key
Results
OKR
F
A
S
T

TODAY

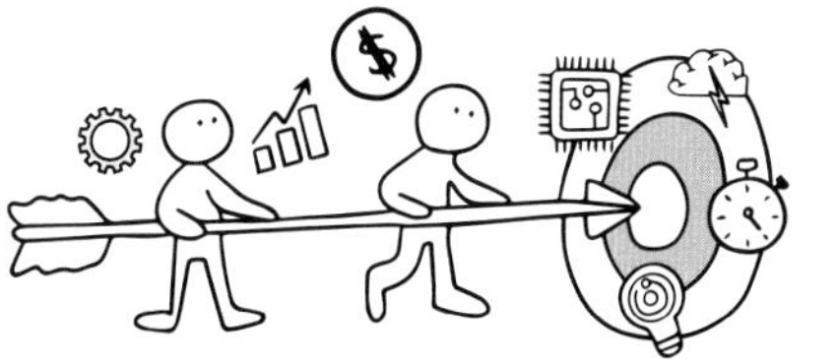

FUTURE

Management by Objectives (MBO)

Balanced Scorecards & S.M.A.R.T Goals

Objectives & Key Results (OKRs) & FAST Goals

Advanced performance measurement and team-of-teams coaching system with a strong focus on value creation, entrepreneurial thinking, and self-directed learning for innovation teams

1960 — 1980 — 2000 — 2020 — 2040 — TIME

Advanced OKRs fit well in many organizations' transformative exploration into a new mindset, where first the appropriate problem is identified and second the solution is developed. Design thinking, implementation via sprints, the concept of minimum viable metrics, and the agile toolbox support the desired outcomes. For example, up until today OKRs have been also a great addition in many projects for building design thinking capabilities and introducing agile methods, as they are set and evaluated in a regular rhythm that is much more frequent than that of traditional organizational or team goals. OKRs also help to accelerate the transition from traditional structures to a networked organization, working in team-of-teams structures. Building on the core concepts of this book, modern measurement systems go a step further and are focused on EXPLORE and EXPLOIT. Concepts like EXPLORE internalize even more that the teams define their objectives independently. Besides Moonshot and Roofshot objectives, teams have the corresponding freedom to explore new problems themselves, to be creative and to search for new market opportunities in an explorative and aspirational mindset based on customer insights.

With these objectives, individuals and teams can be additionally inspired to get even more into the exploration mode and achieve peak performance. Everest goals often start with the question "What if...," and the results very often represent achievements far beyond normal success. But Everest objectives are not just fantasies or dreams. They have special characteristics that actually motivate spectacular achievements, breakthrough innovations, and the creation and implementation of black-ocean strategies. Everest goals, however, require even higher degrees of freedom, knowing the appropriate problem, but not the solution at the beginning of the journey. Through iterations, and with a lot of exploration, they have to be elaborated by the teams. Some organizations leap directly into the so-called FAST goals. The acronym FAST means frequently discussed (faster and from the bottom up), ambitious (difficult but not impossible to achieve), specific (concretized by measurable milestones), and transparent (shared within and between teams). FAST goals might be a first attempt toward team coaching and goal setting that allow improvement in various areas at the same time and emphasize continuous discussions of objectives to adapt in an unpredictable and complex environment.

Some companies use the term Aspirations and Key Insights (AKIs) in the context of OKRs for innovation teams to highlight the purpose in EXPLORE.

Another key difference between modern and traditional measurement systems, such as balanced scorecards, is transparency. Modern measurement systems should be fully transparent, connect teams, and break silos within the organization, as well as encourage collaboration and co-creation with other companies. However, with the introduction of such systems, it is also observed that employees often feel exposed when sharing their thoughts on what is most important and how they think they can contribute most effectively to the overall goals of their team and the organization as a whole. By applying the appropriate design thinking tools and facilitation, in many cases support can be provided to define the problem statement and set the right level for short- and medium-term team/individual objectives and key results for all the Moonshot, Roofshot, or Everest objectives. Within the framework of a moderated process, it is possible to introduce the teams to transparency step by step.

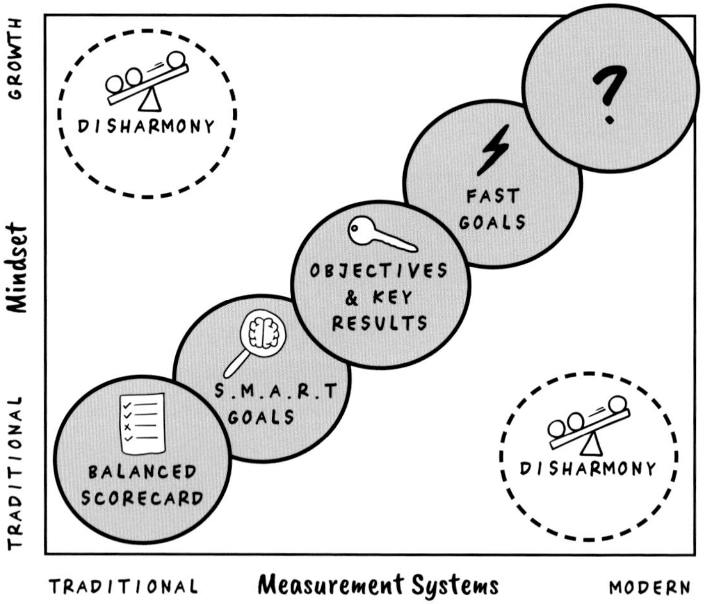
GROWTH
Mindset
TRADITIONAL
DISHARMONY
?
FAST GOALS
OBJECTIVES & KEY RESULTS
S.M.A.R.T GOALS
BALANCED SCORECARD
DISHARMONY
TRADITIONAL
Measurement Systems
MODERN

Modern performance measurement tools are based on objectives that require teams to leave their comfort zone, pull more levers, get the right resources to go to the moon, or climb Mount Everest, and thus be motivated to take bold steps to achieve the goals. Moonshot objectives in particular are often seen as unattainable, at least initially. However, these goals are what really push teams to do things they didn't initially think were possible. Finally, it is the use of Moonshot, Everest, and Roofshot objectives that make a governance approach a powerful tool for measuring the progress of any initiative. Unlike Moonshot or Everest objectives, Roofshot objectives are goals that the team wants to achieve 100 percent of the time. They are still hard objectives, but they are achievable. True Moonshots are not usually achieved 100 percent, but an achievement rate of 60–70 percent is a very decent result. Everest objectives generally provide more room for creative solutions and they are clear, engaging, convincing, exciting, stimulating, and impassioned at the same time. The different scoring systems are explained on page 274.

	Moonshot	Roofshot	Everest
CHARACTERISTICS	Stretch targets beyond the threshold of what the team thinks is possible.	Still hard goals, but the team likely did something similar before and thinks it is achievable.	Goals that support team-of-teams by enhancing positive relationships because it is rare to achieve them alone.
SUCCESS CRITERIA	Success is achieving 60–70 percent.	Success is achieving 100 percent.	Success is achieving something new to the customer and the market.
EFFECTS	Strong promoter for innovation and creative solutions.	Provides structure and alignment for interdependent goals.	Individuals develop better for having engaged in its pursuit.
APPLICABLE CONTEXT	Stretch goals with ambitious target to hit at the critical moment. Mostly seen in exploring new market opportunities.	Regular goals for more operational targets and gains in effectiveness and efficiency. Mostly observed in exploiting existing market opportunities.	Special goals that require supreme effort but also motivate learning and wisdom as the team conquers and achieves them.
SPLIT BETWEEN AMBITIOUS AND ACHIEVABLE	One in three objectives or key results should be related to Moonshot.	Two of three objectives or key results should be more related to Roofshot.	In fast changing and dynamic environments, Moonshots might be replaced with one of three Everest objectives.

THREE MINDSHIFTS IN APPLYING OKRs

Three Shifts to Grasp the Topic Quickly for the Perspectives of Team Members and Teams

For a quick introduction to this chapter on modern performance measurement systems, the three most important changes from the perspective of employees are briefly discussed before presenting the most important tips and tricks for the implementation of OKRs, including guidance for their moderation and formulation.

New systems for performance objectives, for example, based on the idea of OKRs or FAST Goals, require a different behavior from employees, which in turn requires a shift away from the mere processing of activities to actions that create a true impact **(Shift #1)**. Another important element is the ability to develop trust in the work of other teams and even individuals. This means, trust that all teams are intent on achieving the overarching goals (North Star) and along the way, continuously reflect and adjust the path to the defined objectives **(Shift #2)**. The mindset and the work of the individual teams are based, for example, on the fundamentals of design thinking. This mindset (see page 64) makes it possible, especially in innovation projects, to create initial prototypes very close to the customer needs with few resources. It also enables teams to test potential solutions in the market as prototypes and first minimum viable products (MVPs) that are good enough to engage and learn from the customer **(Shift #3)**. Agile customer development and the courage to show potential solutions to the customer at an early stage often means that team members need to get out of their comfort zone. This requires initial guidance, encouragement, and support to make this very important element part of the transformation.

1 Shift: From ACTIVITY to IMPACT

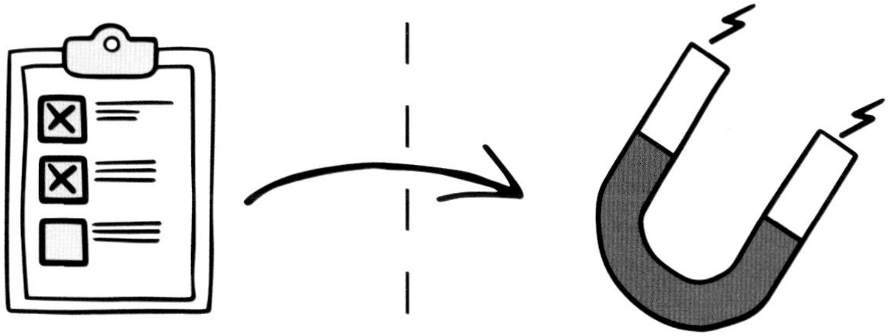

The artificial operational hustle and bustle, often enriched with many unnecessary tasks and uninformative metrics, prevails in many organizations. This approach is replaced by a powerful new perspective focused on real impact.

2 Shift: From CONTROL to TRUST

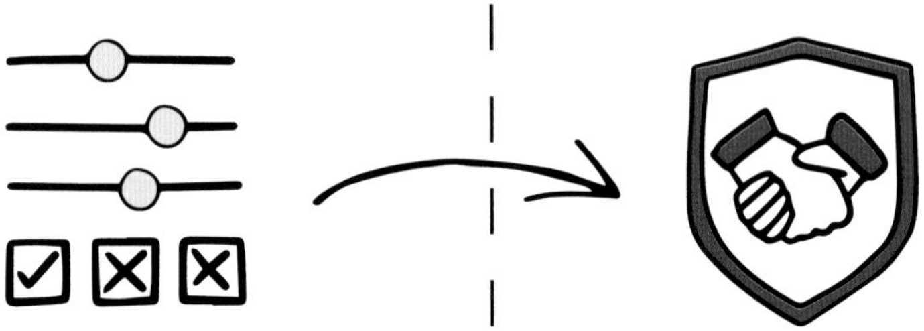

In classical hierarchical organizational models, control, supervision, and the tendency to micro-manage are part of the corporate culture. In team-of-teams structures, the focus is on mutual trust, the transfer of responsibility, and the independent and autonomous work of the teams in fulfilling the objectives.

3 Shift: From PERFECT to GOOD ENOUGH

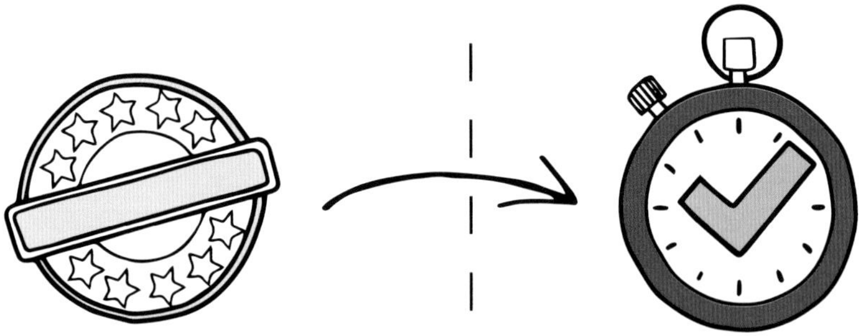

A classic approach to product and service development, with a comprehensive declaration of all current and future requirements before implementation, is replaced by an iterative approach, from the exploration of customer needs through initial prototypes, MVPs and MVEs.

How to Embed Objectives for Innovation Teams

Setting the appropriate objectives and applying modern performance measurement tools should be embedded in the transformation of the organization. Five dimensions for success can be derived from practical experience. Two of these are a strategy based on customer needs and an appropriate mindset that allows organizations to apply varying design lenses according to their characteristics (EXPLORE or EXPLOIT). Two more are an organizational structure that allows interdisciplinary teams to work together across existing business silos, and the fact that each team knows its role in the process so that teams can make the right decisions to achieve goals. Finally, using necessary enabler technology to achieve the objectives and the efficient operationalization of initiatives.

In general, all teams should have visible expectations of their tasks based on the objectives and continuously measure the performance to adjust the respective actions accordingly. Team-of-teams structures can also reduce the waiting times for decisions and coordination, since the decision-making path is simply shorter and closer to the customer and their needs. Additionally, the transformation and team effort itself can be measured on core elements related to operational performance (e.g., FTE cost reduction), employee engagement, or well-known customer satisfaction indicators (e.g., net promoter score, customer effort score), all the way up to financial measurements.

It is important that the measurement system fits the speed of the transformation, and the chosen approach fits the culture of the company. Google's OKR system is too often used as a blueprint, and organizations find that there are hardly any Moonshots that allow them to try radically new ways, not to mention the freedom to let teams work on Everest objectives.

Especially in traditional hierarchical organizations based on command and control, there is little room for "What if..." and "How might we..." (HMW) questions or "Imagine that ..." statements (see page 287). Most of the time there are no aspirational objectives in classical structures, but only the mandatory goals expressed in classical KPIs of sales and financial performance measurement. Another observation is that in classic organizations the well-intentioned objectives are 100 percent delegated downwards to individual employees and their teams. With such a culture, teams are left with little freedom to make team decisions and individual contributions to achieving the North Star, nor to define their own key results. As a rule of thumb, at least 50 percent of objectives should originate from the bottom up, and supervisors should motivate employees to collaborate across teams for all projects. Supervisors, up to the top management level, are part of the team in the implementation of the objectives and not only the recipients of the finished concepts and solutions.

In addition, traditional organizations find it difficult to dynamically adapt objectives to new factors and conditions. However, it is part of a modern measurement system to adjust or discard team objectives if they are no longer appropriate or feasible. Modern measurement also includes a culture in which failure is perceived as a learning opportunity. Openness and transparency are important for mutual success, and the teams should ultimately be intrinsically motivated to find the right problem and realize the best solution for the customer.

VISION

A shared strategy and purpose based on customer needs.

MINDSET

An open-minded workforce with the ability to learn, reflect, and grow.

ORGANIZATION

A team-of-teams structure that allows radical collaboration.

PROCESS

Iterative and agile create, build, and decision cycles.

TECHNOLOGY

Existing and game-changing enabling technologies.

Typical Impact of Higher Employee Engagement with OKRs, Building Design Thinking Capabilities, Increasing Customer Satisfaction, and Improved Operational Performance

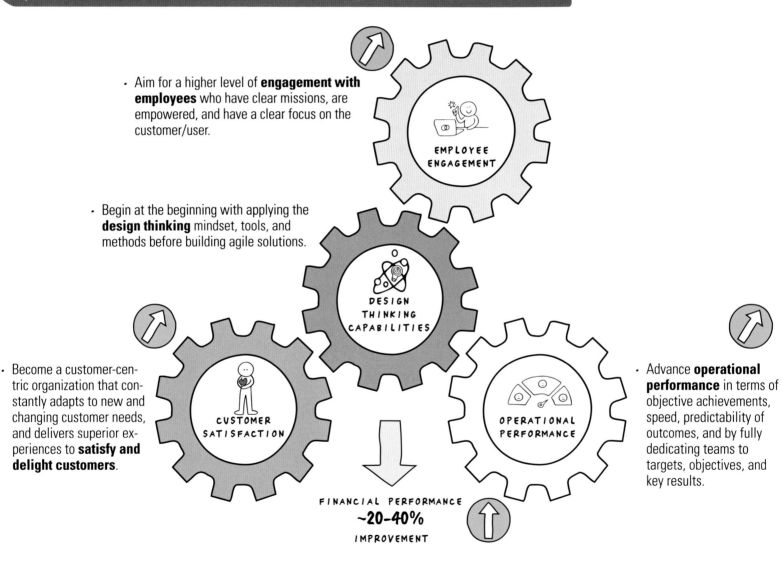

- Aim for a higher level of **engagement with employees** who have clear missions, are empowered, and have a clear focus on the customer/user.

- Begin at the beginning with applying the **design thinking** mindset, tools, and methods before building agile solutions.

- Become a customer-centric organization that constantly adapts to new and changing customer needs, and delivers superior experiences to **satisfy and delight customers**.

- Advance **operational performance** in terms of objective achievements, speed, predictability of outcomes, and by fully dedicating teams to targets, objectives, and key results.

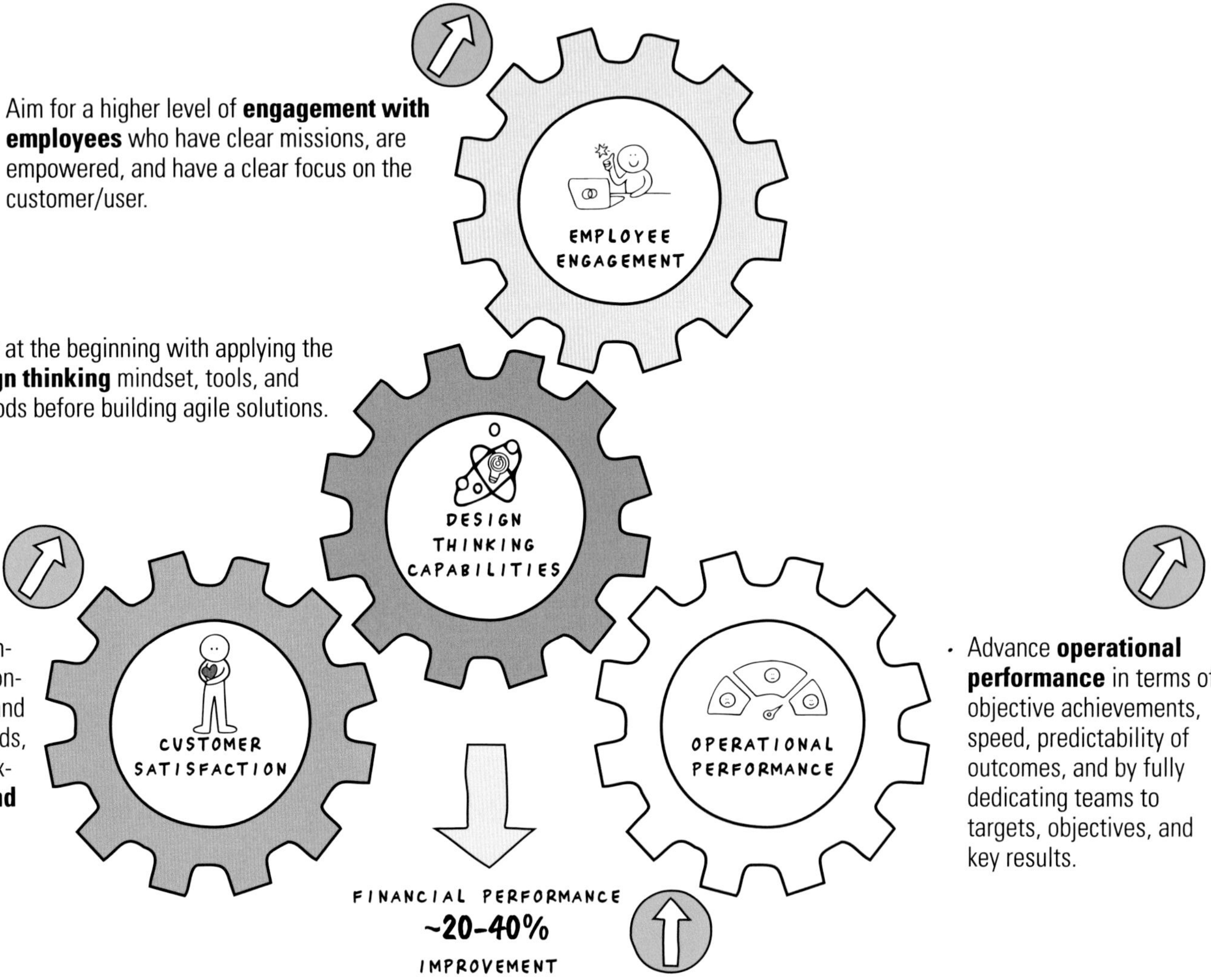

EMPLOYEE ENGAGEMENT
DESIGN THINKING CAPABILITIES
CUSTOMER SATISFACTION
OPERATIONAL PERFORMANCE
FINANCIAL PERFORMANCE
~20-40%
IMPROVEMENT

The Importance of Understanding Impact

The best measurement system will not be effective if the core concept of impact is not understood by leaders, teams, and individuals. Making an impact is not isolated to **WHAT** everyone in the organization is doing. It goes together with **HOW** everyone does it and **FOR WHOM**. These three elements are closely linked and make a big difference in the definition of objectives compared to more traditional measurement systems.

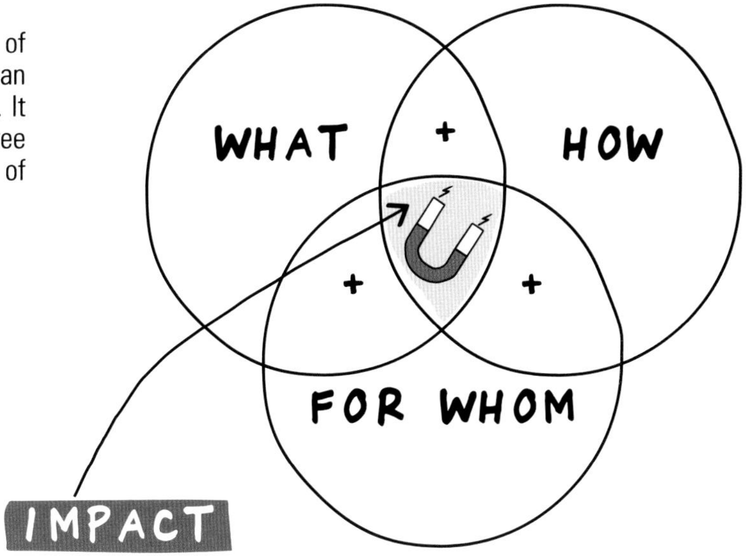

WHAT
HOW
FOR WHOM
+
+
+
IMPACT

Impact is not just...

- goal-oriented and rigid work,
- being present and delivering good results,
- achieving the individual year-end goals, and
- delegating unexpected challenges to others.

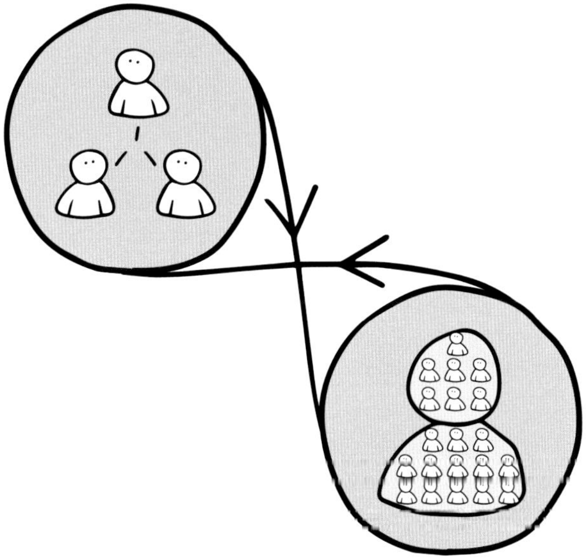

...it feels more like

Allowing radical collaboration
Collaborate in an impactful way and declare your behavior.

Positively affecting each other in the team/organization
Make others successful because of an individual contribution.

Overcoming adversity and being a role model
Each individual responds to challenging circumstances and learns lessons from them.

Be goal-oriented and smart at work
Focus on the right priorities and ask each other for feedback by putting the team's success ahead of your own.

Objectives Are Not Task Lists

The objectives should be defined by the teams, which are responsible for the tasks. In contrast with the traditional approaches, they do not have the prescribed path of being cascaded and delegated from top to bottom in the organization. The strategy and guiding principles from the board of directors provide the basis for the North Star to which teams align. The strategy should also communicate the common purpose and a shared vision based on customer needs for the entire company. Thus, the North Star is critical for transformation and radical collaboration among teams as it informs all decisions and missions and provides a common language for the entire organization. However, it is often observed that the vision is not clearly derived from the customer needs. In this case, the teams are again responsible for exploring the needs themselves within the first mission to be completed based on the team objectives.

Traditionally, cascading is nothing more than cutting and pasting goals from the top to teams. Aligning for impact means that teams and individuals consider where they can have the greatest impact on priorities.

VISION
MISSIONS
GOALS
STRATEGIES
TACTICS
ACTION PLANS

1. ALIGN THE VISION WITH THE OBJECTIVES.
2. DECIDE ON WHICH OBJECTIVE TO COLLABORATE.
3. BREAK THE OBJECTIVES DOWN IN MISSIONS.
4. SET KEY RESULTS FOR EACH MISSION AND START WORKING ON IT.
5. DELIVER KEY RESULTS.
6. DEFINE TASKS.
7. PLAN YOUR ACTIVITIES.

The Six Golden Rules for Objective Setting

1. Keep It Simple:
Objectives should be memorable and easy to understand.

2. Be Ambitious:
Objectives are meant to be bold and ambitious, and should be challenging for the team.

3. Be Transparent:
Keep all objectives in a publicly accessible folder (e.g., SharePoint) so all team members can access them.

4. No Tactics:
Avoid turning objectives into a to-do list. They should add value and inspire learning rather than efficiency in completing tasks.

5. Track Regularly:
Track your teams' objectives on a weekly basis. This approach is not about simply reporting back, but reflecting on what is working, where the teams are getting stuck, and what the team can do differently.

6. Stay Agile:
More frequent adjustment (e.g., quarterly) of objectives is critical for innovative teams, as customer needs and the environment change quickly.

TIPS AND TRICKS FOR IMPLEMENTING OKRs

How to Implement a Revised Performance Measurement System

The central question of **WHAT** the performance measurement system should be and **HOW** it should evolve is often driven by key desires for change and associated goals.

In more advanced organizations, individual objectives are dispensed entirely, and objectives are measured only at the team level. Practical application proves that this approach has a very strong effect on cultural change and is highly recommended.

The aspirations usually range from a desire for more focus; the prevention of silo effects; greater transparency of strategic goals, initiatives, and growth areas; to the establishment of a revolutionary team-of-teams culture. Thus, the initial impetus for a more modern performance measurement system comes primarily from the top level of an organization along with the desire of the teams to bear more responsibility. Both must be a committed part of a realignment. All variants of performance measurement systems, including the well-known OKRs, are not a new way to delegate tasks for management, but to become an active part of the team!

In addition, it is recommended not to stick to organizational silos. Working beyond departmental boundaries and future alignment on objectives and key results is an important part of transformation and change for many organizations. Thus, it should be emphasized that, especially in a transformation phase of performance measurement systems, the teams that have come together for the initiatives should define their objectives before the focus and extension are put on the individual objectives and respective key results.

The best way to learn about a new performance measurement system is to start linking objectives to real-world projects and initiatives.

It is especially important for teams to know their purpose and how to align their activities with a larger aligned goal. Core questions include:

- **Why does our team exist and what is the purpose of our work?**
- **What is the most important objective the team should focus on (in the short, medium, and long term)?**
- **What do we need to do and what do we require to achieve the objectives as a team?**

Depending on the team, the problem, and the initiatives to be accomplished, the team's objectives can have an outward or inward orientation as the object of measurement.

Typical internal objectives relate, for example, to improving the employee experience as part of a realignment of interaction with employees from the initial contact and beyond the last working day. Objectives can also focus on creating a greater sense of *we* and aim to achieve a bowl of trust.

External objectives often refer to the expansion of market areas or collaborating with other actors in a business ecosystem initiative. External objectives also can refer to the acquisition targets of certain customer segments or the expansion of features of existing products and services on the market.

An example of OKRs related to customer satisfaction during MVP testing is provided on page 274. The example from a real project also shows the challenge of focusing on the outcomes related to the value the team should create and the effect of the activities. Well-set key results are quantitative lead metrics that are accepted by the stakeholders and influenced by the work performed from the team. The Cheat Sheet for Formulating OKRs (see page 275) is very helpful for teams and individuals setting objectives and key results for the first time.

OKR

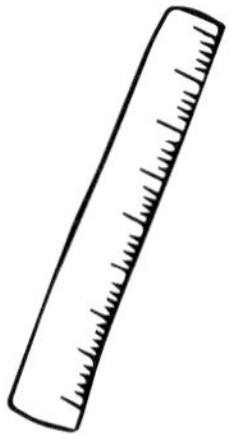

KEY QUESTIONS TO BE ANSWERED DEFINING CONCRETE OBJECTIVES

- **What is the ultimate goal?**
- **What do we want to achieve?**
- **What will have changed when we reach the goal?**

The objective is the description of what the company, team, or individuals want to achieve together. It is in most cases and based on a clear direction.

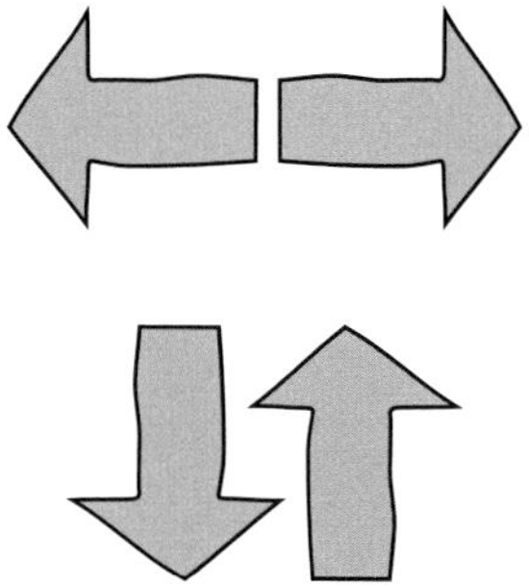

DEFINITION OF THE DISTANCE TO THE DESIRED OUTCOME (SHORT, MID-, OR LONG TERM) BASED ON KEY RESULTS

- **How can it be determined whether the activities contribute to getting closer to the goal?**
- **How can progress be measured?**

The key result is basically metrics with a starting value and a target value that measures the progress toward a short-, mid-, or long-term objective. Key results should not relate to specific solutions or specific tasks.

WORKING ON INITIATIVES AND PERFORMING ACTIVITIES

- **What is concretely done within an initiatives to progressively contribute to the objective and to the overarching vision?**

A specific initiative is based on a set of activities. It describes the work the team has to perform to reach a short-, mid-, or long-term objective. The activities and the initiatives should have a positive impact on the key results. Related to the team's efforts regarding the key results are the following: Keep it, leave it, or change it. This means that if the metrics do not improve, the approach must be adjusted, or the activities changed.

Defining OKRs Based on Observed Situation

Example

Situation: During MVP testing, it becomes apparent that customer satisfaction with the product experience is low. The team wants to increase customer satisfaction and establish a data-driven basis of possible measures.

Objective:

We know which critical experiences and functions are highly relevant to customers and influence their satisfaction when engaging and using the offering.

The objective is formulated from the customer/user point of view.

Key Result #1:
Conduct an extraordinary online customer survey which generates at least 100 strong customer/user statements to develop the top three actions that will improve the customer experience today.

Based on two outcomes, the potential actions to improve customer satisfaction are clear.

Key Result #2:
Perform 10 in-depth interviews with test customers who indicated a low level of customer satisfaction in order to validate the hypotheses made for the development of the prioritized actions.

Validating the hypotheses based on customer feedback is considered an outcome for the team.

Key Result #3:
Define three meaningful minimum viable exploration metrics that allow constant measuring of prioritized actions and periodically over a quarter checks if there is a correlation between actions and increased customer satisfaction.

Monitoring the actions implemented is key for the team to prove evidence of measurements.

In the Practical Application, It Is Recommended to Apply the Principles of OKR Setting in the Best Possible Way:

- The objective should be a qualitative, outcome-oriented, ambitious, and inspiring goal that describes what needs to be achieved.
- The corresponding key result is quantitative and describes how we will get to the objective. It also clearly states whether the objective has been achieved.
- Both objectives and key results describe the outcome, the benefit or value contribution, and not the output, or tasks.

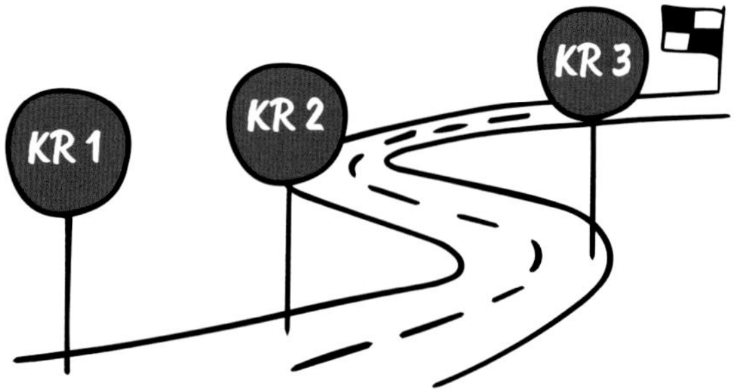
KR 1
KR 2
KR 3

Cheat Sheet for Formulating OKRs

Template download

CUSTOMER / USER

Who Is the Customer / User?

(Based on Personas or Customer Segments)

INSIGHT, ASPIRATION, FORESIGHT

BASED ON ______________________________

(foresight, insights, customer feedback)

For example:

- Ethnographic research on Generation Z reveals that TikTok is the predominant search engine.
- Since 2022, data shows a high demand for travel experiences in the Metaverse.
- A global shortage of gas and oil will boost the demand for renewable and alternative energy sources by 2040.

EVIDENCE, INSIGHTS, BELIEFS, BETS

WE BELIEVE ______________________________

(actions)

WILL RESULT IN ______________________________

(outcome)

WE KNOW/ASSUME THE TEAM, ORGANIZATION, OR COMPANY WILL BE SUCCESSFUL WHEN THE OUTLINED KEY RESULTS ARE REACHED

____

(see below leading or lagging indicators)

LEADING INDICATORS

KEY RESULTS THAT PREDICT SUCCESS

For example:

- Double the number of users every quarter.
- Promote all offerings via ecosystem actors quarterly.

LAGGING INDICATORS

KEY RESULTS THAT SHOW THE EFFECT

For example:

- Increase market share from 30 to 45 percent.
- Improve NPS from 50 to 70.

CONFIDENCE LEVEL TO REACH OKRS

Scoring between 50 and 100 percent.

DOWNLOAD TOOL
www.design-metrics.com/en/cheat-sheet

Different Scoring Systems for Measuring Performance

The respective key results should ideally be based on a coordinated and pre-defined scoring system. Different approaches can be observed in corporate practice. Mostly, the focus of the selection is on simplicity, accountability, outcomes, or the desire for radical innovations. It is recommended to make clear and transparent what is to be achieved with the scoring system and why it is used in the current phase. Depending on the level, the confidence interval can be set at 50 percent, at stretch, at 10 percent, or up to 90 percent for objectives that require a higher level of commitment. The maturity of the definition of objectives and key results often reflects the development status of the associated measurement system. Without the appropriate deployment parameters and a clear North Star, the implementation of an appropriate scoring system will only be successful to a limited extent. Similar to the design thinking capability building programs introduced in the 101 section, it has proven useful to start with a smaller set of pilot teams, and to anchor the scoring system well and appropriately before scaling it and the final setting of the measurement system.

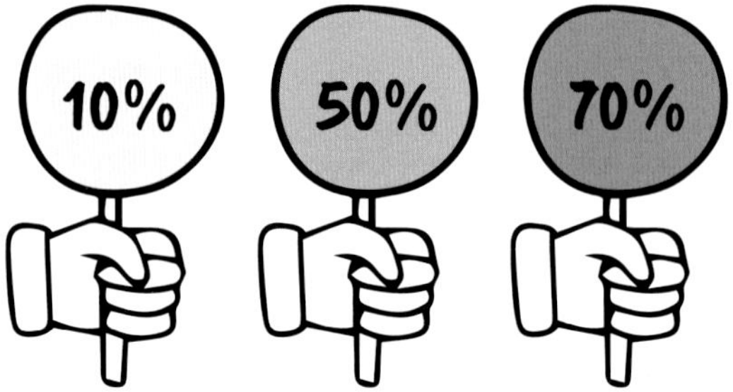

10%
50%
70%

	Moonshot	Roofshot	Everest
MAIN FOCUS	Outcomes	Simplicity	Radical innovation
BENEFITS	Possibility for establishing stretch thinking	One very standardized way of scoring	Appropriate to score in a dynamic environment
CONSTRAINTS	Confidence level/aspirational target level	Difficult to score in the middle of the cycle	High energy and engagement level needed to score
CONFIDENCE LEVEL/ ASPIRATIONAL TARGET LEVEL	⟹ **Set, for example a 10 percent confidence level, for stretch, 50 percent confidence level to target.**	⟹ **Set, for example, a 90 percent confidence level**	⟹ **Set a confidence level suitable for the initiative, for example, a an aspirational target level of 70 percent**

FACILITATING PERFORMANCE OBJECTIVES

How to Coach and Assist in the Formulation of Objectives & Key Results

It is important to clearly communicate **WHAT** the new or adjusted measurement system is intended to do and for it usually to exist in parallel with the ubiquitous individual measurement criteria that are bonus relevant. Another challenge is often inherent in very large organizations that are decentralized in many respects. Here, regional targets and indicators are added to the OKRs and individual pay-related components. All of this can quickly become overwhelming. However, the journey usually starts much earlier. A clear corporate strategy that provides the teams with the guidelines for formulating their own objectives is central to the establishment of a measurement system.

The objectives at the leadership level are also important, as they send out a signal and also contribute to the commitment of the entire organization to the respective system and structure to measure. Company-level objectives should have the business goals in mind, preferably start with a verb (optimize, improve, redesign, increase), and be qualitative in nature. For example, based on a formulated business strategy, the objective could be: Reduce CO2 emissions! A coach or facilitator should immediately follow up with the question "Why now, and what evidence is there for this?" This combined question helps not only to get superficial arguments, but also to retrieve the facts. These, in turn, help to define the key results. Depending on the level of ambition, these tend to be short, medium, or long term. A CO2 reduction of 10 percent may be possible in the short term, whereas CO2 neutrality has a rather longer-term character. In most cases that have an increased complexity and require new and previously unknown solutions, the teams must also be given the necessary freedom to leave the existing creative framework.

In the OKR toolkit (starting on page 286), a possible journey of envision, focus, create, define, mobilize, and (re-)define is described via divergent and convergent thinking, which are particularly suitable for innovation in the context of this book.

Important Steps from Envisioning to Measuring Key Results:

1. **Envision** a shared view of the long term directed to the organization and the team-of-teams.
2. **Focus** on strategic priorities for which a potential team has the opportunity for the greatest impact.
3. **Create** the Moonshot, Roofshot, or Everest objectives with a bold statement and describe missions with objectives to be completed.
4. **Define** key results for each objective or cluster and define success for each mission.
5. **Mobilize** skills and capabilities needed to complete missions for distinct phases of the design cycle (understand, create, implement, lead) to ensure that the critical mass of resources aligns to the missions.
6. **(Re)define** first key results to accomplish first objectives based on activities performed.

QUICK CHECKLIST THAT ENCOMPASSES TEAMS AND INDIVIDUALS

- ✓ The objective should be ambitious and inspire the team.
- ✓ Each objective should have two to five key results.
- ✓ The key result should be measurable and specific (numbers, currency, etc.).
- ✓ Avoid vanity metrics that just make the team look good (number of ideas, likes, page views, etc.).

Communicate Clearly the Benefits for Employees, and Walk the Talk

Depending on the company and organization, 70 to 95 percent of employees working in a company with a more traditional measurement system and organizational structures still do not understand or know the strategy of the company they work for and how their work contributes to the larger goals.

KEY QUESTION

How can the leadership team ensure that everything the teams and employees do actually contributes to the implementation of this strategy?

Modern measurement systems should help make the company an attractive place to work and have a clear path to success, and they should help teams and individuals consistently achieve their goals.

These systems help companies become a great place to work by giving teams more autonomy. They put people in charge of goals without dictating how they should achieve them. Plus, everyone in the company can see where the company wants to go and how their work contributes to that.

Improved measurement systems and a new way of defining objectives will help provide a clear path to success by offering more clarity about the strategic priorities in the company and making them transparent to the entire organization. This allows everyone to focus on what is really important.

Teams are able to consistently achieve their goals (depending on the definition of success for Roofshot, Moonshot, and Everest goals) by following a simple structure. This structure is based on real data, so teams can ensure they are actually driving the business forward.

Example

KEY POINTS EXPLAINING THE TRANSITION TO A NEW PERFORMANCE MEASUREMENT SYSTEM

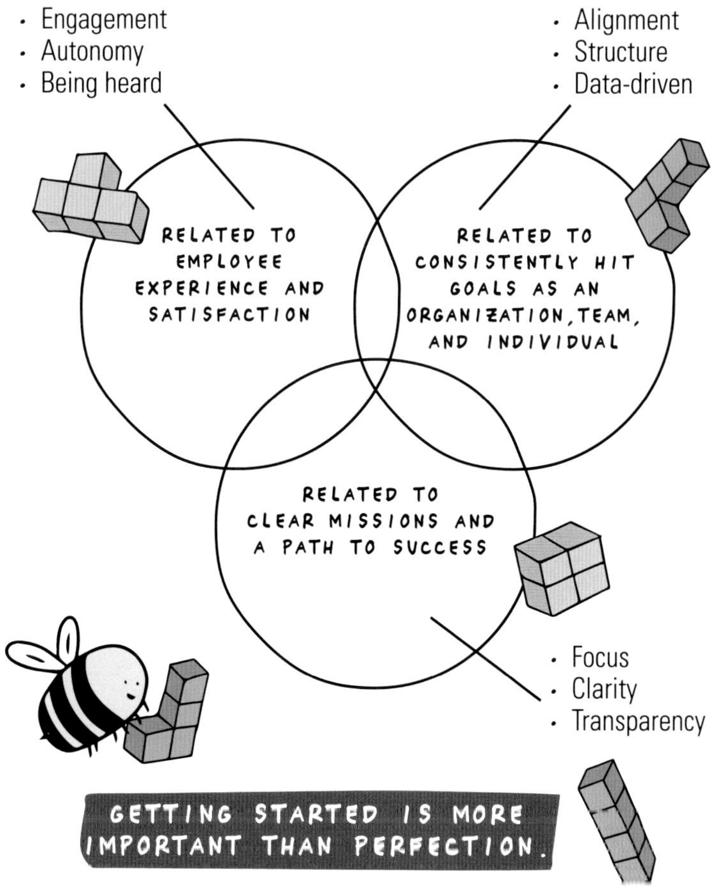

- Engagement
- Autonomy
- Being heard
- Alignment
- Structure
- Data-driven
RELATED TO EMPLOYEE EXPERIENCE AND SATISFACTION
RELATED TO CONSISTENTLY HIT GOALS AS AN ORGANIZATION, TEAM, AND INDIVIDUAL
RELATED TO CLEAR MISSIONS AND A PATH TO SUCCESS
- Focus
- Clarity
- Transparency

GETTING STARTED IS MORE IMPORTANT THAN PERFECTION.

Use of Metaphors and Simple Comparisons to Explain Different Levels of Ambition

It is very important that the teams and individuals get a good sense of the ambition at hand. Easy-to-imagine metaphors and comparisons help to classify and make leveling part of the team's discussion:

Imagine the ambition relates to the North Star of becoming the best sailing team in the world. One possibility to prove this is for the team to win the America's Cup. The Cup is the oldest and one of the most prestigious sailing competitions in the world, with a history of over 170 years.

Example

Moonshot and Everest objectives are usually audacious, inspirational, and focused on an area of great impact to the team members, the team, and the organization.

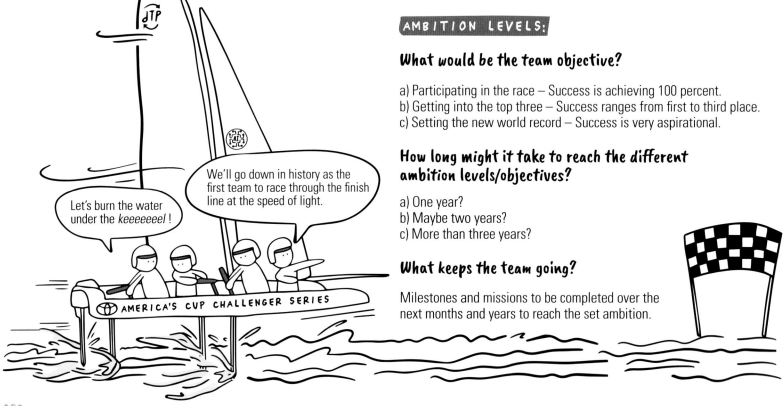

dTP
Let's burn the water under the *keeeeeeel* !
We'll go down in history as the first team to race through the finish line at the speed of light.
AMERICA'S CUP CHALLENGER SERIES

AMBITION LEVELS:

What would be the team objective?

a) Participating in the race – Success is achieving 100 percent.
b) Getting into the top three – Success ranges from first to third place.
c) Setting the new world record – Success is very aspirational.

How long might it take to reach the different ambition levels/objectives?

a) One year?
b) Maybe two years?
c) More than three years?

What keeps the team going?

Milestones and missions to be completed over the next months and years to reach the set ambition.

How to Coach Teams and Individuals on a Regular Basis

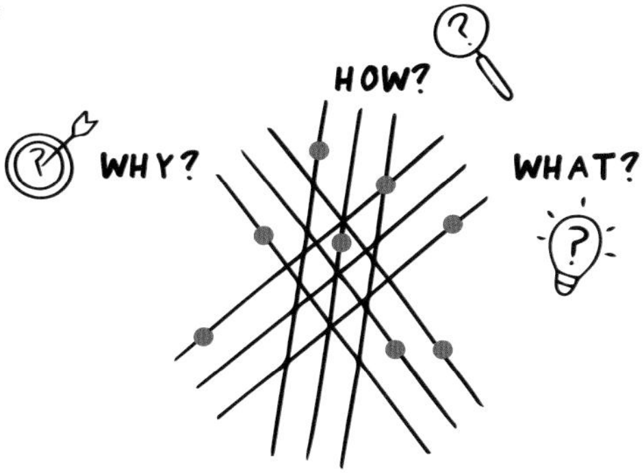
HOW?
WHY?
WHAT?

Regular check-ins initiated by coaches and management members with the teams and team members provide information on whether the overall and desired objectives are being achieved. The reflective questions usually relate, for example, to the impact, focus, communication, desired transparency, or learning success (depending on the organization).

IMPACT Why have the key result(s) impacted your team?	• How did the team inspire and encourage those around you? • How did the team explore bold ways to power the purpose?	• What will your team do in the future to engage more teams and to live our purpose?
FOCUS Why has the focus changed and why are the new objectives of higher priority?	• How did the team earn the trust and belief of others in doing what's appropriate? • How did the team stop and delegate less important tasks, projects, and initiatives?	• What will your team do in the future to focus on the important things and not waste time on trivialities?
COMMUNICATION Why was it important to set objectives on different levels and across organizational communication?	• How did the team share insights from exploring the problem space and/or ideas generated in the solution space? • How did the team embrace feedback to grow personally or as a team?	• What will your team do in the future to align to the overall objectives, sharing knowledge, and success stories?
TRANSPARENT Why does transparency get us further than silo thinking?	• How did the team provide and seek for clarity? • How did the team remove barriers for other teams and/or hold yourself or the team accountable?	• What will your team do in the future to continue to serve others and to own teams' actions?
ORGANIZATIONAL LEARNING Why did the team learn the most from this key result, and is it a benchmark for others?	• How did the team seek opportunities to experiment and learn? • How did the team contribute to the collective success across all teams?	• What will your team do in the future to continue being self-aware, reflective, and open to learn?

High-Level Visualization of Key Results at Three Levels: Company, Team, and Individual

Temporary worksheets can be used for the initial definition of the respective objectives and key results. In particular, it seems important that individual missions and short-term objectives in design thinking and innovation projects can also change depending on the state of knowledge, since often the exploration of the problem space brings completely new insights that have an influence on the objectives and key results.

The longer-term objectives can also change depending on the teams' degrees of freedom, especially with Everest objectives. In organizations where individual objectives continue to be set, these must also be made transparent and visualized. There are also several applications and tools on the market that allow tracking key results over the short, mid-, and long term across the organization and making them transparent on dashboards.

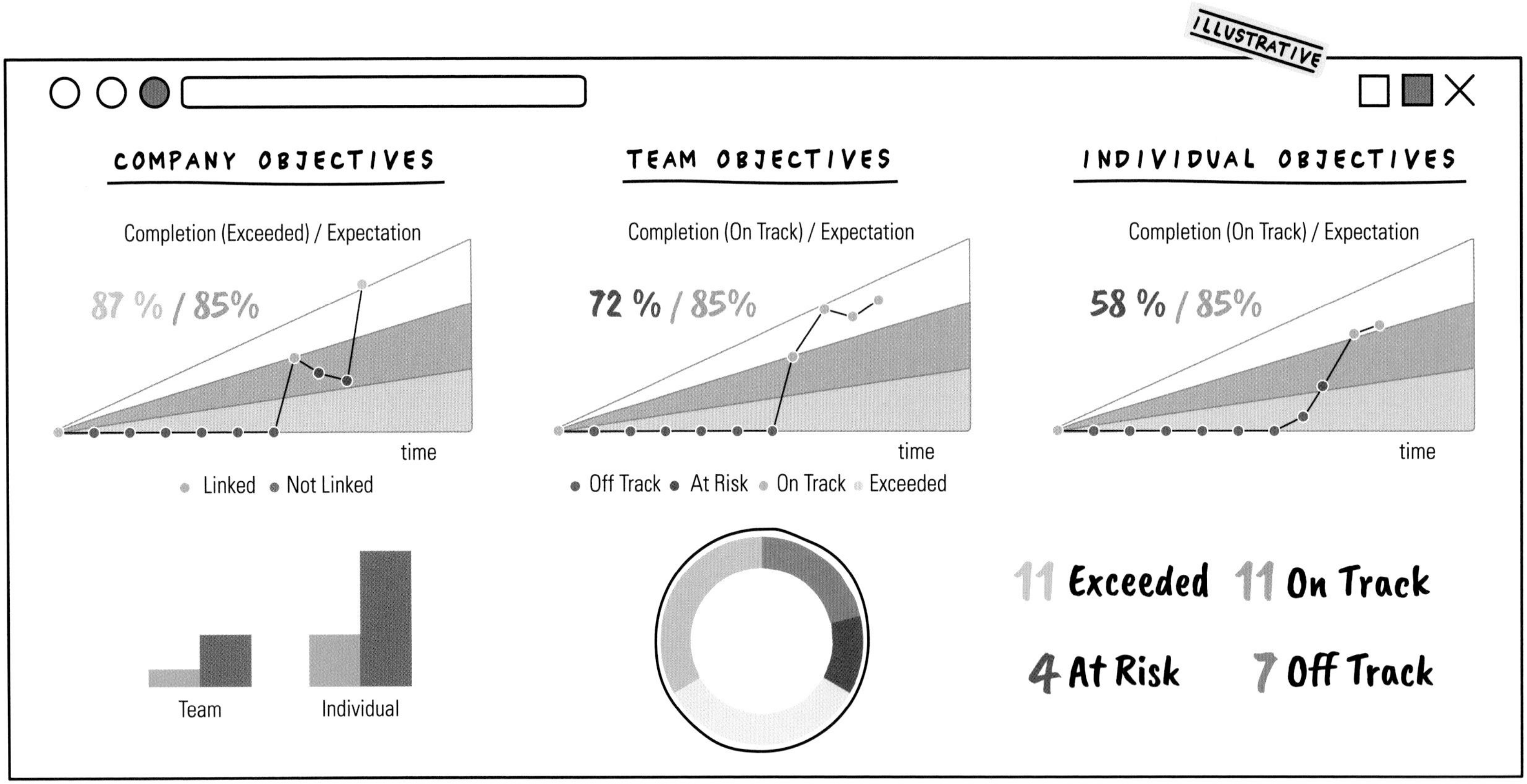

ILLUSTRATIVE
COMPANY OBJECTIVES
Completion (Exceeded) / Expectation
87 % / 85%
time
Linked
Not Linked
Team
Individual
TEAM OBJECTIVES
Completion (On Track) / Expectation
72 % / 85%
time
Off Track
At Risk
On Track
Exceeded
INDIVIDUAL OBJECTIVES
Completion (On Track) / Expectation
58 % / 85%
time
11 Exceeded
11 On Track
4 At Risk
7 Off Track

A Typical OKR Cycle

The North Star and strategic choice are the basis for setting the priorities before the next cycle. Usually in established systems, the alignment starts before the end of the quarter with informal priority discussions. Based on a retrospective/lessons learned, the objectives, and key results for the next quarter are set by leadership team.

Teams and individuals review objectives and key results, provide feedback, and align team and individual objectives and key results. The monitoring process is based on weekly check-ins and mid-cycle reviews. The achievements are based on the set scoring system, for example 70 percent achievement (see different scoring systems on page 276):

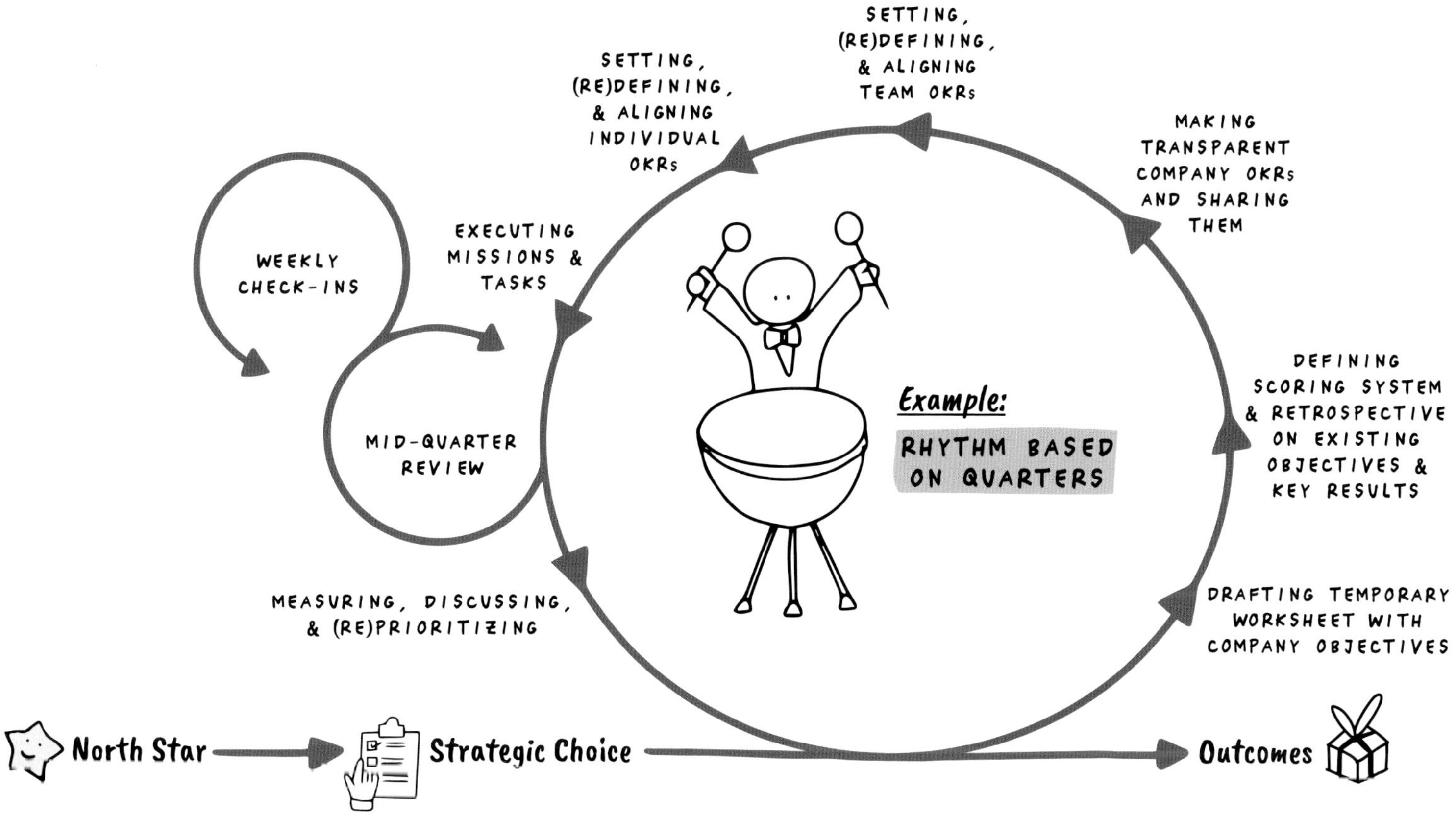
SETTING, (RE)DEFINING, & ALIGNING TEAM OKRs
SETTING, (RE)DEFINING, & ALIGNING INDIVIDUAL OKRs
MAKING TRANSPARENT COMPANY OKRs AND SHARING THEM
WEEKLY CHECK-INS
EXECUTING MISSIONS & TASKS
MID-QUARTER REVIEW
Example:
RHYTHM BASED ON QUARTERS
DEFINING SCORING SYSTEM & RETROSPECTIVE ON EXISTING OBJECTIVES & KEY RESULTS
MEASURING, DISCUSSING, & (RE)PRIORITIZING
DRAFTING TEMPORARY WORKSHEET WITH COMPANY OBJECTIVES
North Star
Strategic Choice
Outcomes

To the Point!

It is important to make the benefits of a new or adjusted measurement system clear to everyone. This includes higher engagement, more autonomy, better alignment, and focus on priorities.

Objectives are action-oriented, inspirational statements that connect the work with purpose. The objectives simply answer the question "What do we want to do?"

Key results are answering the question "How do we know we are making progress?" Key results provide clarity on accountability and are aligned, result based, measurable, challenging, and mostly leading indicators.

Outcome metrics, for example, refer to behavioral change, customer satisfaction, referrals, or lean process metrics. The result is usually measured in financial indicators based on profits, sales results, market share, and throughput.

OKR TOOLKIT

Create Objectives for Design Thinking & Innovation Teams

There are many ways of creating the objectives together with the team. The central element is to align the objectives with the strategic priorities of the company or business unit. This ensures that the objectives have the greatest impact. It has proven useful to accompany the identification process with facilitation and to include well-known methods from design thinking in order to work on the appropriate problems and not to jump directly into the implementation of predefined solutions and existing ideas. The OKR toolkit provides different templates and procedures which support facilitators and teams in defining OKRs. However, the tools must be adjusted to the context and goals of the measurement system.

Qualified teams with experience in defining objectives can usually realize OKR definition in a single workshop. Teams working with objectives for the first time usually need several interactions and workshops until the team objectives with the respective first missions to be worked on are ready for concrete actions.

Possible Steps in Effective Work with OKRs

1. **Envision** a shared view of the long term directed to the organization and the team-of-teams.
2. **Focus** on strategic priorities for which a potential team has the opportunity for the greatest impact.
3. **Create** the Moonshot, Roofshot, or Everest objective with a bold statement and describe missions with objectives to be completed.
4. **Define** key results for each objective or cluster and define success for each mission.
5. **Mobilize** skills and capabilities needed to complete missions for distinct phases of the design cycle (understand, create, implement, lead) to ensure that the critical mass of resources is aligned with the missions.
6. **(Re)define** first key results to accomplish first objectives based on activities performed.

These steps will be explored on the following pages.

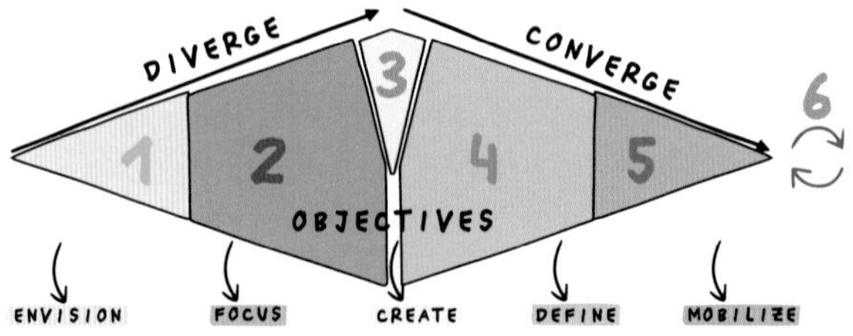
DIVERGE
CONVERGE
1
2
3
4
5
6
OBJECTIVES
ENVISION
FOCUS
CREATE
DEFINE
MOBILIZE

Suitable methods & tools

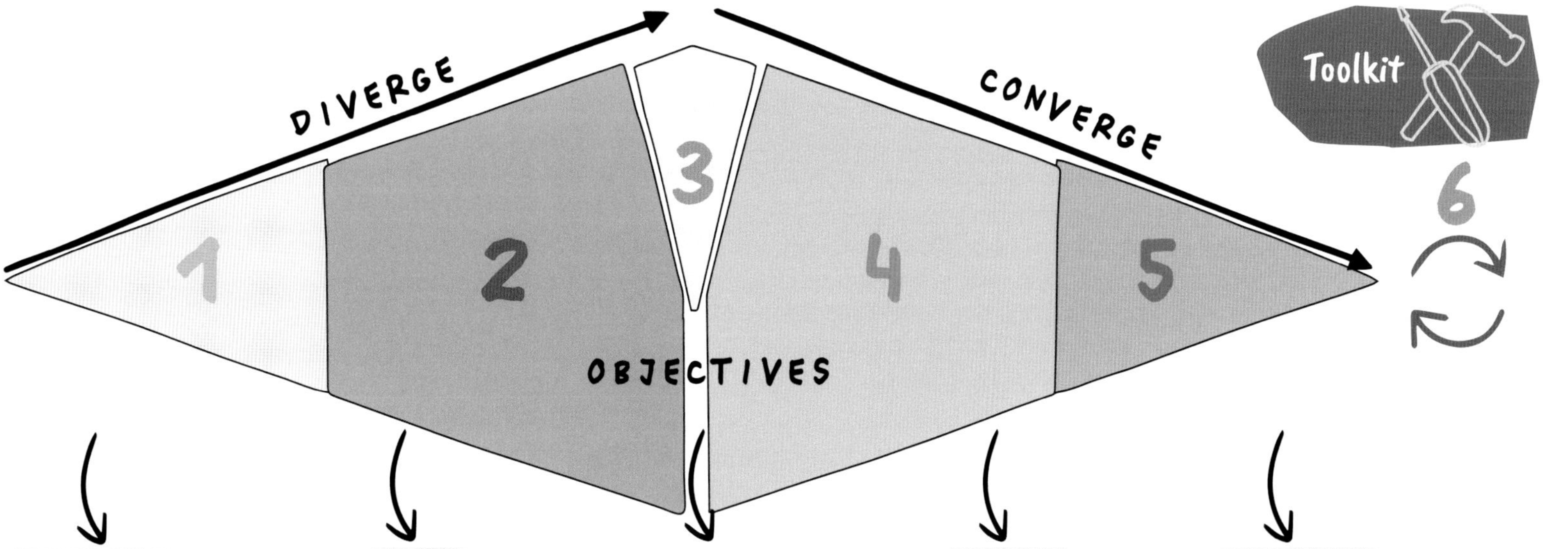
DIVERGE
CONVERGE
Toolkit
1
2
3
4
5
6
OBJECTIVES

ENVISION

Synthesize a shared view of the long term directed to the organization and the teams.

- What are the plausible futures of your industry, business model, or domain?
- How does the future look in 5, 10, or 15 years?
- What are the new, next, and future social, technological, economic, environmental, and political trends?

FOCUS

Specify strategic priorities for which a potential team has the opportunity for the greatest impact.

- What kind of objectives are needed to achieve the priorities (Moonshot, Roofshot, or Everest objectives)?
- Which objectives at the corporate, team, or individual level bring the greatest benefit for the organization?

CREATE

Break initiatives down into missions to accelerate the path toward that long-term destination over the next 12 months.

- Which of these objectives align with the organization's key priorities?
- Which mission must be started now to secure the envisioned future?

DEFINE

Define key results for objectives or cluster and define success for each mission.

- How is success defined for each mission?
- What are the design principles for the missions?
- Who should work on which mission?
- How much time and effort are needed?
- Which kind of capabilities are needed?
- What are potential constraints in achieving the mission?

MOBILIZE

Ensure there is a critical mass of resources aligned with the key missions and the objectives are communicated transparently in the organization.

- Is there an action plan with clear milestones and key results defined?
- Are all stakeholders, team members, and supporting entities informed about the mission, objectives, and key results?

(RE)DEFINE

Adapt or redefine objectives to new factors, key results performance, and/or other conditions, if necessary.

Envision the Future: Strategic Foresight

In order to set the appropriate level for objectives, different horizons of forward thinking might be applied. For Moonshots, it is often necessary to apply, for example, a strategic foresight methodology. Strategic foresight is looking 10+ years into the future and asking, "What will our customers value? What kind of products and services will our customers need, and how can we best serve them?" These questions are different from the ones we usually ask in the context of two to three years of strategic planning, "How can we do this a little better, faster, or more efficiently? How can we improve or expand upon what we're currently providing to our customers?" Strategic foresight means better addressing volatility, uncertainty, chaos, and ambiguity (also known as VUCA).

Strategic Foresight... Helps the Team

- to gain a better understanding of future customer behavior.
- to make the right technology and architecture decisions.
- to create future scenarios that are plausible and relevant.
- to explore different types of disruption (positive and negative disruption with respect to the status quo).
- to scan for signals of change, identify patterns.
- to look back to see forward.

Procedure and Template

- Frame the industry, ecosystem, business model, or domain.
- Apply different techniques for the creation of an aspirational future scenario or the definition of a problem statement. Common techniques are **"What if..."** and **"Imagine that..."** statements or **"How might we..."** questions.
- Start to envision the future, consider strategic guidelines, and make inferences to the already existing vision/North Star.

Envisioning the best possible future is the first step in developing the solutions that will define that future.

Guidance to Talk about the Future

What you (for)see is what you get:
Do better, be more effective, and find a different approach to the solution of a problem.

Take a different perspective:
Analyze the past to be able to shape the future actively.

Think in opportunities:
Identify current growth opportunities and explore them by understanding changed customer needs of the future.

Find solutions that matter:
Find the right way to discover innovations by applying an EXPLOIT and EXPLORE mindset.

Create powerful interdisciplinary teams:
Identify leadership and develop talents that are needed to transform organizations and measurement systems and to collaborate.

Communicate the vision (North Star):
Develop a clear vision, so the teams will be able to survive on their own with a mix of top-down and bottom-up objectives and key results.

Actively shape the future:
Create team objectives and turn missions into actions which can influence tomorrow.

Spark New and Different Directions for the Future with "What if...", "Imagine that...", and "How might we..."

The next few pages describe three techniques from the *design thinking toolbox* that help to look into the future and at the same time initiate the discussion with the team about the problem space and potential solutions. The terms Moonshoot, Roofshot, and Everest exemplify the dimension, scope, and purpose of the objectives. The Envision OKR Canvas (see page 293) supports the brainstorming and documents potential objectives, purposes, and first ideas for leading and lagging indicators. It provides at the same time flexibility to apply different levels of envisioning.

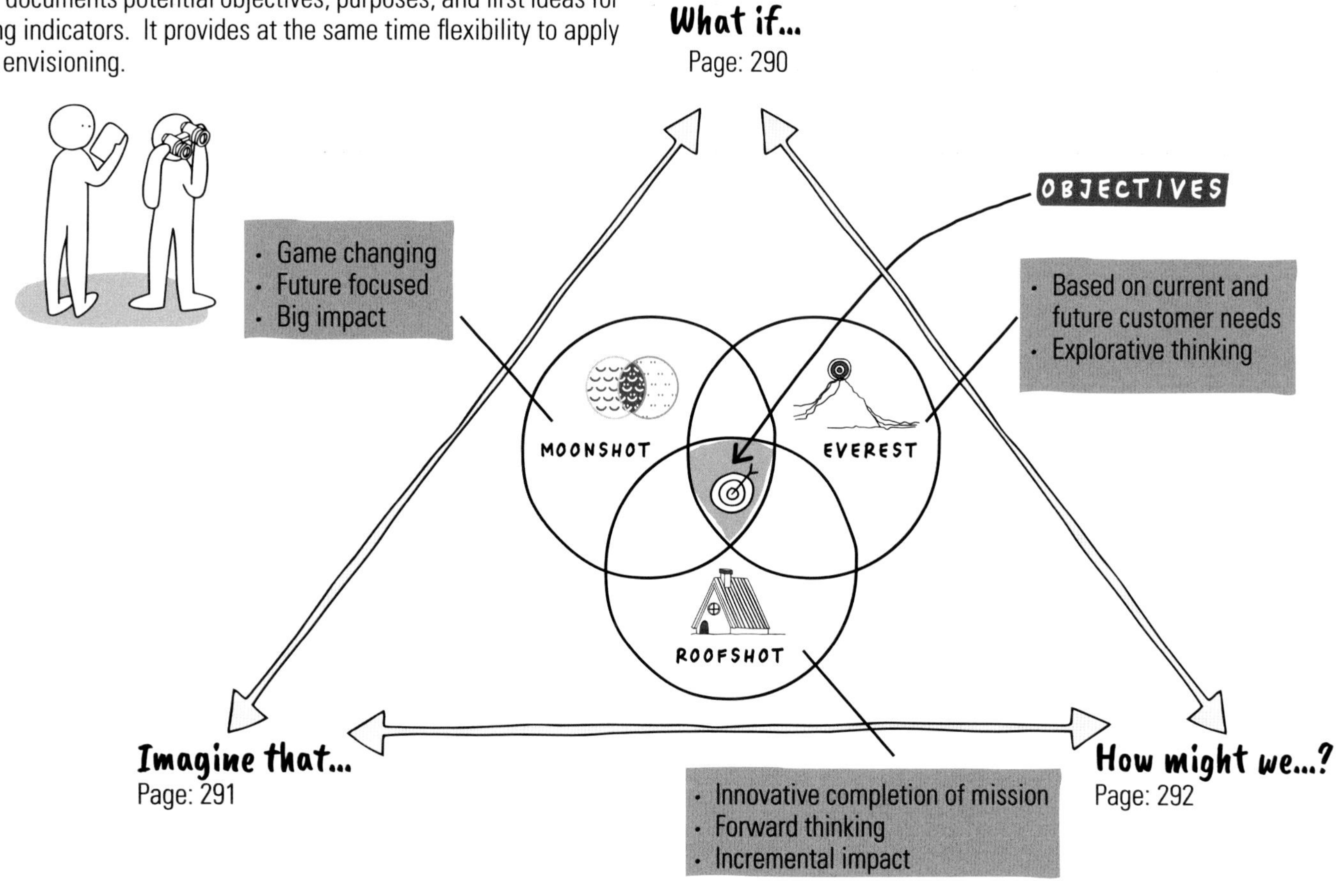
What if...
Page: 290
OBJECTIVES
· Game changing
· Future focused
· Big impact
· Based on current and future customer needs
· Explorative thinking
MOONSHOT
EVEREST
ROOFSHOT
Imagine that...
Page: 291
How might we...?
Page: 292
· Innovative completion of mission
· Forward thinking
· Incremental impact

What If...

From the *design thinking toolbox,* a well-known technique is the "What if..." question, which helps to leave the existing framework and to open up for something new. It is a relatively simple but powerful process to explore potential upside and downside possibilities. The question is often the starting point for divergent thinking and later for identifying the appropriate problem to develop new market opportunities that are real game changers. "What if..." sparks new and different ideas and provides directions for the future, which especially helps to envision Moonshot and Everest objectives.

What if... Helps the Team

- to provoke bigger and out-of-the-box thinking.
- to anticipate possible opportunities, risks, and bets on the future.
- to span the world and open up all conceivable possibilities.
- to explore options that lie away from the day-to-day operations of the business.

Procedure and Template

- Create an initial list of "What ifs" to bring to the workshop.
- Ask, "What are three 'what if' questions we could ask in the current business situation?"
- Establish potential criteria for how the outcomes will be prioritized (e.g., impact on customer/user, time to fix).
- Collect additional "What if" questions from the team or aim for boldness by asking, "What would we do next if we had no limitations?"
- Constantly reflect which questions will deliver the most value.
- Determine what it will take to execute the envisioned future.

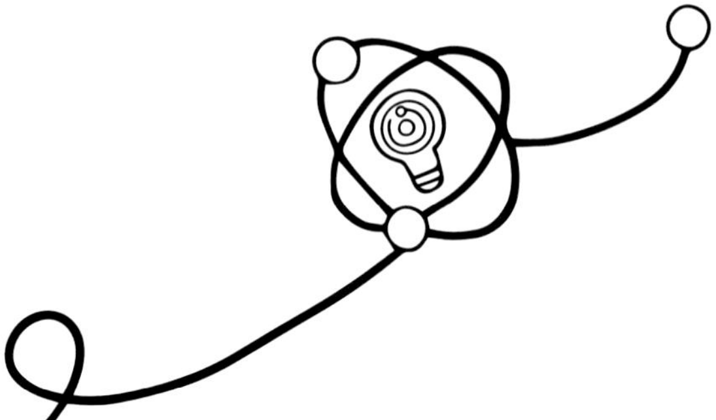

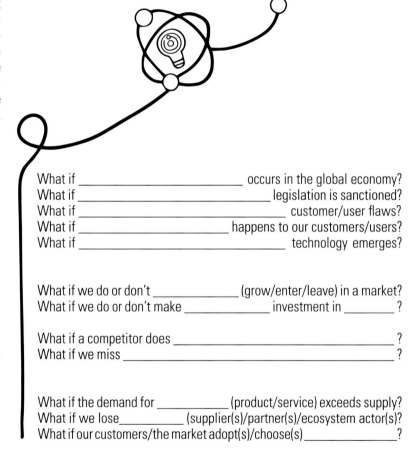

What if ______________________ occurs in the global economy?
What if ______________________ legislation is sanctioned?
What if ______________________ customer/user flaws?
What if ______________________ happens to our customers/users?
What if ______________________ technology emerges?

What if we do or don't ____________ (grow/enter/leave) in a market?
What if we do or don't make ____________ investment in ________ ?

What if a competitor does ______________________ ?
What if we miss ______________________ ?

What if the demand for ____________ (product/service) exceeds supply?
What if we lose ____________ (supplier(s)/partner(s)/ecosystem actor(s)?
What if our customers/the market adopt(s)/choose(s) ____________ ?

Imagine That...

Every dream, strategy, or need includes visualization. A very powerful technique is to start with "Imagine that...", which leads also to focusing, converging, and creating action: it makes a potential future or idea feel somehow real. "Imagine that..." is a verbal cue intended to provoke bigger thinking, which is why it is invaluable in a practice of building, for example, Moonshots. A visual image of the future is an important element of envisioning. After all, the picture of the future allows us to better memorize all the fragments and the hierarchy between them.

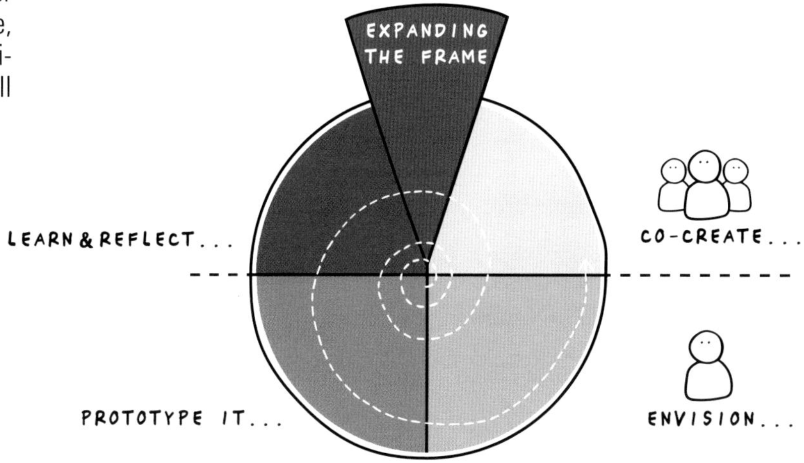

EXPANDING THE FRAME
LEARN & REFLECT...
CO-CREATE...
PROTOTYPE IT...
ENVISION...

Imagine That... Helps the Team

- to focus on core aspects of the desired future.
- to converge from the status quo.
- to create actions for the first initiatives.
- to feel real with the envisioned future.
- to provoke bigger thinking.
- to share an image of the future with the wider team.

Procedure and Template

- Start to imagine the world from multiple perspectives with the team.
- Apply reframing exercises or take the perspectives of customers/users, colleagues, competitors, or other stakeholders.
- Apply symbolic and visual communications to share the potential future or idea with the team.
- Allow creative, subjective, and emotional alternatives to the analytical logic which might be predominantly applied in the team.

Imagination is more important than knowledge. Knowledge is limited. Imagination encircles the world. – Albert Einstein

How Might We...

The "How might we ...." question is also a great starting point for defining an initial problem statement or for creating new point of view based on collected insights and research. The HMW question opens the exploration space to a range of possibilities. In the context of creating objectives or missions, the techniques encourage discussion within the team. Together with asking a minimum of five *whys* and five *hows*, based on the initial thoughts about the topic, the appropriate starting point might be defined. Asking for the *why* is also very powerful for linking purposes to objectives. Both ways of asking are also part of the Envision OKR Canvas (see page 293), which is for most teams the first brainstorming attempt for long-term objectives and first ideas of leading and lagging measures.

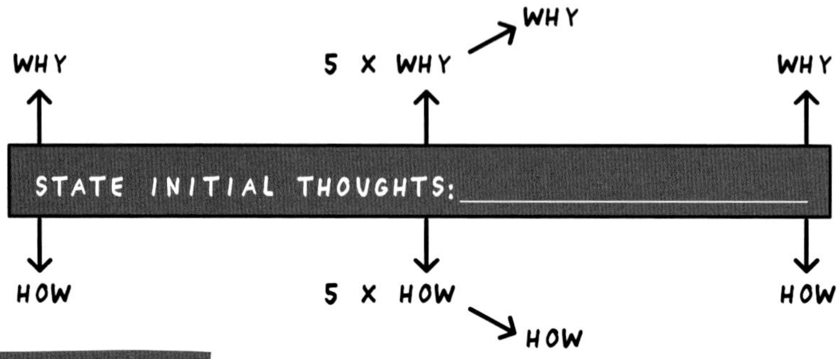

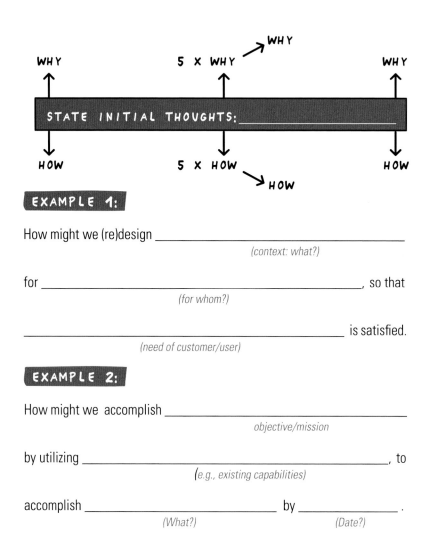

WHY
5 X WHY
WHY
WHY
STATE INITIAL THOUGHTS:
HOW
5 X HOW
HOW
HOW

"How Might We..." Helps the Team

- to set aside prescriptive beliefs.
- to explore a variety of endeavors.
- not to jump directly into solution.
- to team-up with team members with different thinking preferences and T-shaped profiles.

Procedure and Template

- The exploration of WHY and HOW is important for the formulation of the "How might we...." questions and it helps at the same time to envision OKRs.
- The "How might we...." questions can be applied to transform customer needs into a real design challenge (see example 1).
- In the context of formulating OKRs, the "How might we...." question is also helpful in a later phase. It is often applied to investigate how an objective or mission can be accomplished (see example 2).

EXAMPLE 1:

How might we (re)design ______________________________

(context: what?)

for ______________________________, so that

(for whom?)

______________________________ is satisfied.

(need of customer/user)

EXAMPLE 2:

How might we accomplish ______________________________

objective/mission

by utilizing ______________________________, to

(e.g., existing capabilities)

accomplish ____________________ by ______________ .

(What?) *(Date?)*

Template download

ENVISION OKR CANVAS

1 ENVISION NORTH STAR
State and describe the long-term vision or high-level company-wide North Star (for example based on one metric that matters most).

2 STRATEGIC GUIDELINES AND SCOPE FOR ACTION FOR THE TEAM
Describe any boundaries and defined scope of actions for the organization.

3 SELECT AND APPLY TECHNIQUES FOR ENVISIONING
Apply the appropriate techniques to explore the future and to describe the problem statement.

WHAT IF... Questions	**LONG-TERM OBJECTIVES** Aspirational qualitative statement	**WHY?/HOW?** Ask why/how multiple times to explore and adjust the problem space.	**LEADING/LAGGING MEASURES** Brainstorm and describe first high-level activities/milestones or/and outcomes/impact in the context of the objective and purpose.
IMAGINE THAT... Statements	**LONG-TERM OBJECTIVES** Aspirational qualitative statement	**WHY?/HOW?** Ask why/how multiple times to explore and adjust the problem space.	**LEADING/LAGGING MEASURES** Brainstorm and describe first high-level activities/milestones or/and outcomes/impact in the context of the objective and purpose.
HOW MIGHT WE... Questions	**LONG-/MID-TERM OBJECTIVES** Aspirational qualitative statement	**WHY?/HOW?** Ask why/how multiple times to explore and adjust the problem space.	**LEADING/LAGGING MEASURES** Brainstorm and describe first high-level activities/milestones or/and outcomes/impact in the context of the objective and purpose.

4 ENVISION
What is, for example, the bold and ambitious more-than-three-year Moonshot or Everest objective?
How is it linked to the overall vision and North Star of the company or organization?

DOWNLOAD TOOL
www.design-metrics.com/en/envision

Guiding Principles for Envisioning:

Improving Companies' Performance and Resilience, for Example, by a Factor of More Than Five

The first step, envisioning, was important because thinking long term is a state of mind. Especially when it comes to innovation, the mindset influences the decisions and attention teams give to future market opportunities. When the entire organization is focused on a North Star, it has a long-term mindset, thinking in terms of years, not project weeks or shareholder quarters, and achieves the greater results it seeks. When teams think this way, it has the advantage that everyone in the organization can accept that change will take time in the future, so efforts are put into perspective by teams. Another benefit is that a long-term perspective helps anchor the vision in the future. Questions like "What if..." and "Imagine that…", aim at what could be, rather than using today's perceptions as roadblocks. The following examples of guiding principles might serve as an inspiration to allow Moonshot and Everest thinking for the respective teams:

Shift the approach from improving the current offering to solving problems that matter for the customer/user.

Apply strategic foresight (> 10 years) and try to anticipate the new and future customer behaviors and needs.

HI! HELP ME GET TO THE MOON.

Use re-framing to redefine the nature of competition using "What if..." and "Imagine that..."

Change the perspective and questions of all critical assumptions with regard to potential solutions and customer needs.

Focus on a North Star that is ambitious, achievable, measurable, and within the limits of the organization or different teams.

Introduce a new worldview to ensure that myopic beliefs and expert opinions do not interfere with judgment.

Focus: Be Mindful about Purpose, Priorities, and Greatest Benefit

When the North Star, potential future, or mandate to think beyond existing strategic boundaries are clear to everyone, it is the task of the design thinking or innovation teams to focus on the appropriate objectives and ambition levels in order to work purposefully with their activities on the envisioned Moonshot or Everest objective. This requires focus. The *Team Collaboration Canvas*, which is outlined on page 297, is a great supplementary tool to synchronize the team members for the create phase of OKRs and later during team collaboration.

Focus Helps the Team to Answer the Questions:

- What kind of objectives are needed to achieve the priorities? (Moonshot, Roofshot, or Everest objectives)
- Which objectives at the corporate, team, or individual level bring the greatest benefit for the organization?
- How to break long-term objectives in mid- and short-term objectives?

As a general rule of thumb, each team should have no more than three to five objectives at any one time. The quarterly review of each OKR set enables a switch to more valuable goals when changes occur, such as those caused by crises, new insights from interaction with potential customers/users, due to measurements and applied minimum viable exploration metrics, or just market shifts that are natural in today's dynamic environment.

At this point it is also important to note again that key results are not classic KPIs. On the one hand, key results have a purpose and answer the question of whether the team/individual made measurable progress on a specific objective by a defined date. A KPI, on the other hand, can be anything and any metric is often mistakenly considered a KPI. In summary, key results are defined in the context of an objective, originated by the team or individuals, are visible to all employees, aim for alignment across teams and toward the North Star, and inform predominantly near and mid-term focus and prioritization. In addition, objectives and key results will change with the cycle of reflection and progress over the design cycle.

Team & Personal Checklist for Objectives and Key Results

Objectives:

- ☐ Aligned to North Star and strategic choice
- ☐ Directional to a defined objective
- ☐ Reachable, ambitious, or very bold
- ☐ Time bound
- ☐ Do not have too many

Key Results:

- ☐ The objectives are achievable
- ☐ Measurable
- ☐ Not a classic KPI
- ☐ Not a task or to-do
- ☐ Do not have too many

Keep in Mind:

- You should always be aware if the intention of all objectives (team and individual) is to create alignment to the priorities of the company, team, and beyond.
- To do so, all objectives should be public to all peers — everyone has access to everyone else's objectives.

Objectives Styles: Moonshot versus Everest

The focus of objectives and key results for innovation and design thinking teams is the realization of innovation and new market opportunities as well as imaginative and inspiring thoughts about the future, which will fundamentally (positively) change our society, actions, and behaviors. Analogies such as going to the moon (Moonshot) or climbing Mount Everest (Everest) are suitable for many projects aiming for radical innovation. Well-known examples of moonshots include self-driving cars, the realization of the Internet from space, CO2 neutrality, and cell-gene therapy with new success-based business models in medicine. Moonshot objectives are mostly based on the future-back approach (e.g., how the future can be accomplished through foresight). So it is not scenario planning in the narrow sense, but as designed for big impact, including what will significantly change the value proposition, and also which adjacent markets a company wants to edge into. When it comes to realization, the design thinking and innovation teams are guided by the envisioning principles presented on page 288. This creates space for new ideas and agile customer development. In actualizing a big moonshot idea, the teams focus on starting with smaller missions, with the ultimate big, bold objective "of landing on the moon" in the long term.

With the Everest objectives, the goal is closer and already tangible, yet not easy to achieve. It takes courage for an Everest objective in "expedition style." Everest objectives are often engaging, compelling, exciting, stimulating, and passionate. An Everest objective is visionary, and it leaves individuals better for having engaged in its pursuit. However, Everest as a metaphor also represents the harsh, ever-changing landscape of business, and the need to focus on big objectives. Everest objectives serve in many respects as a guide to fine-tuning team performance dynamics and relating lessons learned on the treacherous slopes of the world's highest peak in sub-zero temperatures. These lessons should be compellingly applied to modern approaches for networking organizations.

A powerful tool for aligning teams around Everest and Moonshot objectives is the Team Collaboration Canvas. It is designed to get to know the various team members, document team objectives, and find alignment with the purpose of future activities as a team. The canvas also provides space for defining principles and action items to help orchestrate collaboration and assign responsibilities.

A Moonshot Objective Motivates Teams:

- to complete first missions of longer transition (e.g., more than three years,) with big impact on the world.
- to get inspired to create extraordinary solutions.
- to work on a future with a reasonable chance of success.
- to be part of a meaningful break from the past.
- to tap into new and emerging technologies.

An Everest Objective Motivates Teams:

- to work better and more closely together, as an Everest objective is rarely achieved alone.
- to develop a deeper inner commitment to an initiative.
- to make a high commitment and go the extra mile.
- to embark on a new path of learning.
- to pioneer with a new product or service in the market (analogous to a first ascent of a mountain).
- to redefine the deeper purpose of work and values.

Failure is a part of success, for both Moonshot and Everest objectives, if the team strives to achieve remarkable results.

Template download

TEAM COLLABORATION CANVAS

NAME OF TEAM:

T-SHAPED PARTICIPANTS & THINKING PREFERENCES — 2

What T-shaped profiles are available in the team?
How are the thinking preferences distributed in the team?
What are the additional skills needed?

TEAM GAOLS & OBJECTIVES — 3

What do we want to achieve as a high-performing team?
What are our mid- and long-term team objectives?
What are good time-bond measurements and metrics?

TEAM VALUES — 4

What is the team standing for?
What are common values?
What values are of paramount importance for reaching the team objectives?

PURPOSE — 1

Why are we doing what we do as a team? What is the overall purpose and potential impact?

INDIVIDUAL GOALS & OBJECTIVES (if applicable) — 5

How are the individual objectives aligned to the team objectives?
Which individual goals are in conflict with the team objectives?

INDIVIDUAL EXPECTATIONS — 6

What does each of the team members need to be successful?
What are the expectations regarding collaboration?

PRINCIPLES AND ACTION POINTS — 7

What are the design principles?
How to communicate? How are decisions created?
How are the execution and measurement organized?
How are lessons learned, weaknesses, and development areas of the team and single team members addressed?

DOWNLOAD TOOL
www.design-metrics.com/en/collaboration

OKR
The way up is through prioritized missions, which can be measured as part of OKRs.
NAME & TIMELINE
OBJECTIVE 1
KEY RESULT
KEY RESULT
OBJECTIVE 2
KEY RESULT
KEY RESULT
EVEREST OBJECTIVE
MOONSHOT OBJECTIVE
MISSION 3
MISSION 2
MISSION 1
HI! HELP ME GET TO THE TOP.
To the Galaxy

Create: Team and Individual Objectives

Working at different levels for the short and long term with the North Star in mind gives the teams orientation toward the implementation of necessary missions on the way to creating impact. Particularly with the desire for radical innovations, the scoring system is suitable for setting, negotiating, and, at the same time, aligning objectives of high importance. Through the freedom, transparency, and the perception of the teams at eye level, it is possible to increase team commitment and start the creative process. The breakdown into prioritized missions has proven to be a good intermediate step that helps teams to begin to focus on what is essential for success. The Moonshot/Everest Canvas for Innovation Teams (see page 300) helps (if necessary and desired) to break down the vision into more executable and prioritizable missions.

For innovation and design thinking teams working on the next big ideas, it is appropriate to speak about missions to complete, learning in conjunction with performance settings and the application of OKRs, improving and building new capabilities at the same time.

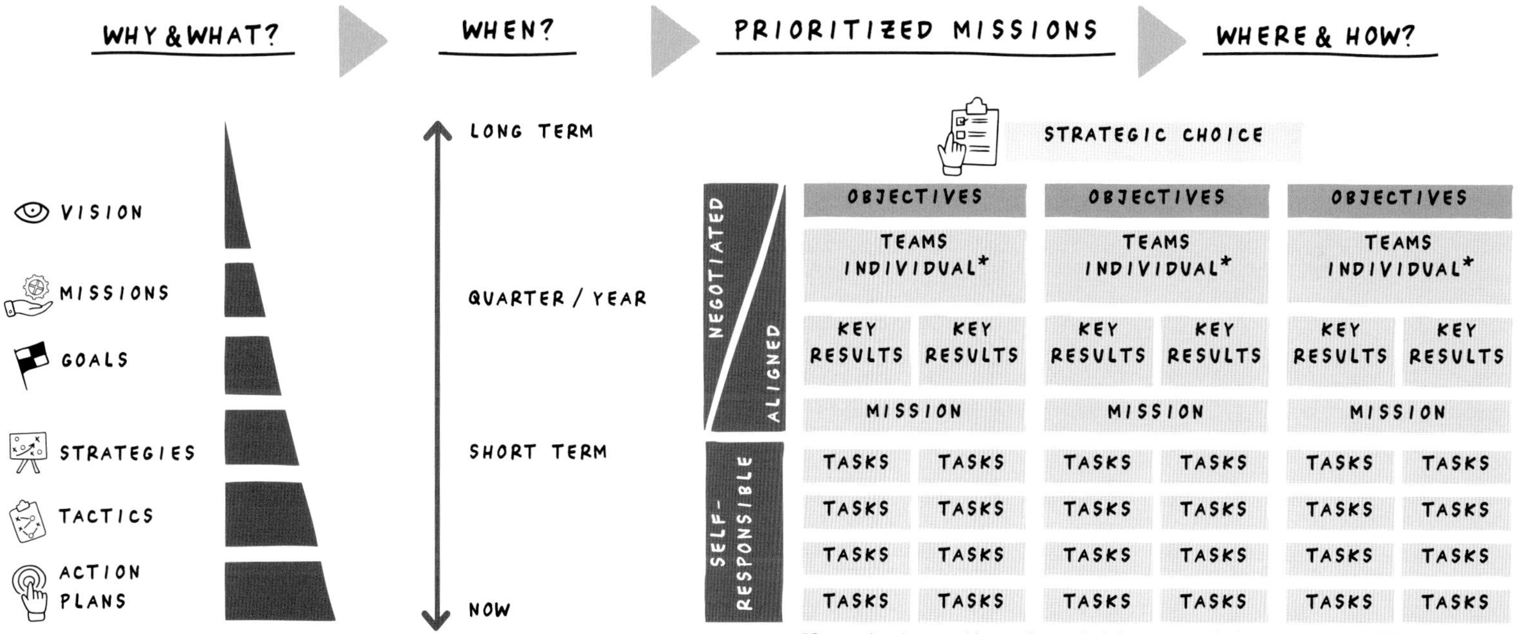

WHY & WHAT?
WHEN?
PRIORITIZED MISSIONS
WHERE & HOW?
VISION
MISSIONS
GOALS
STRATEGIES
TACTICS
ACTION PLANS
LONG TERM
QUARTER / YEAR
SHORT TERM
NOW
STRATEGIC CHOICE
NEGOTIATED
ALIGNED
SELF-RESPONSIBLE
OBJECTIVES
TEAMS INDIVIDUAL*
KEY RESULTS
MISSION
TASKS

*Concerning the actual innovation work, it is recommended not to agree on individual objectives.

MOONSHOT /EVEREST CANVAS FOR INNOVATION TEAMS

Template download

ENVISION
What is the bold and ambitious three-plus-year Moonshot and one-to-three-year Everest objective?
Describe how the ambition will look like in the future.

1

BREAK INITIATIVES INTO MISSIONS
What will the team probably realize over different time periods?
What are the concrete missions the team is able to achieve within 12, 24, 36+ months that will keep the Moonshot/Everest objective on schedule?

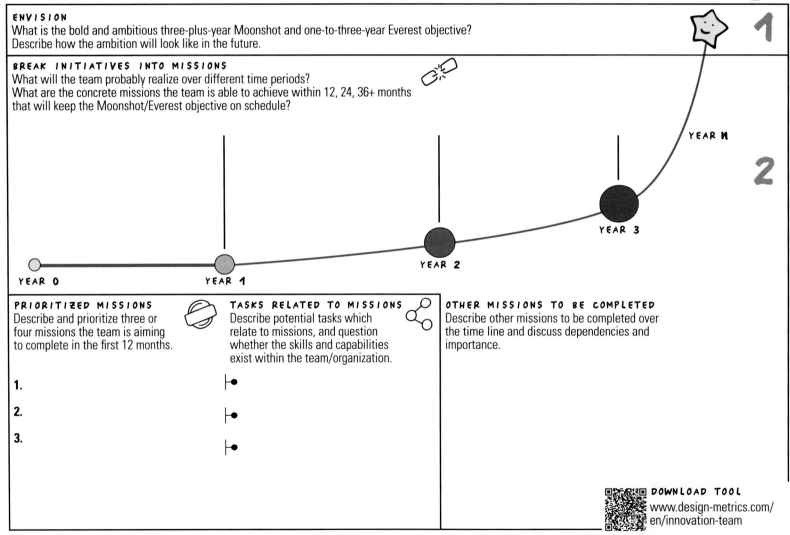

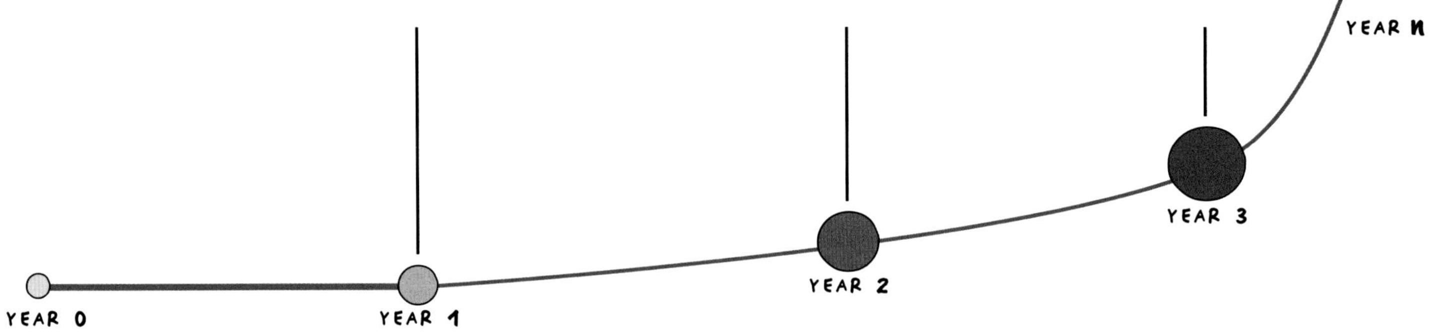
YEAR n
YEAR 3
YEAR 2
YEAR 0
YEAR 1

2

PRIORITIZED MISSIONS
Describe and prioritize three or four missions the team is aiming to complete in the first 12 months.

1.

2.

3.

TASKS RELATED TO MISSIONS
Describe potential tasks which relate to missions, and question whether the skills and capabilities exist within the team/organization.

OTHER MISSIONS TO BE COMPLETED
Describe other missions to be completed over the time line and discuss dependencies and importance.

DOWNLOAD TOOL
www.design-metrics.com/en/innovation-team

Prioritization of Assignments and Activities

If team and individual goals are provided with overall objectives and degrees of freedom, it is equally important to take care about the other tasks and activities. New objectives need resources and time, and so the resulting question is what can be delegated to gain more time for the objectives with a higher priority, or which have become superfluous. Innovation work also means stopping things or delegating things. A simple prioritization matrix, as illustrated here, helps to make transparent which activities are being stopped or delegated, and where the team is currently engaged with its first missions to complete, and where it will be engaged in the future based on the set objectives.

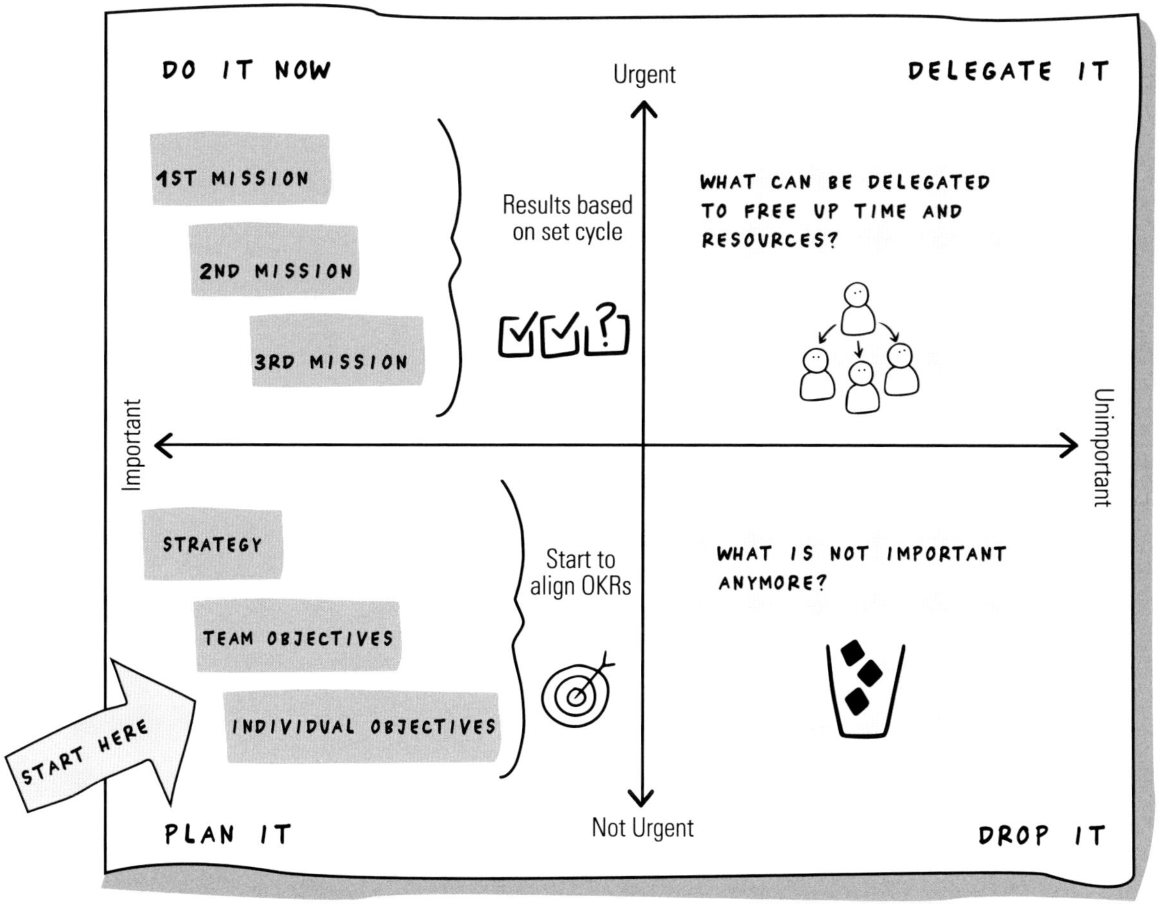

DO IT NOW
Urgent
DELEGATE IT
1ST MISSION
2ND MISSION
3RD MISSION
Results based on set cycle
WHAT CAN BE DELEGATED TO FREE UP TIME AND RESOURCES?
Important
Unimportant
STRATEGY
TEAM OBJECTIVES
INDIVIDUAL OBJECTIVES
Start to align OKRs
WHAT IS NOT IMPORTANT ANYMORE?
START HERE
PLAN IT
Not Urgent
DROP IT

Define: Cluster and Success

Template download

There is often no reference point in the definition of the key results for a specific mission or objective. For this reason, an intermediate step is sometimes helpful for teams. A good option is to define success together in the team or as part of the alignment process with other teams or the leadership board. Depending on the project, this can be done at the customer level, during collaboration between the teams or for the initiative itself. The exchange and dialog help in the alignment and definition of key results.

Working with success criteria helps the team to facilitate discussions about OKRs.

Defining Success Helps the Team:

- to start a discussion about what matters.
- to align with other teams.
- to identify independencies and opportunities with joint objectives and key results with other teams.
- to steer both discussions about external and internal objectives and key results.

Procedure and Template

1. Decide if success should be defined on the objective or on a concrete mission.
2. Brainstorm success criteria with the team.
3. Cluster the success criteria into major topics or points of views based on the customer, team, or project perspective.
4. Select the top three success criteria, for example, per mission.
5. Apply the criteria as basis for further development of short-term key results.

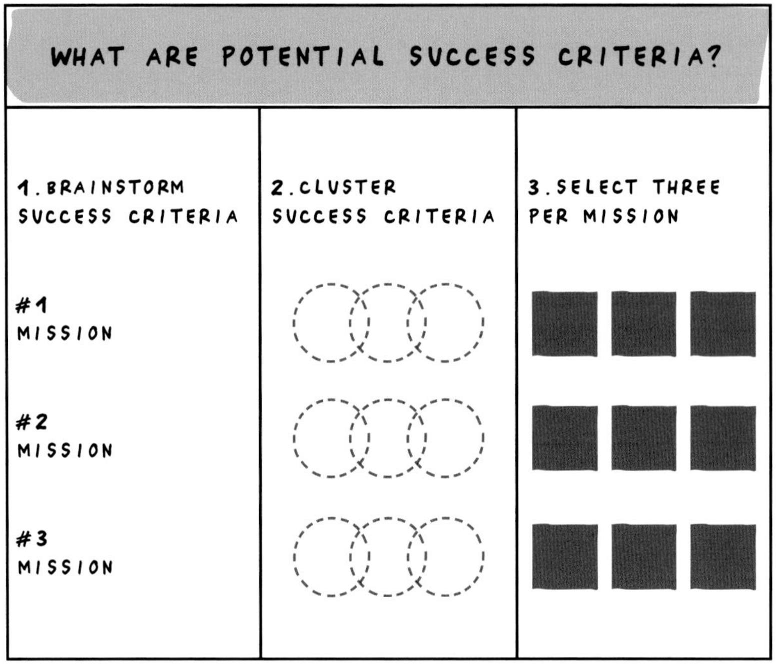

WHAT ARE POTENTIAL SUCCESS CRITERIA?
1. BRAINSTORM SUCCESS CRITERIA
2. CLUSTER SUCCESS CRITERIA
3. SELECT THREE PER MISSION
#1 MISSION
#2 MISSION
#3 MISSION

DOWNLOAD TOOL
www.design-metrics.com/en/success-criteria

How to Draft Key Results

For most teams, describing the key results is the biggest challenge. The OKRs Documentation Template (page 304) provides the most important parameters to define the key results. As a mental model, it helps to follow a defined structure for the description of a key result that addresses the four most important questions:

At ______________________________(What end date ?),

we know that the objective is achieved because ______________

______________________ (What outcome?).

The underlying metrics are based on______________________

(What happens regarding the progress?).

The key result makes the objective measurable based on__________

__________________ (How to make the objective measurable?).

Procedure of Drafting Key Results

The key results follow the key principles of defining measurements and metrics, the achievement of an objective within a project, initiative or mission. Most of the key results should be leading indicators to measure progress and lagging indicators to guide the broader execution. For teams new to design thinking, adding the declared behavior provides additional guidance to the broader team and confidence in working on the missions.

It is important that each key result adds value and all of them are contributing somehow directly or indirectly to the North Star. The key result may vary across the measurement of quantity to quality, depending on the desired outcomes.

Team Collaboration Agreement

In addition to the sign-off of the OKRs, it is recommended to define within the working agreement the following roles and responsibilities of the team:

- ✓ Who is the OKR representative in our team?
- ✓ What is the team's check-in/review cadence?
- ✓ Who is the team's key results maintainer?
- ✓ Who from the team is responsible for maintaining the progress measurement (results maintainer)?
- ✓ Who from the team attends the check-in?
- ✓ How will the team address the daily business and ways of collaborating?
- ✓ How will the team celebrate achievements and lessons learned after the check-ins?

It helps if all team members sign their agreed team OKR template together at the end.

Set it, sign it, and start collaborating!

Template download

OKRs DOCUMENTATION TEMPLATE

VISION / NORTH STAR What is the long-term vision (three to five years)? How will you define the one metric that matters most to the North Star?	**NAME OF THE TEAM:**
STRATEGIC GOALS / PRIORITIZED MISSIONS FOR THE INITIATIVE Why are we doing what we do as a team? What is the overall purpose and potential impact?	**TYPE & SCORING:** For example: Moonshot, Roofshot, Everest

OBJECTIVES What does the team want to do/achieve?

OBJECTIVE 1	OBJECTIVE 2	OBJECTIVE 3

KEY RESULTS How would the team like to measure progress?

KEY RESULT 1:	OWNER:	KEY RESULT 2:	OWNER:	KEY RESULT 3:	OWNER:
	START VALUE:		START VALUE:		START VALUE:
	END VALUE:		END VALUE:		END VALUE:
KEY RESULT 1:	OWNER:	KEY RESULT 2:	OWNER:	KEY RESULT 3:	OWNER:
	START VALUE:		START VALUE:		START VALUE:
	END VALUE:		END VALUE:		END VALUE:
KEY RESULT 1:	OWNER:	KEY RESULT 2:	OWNER:	KEY RESULT 3:	OWNER:
	START VALUE:		START VALUE:		START VALUE:
	END VALUE:		END VALUE:		END VALUE:

NOTE
Descriptions, labels, questions, important stakeholders, benefiting (parent) objective, resources, skills needed

DOWNLOAD TOOL
www.design-metrics.com/en/okr-documentation

Mobilize: Reflect on Skills & Capabilities

Template download

The actual work for the teams starts right after the alignment of the missions, objectives, and key results. However, in many innovation projects additional skills and capabilities are needed to start working on the mission to complete. A short reflection on existing and missing skills and capabilities helps the teams to plan ahead. The most important question is about what skills and capabilities are key to start a specific mission. Skills are usually somewhere in the organization or with external service providers. For example, if it is needed to understand the customer better, ethnographic research (observe and interact with customers in their real-life environment) or insights based on neuroscience must be included to build up evidence for testing critical assumptions.

The Reflection on Skills and Capabilities Helps the Teams:

- to understand what skills are present in the team.
- to analyze what is needed to successfully complete the first mission.
- to initiate a planning of new skills and abilities in case they are needed recurrently.
- to reinforce with team members from the organization who have the desired skills.

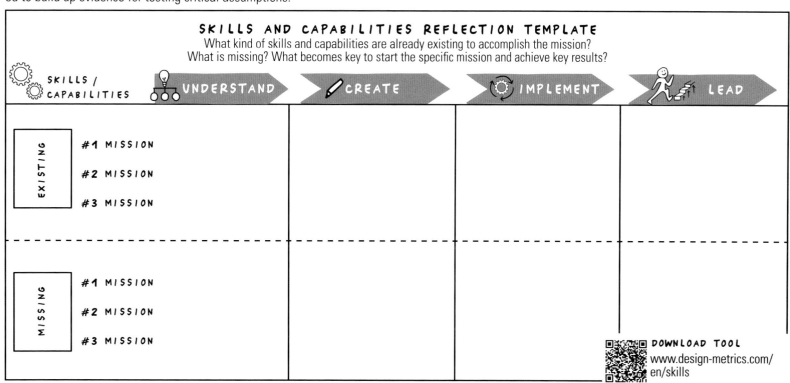

SKILLS AND CAPABILITIES REFLECTION TEMPLATE

What kind of skills and capabilities are already existing to accomplish the mission?
What is missing? What becomes key to start the specific mission and achieve key results?

SKILLS / CAPABILITIES		UNDERSTAND	CREATE	IMPLEMENT	LEAD
EXISTING	#1 MISSION #2 MISSION #3 MISSION				
MISSING	#1 MISSION #2 MISSION #3 MISSION				

DOWNLOAD TOOL
www.design-metrics.com/en/skills

Cultivate Curiosity & Experimentation

In design thinking and innovation initiatives, first missions are of paramount importance to mobilize an interdisciplinary team (see page 84) and to encourage work across teams (see team-of-teams, page 85). Based on the previous reflection on skills and capabilities, new team members might be needed. The Mobilize & Communication Canvas (see page 307) supports spreading the key information to new team members and updating existing team members about aligned OKRs. However, objectives and key results have also the aim of cultivating curiosity and experimentation. For new team members with diverse skills, the associated mindset related to design thinking and lean start-up might be new. For this reason, accompanying measures and coaching in the development of design thinking skills and capabilities become of paramount importance (see page 92).

Customized Design Thinking Training and Coaching Help the Team:

- to unlearn and challenge assumptions.
- to detach from daily operations and be in the moment.
- to combine internal and external perspectives to the connections.
- to create a starting point for building bold ideas from real insights.
- to merge learning through seeing, dialogue, and experiencing.
- to lead to distinct investment choices in line with a strategic narrative.

Procedure in Accepting, Learning, and Applying Design Thinking

The best way of learning design thinking is to apply the mindset, tools, and methodology with experienced design thinking coaches or facilitators. The coach or facilitator supports the teams in applying the appropriate tools and encourages them to leave their comfort zone. Techniques related to framing the right question, seeking inspiration in the real world, ideation, making ideas tangible, testing, learning, and sharing the story are core parts of understanding the customer problem and realizing new market opportunities.

Strategies that are directly linked to solving a problem have the best chance of performing.

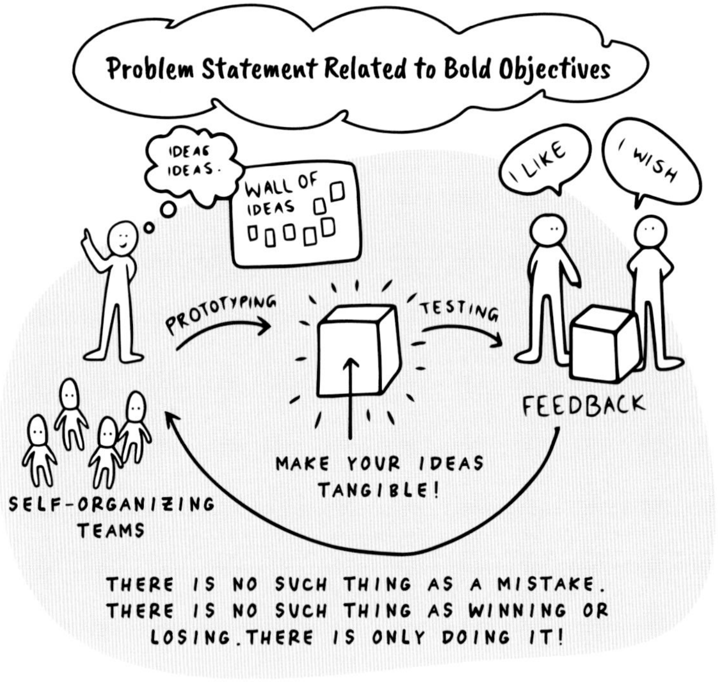

Problem Statement Related to Bold Objectives
IDEAS IDEAS.
WALL OF IDEAS
I LIKE
I WISH
PROTOTYPING
TESTING
FEEDBACK
MAKE YOUR IDEAS TANGIBLE!
SELF-ORGANIZING TEAMS
THERE IS NO SUCH THING AS A MISTAKE. THERE IS NO SUCH THING AS WINNING OR LOSING. THERE IS ONLY DOING IT!

Template download

MOBILIZE & COMMUNICATION CANVAS

6

SHARE AND COMMUNICATE THE SHORT-, MID-, AND LONG-TERM OBJECTIVES
What is the team expecting to achieve?

SHARE AND COMMUNICATE THE EXPECTED OUTCOMES
What outcome is expected from the team?

EXPLAIN THE PROBLEM STATEMENT/ASPIRATIONAL STATEMENT
Explain WHY it is worth working as a team on the objectives.

1

OUTLINE CRITICAL ASSUMPTIONS
Elaborate or communicate, which assumptions are critical and have the highest priority to complete in the mission?
Make transparent how frequent the progress of verifying measured assumptions is.

2

MAKE KEY RESULTS TRANSPARENT
Define with the team and communicate after each check-in which leading and lagging indicators will be used to determine if the mission is progressing toward the outcome.

3

SHOW VALUE FOR CUSTOMER / USER
Provide a good understanding about the value for the customer/user. Connect the dots and show the link between customers/users, the key value measures, and the overall outcome?

4

5

EXPLAIN COSTS AND AVAILABILITY OF BUDGETS
Calculate the expected investment to deliver the outcome.
Divide the cost between CAPEX and OPEX.

HIGHLIGHT FUNDING MECHANISM
Make explicit who the sponsor of the initiative is and where the funding is coming from. Make clear from the beginning if the funding provided is based on the achievement of missions or is for achieving the long-term and bold objective.

DOWNLOAD TOOL
www.design-metrics.com/en/mobilize

Redefine: Measure, Discuss, & (Re)prioritize

Objectives and key results are not set in stone. They can change during finalization, dissemination, and regular check-ins with the teams. Experienced teams usually deal with this uncertainty and convergence gracefully. Teams that are working with objectives and key results for the first time are advised to keep in mind the target of the measurement system. The Redefine OKR Canvas (see page 309) provides assistance with documenting the achievements/lesson learned and summarizes the adjusted or re-prioritized OKRs for the team and the larger organization. The adjustments should be made known immediately and should be transparently available.

The Redefine Phase Helps the Team:

- to reflect if the key result will really inspire team to take the appropriate actions, or whether it represents an operational task list.
- to verify if the descriptions of the objectives and key results are clear to the team and other teams within the organization.
- to verify that the progress is measurable and provides good evidence.
- to test if the objective is bold and reflects aspirational thinking, for example.
- to ensure that the appropriate scoring system is applied for the objective or a first mission.

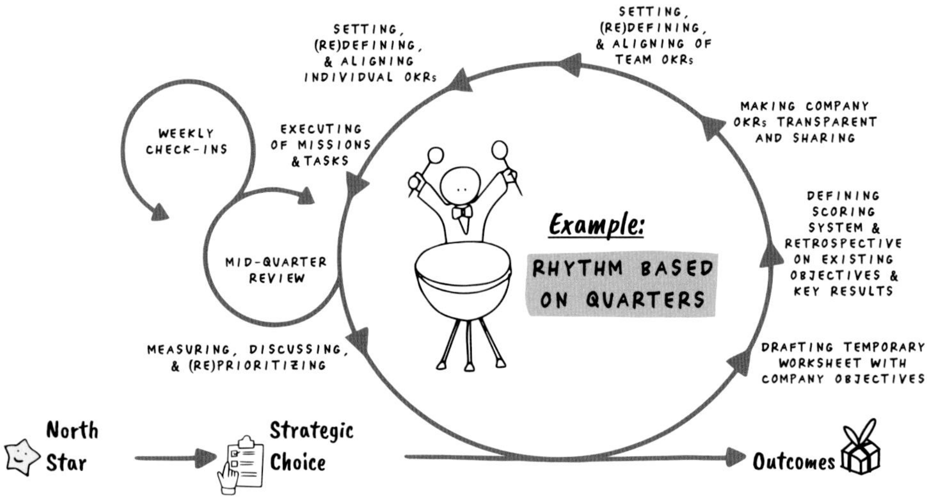
SETTING, (RE)DEFINING, & ALIGNING INDIVIDUAL OKRs
SETTING, (RE)DEFINING, & ALIGNING OF TEAM OKRs
MAKING COMPANY OKRs TRANSPARENT AND SHARING
WEEKLY CHECK-INS
EXECUTING OF MISSIONS & TASKS
DEFINING SCORING SYSTEM & RETROSPECTIVE ON EXISTING OBJECTIVES & KEY RESULTS
Example:
RHYTHM BASED ON QUARTERS
MID-QUARTER REVIEW
MEASURING, DISCUSSING, & (RE)PRIORITIZING
DRAFTING TEMPORARY WORKSHEET WITH COMPANY OBJECTIVES
North Star
Strategic Choice
Outcomes

Check-in

The regular check-in helps to review progress and see to what extent adjustments need to be made. For the check-ins, it has proven useful to go from the rough to the detailed, as in a retreat, and to create an action plan at the end that enables the teams to act or encourages them in their activities. In general, all OKRs should be reflected on briefly. In general, it is recommended to spend more time on those with a low confidence level.

Procedure and Template:

Step 1: Current progress of the team

- How did the team do on recent activities to drive the key result?

Step 2: Predicted score based on agreed scoring system

- What is the current score, and what is the confidence level of the team to achieve the result?

Step 3: Potential roadblocks/constraints

- What is preventing progress?
- Which problems are solvable, and which facts have to be accepted?

Step 4: Concrete adjustments, measurements, and action plans

- What needs to be adjusted?
- What are the next steps?
- Who can help?

Template download

REDEFINE OKRs CANVAS

DEFINED OBJECTIVES & KEY RESULTS (SHORT / MID-/ LONG TERM)
What is the set of OKRs relevant for the check-in?

REDEFINED OBJECTIVES & KEY RESULTS (SHORT / MID-/ LONG TERM)
What is the redefined set of OKRs for the team to work on?

NOW: DISCUSSION ABOUT CURRENT PROGRESS
What did the team do in recent activities to drive the key result?

MEASURE SCORE
What is the current score and what is confidence level of the team to achieve the result?

ADJUSTMENTS NEXT CYCLE REPRIORITIZE OKRs
What are the implications of the current progress in the next cycle?

ONE-YEAR OKR ADJUSTMENTS ONE YEAR
What are the implications for the OKRs set for the first year?

LONG-TERM OKR ADJUSTMENTS
What are the implications for the long-term OKRs?

ACHIEVEMENTS/LESSON LEARNED
What has the team achieved?
What are the lessons learned at the initiative and team level?

ROADBLOCKS/CONSTRAINTS
What is preventing progress?
Which problems are solvable and which facts have to be accepted?

ADJUSTMENT ACTION PLANS
What needs to be adjusted?
What are the next steps?
Who can help?

ADJUSTMENT ACTION PLANS
What needs to be adjusted with regard to the one-year objectives ?

ADJUSTMENT ACTION PLANS
What needs to be adjusted with regard to the long-term objectives ?

DOWNLOAD TOOL
www.design-metrics.com/en/re-defining

Ways of Working & Cultural Reflection

Building a measurement system is a challenging task. Especially in traditional companies where command and order prevails, or where management spends most of its time monitoring micro tasks, it is necessary to constantly reflect on the behavior and culture. The cultural change needs to bring a new understanding of the fact that the objectives were not introduced to perform a rating at the end of the year, and that a universal KPI is the measure for this. With OKRs, the engagement of every team member is bigger than numbers. If the organization tends to act chaotically, it is often due to the lack of a North Star.

Good strategies create a path for action based on a high common alignment. A strategy without a plan of action is inherently incomplete.

The organization should reflect regularly on two key questions:

- **How do individual team members and management feel in the environment of the adapted performance management system?**
- **Have ingrained behaviors been discarded?**

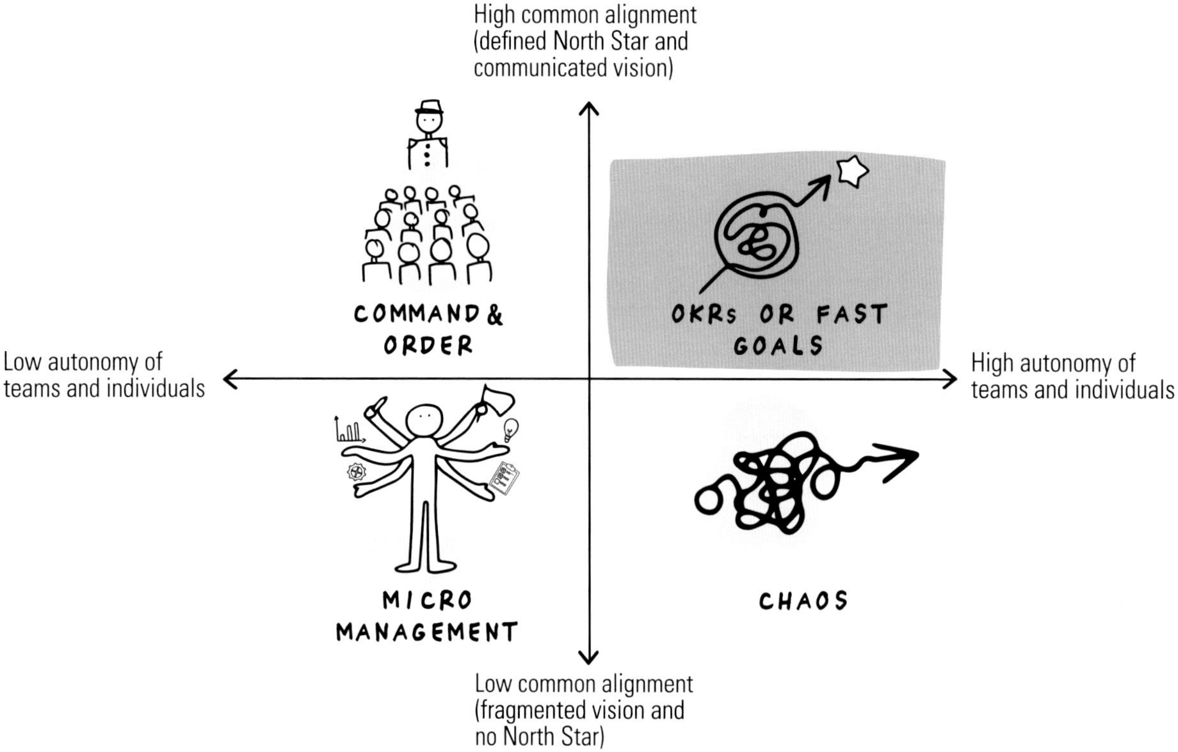
High common alignment (defined North Star and communicated vision)
COMMAND & ORDER
OKRs OR FAST GOALS
Low autonomy of teams and individuals
High autonomy of teams and individuals
MICRO MANAGEMENT
CHAOS
Low common alignment (fragmented vision and no North Star)

No organization has more collective experience implementing OKRs than Alphabet. Since 1999, the company has impressively demonstrated how exponential growth can be realized through the appropriate culture and measurement systems.

Example

INSPIRING

Every OKR is drafted as a sophisticated stretch goal for Everest and Moonshot objectives.

CREATING IMPACT

The successful achievement of an objective must provide clear value for Alphabet.

COLLABORATING

Goals are created 60 percent from the bottom up, and the autonomous teams are synced vertically and horizontally.

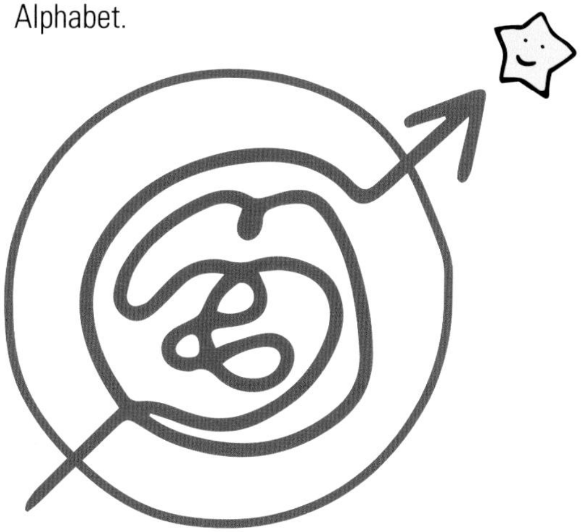

PROVIDING FEEDBACK

OKRs are drafted by considering feedback from peers acting as OKR facilitators and who are visible to the entire workforce.

FOCUSING

Each quarter, a maximum of five objectives with four verifiable key results each is practical to stay focused.

Google

40 employees applied OKRs at Google for the first time.

OKRs have helped lead us to 10x growth, many times over. They`ve kept me and the rest of the company on time and on track when it mattered the most.
- Larry Page, Co-founder of Google

Alphabet

140,000 employees of Alphabet still applying OKRs in combination with other innovation metrics.

1999 → 2022

THE FUTURE OF DESIGN THINKING & DATA-DRIVEN INNOVATION

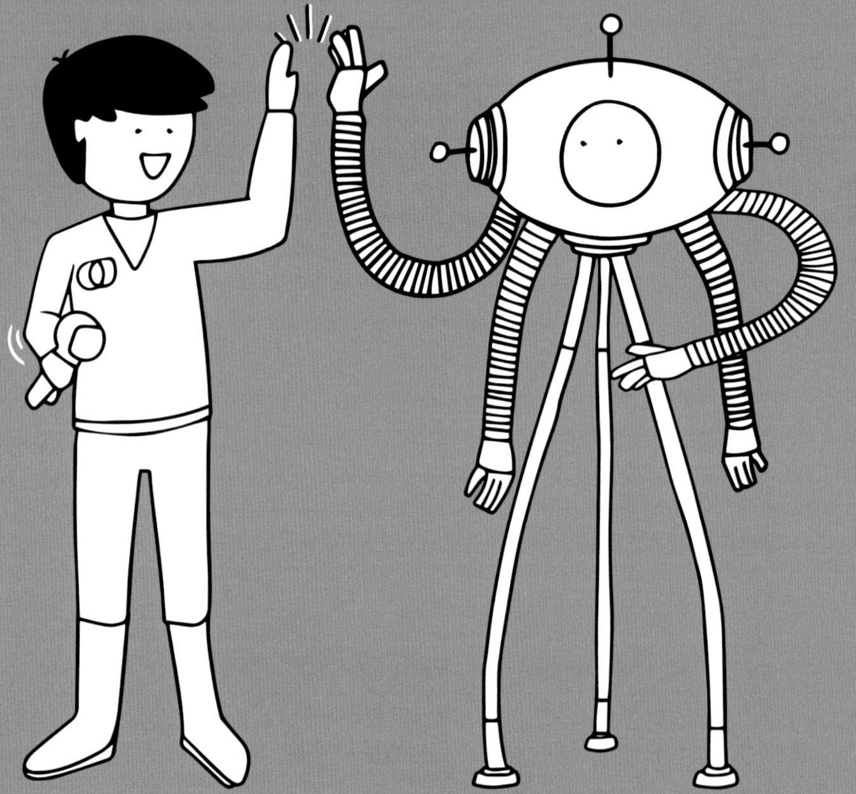

WHY? HOW? WHAT?

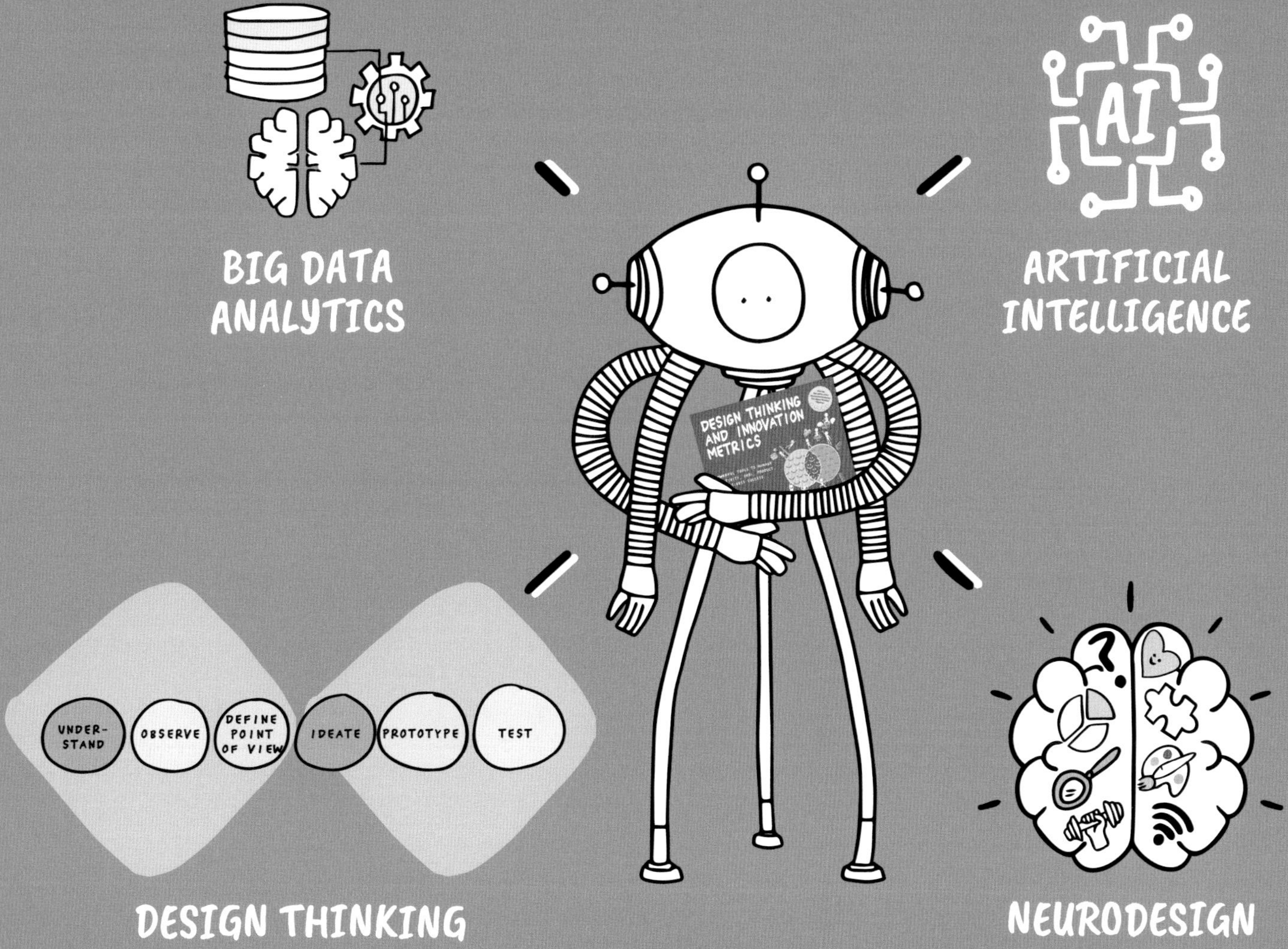
BIG DATA ANALYTICS
ARTIFICIAL INTELLIGENCE
AI
DESIGN THINKING AND INNOVATION METRICS
UNDER-STAND
OBSERVE
DEFINE POINT OF VIEW
IDEATE
PROTOTYPE
TEST
DESIGN THINKING
NEURODESIGN

Creating Impact with Advanced Tools

In this part of this book, a brief introduction to the world of data-driven innovation is outlined. The disciplines discussed here have become already an indispensable part of the toolbox of innovation teams and integral part of measuring innovation and success. The new approaches also help to better understand the problem space, take experiments to a new level, and ultimately design better solutions for customers/users. In the two 101s, the basis for this has been laid: Design thinking, measurements, and the basic concepts of analytics and statistics are central elements needed for this purpose. Leading companies as well as world-class universities provide training to employees and students in a combination of data science and design thinking.

Why Design Thinking and Data Matter

The ability to support the creative process with data, and to continuously collect intelligent data, is the basis for data-driven innovation and evidence-based development of new products, services, and experiences. At the same time, data, ML, and AI provide the future for building powerful measurement systems. While new language processing AI models help to engage with customers.

Data-driven innovation helps to enhance our intuition in many directions. Design thinking and data sciences support each other.

The terms and fields of data-driven innovation are not always clear. Rather, the individual disciplines overlap and serve each other. In the context of this book, data-driven innovation requires the intersection of design thinking and data science. Data scientists, for example, use computer science and mathematics, although the distinction between the two terms is blurred. The two are not mutually exclusive. For instance, AI is defined as computer science, but uses statistical methods as part of its process. Another important discipline outlined in this part of the book is neurodesign. Neurodesign is a design paradigm in which design thinking teams and cognitive neuroscientists work as a team, using knowledge of what makes the human brain tick to understand how creativity and design thinking are measured. They also seek to learn how to better design products and services by understanding customers and their brain activities associated to a potential product, service, or experience.

Data-Driven Innovation Has Become Part of Creating, Delivering, and Capturing Value

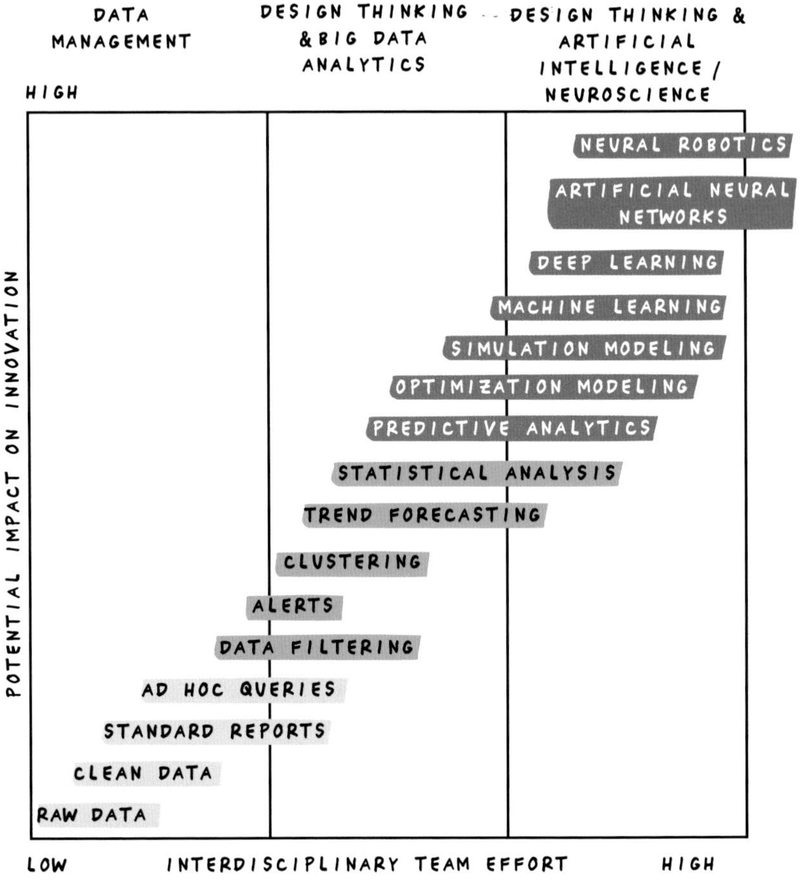
DATA MANAGEMENT
DESIGN THINKING & BIG DATA ANALYTICS
DESIGN THINKING & ARTIFICIAL INTELLIGENCE / NEUROSCIENCE
HIGH
POTENTIAL IMPACT ON INNOVATION
NEURAL ROBOTICS
ARTIFICIAL NEURAL NETWORKS
DEEP LEARNING
MACHINE LEARNING
SIMULATION MODELING
OPTIMIZATION MODELING
PREDICTIVE ANALYTICS
STATISTICAL ANALYSIS
TREND FORECASTING
CLUSTERING
ALERTS
DATA FILTERING
AD HOC QUERIES
STANDARD REPORTS
CLEAN DATA
RAW DATA
LOW
INTERDISCIPLINARY TEAM EFFORT
HIGH

AI and machine learning for creativity and innovation are very different from the established areas where AI has replaced traditional operational tasks.

The Future Fusion of Disciplines

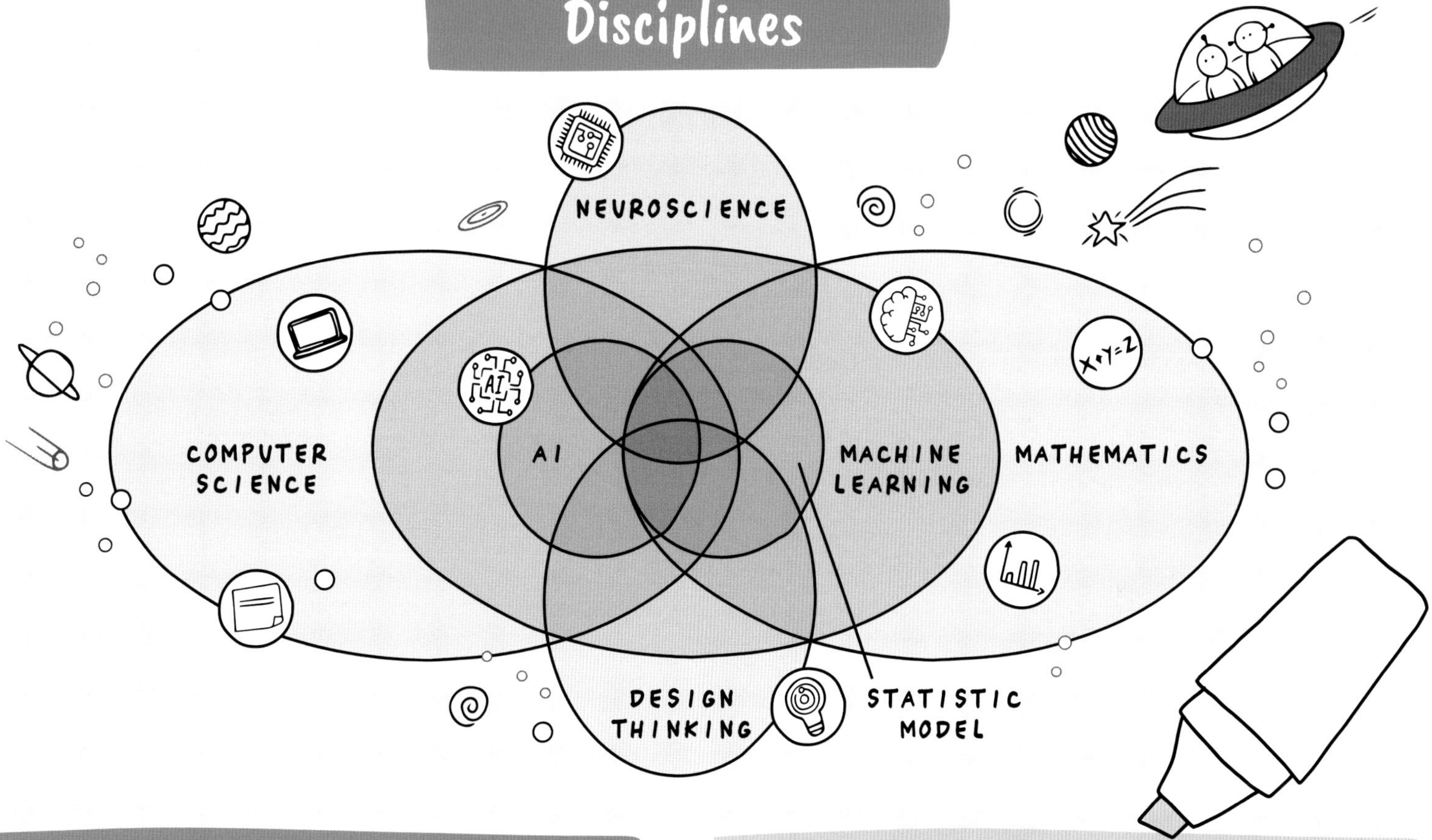

NEUROSCIENCE
COMPUTER SCIENCE
AI
MACHINE LEARNING
MATHEMATICS
DESIGN THINKING
STATISTIC MODEL

- Data-driven innovation includes computer science, mathematics, statistical models to perform machine learning, AI, and neurotechnologies.
- Data provides insights to understand the WHAT and, increasingly, through data from neuroscience, the WHY.

- Design thinking is applied to support the data scientist to find the appropriate starting point and develop AI solutions, or it is applied in combination with design thinking and data analytics to create better evidence-based solutions.
- Design thinking helps to understand the WHY in many directions.

WHY It Matters for Measuring

As part of the iterative evolution of measurement systems, companies are increasingly applying data analytics and AI.

Organizations and teams that have progressively professionalized their innovation measurement systems from IMS 0 to IMS 2 or IMS 3 are now applying predictive analytics such as machine learning together with leadership knowledge and expertise from interdisciplinary teams to better understand cause and effect, and to validate strategic and tactical actions. In many ways, those companies have also mobilized their employees across all departments and functions to get acquainted with data analytics skills and competences. On top, for adequate measurement in EXPLORE and EXPLOIT, technology is increasingly becoming a key competitive differentiator. It can be observed that the use of data analytics and AI also has impact on more advanced organizations in their general behavior, trust in machine-generated indicators, and the overall culture of companies.

Predictive analytics can do more than just improve the achievement of the North Star. It can also create new indicators of success and change the way companies define performance.

Developing, connecting, and tracking a mix of analytically identified measures, and those determined bottom-up by teams, represents a fundamentally new organizing principle to align behavior with the strategic reasoning and responsibilities of the design thinking and innovation teams for innovation measurements systems.

The future form of alignment not only transits from EXPLORE to EXPLOIT between the company and the team objectives, but it also adds another dimension, that of predictive alignment with the teams and leadership.

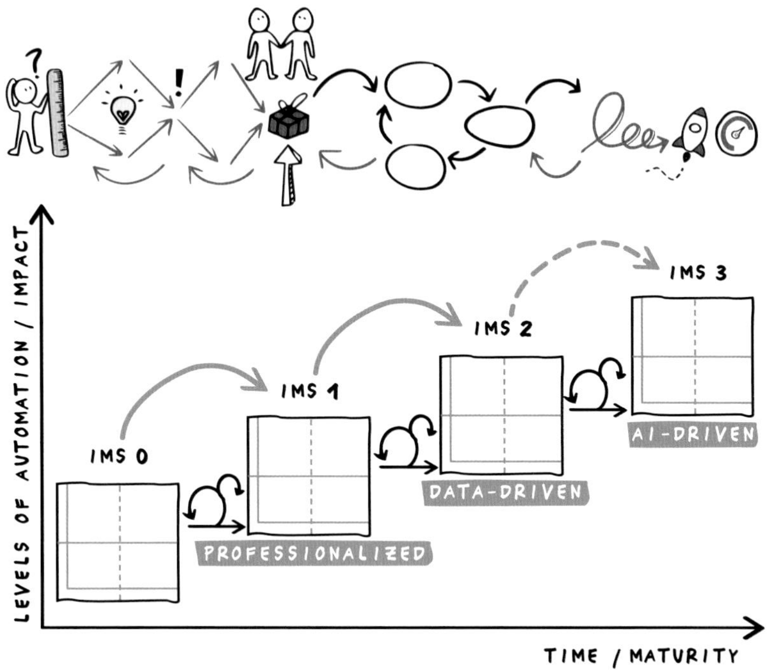
LEVELS OF AUTOMATION / IMPACT
TIME / MATURITY
IMS 0
IMS 1
IMS 2
IMS 3
PROFESSIONALIZED
DATA-DRIVEN
AI-DRIVEN

Companies which already focused on customer behavior and applied data-driven innovation tools are managing the transition much faster than other companies, which need longer to change measurement systems and the underlying mindset. In addition, new technologies (data analytics, AI, and neuroscience) can be used to shape entirely new leading indicators that not only determine future customer behavior, but also open up entirely new ways of creating value for customers, companies, and entire business ecosystems alike.

HOW to Apply It to Unlock Value

Design thinking and innovation work applies real-time information processing based on in-depth interviews with potential customers/users, moving beyond the traditional ethnographic research methods, as part of a data-driven and neuro-led work routine. For example, design thinking teams receive custom-curated streams and stories, each tuned to interests of a particular innovation topic or growth initiative. Instead of collecting insights from potential customers/users manually, a machine is connecting and interacting with thousands of them at the same time—then summarizing their insights, generating new points of view, and even testing them in robot labs. Innovation teams are increasingly delegating ground innovation work to machines, leaving them space to apply higher levels of problem-solving competence, creativity, and innovativeness.

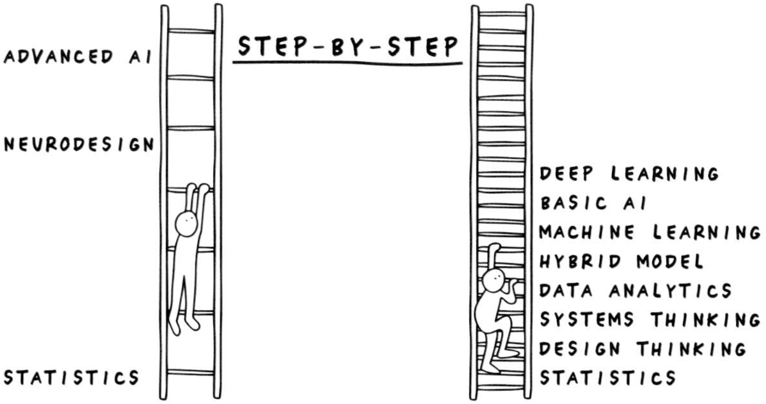
ADVANCED AI
NEURODESIGN
STATISTICS
STEP-BY-STEP
DEEP LEARNING
BASIC AI
MACHINE LEARNING
HYBRID MODEL
DATA ANALYTICS
SYSTEMS THINKING
DESIGN THINKING
STATISTICS

However, this journey started only a few years ago, and innovation teams globally are learning step by step how to apply these new data and neuro-related technologies to manage and measure innovation work. Many possibilities in the field of AI are imaginable but have not yet been realized. Likewise, the exploration of our brains with the future tools of neuroscience still has a lot of potential. It is important to be open to these new possibilities and to link the different views and approaches in the best possible way.

WHAT Tools, Methods, and Mindsets Are Required to Become Successful?

To innovate far beyond today's imagination, design thinking and innovation teams need intelligent tools, very much like the tools that have been built to overcome physical limitations. With the application of neurodesign and the efforts to understand how the brain behaves when composing exceptional ideas, a new era of augmenting our intellectual abilities with stimulation has started.

Neuroscience can also be considered the root of today's artificial intelligence. It is becoming increasingly apparent that neuroscience can enrich many topics related to artificial intelligence, for example, by creating better neural network architectures. New data points will provide the basis for advanced tooling and more automated predictions of potential solutions, as well as innovation and business success.

Neuroscience reveals the truth of design thinking, or simply the satisfaction of formal expectations that are still not completely understood by design thinking and innovation teams.

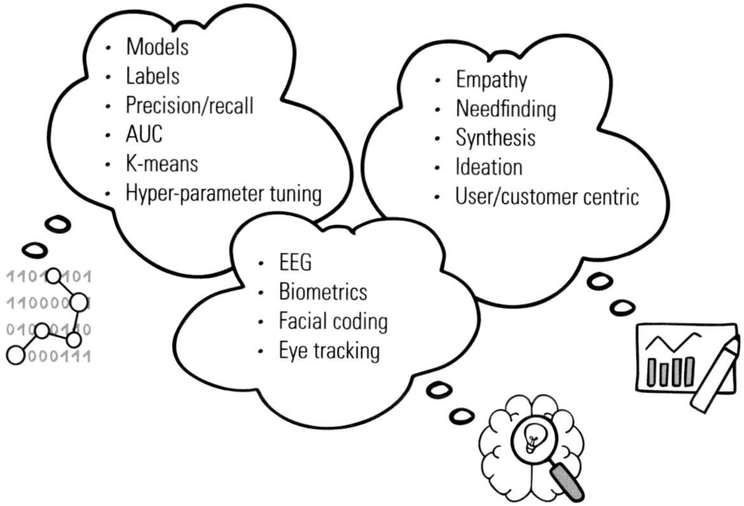
• Models
• Labels
• Precision/recall
• AUC
• K-means
• Hyper-parameter tuning
• Empathy
• Needfinding
• Synthesis
• Ideation
• User/customer centric
• EEG
• Biometrics
• Facial coding
• Eye tracking

301

DESIGN THINKING & DATA ANALYTICS

11011101
11000011
01000100
10000111

Applying the Hybrid Model

Overcoming the Lack of Strong Evidence

The hybrid model was the first attempt to combine big data analytics and design thinking to solve wicked and ill-defined problems. At the time the hybrid model was developed, the term "data scientist" had not yet been created, and the tools and methodology were limited to analyze, process, and visualize data. In the meantime, working with smaller but more diverse and complex data sets and combined data lakes has continuously advanced, and today machine learning (ML) technology and artificial intelligence (AI) have impacted how insights for our design challenges and projects are collected, interpreted, and presented (see Design Thinking & AI, page 341). Advanced data visualization techniques provide the opportunity to present outcomes in such a way that, for all team members and business leads, explainable data analytics are accessible.

Data analytics has become key to elaborate on the WHAT while design thinking is still the main cornerstone to find out about the WHY.

Another trend to observe is the application of technology-aided design and business decisions based on data science. However, even the most advanced technology is still not able to replace the ability to show empathy or tap into human intuition and experience. Over the last decade the hybrid model has proven to generate better market opportunities, and the combined mindset has found acceptance in interdisciplinary teams supported by design thinking and data science experts.

As AI-powered customer/user interfaces become more professionalized, more valuable data sets will be added. Touchpoints, customer journeys, and interactions will become more personalized and behave more like human interactions. The combination of data analytics and design thinking is ideally suited for the design and implementation of such projects.

In the combined application of design thinking and big data analytics, decisions are based on what future customers really want, and data can be used to validate these assumptions. Together they ensure that risky bets are not made based on instinct only, but also on evidence.

From practical experience and various studies since 2005, it could be inferred that data science can help designers think outside the box, providing relevant inspiration and alternative points of view. Moreover, design thinking is applied for data analytics, to identify, for example, the appropriate questions for data analytics. The hybrid model has enhanced design thinking with data analytics and become a joint process framework, toolkit, and mindset for solving problems.

DATA SCIENTIST:
THE SEXIEST JOB
IN THE WORLD!
DATA
GIVES YOU THE
WHAT
BUT HUMANS
KNOW THE
WHY
1 1 0 1 1 1 1 0 1 0 1 0 0 1 0 1 1 1 0 1 0 1 1 0 1 0

DESIGN THINKING & DATA ANALYTICS

[For a purposeful and successful collaboration of data scientists and design thinking teams, it has proven to be beneficial to work with a mindset that reflects the work of the hybrid model.]

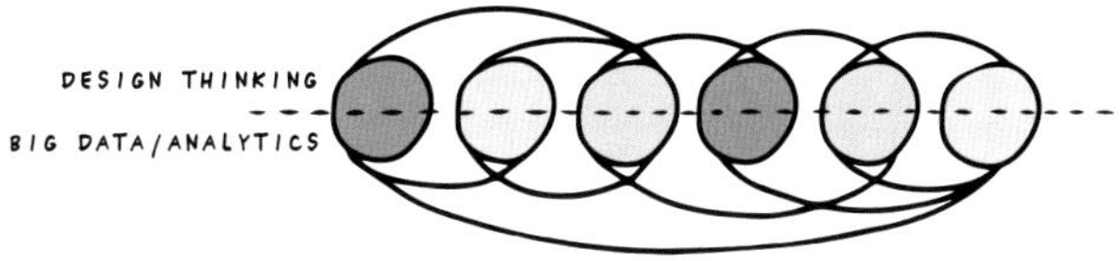
DESIGN THINKING
BIG DATA/ANALYTICS

- Combining the toolbox of data analytics and human-centered design
- Increasing the efficiency and effectiveness of the innovation process
- Combining the deep customer insights and the deep learning from data
- Generating synergies by combining the rather qualitative aspects of design thinking and the quantitative methods of data science

- Understanding the customer from a different perspective
- Improving the decision confidence
- Smoothing, validation, and alignment of insights
- Validating assumptions and justified inferences
- Proofing of solutions based on intuition and enhanced evidence
- Contextualizing data in order to link data insights with stories
- Accelerating the iterative problem-solving process by applying data analytics tools and design thinking methods

- Design thinking for data analytics
- Data-enhanced design thinking
- Hybrid model (combined process of design thinking and data science, based on hybrid mindset and toolkit)

Future application of technology-aided design and business decisions based on data science, AI, ML, and in combination with neurodesign.

2005 2010 2020 2040

THE HYBRID MINDSET

Combine an **analytical and intuitive way of thinking** in a holistic approach.

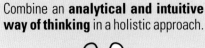

Combine human and data insights over the entire design cycle (e.g., development of points of view).

Accept uncertainty at the beginning of the design challenge and interpret correlations in the context of the user.

Apply an **optimistic mindset** and perform iterations throughout the design cycles for both thinking states (design thinking and data analytics).

Reflect constantly on the approach and **enhance the data capabilities step by step** with new and enhanced tools and methods.

Encourage the team to be **interested in the unknown** and create clarity through a **creative and analytical approach.**

Create an environment that allows you to merge expertise and mutual **inspiration from human insights and visualized data** from T-shaped team members.

Generate stories from data, experiences, customer feedback, and innovation metrics, and mirror the results in prototypes and experiments.

Learn and expand the toolbox of creative and analytical tools constantly. Experiment with new measurements, technology, and ways of generating evidence.

The Symbiosis of Thinking Attitudes

With the application of the hybrid model, design challenges with interdisciplinary teams of data scientists and design thinkers are being pursued in the exploration of the problem space and the development of complex solutions. Collaboration between design thinkers and data scientists is not only profitable, but often very challenging for both groups. A good understanding of each other's work is therefore very important, especially how the disciplines can support each other to generate better solutions in the end. Experience shows that mixed teams ultimately achieve better results.

For data scientists the application of design thinking is often an important starting point for the analysis of data. Asking the right questions is central and design thinking helps to get to the bottom of things with tools like the Question-Analyze-Builder (see for example The Design Thinking Toolbox, page 111) before building data models and performing analysis. Data scientists also start with assumptions that must first be validated and gradually provide clarity about the problem to be solved. The diagram on the right is self-explanatory; the use of design thinking gradually improves the quality of the questions while improving the evidence. However, design thinking teams can also ask purposeful HMW questions about data and enabler technologies in the context of problem solving, as the following examples show.

Data-driven innovation allows design thinking teams to break new ground and ask:

How might the available data provide new insights about the customer needs or, more broadly asked, how can they create value for customers/users?

More advanced teams, in collaborating with data scientists for problem solving, formulate questions regarding enabler technologies:

How might AI/ML help us to solve the problem in a unique way for existing scenarios with applications providing enough data points?

In addition, the combination potentially offers a wide field of tech-driven realizations of market opportunities by developing new models based on HMW-questions related to:

How might we leverage AI/ML for customer journeys where good outcomes are clear, but rules are not?

Among other things, the hybrid model uses design thinking methods to identify customer needs and find new solutions for data-driven challenges using creative methods. The model also includes structured and established methods for developing data solutions, a mindset that allows teams to collaborate efficiently, act flexibly during the development process, and work quickly toward the set objectives.

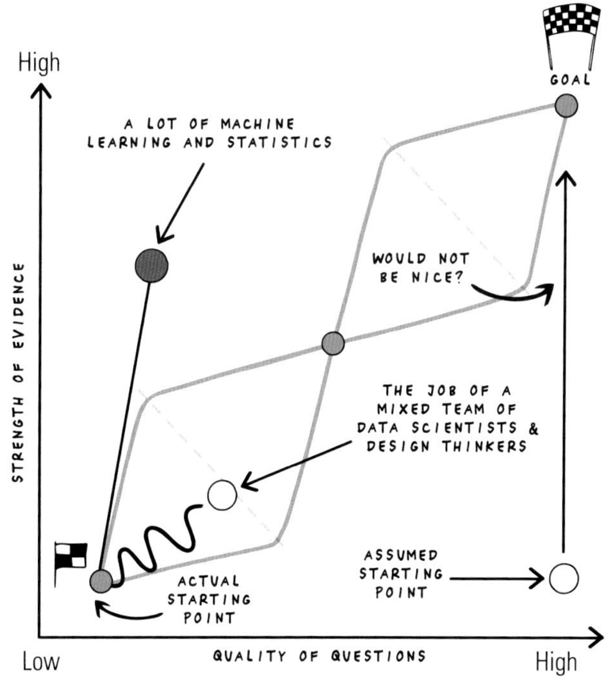

GOAL
High
A LOT OF MACHINE LEARNING AND STATISTICS
WOULD NOT BE NICE?
STRENGTH OF EVIDENCE
THE JOB OF A MIXED TEAM OF DATA SCIENTISTS & DESIGN THINKERS
ASSUMED STARTING POINT
ACTUAL STARTING POINT
Low
QUALITY OF QUESTIONS
High

Over the entire design cycle the appropriate enabler technologies help to create, deliver, and capture value.

Creating Value with an Extended Toolbox

The hybrid process (see page 325) aims to bring design and design thinking into the world of data science and vice versa. Different types of data mining and other methods from the data science toolbox can be combined with design thinking throughout the Double Diamond. In the problem space, the focus is on building empathy with exploration of pains, gains, and jobs to be done of customers/users. The combination of data, in-depth interviews, and observations can also help to make hidden insights and patterns visible and ultimately improve the quality and evidence of the insights.

In the context of opinions, feedback, changes, and other dynamics, data mining aims to extract and analyze customer/user responses. The analyses range from simple queries of current satisfaction with a service or product to influences of social media on current and future offerings. For optimization and redesign, for example, comments about products on marketplaces can be extracted and tested for correlation with ratings using factor analysis.

Other data points arise from the analysis of unconscious information from the human brain (see Neurodesign, page 360). Physiological condition mining, for example, trains models to decode customer/user signals. Here, the ML first learns from a data set and then predicts responses from new data collected from customers/users. In a laboratory setting, these signals can come from electrocorticography (ECoG), electroencephalography (EEG), patch clamp recordings, and also from more accessible means, such as audio and video analytics. For example, video analytics and eye tracking can be used to track eye movement to measure attention. Such measurements are particularly easy to perform in virtual settings where customers/users are already using VR devices.

The kinetics can also be used to analyze the communication in a consultation or sales talk and gain valuable insights. Likewise, in a combination with the technologies described above, QR codes with product information or RFIDs can help to better understand the actions of customers in their interaction with products and services.

At the process level, data points can also provide important information. For example, data can be collected analogously to the current customer journey or at specific touchpoints. In the same way, data can be collected proactively in order to better understand what the current customer ecosystem journey looks like. Comparing the actual journey with the journey designed in the business ecosystem can help prioritize actions and lead to ideas on how to improve the overall customer experience or radically redesign the value proposition. In various projects, data on specific traffic congestion and frequency of critical infrastructure to movement patterns have also generated new insights.

In the context of the point of view (PoV), the insights of the data scientists and the insights of the design thinkers often come together. Additionally, data mining can help to neutrally examine and process the gained insights and give corresponding predictions about the relevance. In this context, the hybrid model not only helps in exploration, but also in analyzing large amounts of data, for example, to identify previously undiscovered patterns.

With the application of generative models, alternative views can be generated quickly and efficiently. Based on previous design thinking challenges, new prototypes and variants are generated. This helps the team to develop a large number of alternative prototypes and simulations. This kind of prototyping is used for wireframe sketches in digital interactions. User testing (A/B testing) also makes use of generative models that, in conjunction with eye tracking, can simulate where the human gaze tends to orient itself for certain prototypes. The clusters that are generated from the data often provide insights into new categories that were previously unseen by design thinking teams. In the last decade, many ways have also emerged to visualize data and share the stories from data with the team and other stakeholders.

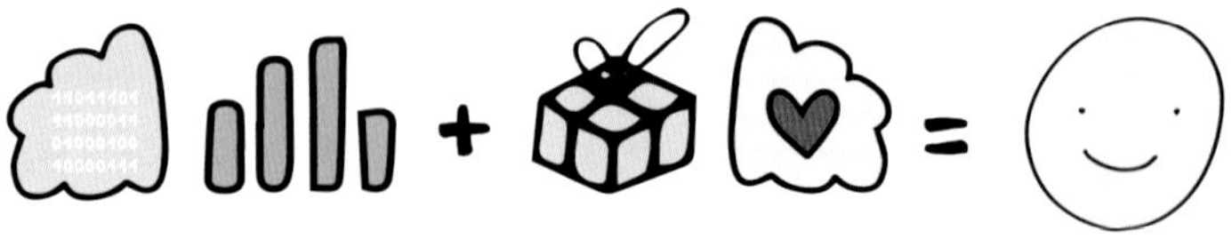

The Hybrid Process: Design Thinking & Big Data Analytics

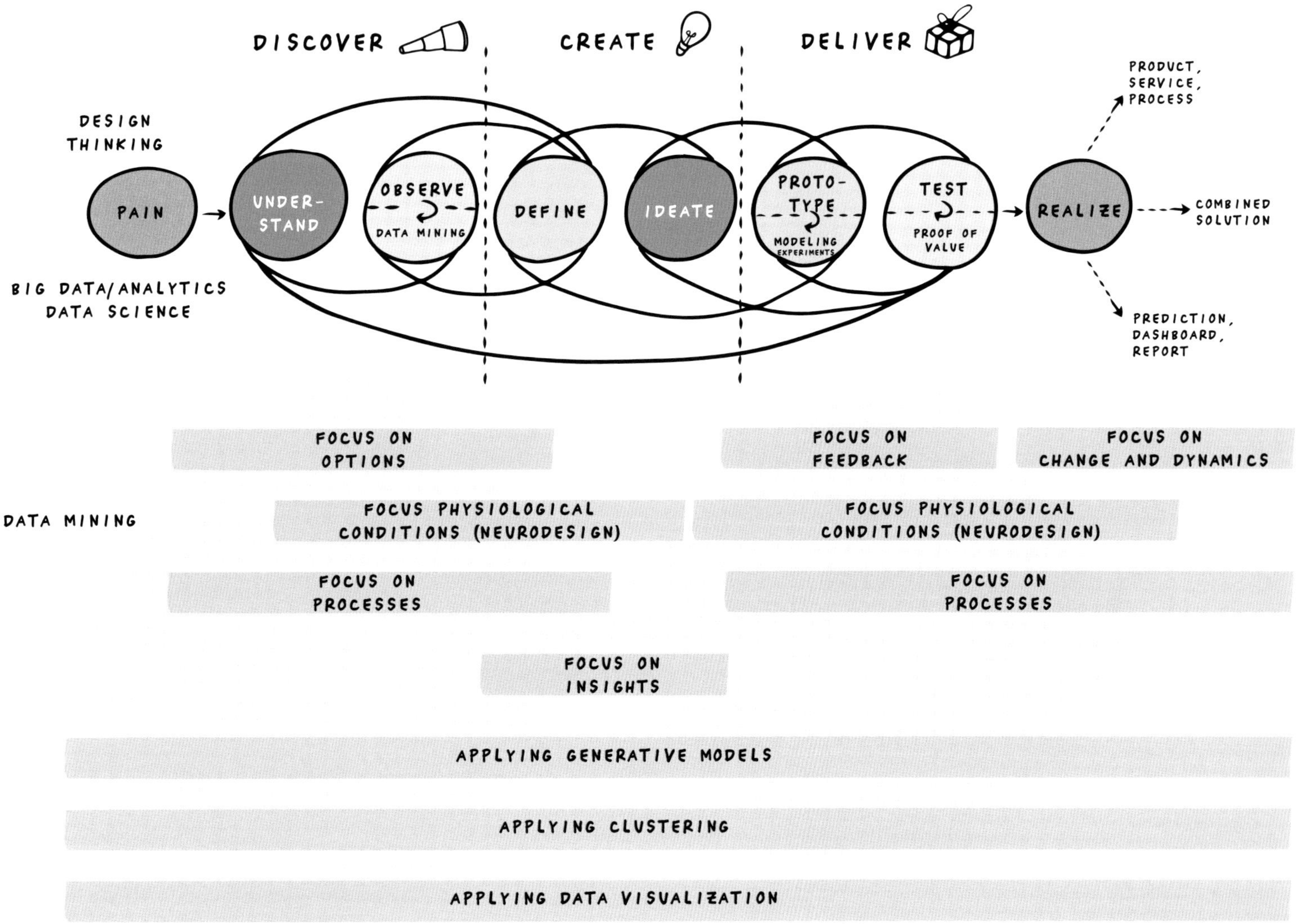
DISCOVER
CREATE
DELIVER
DESIGN THINKING
BIG DATA/ANALYTICS DATA SCIENCE
PAIN
UNDER-STAND
OBSERVE
DATA MINING
DEFINE
IDEATE
PROTO-TYPE
MODELING EXPERIMENTS
TEST
PROOF OF VALUE
REALIZE
PRODUCT, SERVICE, PROCESS
COMBINED SOLUTION
PREDICTION, DASHBOARD, REPORT
FOCUS ON OPTIONS
FOCUS ON FEEDBACK
FOCUS ON CHANGE AND DYNAMICS
DATA MINING
FOCUS PHYSIOLOGICAL CONDITIONS (NEURODESIGN)
FOCUS PHYSIOLOGICAL CONDITIONS (NEURODESIGN)
FOCUS ON PROCESSES
FOCUS ON PROCESSES
FOCUS ON INSIGHTS
APPLYING GENERATIVE MODELS
APPLYING CLUSTERING
APPLYING DATA VISUALIZATION

How to Evolve from Data Experiments to Thrive with the Hybrid Model

Data science tools and methods can support activities in various phases across the design cycle and accelerate the exploration of the problem space and the solution space. The appropriate interweaving of design thinking, strategy, and data science tools and methods are now an integral part of innovation work. The Data Ideation Canvas supports with powerful questions the approach of integrating existing and new data points.

However, even when applying the hybrid model, there is no boilerplate recipe. To realize the best market opportunities, it is important to understand how qualitative and quantitative methods complement each other throughout the design cycle when teams come together, and how insights are shared to ultimately realize true innovation. Above all, it is also important that design thinking and innovation teams understand the entire process of data analysis and operationalization (see Data Analytics Process on pages 228–234).

Data science enables innovation teams to interact with user data in new ways. The tools are a means to an end for collecting and analyzing new types of user information and adding statistical power and validation to the design thinking work. Ultimately, the hybrid model is about understanding the user/customer even better and, on the other hand, using modern tools to perform, for example, automatic audio transcription to pattern recognition in clustering. In practice and from working with different teams, there are also different levels of maturity with respect to the hybrid model. These range from teams that are only gaining initial experience with a combined approach (EXPERIMENT) to teams that have established the approach and work with data across the board (THRIVE).

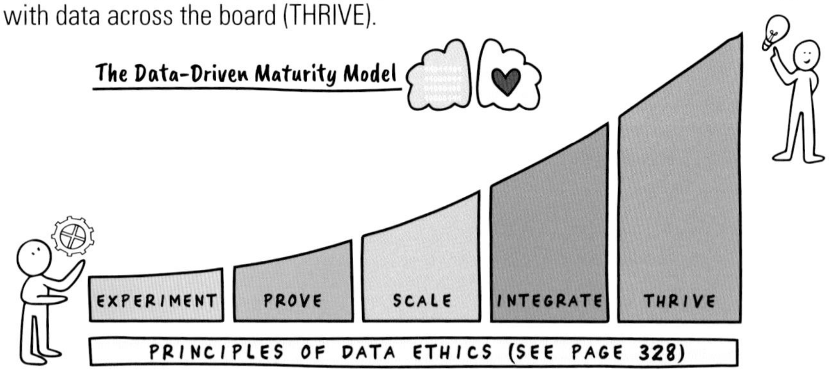
The Data-Driven Maturity Model
EXPERIMENT
PROVE
SCALE
INTEGRATE
THRIVE
PRINCIPLES OF DATA ETHICS (SEE PAGE 328)

Data Science Tools Added over the Design Cycle

Phase: Understand
- Digital process mapping
- IT architecture cross check
- Trends/correlations/quantitative insights

Phase: Observe
- Video analytics
- Heatmaps of customer/user interaction
- Mouseflow and live chat analytics tools

Phase: Point of View
- Correlations (e.g., between pain points and the types of users)
- Bipolar factors
- Different scenarios (matrix, stories, customer)

Phase: Ideate
- T-SNE, neural networks, PCA
- Hyperparameter optimization
- Digital process re-design
- Intelligent clustering of idea

Phase: Prototype
- Behavioral models
- Animation objects, flow components
- Data acquisition strategies

Phase: Test
- Deepfake
- Large-scale testing
- Measuring outcomes/acceptance/churn

Template download

DATA IDEATION CANVAS

1 DISCOVER	2 CREATE	3 DELIVER
EXISTING DATA ASSUMPTIONS What data is available? What data can be acquired? What data can be accessed on open data platforms?	**APPLIED DATA** Which of the existing or previously collected data can be utilized for a function, experience, or prototype?	**NEW DATA** How active is the customer/user? What is the churn rate? What is the active time the user is engaging? How often is the customer/user using the product/service?
RESEARCH What trends, mega trends, and scenarios of the future are relevant?	**PROTOTYPES (DIGITAL TANGIBLE)** How is the prototype visualized? What are the critical functions and experiences? Is the idea/prototype desirable, feasible, and viable?	**FINAL PROTOTYPE** What kind of data is collected with the solution? How is the experience/value proposition improved? Is there new data to expand the service, product, or entire value proposition? Over which channels is the customer/user engaging? Which are the most important touchpoints for the customer/user?
INSIGHTS FROM DATA & OBSERVATION Which insights provide new physical or emotional learnings about the customer/user, situation, or context of the problem statement? What kind of insights derive from data?	**ADDITIONAL DATA INSIGHTS** What data is missing? How to collect, acquire, or obtain additional data? How much data is the customer/user willing to share?	**MEASUREMENTS & METRICS** How is the impact of the hybrid team measured? What are the efficiency gains over the entire design cycle applying the hybrid model? What measurements and metrics might apply in similar projects or are part of the innovation measurement system?
PERSONA USER FROM DATA & EMPATHY MAP Who is the customer/user? Pains/gains? Jobs to be done? Use cases? Influencers? What data provides evidence about personas or preferences?	**ADDRESSED NEEDS FROM DATA & EXPERIENCE** What customer/user needs are solved with data, automation, or additional information?	**IMPACT ON CUSTOMER / USER SATISFACTION** What is the customer/user satisfaction? What is the impact on the net promoter score? How often is the customer/user recommending the product, service, or value proposition?

DOWNLOAD TOOL
www.design-metrics.com/en/data-ideation

How to Consider Data Ethics

A major challenge in the context of data-driven innovation is the handling of regulations and ethical principles with regard to the use of data for innovation work or the customization of offerings or experiences. On a positive note, many countries and economies have established clear rules and guidelines in recent years that help organizations stay within the right framework. However, ethics and privacy also represent central components that are highly important for shaping a data-driven culture. Many companies have set the ambition that everyone in the organization embrace data, and thereby leverage the potential of new market opportunities. Employees are expected to acquire skills that allow them to work with internal and external data as a high priority. At the same time, the known silos in organizations are to be eliminated and data made available accordingly, as it has been realized that data is an important asset to deliver competitive advantages. Some companies are making data the core business strategy, the main success factor for developing future business success, and going beyond just using data to explore the problem space, new ideas, and innovation.

The principles defined, for example, with the help of the Data Ethics Canvas (see page 329), support the hybrid innovation teams and every single team member to work toward the agreed frame of values and purpose. Typical themes for principles are ranging from privacy to the promotion of human values (in the respective cultural setting).

EXAMPLES OF VALUES AND PURPOSE	THEME
• Control over the use of data based on relevant data laws • Ability to restrict processing, building consensus, or erase data	Privacy
• Verifiable and replicable with governance via monitoring bodies • Ability to evaluate, audit, and create impact assessments	Accountability
• Technical robustness of data models, protocols, and algorithms • Security by design and predictability	Security & safety
• Transparency about data and algorithms (e.g., open source) • Notification in the event of decision making of AI	Explainability & transparency
• Prevention of bias; striving for fairness and equality • Application of representative and high-quality data	Non-discrimination & fairness
• Human agency and oversight for automated decisions • Ability to opt out of automated decisions	Human control of technology
• Access to technology; human values and human flourishing • Focus on environmental and societal well-being	Professional responsibility

DATA ETHICS

- The principles of data ethics, including AI and neuroscience, ensure that data-driven innovations have the greatest possible impact.
- Data ethics is part of the entire development cycle from problem identification, implementation, and ecosystem realization to scaling.
- It is also necessary because the expectations of customers/users have also evolved, and they expect that an offering knows who they are and what they need.
- It is the foundation for good evidence-based innovation work and part of the hybrid mindset.
- It is also part of a data-driven leadership culture and well-defined governance structures across corporate boundaries.

Template download

DATA ETHICS CANVAS

1 PRINCIPLES	2 DESIGN CYCLE	3 IMPLEMENTATION
GENERAL PURPOSE What is the core purpose for using data?	**DATA SOURCES & RIGHTS** What are the data sources? Internal or external data? Who owns the data?	**ACTIONS** Which steps, considerations, and compliance clarifications are needed before implementing?
GENERAL VALUES What values apply for our company or region?	**DATA LIMITATIONS** What are the limitations of the data source? What is influencing the outcomes (e.g., data or timeline gaps, omissions, general bias in data)?	**ENGAGEMENT** How is the customer/user informed about data collection and uses of data? What is the mechanism to erase data or other actions required in conjunction to the product, service, or experience?
GENERAL POLICIES & LAWS What data protection legislation applies? What sector specific data sharing policies are to be considered?	**DATA SHARING** Is the data shared with other departments, regions, or companies? What is the policy for data sharing?	**ENHANCEMENT** What data points help to improve the product, service, or experience? How might the data set help to expand the core value proposition? How can the data be combined to obtain better insights from data driven innovation?
DEFINITION OF CORE PRINCIPLES What are the defined core principles working with Data, AI, and the application of neuroscience? Are the principles communicated with the wider organization, the core team, as well as the stake-holders (inclusive customers/users)?	**MINIMIZING NEGATIVE IMPACT** What measures can be applied to reduce any limitations with regard to the data? How is the potential negative impact monitored? What benefits are related to the actions applied in the specific design challenge?	**IMPACT OF DATA** What is the impact of the data? For example: - Better quality of service? - Shorter development cycles? - Better customer engagement? - New revenues?

DOWNLOAD TOOL
www.design-metrics.com/en/data-ethics

Example of Applying the Hybrid Model

For many wicked problems, the hybrid model is outperforming the single use of big data analytics or design thinking. Especially in this situation, the customers/users are looking for a suitable solution which fits into the core activities they are performing. The design challenge stated about preventing ski accidents on slopes has been given to pure data teams, professional design thinking teams, as well as mixed innovation teams working in a hybrid setting over 10+ years. As a result, data scientists tend to create solutions based on many data points, integrating the information in augmented realities, and inform ski and snowboard riders constantly about potential danger zones, high density on slopes, weather conditions, and much more information. However, most of the ski and snowboarders are not enjoying the purely data-driven and digital solution, causing an information overflow on a nice day out skiing.

On the opposite side, design thinking teams tend to make the wrong assumptions about data. Assuming for example that more accidents are happening on ski slopes with bad weather conditions (e.g., heavy snowfall and fog). In addition, observation and deep interviews with snow sports enthusiasts alone did not lead to any superior solution to prevent accidents. Also going into pure neuroscience paths to test different campaigns for preventing accidents resulted in more attention at the targeted audience but did not help to reduce the number of accidents. The best results have been achieved in the creation of a solution which included both the power of ethnographic research and insights from data, topped-up with learnings from neuroscientific research.

Design Challenge from an Accident Insurance Company:

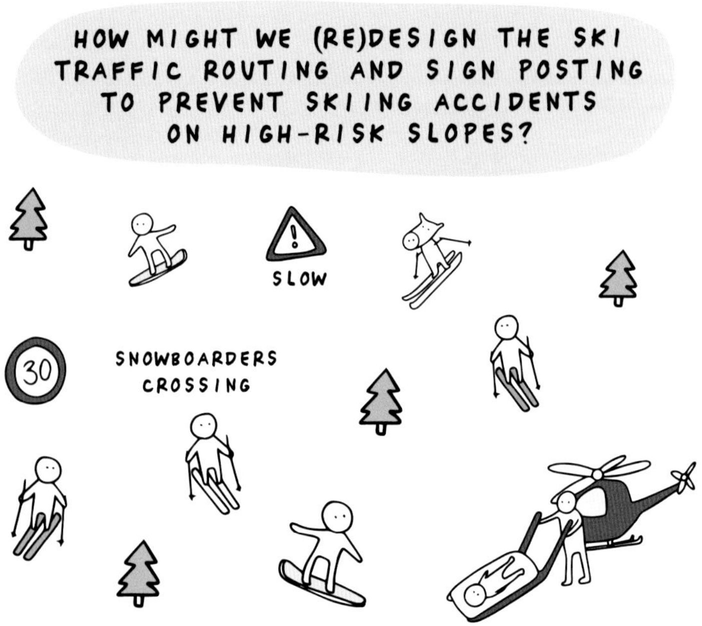
HOW MIGHT WE (RE)DESIGN THE SKI TRAFFIC ROUTING AND SIGN POSTING TO PREVENT SKIING ACCIDENTS ON HIGH-RISK SLOPES?
SLOW
30
SNOWBOARDERS CROSSING

Creation of a Mixed Innovation Team with Data Scientist and Experts in Ethnographic Research

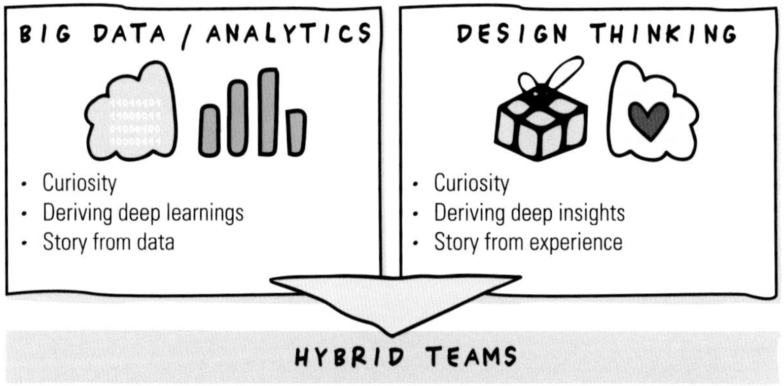
BIG DATA / ANALYTICS
• Curiosity
• Deriving deep learnings
• Story from data
DESIGN THINKING
• Curiosity
• Deriving deep insights
• Story from experience
HYBRID TEAMS

The design challenge set was worked on by over 300 different teams in a wide variety of constellations over a period of 10 years. This has allowed the impact of the hybrid model to be measured and the tools and methods to be progressively developed.

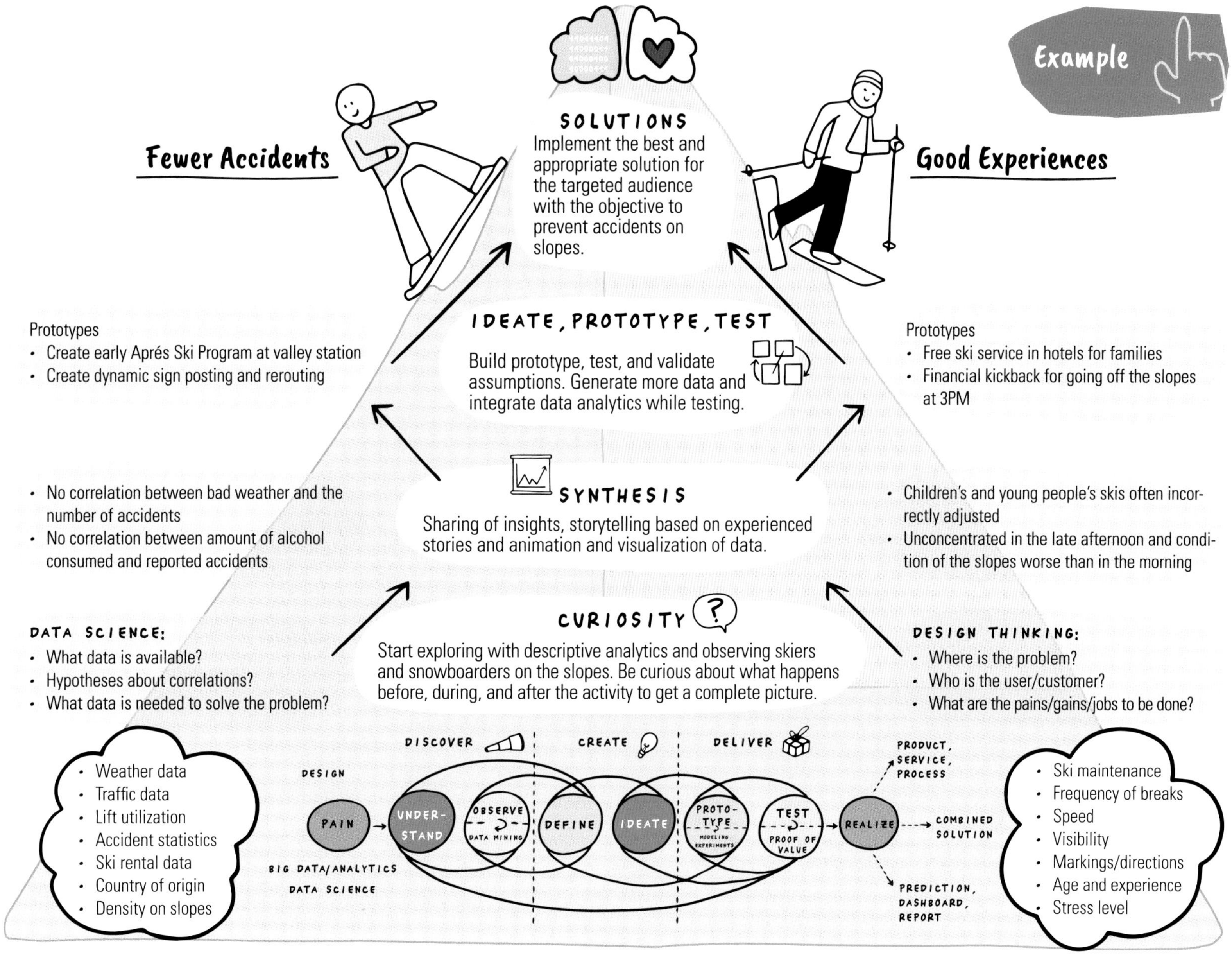
Example
Fewer Accidents
Good Experiences
SOLUTIONS
Implement the best and appropriate solution for the targeted audience with the objective to prevent accidents on slopes.
IDEATE, PROTOTYPE, TEST
Build prototype, test, and validate assumptions. Generate more data and integrate data analytics while testing.
Prototypes
• Create early Aprés Ski Program at valley station
• Create dynamic sign posting and rerouting
Prototypes
• Free ski service in hotels for families
• Financial kickback for going off the slopes at 3PM
SYNTHESIS
Sharing of insights, storytelling based on experienced stories and animation and visualization of data.
• No correlation between bad weather and the number of accidents
• No correlation between amount of alcohol consumed and reported accidents
• Children's and young people's skis often incorrectly adjusted
• Unconcentrated in the late afternoon and condition of the slopes worse than in the morning
CURIOSITY
Start exploring with descriptive analytics and observing skiers and snowboarders on the slopes. Be curious about what happens before, during, and after the activity to get a complete picture.
DATA SCIENCE:
• What data is available?
• Hypotheses about correlations?
• What data is needed to solve the problem?
DESIGN THINKING:
• Where is the problem?
• Who is the user/customer?
• What are the pains/gains/jobs to be done?
• Weather data
• Traffic data
• Lift utilization
• Accident statistics
• Ski rental data
• Country of origin
• Density on slopes
DISCOVER
CREATE
DELIVER
DESIGN
PAIN
UNDER-STAND
OBSERVE
DATA MINING
DEFINE
IDEATE
PROTO-TYPE
MODELING EXPERIMENTS
TEST
PROOF OF VALUE
REALIZE
BIG DATA/ANALYTICS
DATA SCIENCE
PRODUCT, SERVICE, PROCESS
COMBINED SOLUTION
PREDICTION, DASHBOARD, REPORT
• Ski maintenance
• Frequency of breaks
• Speed
• Visibility
• Markings/directions
• Age and experience
• Stress level

Applying Predictive Analytics for Enhancing Innovation Metrics

Data analytics also helps in the extension of the innovation measurement system (IMS). As already presented in the Measuring Toolkit (see pages 224–255), two core areas of statistics and analytics help in the formulation of operationalized measurement systems and in the extension and alignment of such systems toward predictive and prescriptive analytics and beyond.

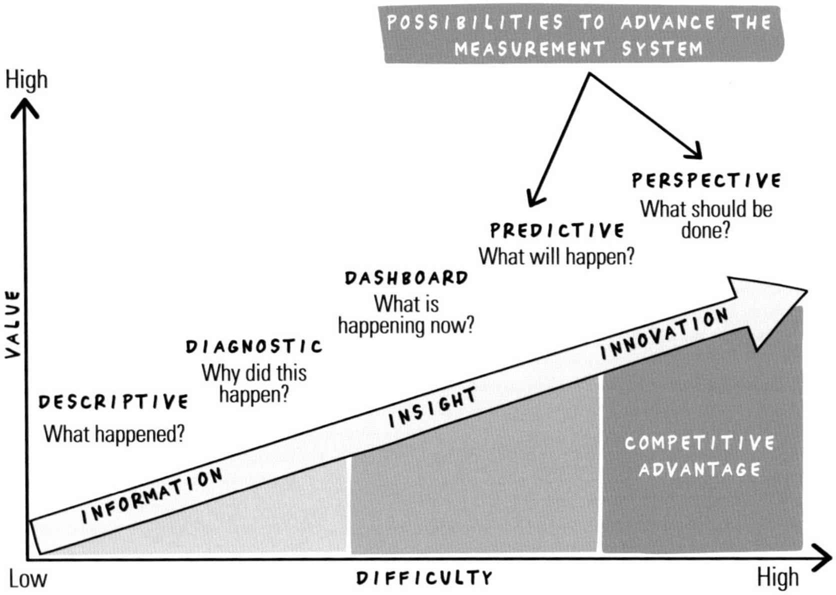

POSSIBILITIES TO ADVANCE THE MEASUREMENT SYSTEM
High
VALUE
PERSPECTIVE
What should be done?
PREDICTIVE
What will happen?
DASHBOARD
What is happening now?
DIAGNOSTIC
Why did this happen?
DESCRIPTIVE
What happened?
INFORMATION
INSIGHT
INNOVATION
COMPETITIVE ADVANTAGE
Low
DIFFICULTY
High

With new and more sophisticated measurement methods, as well as a deeper understanding of the factors, the behavior of the respective teams is changing and the continuous cycle of measure, evaluate, and operationalize is accelerating. The basic principle of the learning loops remains the same in the predictive procedures, but the test of the respective hypotheses is not performed backwards, but based on predictive, prescriptive, and machine learning principles.

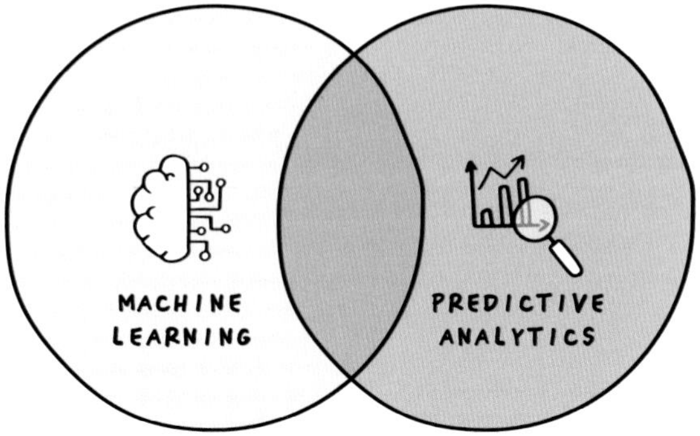

MACHINE LEARNING
PREDICTIVE ANALYTICS

The development of metrics is based on the initial identification of the problem, and in the second step, which is crucial for this approach, new metrics are created from the data instead of defining or adapting metrics based on assumptions. These data-based metrics are used again for analysis of the targeting purposes to achieve the desired effect in terms of EXPLORE or EXPLOIT. The data enables new ways of looking at the problem that go beyond the imagination and the current prevailing assumptions.

Predictive analytics for enhancing innovation metrics is a powerful tool for rethinking a variety of assumptions related to measurement. For example, incorporating predictive alignment changes assumptions and metrics that align organizational behavior with North Star, company, and team objectives. The combined efforts of a hybrid team of design thinkers and data specialists help organizations not only ask the right questions, but also identify the right data flows and create governance systems so that the data required for key metrics has consistent significance across networked organizations.

However, it is still needed to have in a world of data analytics a well-managed transition from EXPLORE and EXPLOIT as well as an alignment and governance between the predictive analytics applications and the respective performance measurements.

How to Apply Predictive Analytics Applications

In practical implementation, there are several things to keep in mind. These include the principles listed in Measuring 101 (see page 158), as well as other specifics that should be considered about culture, alignment, and responsibilities.

The Expansion of Accountability to Include Data

In applying predictive tools and methods, the design thinking and innovation team will be accountable not only for their performance on a metric or key result, but also for how well the measurement system itself is able to test existing assumptions and formulate new hypotheses. Developing an analytics capability that enables testing and learning is in itself a challenging task for many organizations, but it is critical to support this with expertise and data literacy from team members.

Teams, including their respective leaders, must take responsibility for the development of innovation measurement systems. It becomes mandatory for them to set aside their deadlocked assumptions about measurements and understand the factors that influence metrics so they can generate and adapt to a set of metrics. Especially for less data-driven companies, this requires the development of new analytical capabilities and processes, the willingness to continuously adapt the measurement system, as well as the openness to new behaviors.

In the formulation of the metrics, the hybrid approach is suitable because it brings in the expertise of business intelligence executives and analysts as well as data scientists in addition to the design thinking expertise.

The required skills reach from probability and statistics, structured query language, statistical programming (e.g., R or Python), ML, data management, data visualization (e.g., Tableau) to econometrics, which help forecast future trends.

Transparency with Respect to Analytics Tools

The implemented metrics based on predictive analytics require democratic access to advanced analytics tools for teams and relevant stakeholders. The risk of organizational misalignment increases with unequal access to advanced analytics. Unequal access to analytics can reinforce cultural differences, fuel employee dissatisfaction, and undermine the alignment of employee behavior with strategic outcomes. Similar to the application of OKRs (see page 258–284), it is imperative to create the necessary transparency in analytics tools. As design thinking and innovation teams collaborate with many areas of an organization as part of a team-of-teams setting, there is a need for forward-looking alignment for the distribution of analytics tools and expertise across all business functions. This will again allow progress toward these metrics to be measured and their benefits to be continuously assessed.

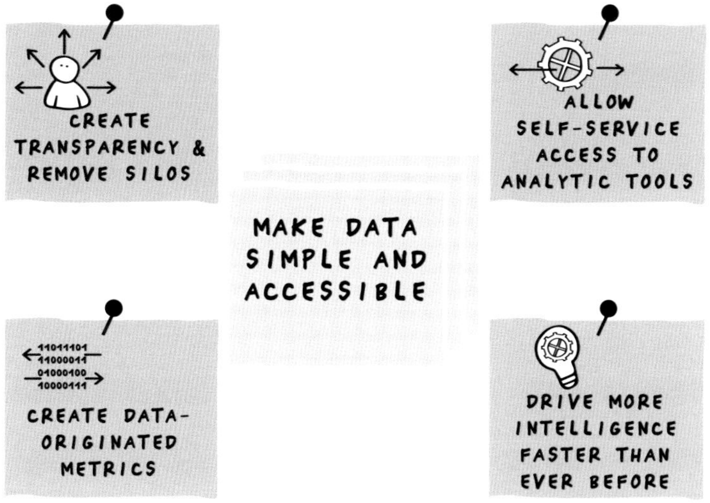
CREATE TRANSPARENCY & REMOVE SILOS
ALLOW SELF-SERVICE ACCESS TO ANALYTIC TOOLS
MAKE DATA SIMPLE AND ACCESSIBLE
CREATE DATA-ORIGINATED METRICS
DRIVE MORE INTELLIGENCE FASTER THAN EVER BEFORE

To the Point!

The combination of data analytics and design thinking accelerates innovation. The hybrid model allows use of data sets and tools in both the problem and solution pace.

Data science can help designers to think outside the box, providing relevant inspiration, alternative points of view, and support with advanced clustering techniques.

Innovation teams are enabled to rapidly prototype new, flexible solutions with, for example, native algorithms, open source, and partner ecosystems while ensuring governance.

Teams and organizations with a high maturity of the hybrid model democratize, co-create, and collaborate with automation, reusable templates, and a common collaborative framework for interdisciplinary teams.

DESIGN THINKING & ARTIFICIAL INTELLIGENCE

AI

Gaining Efficiency in Customer Interactions

As already mentioned in the previous section about the hybrid model, the application of artificial intelligence (AI) gives us completely new possibilities, which is also one of the most transformational technologies that innovation teams currently have at their fingertips. AI helps with gaining efficiency and realizing scale in customer interaction. However, developing sustainable and powerful AI remains a challenge, and design thinking is key to address customer/user needs and emotions in many cases. At the same time, the design thinking mindset supports the adoption of the technology in large organizations, because from identifying use cases to test and implement, design thinking and team-of-teams structures are needed for cross-functional initiatives. The design thinking mindset has helped countless AI projects achieve the interdisciplinary cross-pollination that encourages individuals and teams to ask powerful questions, observe, network, and experiment in an environment that dismantles hierarchies, eliminates vested interests, challenges the status quo, and encourages intelligent risk-taking.

Design thinking also helps to change the awareness that AI projects are not driven by IT, but that the responsibility for innovation and new customer interaction comes from teams in the organization that know the strategies and operational challenges best. A different setting allows the teams to become a networked organization that shares learning across the enterprise, along with costs and risks. In many cases, AI technology supports making the work of every employee more efficient, especially in the recently established hybrid-digital working environment. The human and robot interaction powered by AI is allowing to combine multiple types of work-related application with real-time data analytics. However, the ultimate goal must be that each individual employee is not only empowered by the technology, but that the employees and interdisciplinary teams enable the technology. This allows the current digital-hybrid and future virtual working environment to be configured in such a way that predictive decision making is improved and supports the agile customer development for new or customized products, services, or business ecosystem offerings to customer needs.

Design thinking has evolved and it nowadays provides a powerful mindset to design algorithms that determine the behavior of intelligent systems.

As a result, AI has become a discipline of applied design and computer science that deals with reproducing aspects of human thought and action with computers. The automation of "intelligent" behavior and machine learning are two concepts that need some explanation for their further application together with the concepts of the **design of AI, design thinking of AI, and design thinking with AI**. This distinction is important because, besides building awareness for AI with design thinking, the goal is to design, build, and scale machines that solve problems independently for many at the same time. Areas of application range from marketing to self-service interactions.

If machines are to act human-like, or even more efficiently than humans, in the future, it makes sense to take humans, their interactions, and their behavior as the basis for their superiority. In addition, machines today can analyze data and recognize certain patterns and regularities with the help of self-learning algorithms, and thus learn independently, which is referred to as machine learning (ML). In machine learning, data is linked together, and correlations, inferences, and predictions are made on this basis. The associated automation and tools for evaluating data (as described on page 339) support companies in knowing their customers and target group as precisely as possible, and addressing them personally. This could reduce coverage losses, for example, and is the basis for opti-channel considerations, and it allows mass-customization at scale. The starting point is a simple overview and definition of AI and ML to understand the current and future range of application.

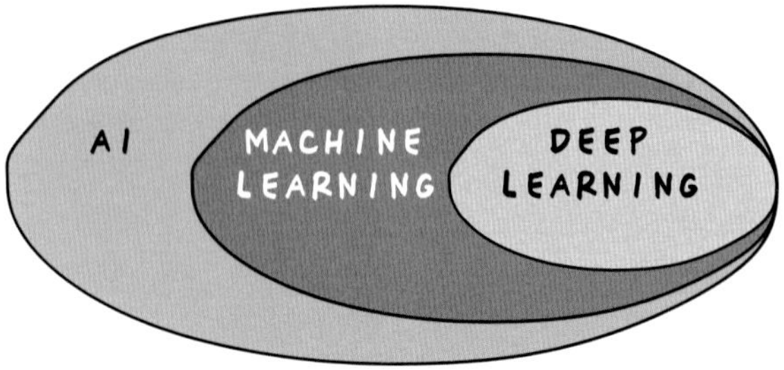
AI
MACHINE LEARNING
DEEP LEARNING

ARTIFICIAL INTELLIGENCE [AI]

[AI is the combined discipline of design and computer science so technology can learn, think, act, and execute tasks in ways traditionally attributed to human intelligence.]

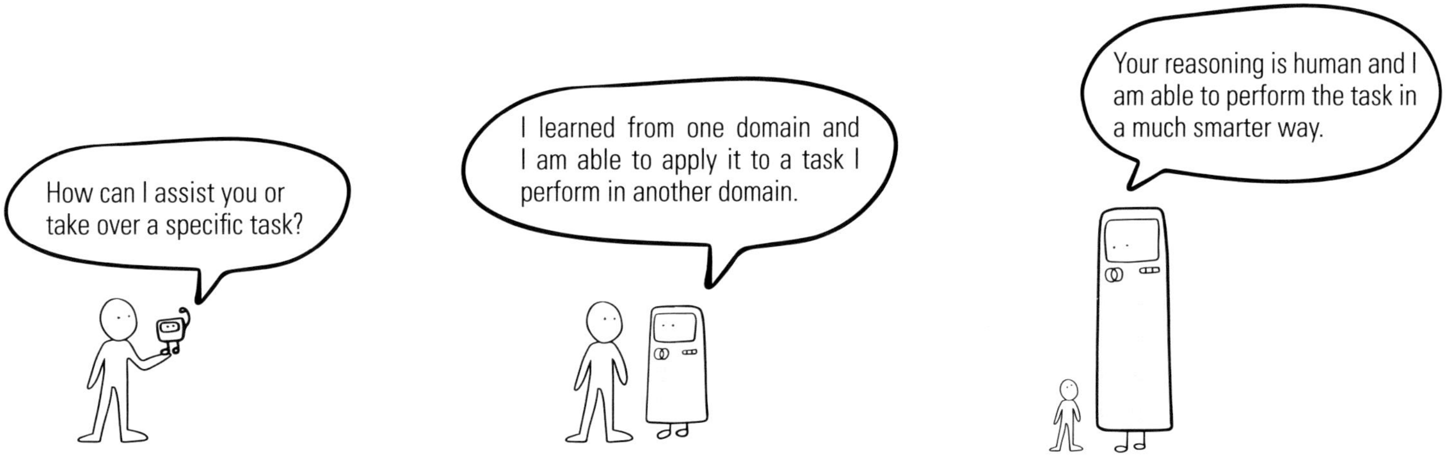

How can I assist you or take over a specific task?
I learned from one domain and I am able to apply it to a task I perform in another domain.
Your reasoning is human and I am able to perform the task in a much smarter way.

MOST COMMON NARROW AI

Most AI applications in use are equal to or surpass human intelligence only in a very specific task.

TODAY GENERAL AI

The next evolutionary step is the development of AI systems that match human intelligence in all domains and tasks.

FUTURE ADVANCED AI

Superior AI systems will surpass human intelligence in the future.

2000 — **2020** — **2040** → **TIME**

Model-Based versus Model-Free Algorithms

The key takeaway from the overview about ML and the types of machine learning is that ML lacks a number of aspects to simulate human learning. All of us are born with a package of intuitions. These early intuitions already act as an existing model for infants to figure out how social and physical interactions work in their environment and can be further extended through reinforcement and conditioning. An ML algorithm must include some sort of bootstrapping software that gives it the same advantage. Nevertheless, it is currently difficult for machine learning to understand social interactions, whereas humans are able to distinguish between friends and enemies from a very young age. From the age of 4 onward, humans develop the ability to build empathy with the people they are interacting with. ML has a hard time making inferences and understanding underlying concepts. However, these are important skills needed to build a causal model, but the models built by ML are mostly correlational. This is what makes human learning so special; it has the ability to learn how to learn and to improve learning. While machine learning is able to learn from its actions and rewards, unlike humans, the same algorithm cannot change its learning strategy. If machine learning is to become more like humans and be applied in superior AI systems, which will surpass human intelligence in the future, AI must definitely be able to achieve the next level of learning.

For the time being, design thinking and AI have become a very powerful combination for solving problems and making decisions. In the first section of 301, it was already explained that data and design thinking are a great combination of providing the WHAT and WHY, which can be linked to model-free learning and model-based learning. See an example on page 340.

Model-based learning is a top-down approach in which the learner begins with a pre-recognized model of the surroundings, and learning is based mainly on planning, predictions, and actions taken according to the known probabilities and rewards of that model. In contrast, in model-free learning, learning does not begin with a model of the environment, but the rewards for each action are learned and stored from the bottom up, and future actions are decided based on what is learned. The theory holds that all humans apply model-based learning when reinforcement is used in a goal-directed manner, whereas a model-free approach leads to the formation of habits. Consequently, we draw the connections between our behaviors and achieving a certain goal, and, while operating in this way, we are building a causal model which can be reused for planning future actions.

Design thinking focuses on understanding how the customer/user is interacting with the environment. In the first two phases of the design thinking process, besides presenting and testing prototypes, design thinking teams build causal models based on observing the consequences of different people's behaviors (vicarious reinforcement). In order to determine causality in the environment, the design thinking team draws inferences, a skill and capability that is very limited in machine learning. Another concept related to the application of ML and types of learning is supervised learning, which considers a model-based learning approach as it relies on a pre-labeled training data set to determine its predictions. In contrast, unsupervised learning, such as clustering, focuses on searching for patterns in the data set (e.g., word associations) that do not build a model; hence it is model-free. Reinforcement learning has both model-free and model-based methods, but it is hard to say which are more dominant. Model-free reinforcement learning (Q-learning) does not have an end goal, but the outcomes of the actions still contain rewards which the program uses to learn in a trial-and-error fashion.

Model-based algorithms = large amount of data (statistically well supported, but computationally inefficient)

VS

Model-free algorithms = less data (computationally efficient, but less statistically supported)

Research shows that heuristics and linear regression are possibly opposites of Bayesian inference. Models built by human learning result in flexible heuristics, whereas the models that arise from machine learning are more complex and rigid.

Application of ML and Types of Learning

MACHINE LEARNING [ML]

[**ML** is the ability for machines to learn and infer from large sets of examples and experiences instead of explicitly programmed rules.]

DEEP LEARNING [DL]

[**DL** is based on an artificial neural network inspired by the human brain, which allows even unsupervised learning.]

REINFORCEMENT LEARNING

Trial & error for maximum reward

Q-Learning

- Consumption
- Reduction
- Policy creation

Model-based value estimation

- Linear task
- Estimating parameters

SEMI-SUPERVISED LEARNING

Model based on a mix of labeled & unlabeled data

Self-trained native Bayes classifier

- Natural language processing

Generative adversarial networks

- Video and audio manipulation
- Data creation

SUPERVISED LEARNING

Model based on labeled data

Decision Tree

- Predictive analysis
- Pricing

Super vector machines

- Image classification
- Comparing performance

Linear regression

- Risk assessment
- Sales forecasting

UNSUPERVISED LEARNING

Model based on unlabeled training data

Apriori

- Sales functions
- Word associations
- Searcher

K-means clustering

- Performance monitoring
- Searcher intent

Example

[How might we (re)design incentives and nudges for commuters to change departure times by 15-20 minutes to avoid traffic jams on highways.]

MODEL-BASED TRAVEL ROUTING AND ETA PREDICTION

&

MODEL-FREE TRAVEL ROUTING AND ETA PREDICTION

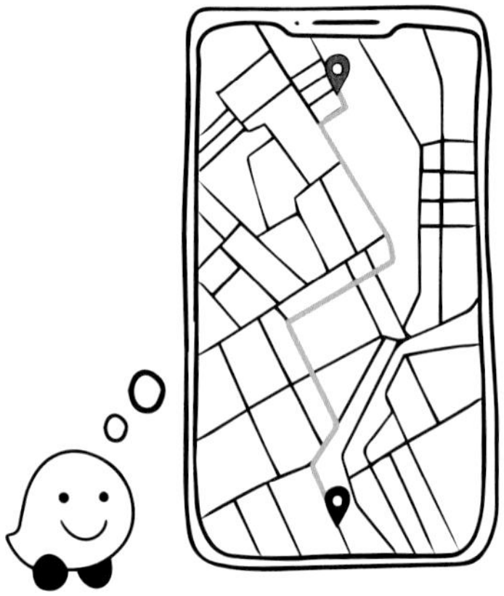

ML solutions @ WAZE:

- Predicting ETA
- Matching riders & drivers (carpool)

Waze Data-Types:

- Big/large scale data
- Low-latency inference (real-time, actual)

- Pains
- Gains
- Job to be done
- Routines

Leave at 7:50 AM and take the highway until exit 3A. Continue on country road 322 to arrive at 8:30 AM.

8 AM = Do not take the highway at this time. It takes 15 minutes longer on Mondays

Waze collects map data, travel times, and traffic information from users to assist travel routing, car pooling, and many other services based on human needs.

Evolution of Combining Design Thinking and AI

In an ideal ML process of machine learning, work is done in a team-of-teams setting. The team includes stakeholders who work closely with the data scientists and domain experts, who bring their respective expertise, tools, methods, and data sets, and are to change the final product/value proposition based on data-driven insights and predictions.

Over the last years, the levels of automation possible for end-to-end machine learning endeavors have evolved significantly. In the early days, ML was entirely manual and required the development and writing of software for the whole process. Automated AI aims for automation from task formulation all the way to result summary and recommendation.

The Three Immediate Effects of Increased Automation:

- Enablement of domain experts to make use of machine learning.
- Increase of data scientists' productivity, because their manual work declines as the levels increase.
- Time savings are realized mainly in data preparation and feature engineering.

Shift: From AI-SUPPORTED IDEAS to AI IDEAS

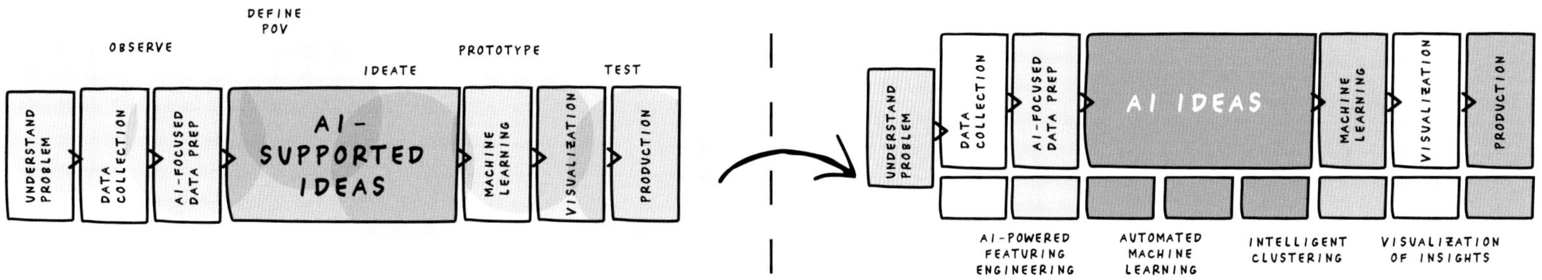

DEFINE POV
OBSERVE
IDEATE
PROTOTYPE
TEST
UNDERSTAND PROBLEM
DATA COLLECTION
AI-FOCUSED DATA PREP
AI-SUPPORTED IDEAS
MACHINE LEARNING
VISUALIZATION
PRODUCTION
UNDERSTAND PROBLEM
DATA COLLECTION
AI-FOCUSED DATA PREP
AI IDEAS
MACHINE LEARNING
VISUALIZATION
PRODUCTION
AI-POWERED FEATURING ENGINEERING
AUTOMATED MACHINE LEARNING
INTELLIGENT CLUSTERING
VISUALIZATION OF INSIGHTS

AI has a lot of potential in the future, but at the moment it is still limited to finding the appropriate problem to solve, asking the relevant powerful questions, or detecting the inefficiencies we created in our processes and systems. In EXPLORE, AI has the potential to augment data insights and delivers unknown unknowns based on the ability to explore millions of feature hypotheses in no time, enhancing the ability to think outside the box, and upending strategy and innovation.

The future applications of AI and machine learning for creativity and innovation are very different from the established areas where the focus is on the design of AI or efforts in leveraging design thinking for creating experiences with AI. In the application and focus related to the design of AI, the well-known standard tools and methods are used to identify, for example, market opportunities for AI. Design thinking for AI works in two dimensions: improving customer experiences with AI and the application of design thinking to facilitate organizational change. The future application of design thinking with AI will be an integrated team effort across the entire design cycle, moving from a hybrid model toward hybrid intelligence. The following pages, which present a short toolkit, are mainly focused on design thinking for AI, and on how design thinking and AI will potentially co-create new products, services, and experiences.

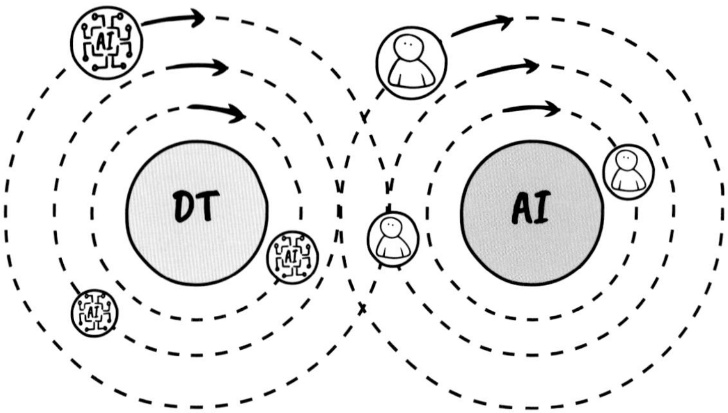

DT
AI

WHO / WHAT	Design of AI	Design Thinking for AI	Design Thinking with AI
DESIGN THINKING	• Understand and observe user/customer needs for AI systems • Identification of opportunities	• Understand, improve, and re-design customer/user experience for AI solutions • Applying design thinking for managing the change process within the organization	• AI co-creates with design thinking team • AI identifies problems to solve and questions to ask • AI creates ideas, tests prototypes, and proposes potential solutions
ML ENGINEERS + DATA SCIENTISTS TEAM	• Creation of system requirements • Decisions on model • Training of model • Calculation cost of errors • Evaluation of the model together with the design thinking team	• Identifying system limitations • Build, test, and integrate user feedback • Machine programming and teaching • Identification of user autonomy • Explainability to user/customer and other data ethics	• Utilizing superior AI systems • Advanced programming languages • Application of learning-based systems and deep code

How to Apply Design Thinking for AI

As highlighted in the overview on page 342, design thinking is predominantly used for AI, and the possibilities and tools would fill an entire book. This section exemplifies the most important activities that are necessary before an AI/ML service is designed and implemented and serves a short toolkit. By using the design thinking mindset, uncertainty can be minimized and a focus on purpose can be realized for the team. However, in the AI age, the challenge might be even bigger than expected because it is necessary to move beyond customer needs alone. Interdisciplinary design teams must develop trust and empathy within the new experiences to create a cooperative relationship between AI and customers/users.

Most of the time, innovation teams have to deal with a whole AI ecosystem consisting of real problems, the benefits for the company, the added value for their customers/users, and the current and future possibilities of ML/AI. Furthermore, it is often observed that AI solutions realize secondary effects in addition to the actual benefits. This means that there is an additional benefit from the interaction and the new data that helps to innovate in a data-driven way or to design even more complex AI scenarios.

As in all applications of the design thinking mindset, it is also recommended in the application of AI to reflect regularly on the project and meta-level. This also includes the ethical discussion of the application of AI up to possible negative consequences for society, the company or the individual customer/user (see Data Ethics Canvas, page 329).

The Design Thinking for AI Canvas (see page 344) supports the respective teams in the document of the outcomes from iterations and data exploration, based on tools from the design thinking, hybrid, and AI toolbox based on needs, approach, and benefits.

Toolkit

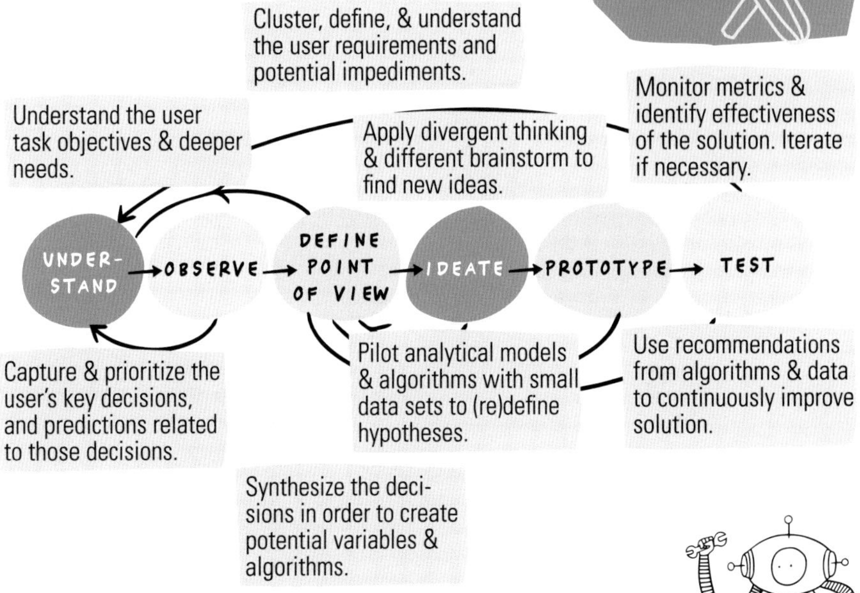
Cluster, define, & understand the user requirements and potential impediments.
Understand the user task objectives & deeper needs.
Apply divergent thinking & different brainstorm to find new ideas.
Monitor metrics & identify effectiveness of the solution. Iterate if necessary.
UNDER-STAND
OBSERVE
DEFINE POINT OF VIEW
IDEATE
PROTOTYPE
TEST
Capture & prioritize the user's key decisions, and predictions related to those decisions.
Pilot analytical models & algorithms with small data sets to (re)define hypotheses.
Use recommendations from algorithms & data to continuously improve solution.
Synthesize the decisions in order to create potential variables & algorithms.

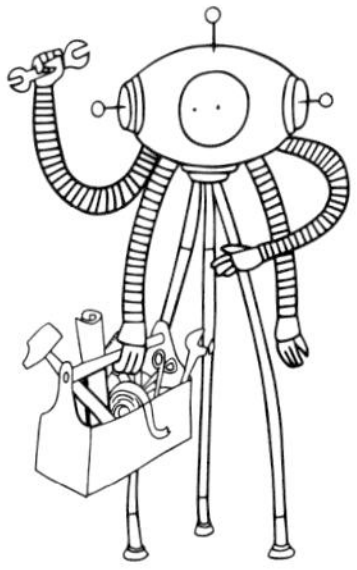

The Following Pages Provide an Extended Design Thinking Toolkit for the Development of AI Solutions.

Preview of Suitable Methods & Tools

Template download

DESIGN THINKING FOR AI CANVAS

1 NEEDS

PERSONA
Who has the problem?
- What are the pains/gains for the customer job to be done (JTBD) and use cases?
- What are the current and future customer needs?

INTERACTIONS
How, when, and where will the customer/user interact with the solution?

DATA ETHICS
What are potential ethical concerns?
How are ethical data concerns addressed?
Is the use of data based on relevant data laws affected?

2 APPROACH

DATA SOURCES, RIGHTS & LIMITATIONS
What are the data sources? Who owns the data?
What are the limitations of the data source?
What is influencing the outcomes (e.g., data or timeline gaps; omissions, general bias in data)?

AI SOLUTION & CUSTOMER/USER EXPERIENCE
What are the core functions of the AI solution?
What are major improvements in the user experience?
Why does the chosen solution/experience bring the the greatest measurable added value?

ECOSYSTEM CONFIGURATION
How is the solution embedded into a wider ecosystem?
What actors are needed to design, build, initiate, and orchestrate?
What data and AI platforms are used?
What partners and suppliers are needed for sourcing data?

3 BENEFITS

AI VALUE ADD
Which steps, considerations, and compliance clarifications are needed before implementing?

EXPERIENCES VALUE ADD
How is the customer/user informed about data collection and uses of data?
What is the mechanism to erase data or other actions required in conjunction to the product, service, or experience?

VALUE ENHANCEMENT
How might the AI solution be used to collect more meaningful data or define meaningful measurements?
How can the experience be expanded or added with additional value-adding functions or interactions?

VALUE PROPOSITION
How does the (re)designed experience bring value to the customer/user? What are the gains for the customer/user through the use of AI?
Examples:
- Proactive
- Higher personalization
- 24/7 availability
- Improved ease of use

DOWNLOAD TOOL
www.design-metrics.com/en/design-thinking-ai

Understanding the User/Customer for AI Scenarios

In the application of design thinking for AI it is central to start from the users/customers and their deeper needs. A customer/user profile (see 101 Toolkit on page 114) helps to identify a persona and use cases, including the jobs to be done. For each AI/ML application, it must be clear which problem is to be solved. In many cases the current customer journey is also helpful to fully understand the interactions of a customer/user with a company.

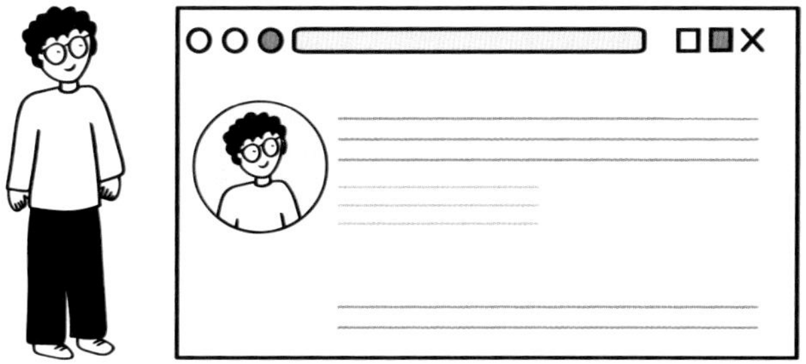

Reflecting on Typical AI / ML Application Scenarios

In many cases AI can be used to realize better, more precise, or personalized interactions for the user/customer. These range from faster recognition of patterns to prevention and detection of faults. The analysis of the customer's/user's intent helps in the matching of the tasks to be fulfilled, use cases, and the possibilities of AI. The central question is:

AI

How do the customers/users and business benefit from the selected application scenarios of AI? What is the additional value for both of them?

Example

Acceleration in Pattern Discovery:
Processing millions of data points to find the elements that have high relevance to the customer/user.

Interaction Enrichment:
Training of AI on frequent requests from customers/users to reduce response times, increase the number of transactions, and make interactions more purposeful.

Active Prevention of Incidents and Disruptions:
Real-time monitoring of systems to mitigate issues that impact the customer experience before they occur.

Trusted Recommendations:
Training of AI on the intricacies of a business relationship so that tailored recommendations are made and customer relationships are strengthened.

Scaling Knowledge and Assisting Learning:
Applying AI/ML as a concept of expertise delivery and demand-driven learning.

Detect Legal, Compliance, and Other Regulatory Changes:
Perform training of AI to quickly understand and clarify impact of relevant regulations and privacy obligations.

Exploring & Selecting Relevant Data

For the first prototypes and tests with the selected application scenarios, the appropriate data is needed, which is either already available in the company or has to be collected. The previously introduced Data Ideation Canvas (see page 327) helps to identify the data available in large companies, which might have not yet been made accessible. All companies have data on operational processes and in the form of user/customer data, which is obtained, for example, from interactions in the customer journey. As already highlighted in conjunction with the hybrid model, there is also public data or data provided by open data initiatives. From the multitude of possible data points, the ones that best fit the application must be used.

However, before the appropriate data points can be used, it must be clear at which point AI can offer real added value. As shown on the next page, an AS-IS customer journey can serve as a basis, which in turn helps the innovation team to improve individual interactions via AI. From this, hypotheses for the application of AI can be formulated:

If.. (data) is integrated/analyzed, it can be predicted/indicated, that .... (hypothesis).

Where AI offers real value, data should be used or actively collected to create a compelling AI experience. The new desired customer journey should be broken down further if possible. This allow us to understand in detail with which inter(action) the customer/user generates data and how the AI interacts and learns. In addition, it is possible to find out which data services arise from this or are required and, above all, what added value arises for the customer and the company as a result. Every interaction, touch point, and pivot of an AI prototype helps the team to:

- define what the customer/user is doing.
- observe which data sources are accessed.
- know how data is analyzed.
- understand how details are made accessible.
- envision how that info will help the customer/user.
- make a decision to and/or move forward.

Critical Issues Related to AI in Customer Interactions:

- Does AI solve a real customer problem?
- Does AI solve the pain points in the interaction?
- Is the customer interaction suitable for AI?
- What can the respective interaction look like?
- Do the sender and receiver understand the respective messages correctly?
- Does this enable the customer to be helped immediately?
- Where do further points of integration into surrounding systems become necessary?
- What is the degree of complexity of the interaction that can be represented?

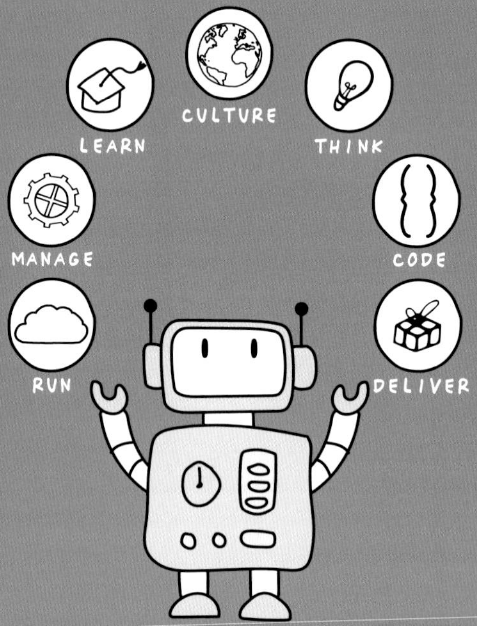
CULTURE
LEARN
THINK
MANAGE
CODE
RUN
DELIVER

(Re)designing Customer Interaction with AI

Template download

Based on an AS-IS customer journey, a mapping can be undertaken that shows where AI adds particularly great value and positively enriches the customer experience and interaction. The Customer Journey AI Template (see below) helps to map and identify the potentials.

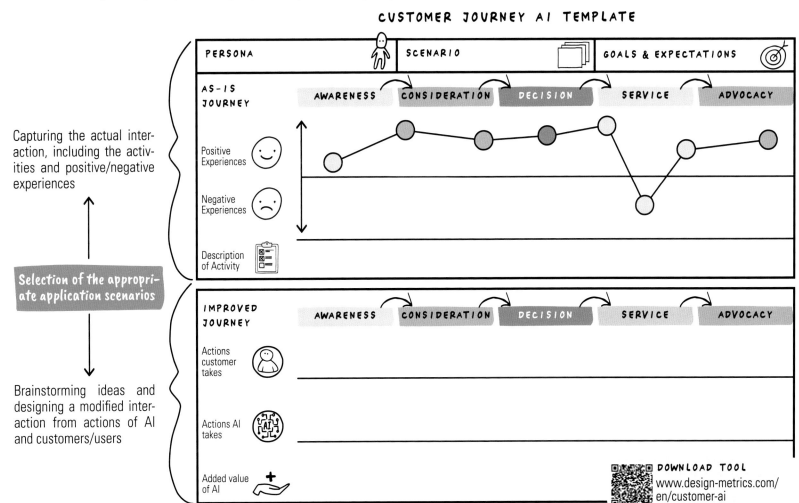

Capturing the actual interaction, including the activities and positive/negative experiences

Selection of the appropriate application scenarios

Brainstorming ideas and designing a modified interaction from actions of AI and customers/users

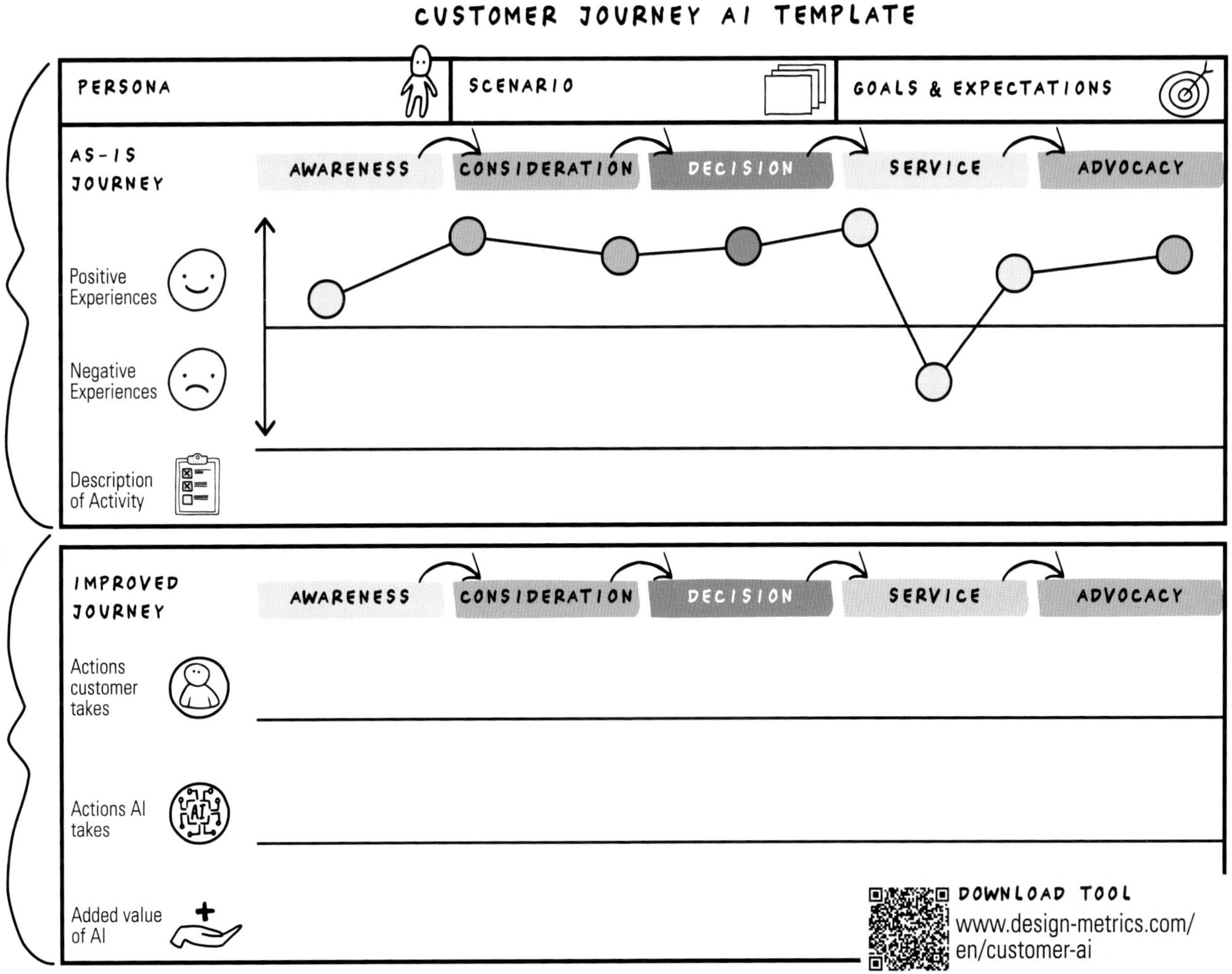

CUSTOMER JOURNEY AI TEMPLATE
PERSONA
SCENARIO
GOALS & EXPECTATIONS
AS-IS JOURNEY
AWARENESS
CONSIDERATION
DECISION
SERVICE
ADVOCACY
Positive Experiences
Negative Experiences
Description of Activity
IMPROVED JOURNEY
AWARENESS
CONSIDERATION
DECISION
SERVICE
ADVOCACY
Actions customer takes
Actions AI takes
Added value of AI
DOWNLOAD TOOL
www.design-metrics.com/en/customer-ai

Ideation for Better Customer Experiences

Based on initial questions in conjunction with the desired outcomes, the ideation part takes place. For example:

How might we apply the power of AI/machine learning to solve the customer/user problem or pain points in order to achieve a specific outcome.

The reference to different creativity techniques for the ideation phase has already been presented in Design Thinking 101. They range from critical item diagrams to dark horse prototypes (see also The Design Thinking Toolbox, pages 187–194). The big pitfall in initial brainstorming for AI solutions is that features are quickly included rather than real customer experiences. At this point, teams should always keep in mind that it's all about ideas for the customer/user experience and not how individual features will be implemented later.

Throughout the design cycle and within the individual brainstorming sessions, it makes sense to cluster, select, and prioritize ideas. The selection and prioritization is necessary to define a starting point for the experiments with the team and to have a deeper discussion about the necessary data.

For example, the selection of critical experiences for the customer/user up to the prioritization of initial ideas has relevance again and again throughout the entire design cycle. Usually, a simple matrix with two axes helps to find priorities and select actions toward the desired outcomes. One possibility is to show the feasibility of the solution on the Y axis and the value of the AI solution for the customer/user on the X axis.

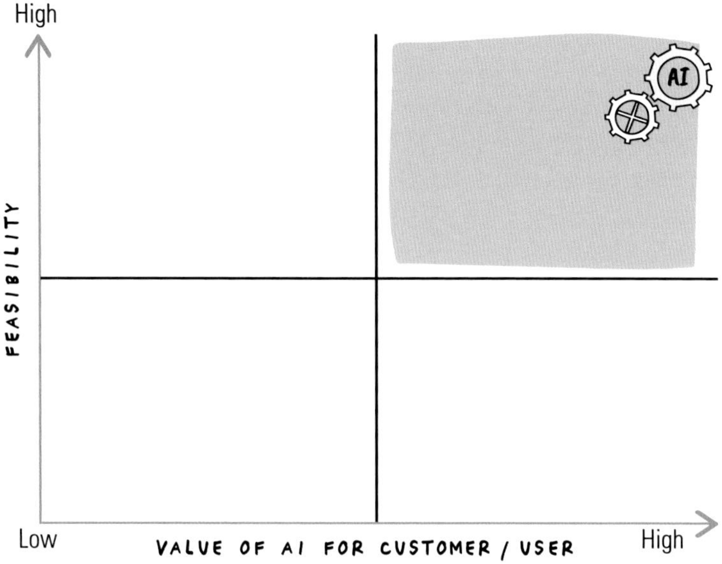
High
AI
FEASIBILITY
Low
VALUE OF AI FOR CUSTOMER / USER
High

Extended design thinking and innovation teams should look for ideas and solutions that are feasible and bring a high value for the customer or, depending on the axis labeling, for the company (e.g., higher conversion rate or cross-selling opportunities). Of course, ideas that are not feasible and do not bring any added value should be avoided for the time being.

Experience (Re)design & Data Selection

Template download

What are the steps the customer/user will perform, or the actions to take based on the proposed AI solution?

Which problem or pain point is solved?

Regardless of the designed experience (e.g., conventional AI solution for ordering food or performing banking transactions), it is also important that the use of AI is a good fit for the solution (AI/solution fit). The (re)designed user experience should be documented, for example, in the previously introduced Customer Journey AI Template, with the aim to outline how customers/users are better served with the new AI-based solution than before. There are two types of prototypes. The first might be not related to real data. Basically, the experience is acted out and human interaction is used to simulate the AI experience. With the application of data and AI tools, the potential output and business value can be measured.

Measuring the Business Value of AI Features in (Re)designed Customer/User Experiences

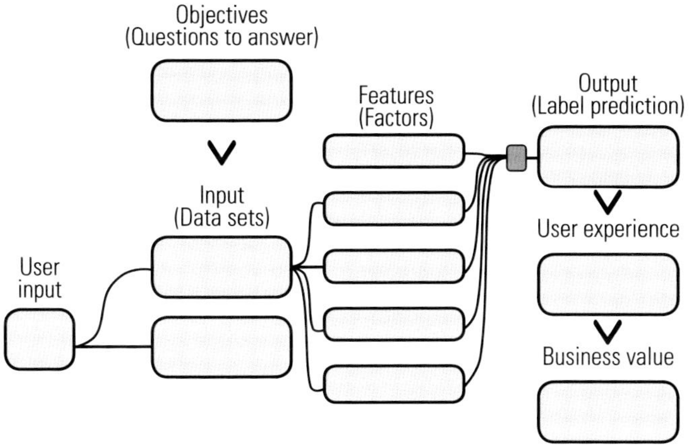

Objectives
(Questions to answer)
Features
(Factors)
Output
(Label prediction)
Input
(Data sets)
User experience
User
input
Business value

For more advanced prototypes with real data, it is necessary to identify the relevant data to train the AI model, as already highlighted on page 326. Again, the Data Ideation Canvas (see page 327) and Data Ethics Canvas (see page 329) might be helpful to identify the data available and data points that need to be acquired or created. In addition, the two templates help to provide evidence that the data sets are balanced, and that security, privacy, and other potential ethical concerns are considered. Data Source Template provides structure in identifying the relevant data sets for the AI solution and distinction between the source of data and how it might be used for the specific scenario or customer/user experience to be (re)designed.

DATA SOURCE TEMPLATE

RELEVANT DATA FOR THE AI SOLUTION				
DATA REQUESTS / SOURCING	SOURCE OF DATA	OWNER OF DATA	PERMISSIONS/ LIMITATION	ETHICAL DATA PRINCIPLES
EXISTING INTERNAL DATA				
EXISTING EXTERNAL DATA				
ADDITIONAL DATA NEEDED				
DATA TO BE CREATED				

DOWNLOAD TOOL
www.design-metrics.com/en/data-source

Prototyping, Testing, & Validating

To create a first functional prototype/MVP and to perform real tests, the appropriate AI tool(s) are needed. Different AI tools might be applied to build the prototype and the entire solution. A first concept for a prototype will show how the AI model will be created to be trained in the interaction with customer/user. The concept will be redefined and improved after the testing and the ongoing validation of the hypotheses. Customer/user testing validates the AI-designed experience based on the needs of the customer/user and the established hypothesis. Based on the design thinking mindset, validation of AI iteration is always performed with the customer/user in mind. If the data is valid, the hypothesis can usually be proven or disproven. It is important to proceed in an iterative way, because, for example, with an appropriate data enrichment, the hypothesis can be confirmed, and the needs of the customer/user can be fully met. Functionality, technology, and AI decision are based entirely on achieving the hypothesis in the context of the customer/user needs, jobs-to-be-done, or defined use cases. The respective prototypes can be measured by appropriate metrics and the collected findings can be visualized. The visualization of the results helps to reflect the results with the team and beyond.

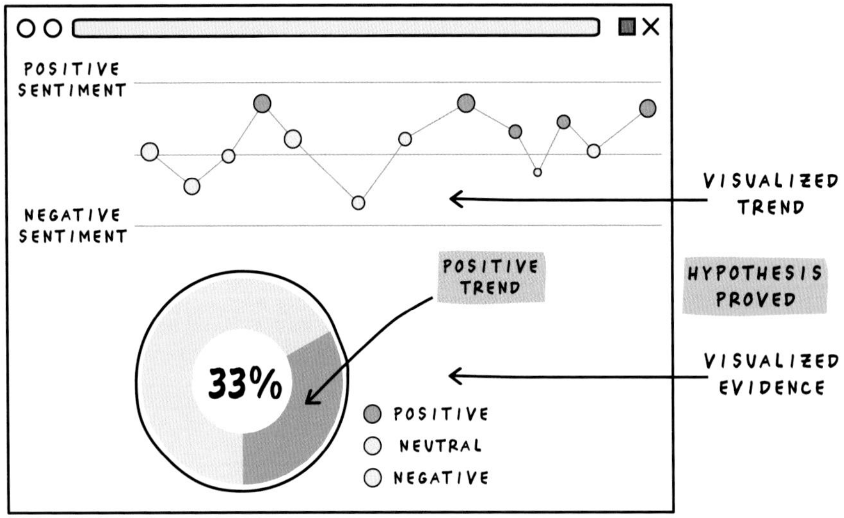
POSITIVE SENTIMENT
NEGATIVE SENTIMENT
VISUALIZED TREND
POSITIVE TREND
HYPOTHESIS PROVED
33%
VISUALIZED EVIDENCE
POSITIVE
NEUTRAL
NEGATIVE

Validating the AI Ecosystem Landscape

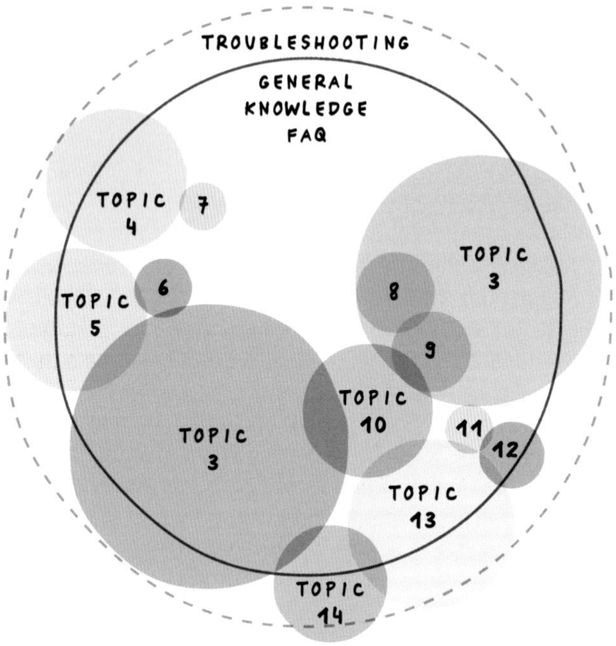
TROUBLESHOOTING
GENERAL KNOWLEDGE FAQ
TOPIC 4
7
TOPIC 3
TOPIC 5
6
8
9
TOPIC 10
TOPIC 3
11
12
TOPIC 13
TOPIC 14

Proper AI development, for example, in the context of conversational AI, needs in many cases also a deep understanding about the ecosystem in which customers and/or employees interact with the organization. An AI ecosystem landscape helps to better understand the initial touch point, which content is needed, and finally how the conversation will create a value-add for customer/user, employees, and for the organization. Design thinking is very valuable at the beginning to understand how touch points, interactions, and conversations relate and link to each other and are supported by systems thinking, which allows for the creation of MVPs for each topic, focusing on a specific path or just specializing on a topic area of high interest for the customer/user. The AI and business ecosystem perspective provides more space for AI innovation than just observing existing customer journeys.

The Third Dimension of Business Ecosystems & AI

The future of AI, however, goes far beyond specific and currently known use cases in which an ecosystem perspective is adopted. In the context of digital transformation, there is also evidence of the third dimension in which the actors open up even further and different platforms and business ecosystem initiatives interact intelligently. This will lead to virtual organizations, companies, and globally decentralized networked ecosystems.

Incorporating intelligent workflows and virtualized processes into broader systems increases the return on investment of virtual collaboration through the emerging ecosystems, digital workflows, and networked organizations. Virtual collaboration makes use of the appropriate value streams that incentivize and invigorate collaboration while connecting ecosystem participants. A key feature of this evolution is data-driven innovation, as described in this book, as well as the openness and transparency of collaboration described comprehensively in the book *Design Thinking for Business Growth*. Virtual and networked collaboration is also accelerating access to new sources of experience, product and service innovation, using of course innovative AI, at various stages of development.

Current prevalent applications, such as how businesses communicate with their customers through virtual assistants, to automating critical workflows and managing network security, are the first prototypes for the broader AI perspective and the opportunity for new market opportunities.

Business ecosystem initiatives that understand data as the fuel for the continuous evolution of the core value proposition and interaction with the customer/user are well prepared for the next steps of digital transformation. The applied models are already enabling ecosystem orchestrators and actors to better leverage data and target their strategic business objectives effectively.

At the end of this mini toolkit, it is important to note and recommend that the well-known tools and methods from the other toolkits and the tools from *The Design Thinking Toolbox* be applied if appropriate to the situation and problem to solve.

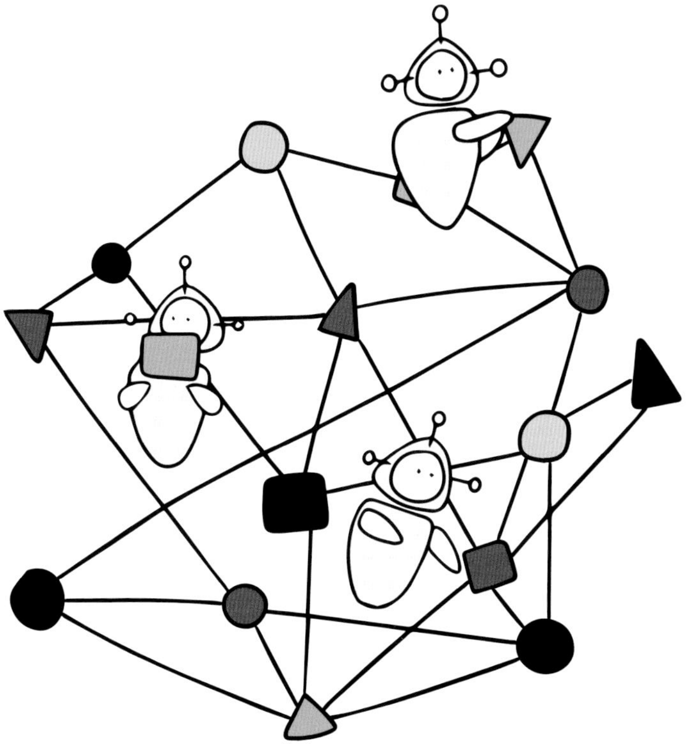

The next level of business ecosystems and AI in the area of virtual collaboration needs profound capabilities and skills in data, new technologies, and business ecosystem design.

The Future: AI as Part of the Design Thinking Cycle

From the presented design thinking and AI toolkit, and the previous examples, it is obvious that there are several reasons why organizations and design thinking teams want to apply AI over the entire design cycle and beyond in the future. Highly volatile and dynamic market environments with changing customer behavior and preference are increasing the pressure of (re)designing and expanding value propositions and offerings. In addition, the global competitive landscape and the orchestration and participation in business ecosystems require from the system to respond and adapt in a timely manner, while new and rival technologies must be considered. At the same time, the availability of information, knowledge, and market insights has intensified and continues to increase significantly. These trends and developments are clear evidence that the basis for competitiveness lies in companies' information and problem-solving capabilities. The intensive exploration of the problem and solution space has been proven over decades to be very useful, but it is also a cost factor in the development of new market opportunities. Thus, it can be argued that the way design thinking and the measurement of the innovation systems are organized today must be challenged by the introduction of AI and machine learning due to their cost advantages in information processing.

AI will support teams in processing the growing amounts of information in an increasingly competitive environment and save costs at the same time.

AI can assist and potentially replace insight collection, storytelling, and decision making over the entire design cycle. AI might find application in all of the design phases (understand, observe, point of view, ideate, test, and prototype) by overcoming mainly information processing constraints and limited search and interaction routines.

AI already outshines the design teams with regard to information processing constraints in the area of ideation and the validation of market opportunities. Deep neural networks are able to process vast amounts of data, which support design teams through the entire cycle by processing a much larger amount of information than is humanly possible. This path of AI is already creating substantial value for companies and entire business ecosystems globally. Enabled by meta-reinforcement learning, related questions concerning how learning can be applied to advance the process of learning will be key to devising algorithms that are able to adapt rapidly to arbitrary new problems.

Example: "Storytelling"
Machine learning methods are used to process raw metric data, into insights that are humanly readable. Out of the analyzed data a summary of actionable and interesting insights in the form of customer stories is generated for teams working in specific innovation fields.

Example: "Deep dive into one specific domain"
AI applications are used to identify options for one solution. For example, interdisciplinary teams at pharma companies train deep domain adaptational neural networks on single-cell RNA genomics data sets to ultimately develop treatments for a virus or disease.

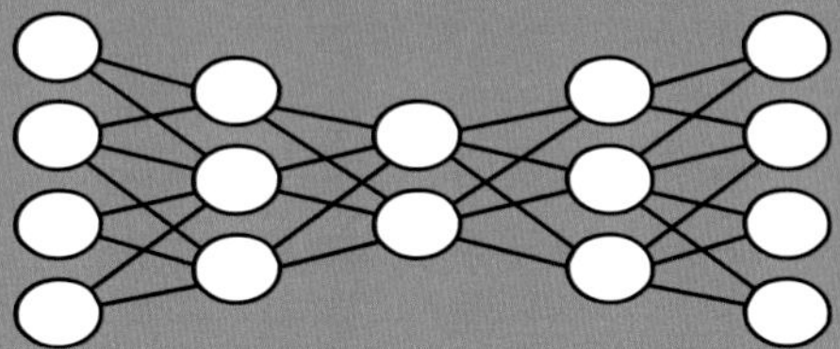

Example: "Process (re)design"
AI applications are trained on process innovations for organizations and specific workflows. For example, process mining can be used to identify processes like order to cash that are suitable for robotic process automation.

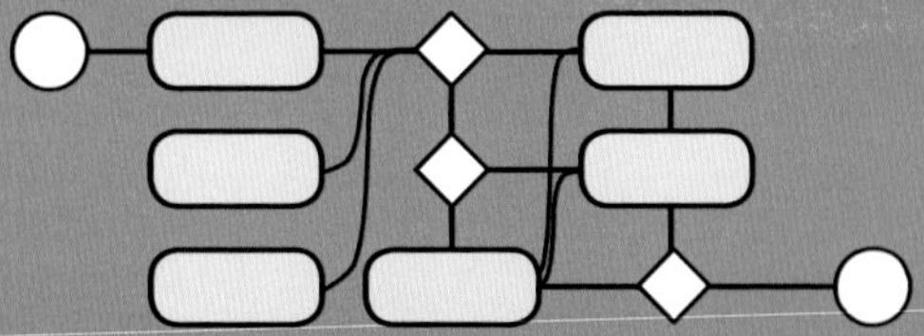

Next Level of AI Performance and Competence

While the first applications of AI are focused on high performance and low competence, developments are increasingly moving toward making complex decisions. The goal is to use AI to act faster and more competently than the human brain in the future.

Superior AI systems will make our smart systems even smarter. Cars will drive themselves, houses will adjust their power consumption, and advanced AI will be able to diagnose even complex medical conditions. In addition, neuro-adaptive AI technologies will become prevalent, enabling the automation of volumetric content, so that holograms will establish themselves as a new form of visual media (see more on neurodesign from page 360 onward). AI will cover more areas of daily life, support creative work, measure innovation, and automate various tasks. Many use cases already exist today with limited functionality and scope, but advanced AI has the potential to unlock significant value.

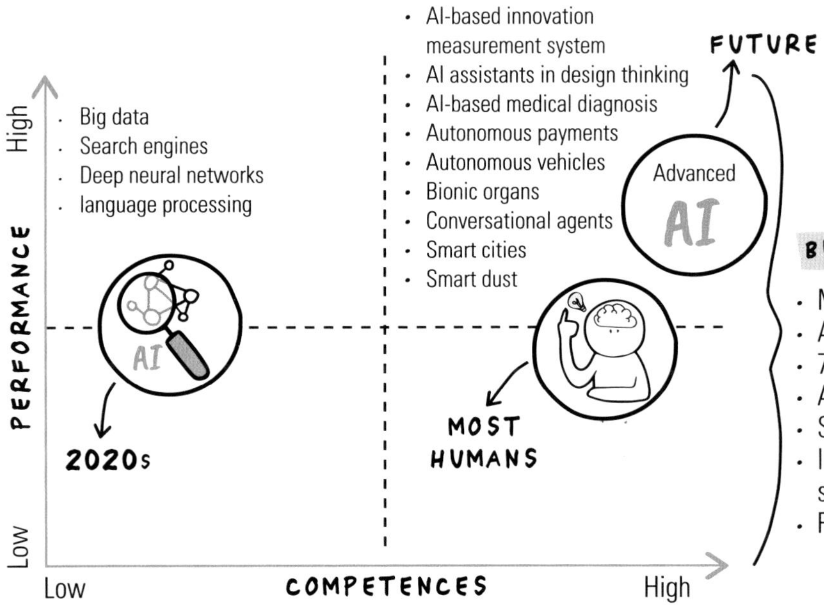
High
PERFORMANCE
Low
Low
COMPETENCES
High
Big data
Search engines
Deep neural networks
language processing
AI
2020s
AI-based innovation measurement system
AI assistants in design thinking
AI-based medical diagnosis
Autonomous payments
Autonomous vehicles
Bionic organs
Conversational agents
Smart cities
Smart dust
FUTURE
Advanced
AI
MOST HUMANS

BUSINESS BENEFITS

- Maximizing channels
- Attracting non-buyers
- 720-degree view on customers
- Automating service integration
- Saving cost and resources
- Increasing efficiency and scalability
- Predicting behavior

CUSTOMER/USER BENEFITS

- Receiving customized information
- Experiencing seamless offerings
- Obtaining personalized services
- Interacting instantly
- Accessing services 24/7
- Great peaking experience
- Simplifying interactions

AI and machine learning for creativity and innovation are very different from the established areas where AI has replaced traditional management. Soon design thinking teams will be expected to work side by side with AI and machine learning algorithms in identifying and selecting market opportunities as well as investigating where to explore and how to exploit.

AI

Integration in Design Thinking

[AI recognizes more problems to solve.]

[AI identifies and evaluates more information for the creation of points of view.]

[AI identifies and evaluates more radical ideas.]

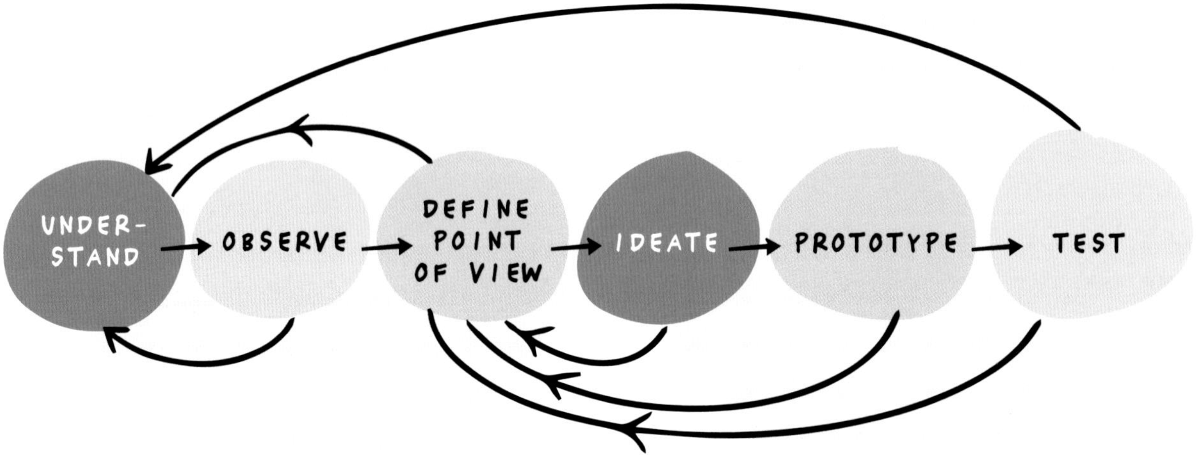
UNDER-STAND
OBSERVE
DEFINE POINT OF VIEW
IDEATE
PROTOTYPE
TEST

[?]

AI will be applied in innovation settings in the future because of the rapid development of AI and machine learning.

New Opportunities for Innovation Metrics Based on AI

The central question in the context of this book is how the development will affect the measurement and management of products, OKRs, innovation, and business success.

Data, AI, and digital science offer new opportunities for the development of valuable and even more meaningful innovation metrics. Research projects with leading universities and the world's most innovative companies are currently exploring innovation statistics as new data sources within and across the extended data ecosystem that will complement the insights in this book on meaningful measurements in the future. Data and analytics from social media sites to customer interaction data in business ecosystems are yielding increasingly valuable insights on agile customer development and innovation success measurement. Through data scraping and analytics tools, it is also possible to gain real-time insights into customer, employee, and innovation activity. However, mapping into innovation categories and success components requires a database on which the algorithms can be built, similar to how real-time and historical data are used today for weather forecasts and phenomena.

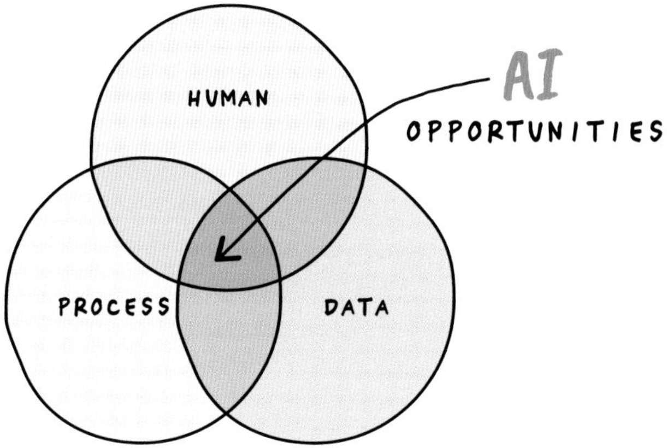
HUMAN
AI
OPPORTUNITIES
PROCESS
DATA

However, organizations must be aware that innovation is complex. Understanding the complexity needs a deep insight into the processes inside and outside organizations, team dynamics, and the motivation of each individual employee to be part of EXPLORE and EXPLOIT in a balanced portfolio. It also requires consideration of how risks are assessed, decisions are made and implemented, and how the rocky roads of internal politics and organizational battles for resources are handled.

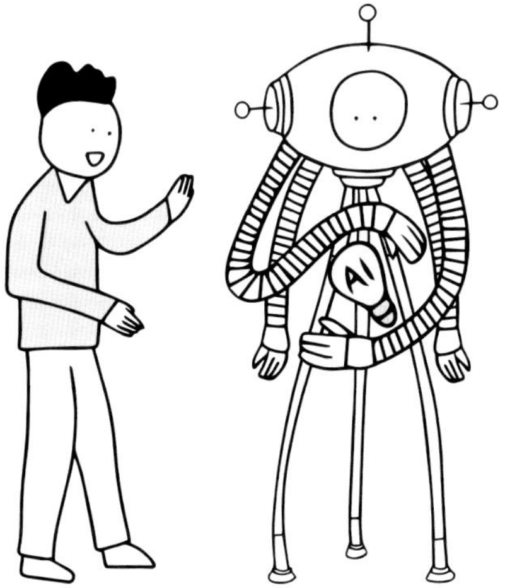
AI

The great value of design thinking for this new opportunity cannot be underestimated. Customer needs and customer problems, which make up most of the modern economy, arise in the current interactions of the customer/user in their natural environment. This means that human-centered design will continue to grow in importance. Especially since many companies in the future will also have to measure themselves against issues such as sustainability, responsible growth, and other environmental-social trends. This requires efficiency in the form of AI and automation as well as AI-supported creativity, a new kind of inspiration, and the human ability to build empathy with the potential customer/user. These form the basis for new metrics that bring together all these components of behavioral psychology, economics, and ecology.

One of the key advantages of AI is that it can be used to make metrics for innovation and business success more forward-looking rather than retrospective.

Applying AI to Understand Organizational Behavior

Example

AI offers the possibility to measure even complex metrics related to corporate culture and employee engagement. The annual employee engagement survey was the standard for many years (passive organizational network analysis). The survey required a lot of time to administer, remind employees to fill it out, prepare reports, and take appropriate actions along the corporate ladder. AI offers a different way of analyzing different indicators allowing a more expansive and more immediate employee engagement culture (active AI enabled organizational network analysis). At the same time the underlying data provides you with the information about potential change agents in the organization and how each of them might have to reach to accelerate change.

WHO IS INFLUENTIAL FOR DRIVING CHANGE?

WHAT ARE THE BEHAVIORS OF HIGH-PERFORMING TEAMS?

HOW CAN A NEW MINDSET IN THE ORGANIZATION BE STIMULATED?

WHO ARE OUR REAL SUBJECT MATTER EXPERTS?

WHICH KIND OF EMPLOYEES ARE MOST AT RISK FOR BURNOUT?

HOW CAN INCREASING THE IMPACT OF THE TEAM-OF-TEAMS INITIATIVES BE MEASURED?

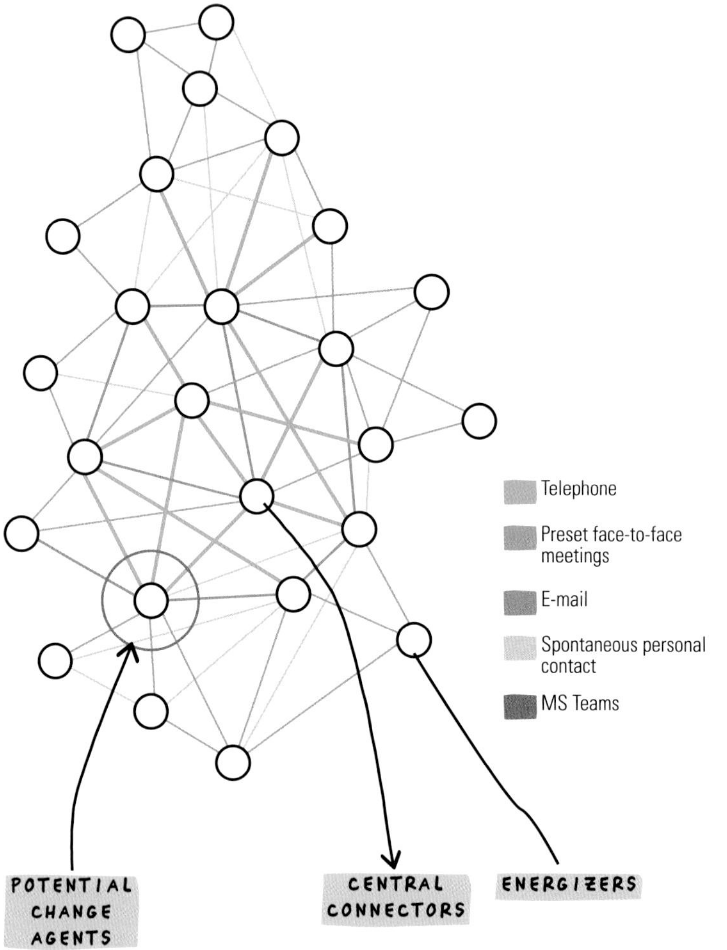
Telephone
Preset face-to-face meetings
E-mail
Spontaneous personal contact
MS Teams
POTENTIAL CHANGE AGENTS
CENTRAL CONNECTORS
ENERGIZERS

The future of innovation measurement systems is comprehensive and analytical. Business and innovation success will depend on how quickly new opportunities are embraced in the future and how quickly employees embrace the new technologies. Teams are increasingly required to apply analytics to every business decision and make data-driven decisions. No longer will finding relevant information be tasked to only specialists. The information will come to the teams, and all team members will be empowered to collaborate and make informed decisions with application of the hybrid model, the application of neuroscience, or with the support of AI.

	Exploit	Explore	Future of Exploit & Explore
ABILITY	Exploiting existing market offerings, processes, and procedures mainly with efficiency gains	Exploring new insights or validating existing assumptions over the entire design cycle	Increased automation of EXPLORE and EXPLOIT to identify or redefine problems, creating ideas and validating prototypes
AI AND ML SETTING	• Overcoming cognitive information constraints • Used for large data sets • Processing many different data sources	• Creating opportunities with the ability to discover new ideas • Supporting design teams and decision makers in specific tasks over the entire design cycle (finding more ideas and solutions)	• Exploring radical innovations and fostering out-of-the-box thinking • Inspiring teams with the creation of new ideas • Exploring new ways of finding the appropriate problem • Addressing new or unknown problems
AUTONOMY LEVEL	Human-designed AI systems	Coexistence of humans and AI as a team	AI systems with a high level of machine autonomy
MATURITY LEVEL (2020s)	High maturity of applications	Medium maturity of applications	Low maturity of applications (mainly research, sandbox experiments, and initial implementations)

To the Point!

AI has become a discipline of applied design and computer science that deals with reproducing aspects of human thought and action with computers.

In AI projects, the design thinking mindset helps to achieve the interdisciplinary cross-pollination that encourages individuals and teams to ask powerful questions, observe, network, and experiment in an environment that dismantles hierarchies, eliminates vested interests, challenges the status quo, and encourages intelligent risk taking.

AI covers more and more areas of daily life, supports creative work for measuring innovation, and automates various tasks.

In the context of this book, AI is the advantage that can be used to make metrics for innovation and business success more forward-looking rather than retrospective.

DESIGN THINKING & NEUROSCIENCE

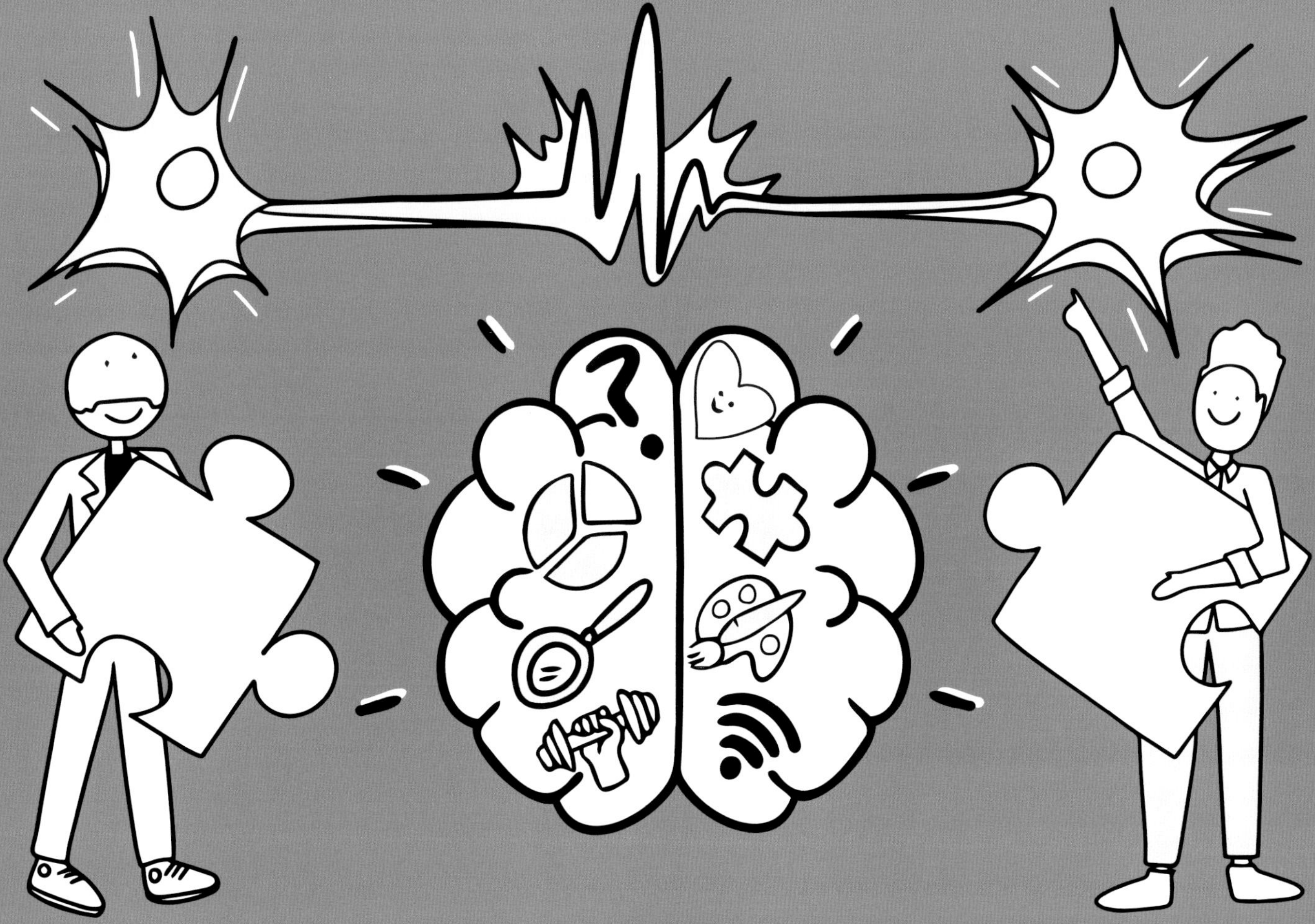

Transferring Brain Patterns into Value

Neurodesign is a design concept in which design thinking teams and cognitive neuroscientists work as a team, using the knowledge of what makes the human brain tick to understand how creativity and design thinking are measured, as well as how to better design products and services by understanding customers and their brain activities as associated to a product or service. In recent years, design thinking teams have become increasingly aware that it is often misleading to merely ask the customer/user for their feedback for a first high-resolution prototype. Measuring the response of the customer/user brain, on the other hand, can in some cases tell us a lot more about perception. Neurodesign, for example, focuses on products and services that incorporate visual, physical, and emotional aspects that the human brain inherently finds appealing.

Over the last years, neurodesign has developed as a special discipline of design thinking. The new paradigm is based on the fundamental principles of human-centered design, which requires radical collaboration across disciplinary boundaries. Active application requires digital enabler technologies as well as knowledge from biology and psychology to explore the relevant and critical problems to explore and to solve. The origin of neurodesign was based on initiatives by Larry Leifer (ME310). He was one of the first researchers at the intersection of design and neuro at Stanford in the 1960s. Later, Bob McKim and Jim Adams developed several practical approaches to creative and visual thinking in design. In the 2020s, the neurodesign research initiative has become more popular at Stanford University and many other leading universities.

The combination is powerful and has some challenges to overcome at the same time. Neuroscience adds value to the design thinking process because it follows, for example, a deductive and inductive reasoning that allows new knowledge creation based on what is already known but with the constraint that the complexity is limited to experiments in the lab environment. For a goal-oriented interaction of mixed teams consisting of design thinkers and neuroscientists, it is necessary that each team is open to the other's approach. This is best achieved when neuroscientists actively participate in design thinking activities throughout the design cycle and design thinkers learn and develop their scientific curiosity and analytical thinking through active engagement in neuroscientific research.

In this final portion of this book, several concepts come together to enable innovation to thrive at the intersection of neuroscience expertise, entrepreneurial vision, and scalable, self-improving technological solutions driven by AI and big data.

DOES HE WANT IT OR SHOULD HE?
ANYTHING YET?
dTP
Mind A
Mind B
Mind C
?

Neurodesign helps to understand humans better. It is the combination of bringing together understanding and transformation or, in the context of this book, making what seems immeasurable measurable.

The Evolution of Neurodesign

NEURODESIGN

[**Neurodesign** emphasizes the importance of understanding customers/users non-conscious behaviors, reactions, and interactions and how it can be applied in designing new or improved products, services, and experiences.]

EARLY DAYS

ADAPTATION

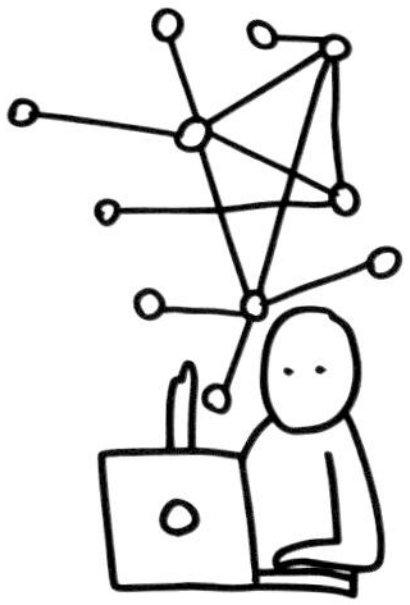

TODAY

FUTURE

Early Days: First research about the intersection of psychology and design and initial application of design thinking and neuro.

Adaptation: Practical models about creativity and visual thinking which are still applied in many design thinking tools.

Today: Neuroscience and design thinking as a combined discipline to quantify the *why* of human behavior.

Future: A new area of human-centered design with strong evidence on real-time data, which is based on the real behavior; seamlessly integrated in physical, VR, and AR environments and experiences.

1960 — 1980 — 2000 — 2020 — 2040 — TIME

Why Is the Combination of Design Thinking and Neuroscience Powerful?

Many innovative companies have recognized the added value of neurodesign, building up their capabilities accordingly and starting first experiments based on mixed capabilities of neuroscience and design thinking. Neurodesign can be used for the development of new products and services on the one hand and, on the other hand, to analyze human actions and motivations in teams and to initiate measures for the transformation of organizations and individual teams. The objective is identical in both areas: to improve products, experiences, and organizations to match the way the brain reacts, creates, and innovates.

The success of the transformation lies in the creation of a new mindset and innovation culture that is aligned with underlying human drivers and thus based on a fundamental understanding of neurodesign. The central task is to identify, recruit, and engage people with their creative perspective, while providing them with the framework, mindset, and appropriate tools to be actionable from the start. To support each new and existing team member, neurodesign can serve as the foundation for creating a set of channels, resources, and tools to tackle any type of design challenge.

Reason Emotion

The entire design cycle, from building empathy, to defining points of view, to ideation, requires a corporate culture and environment based on the neural response of each team member. For example, it needs psychological safety in a team-of-teams setting, where customer/user and team insights are translated into experiments and opportunities to delight customers through new value-added products and services. The step-by-step building of design thinking capabilities (see pages 89–96) shows how building empathy (using a design thinking mindset) can be approached at scale.

Design thinking capability programs are supported by a range of activities that include project-based learning and specialized trainings for facilitators to engage the rest of the organization in solving key customer problems. Design thinking practitioners and facilitators are provided ongoing resources and the full *Design Thinking Toolbox*, and they are also constantly encouraged to use their skills to explore the problem space and drive innovative solutions. Direct business outcomes, appropriate metrics, and financial impact are tracked and made transparent to decision makers and teams on a regular basis.

Research in the field of neurodesign also shows the importance of an environment and culture in organizations where teams can be confronted with radical ideas. In particular, the divergent phase and associations of different approaches activate the brain (i.e., the left frontopolar cortex), which is associated with creativity. In addition, several studies show that the thinking of design thinking teams during the active ideation phase occurs mainly in the parieto-occipital area, an important area for visual function and cognitive control. Likewise, results of neurodesign research show that transparency is enormously important, which in turn leads to better coordination of measurements. The reason for this lies in the anterior cingulate cortex (ACC) of the brain. Here, mixed messages and seemingly contradictory objectives and priorities can be perceived. As a result, flexible thinking and openness are impaired when the brain cannot resolve these dilemmas.

In addition, the neuroscience of intrinsic motivators and personality types are also increasingly being used to assemble teams for design challenges within the organization. Successful innovation teams make their thinking preferences public and arrange teams with T-shaped participants (see page 82). They also know what skills and thinking preferences are needed for each of the steps throughout the design and implementation cycle, right up to scaling and assigning team members to the appropriate tasks based on their skills and strengths.

Reason and emotion work together.

Dimensions of the Brain and Decision Systems

~10%

What someone explains or perhaps thinks

TRADITIONAL CUSTOMER/USER RESEARCH

What someone says

Slow decision system

- Declarative preferences
- Weakly or uncorrelated, for example, when it comes to actions and purchase decisions

THINKING BRAIN

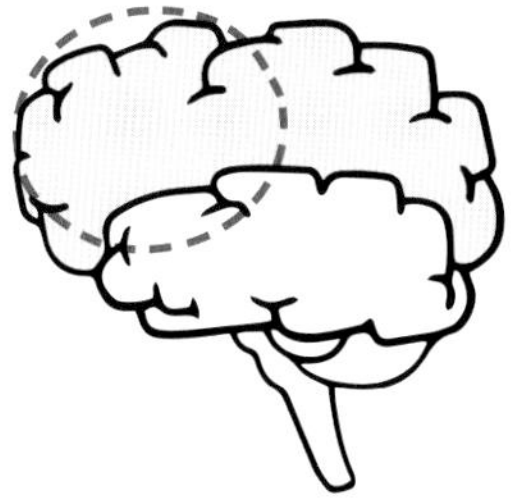

Prefrontal Cortex

LOGICAL
THE FUTURE

+

~90%

Subconscious world for someone, which is based on how and what the brain decides and articulates

CUSTOMER/USER NEUROSCIENCE RESEARCH

What the brain tells one to do

Fast decision system

- 100 percent unconscious
- Survival and sex

EMOTIONAL & DOING BRAIN

Limbic System

EMOTIONAL
THE NOW

What someone is really like

Depends on thickness of deep brain areas

- Correlates with Herrmann Brain Dominance Instrument (HBDI) profile and other personality tests
- Unconsciously modulates humans' decisions

DEEP BEING BRAIN

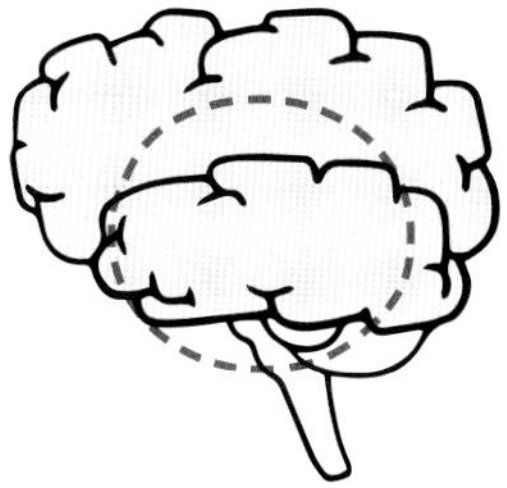

Accumbens

MOTIVATION & FOCUS
ALIGNED WITH
RESOURCES

Stronger Evidence for the WHY!

The field of applications, level of observations, and measurement approaches related with neurodesign are very broad. Within this short introduction the focus is on measuring design thinking activities and applying neurodesign over the entire design cycle to create new products, services, and experiences. With neurodesign, design thinking, and innovation teams have a powerful paradigm to explore more profoundly the WHY. Neurodesign brings a lot of data points, which are evidence-based and not built on limited observations performed predominately in the current settings of applied design. At the same time, design thinking helps to ask the appropriate and powerful questions to be answered by neuroscientists. Over the last decade, neurodesign has already helped tremendously to provide evidence of the power of building design thinking capability build programs in comparison to other training efforts; revealing the secrets of high-performing teams; and the influence on creativity and innovation success triggered by different settings in the physical, hybrid, and virtual spaces. Neurodesign provides the ultimate possibility to include more data points and more objective measures in addition to the human-centered approach and existing measures already applied on various levels in EXPLORE and EXPLOIT. The Neurodesign Project Canvas (see page 365) supports design thinking and innovation teams to align on the initial assumptions, prepare for the experiment set-up, and document results.

WHY?
WHY?
WHY?

Neurodesign helps to question and to reframe the way we did innovation work before. Neurodesign starts with why!

Levels / Measurements/Applications	Measuring Design Thinking Activities	Applying Neuroscience over the Entire Design Cycle
INDIVIDUAL	• Mental stress • Expertise	• Design thinking tasks • Tool preferences
TEAM	• Collaboration, shared goals • Brain synchronization	• Distribution of creativity • Reflection and learning patterns
CUSTOMER/USER	• Engagement level • Mindfulness	• Co-creation patterns • Emotional response insights
ORGANIZATION	• Cross-organizational • Improved tools	• Organizational learning • Engagement levels
ECOSYSTEMS	• Co-evolution of thoughts • Symbiotic behavior	• Customer-led behavior • Data to expand offerings

Template download

NEURODESIGN PROJECT CANVAS

NAME OF NEURODESIGN PROJECT:

1 EMPATHY & ETHNOGRAPHIC RESEARCH

FEELINGS
Collection of sensory inputs to the brain; tapping into subconscious stored memory

WE KNOW FROM MEMORY...

IMAGINATIONS
Applying cognition based on already-acquired knowledge; tapping into experience and/or sensory perception

WE EXPECT...

GENERAL VALUES
Who has the problem?
What are the pains/gains for the customers, jobs to be done, and use cases?
Is the problem related to social activities with other persons?

THE PERSONA MIGHT RESPOND OR TAKE A SPECIFIC ACTION FOR A SPECIFIC SCHEMA, INTERACTION, OR TOUCHPOINT.

CRITICAL ASSUMPTIONS TO BE VERIFIED
Which specific schema, interaction, or touchpoint is most critical?
Which behavior needs verification?
What needs are not understood?
Which environmental setting might have influence?

2 EXPERIMENTAL NEURODESIGN SET-UP

TOOLS
Select the appropriate tools from the neurodesign toolbox (e.g., hyperscanning, heart rate, eye tracking)

SET UP THE EXPERIMENT
What should be measured?
What additional measurements should be applied?
How is the measurement correlating to other measurements?

CAPTURE, ANALYZE, & VISUALIZE DATA
What data was collected?
How do the outcomes fit with existing knowledge and where are there contradictions?
How can the outcomes be visualized and shared with an extended team?

3 LIMITATIONS & RESULTS

LIMITATIONS OF THE EXPERIMENT
What are the potential limitations of applied tools?
How to overcome limitations with different or more advanced experiment settings?
What other social science fields help to overcome limitations? (e.g., social psychology, social anthropology)

RESULTS
What are the main findings?
What is the impact of the findings to the project?
How can the patterns be used in the broader spectrum of machine-readable ontologies, neural networks for better classification, and AI?

DOWNLOAD TOOL
www.design-metrics.com/en/neuroscience

Measuring Design Thinking Activities

Neurodesign has the tools and methods to gain a better understanding of the activities of individuals and teams over the entire design cycle. The results help to improve the design thinking process, to create better tools and methods for the design thinking community, to improve design thinking education, and, consequentially, to advance the outcomes of design thinking in general. For measuring the design thinking activities different measurements can be applied:

1. Measurement of physiological behaviors in performing design thinking.

Typical measurements are eye movements, spectral analysis of heart rate variability, electro dermal activity, as well as standard functional assessment of the autonomic nervous system.

2. Measurement of the connections between cognitive processes and brain activity in performing design thinking.

Typical measurements include an electroencephalogram test that detects electrical activity in the brain, functional near-infrared spectroscopy, and functional magnetic resonance imaging, which depicts brain activity.

3. Measurement of design thinking cognition.

Typical measures are based on reasoning, processes, and patterns; divergent and convergent thinking; design fixation; design creativity; visual reasoning; space co-evolution; and collaboration with cognition tools applied in design thinking.

The three paradigms are examples of exploring the underlying physiological, neurological, and cognitive patterns associated with design thinking. The measurement of design thinking cognition is the most known and developed procedure that aims to detect, for example, design thinking characteristics, measure cognitive behavior, and perform black-box experiments, where multiple conditions are set to perform a design thinking task or to apply a certain design thinking tool.

We should encourage asking, "Where the tools used?" and not so much "Were the tools useful?" – Larry Leifer

Neuroscience reveals the truth of design thinking and offers satisfaction of formal expectations we can still not completely understand.

The main purpose of measuring the design thinking activities should be to generate better insights into each individual's and team's brain, body, and mental activities. This provides the basis for enhancing the individual or team performance, triggering self-reflection, as well as managing better self-regulation in performing or facilitating design thinking.

In many organizations the work of design thinking teams is constantly challenged, and new measurements can help to overcome the bias. Constant feedback to teams and individuals helps to improve collaboration, co-creation, and all activities over the entire design cycle.

The measuring of design thinking activities is advancing more and more with emerging technologies, such as brain–computer interfaces and brain–cognition interfaces, along with the previously introduced availability and use of AI and machine learning (see pages 336–357).

!? – ? = !

? + ! = !?

Paradigms of Exploring Physiological, Neurological, and Cognitive Patterns

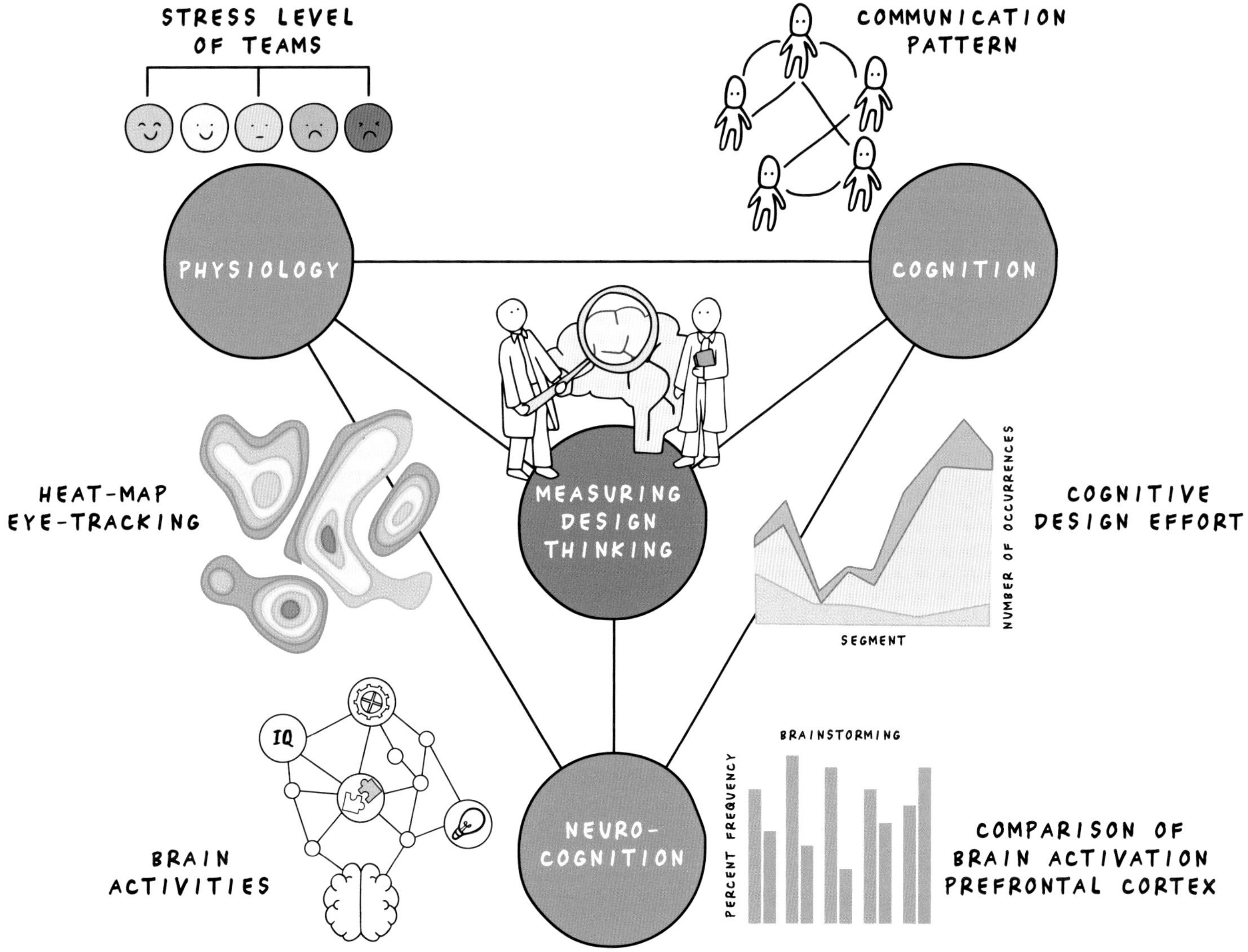
STRESS LEVEL
OF TEAMS
COMMUNICATION
PATTERN
PHYSIOLOGY
COGNITION
HEAT-MAP
EYE-TRACKING
MEASURING
DESIGN
THINKING
COGNITIVE
DESIGN EFFORT
NUMBER OF OCCURRENCES
SEGMENT
IQ
BRAINSTORMING
PERCENT FREQUENCY
BRAIN
ACTIVITIES
NEURO-
COGNITION
COMPARISON OF
BRAIN ACTIVATION
PREFRONTAL CORTEX

Applying Knowledge about the Brain to Design Thinking and Innovation Work

Neuroscience research related to design thinking has validated many assumptions about how the design thinking mindset, process, and tools are applied. In particular, recent studies looking at brain activity in various environments and in interventions of physical activity help us to understand how, for example, creative skills can be best realized.

One most important finding is, for example, that design thinking and creative confidence can be learned (see page 90). Qualitative and quantitative research on brain processes has provided the underlying findings that creating ideas follows specific patterns. Various scientific studies have contributed to building and expanding knowledge about creative processes and better understanding how brainstorms emerge. These studies also confirm that great ideas are not random, but rather the result of a more or less orderly process. Every brain has the natural ability to generate these sudden insights. The good news is that every team member can be creative.

The research also shows that there is no specific part of the brain responsible for creative thinking, but rather several areas of the brain are activated, depending on the nature of the design challenge and the complexity of the problem (e.g., wicked problem, ill-defined or well-defined problem). Both hemispheres of the brain are active in this process. The prefrontal cortex, which is responsible for higher order thinking, has a special function. Scientific research also shows that the brain can be stimulated to think creatively, for example by using a facilitator (see page 75–81) and various creativity exercises.

When the brain first deals with a problem, the prefrontal cortex has the task of evaluating this problem. At this stage, the realization that there is a challenge to be solved occurs. The brain tries to understand what the problem is and checks whether a solution is in sight. In this stage, the brain consciously works to understand, specify, and elaborate the challenge by gathering raw material and finding a solution.

In addition, other brain regions are activated that seem suitable to complete the task. In design thinking, we try to actively disrupt this step of finding a solution quickly from our brain, as experience shows that a more in-depth exploration of the problem space leads to better solutions later on. This responds very well to the motto "begin at the beginning," which introduced the design thinking 101 of this book (see pages 61–62).

By actively and moderately seeking more insights, the brain is fed with inputs to approach the challenge from different perspectives: logical reasoning, trial and error, or even by applying design thinking tools which aim to find analogies. If the brain does not find a solution right away, it is important to keep pushing, as more connections to other brain regions are activated. Failing and making mistakes is an important philosophy in design thinking that also helps—especially in the early stages of creative thinking. Studies show that sometimes, a little frustration combined with emotion is also helpful. In particular, when emotions have a positive effect on memory performance, team members tend to remember the challenge better.

Neuroscience can now explain the phenomenon of why many good ideas come in the shower or, as in the case of Archimedes, arise in the bathtub.

Remarkably, the brain also continues to process the design challenge and to think about solutions as soon as the team members turn their attention to other activities or simply relax. The brain starts to discover new combinations. At this point, the brain begins to change its activation patterns, as seen in the fMRI. It appears as if the brain is changing strategy to determine if other brain regions have something valuable to contribute. Research has shown that it is helpful to relieve the prefrontal cortex from heavy work by doing something that requires less concentration. A nap, walk, jog, or other low-stress leisure activity feels good for one thing, but more importantly for innovation work, the brain continues to work on the design challenge continuously and subconsciously, continuing divergent thinking.

Applying Neurodesign in the Creation of New Products and Services

Even more exciting than using neuroscience to explore creative teamwork is using neurodesign to develop new products and services. Based on the well-known *Design Thinking Toolbox* and other methods for collecting customer insights, such as market data and data analytics, neurodesign complements these activities with the goal of collecting a more complete view of the customer/user. It is important to emphasize here that the combination of design thinking and neuroscience makes the crucial difference. While neuroscience is able to provide deeper and more meaningful insights into the customer/user, it also has its limitations. It is not possible to read minds with neuroscience alone. Based on brain activity, workload and motivation can be measured, but only limited conclusions can be drawn regarding brain activity and the content or experience a test subject is currently exposed to. A larger sample of potential customers/users as test cases correspondingly increases the validity of the data points identified. However, it is also important to consider the data in the appropriate context, as outlined already (see pages 320–333) in the section about design thinking and big data analytics. This seems particularly important in experiments in which only individual functions or features are tested and analyzed. A comprehensive view that takes into account how the individual elements are interrelated is today confirmed from the results of various neurodesign projects over the last few years.

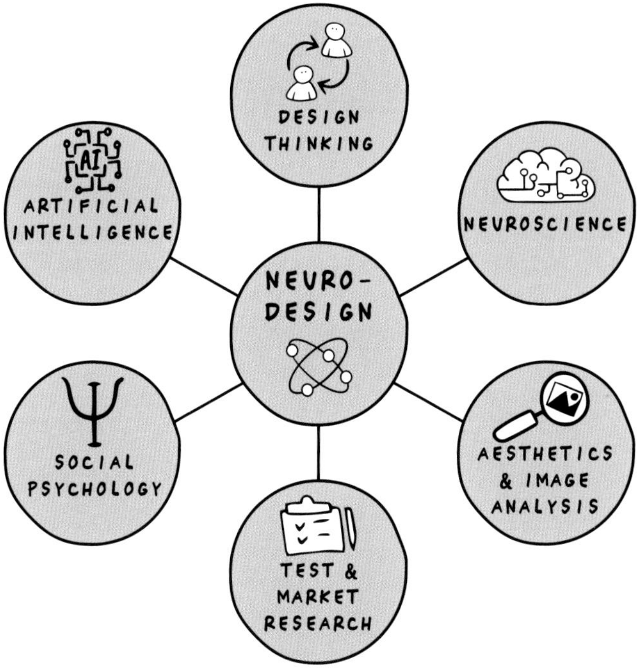
NEURO-DESIGN
DESIGN THINKING
NEUROSCIENCE
AESTHETICS & IMAGE ANALYSIS
TEST & MARKET RESEARCH
SOCIAL PSYCHOLOGY
ARTIFICIAL INTELLIGENCE
AI

By mimicking the brain's process through need finding, iteration, and solution tests, neurotechnologies add value across the design thinking cycle to better fit new products, services, and experiences to human needs.

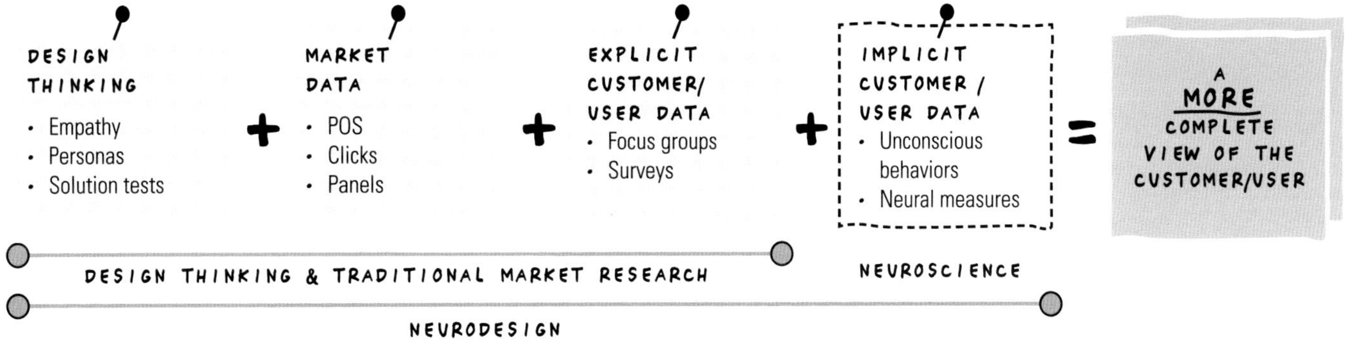
DESIGN THINKING
• Empathy
• Personas
• Solution tests
+
MARKET DATA
• POS
• Clicks
• Panels
+
EXPLICIT CUSTOMER/ USER DATA
• Focus groups
• Surveys
+
IMPLICIT CUSTOMER / USER DATA
• Unconscious behaviors
• Neural measures
=
A MORE COMPLETE VIEW OF THE CUSTOMER/USER
DESIGN THINKING & TRADITIONAL MARKET RESEARCH
NEUROSCIENCE
NEURODESIGN

What Use Cases Exist? What Impact Do These Have on Innovation and Business Success?

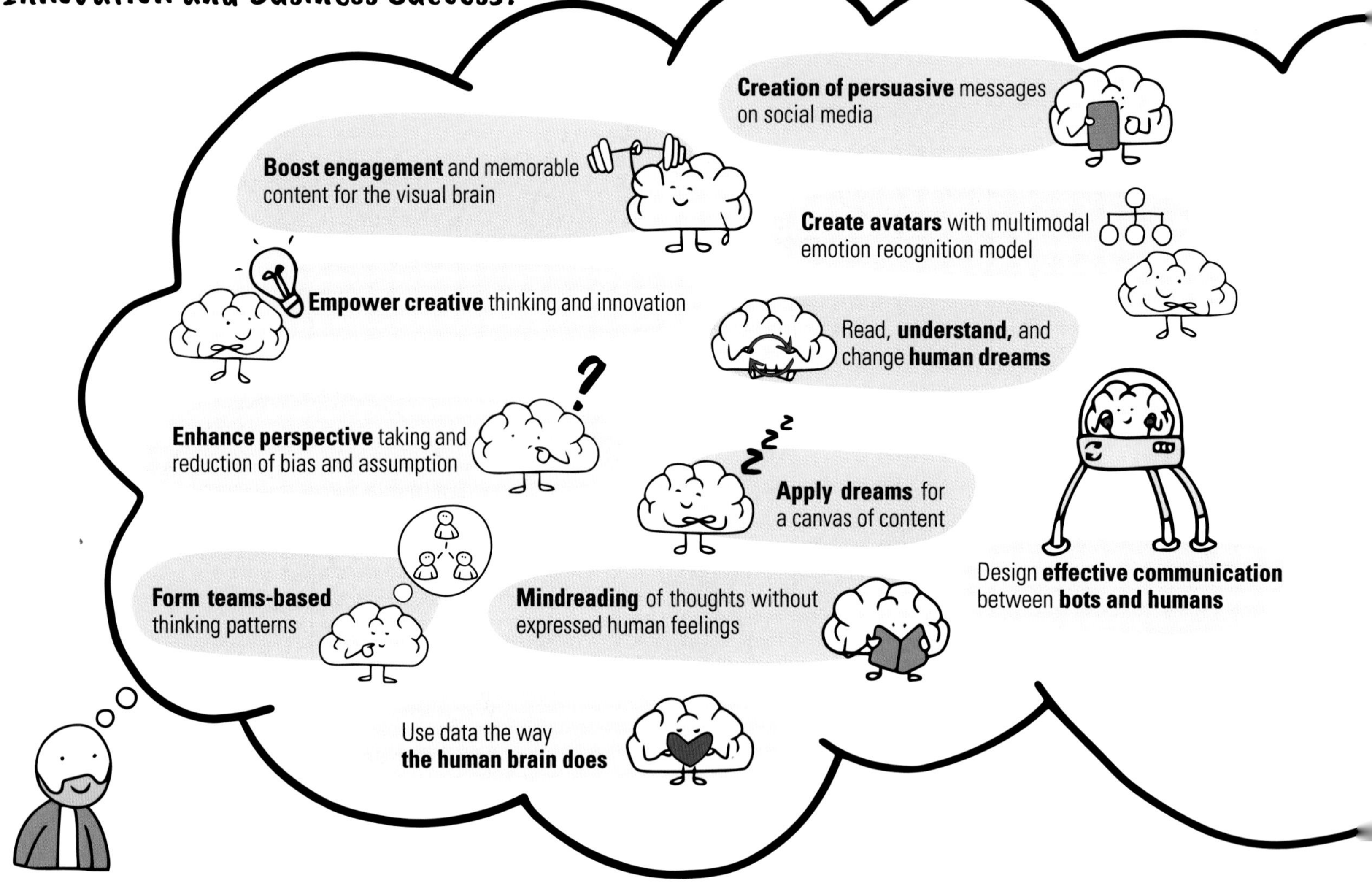

Creation of persuasive messages on social media
Boost engagement and memorable content for the visual brain
Create avatars with multimodal emotion recognition model
Empower creative thinking and innovation
Read, understand, and change human dreams
Enhance perspective taking and reduction of bias and assumption
Apply dreams for a canvas of content
Design effective communication between bots and humans
Form teams-based thinking patterns
Mindreading of thoughts without expressed human feelings
Use data the way the human brain does

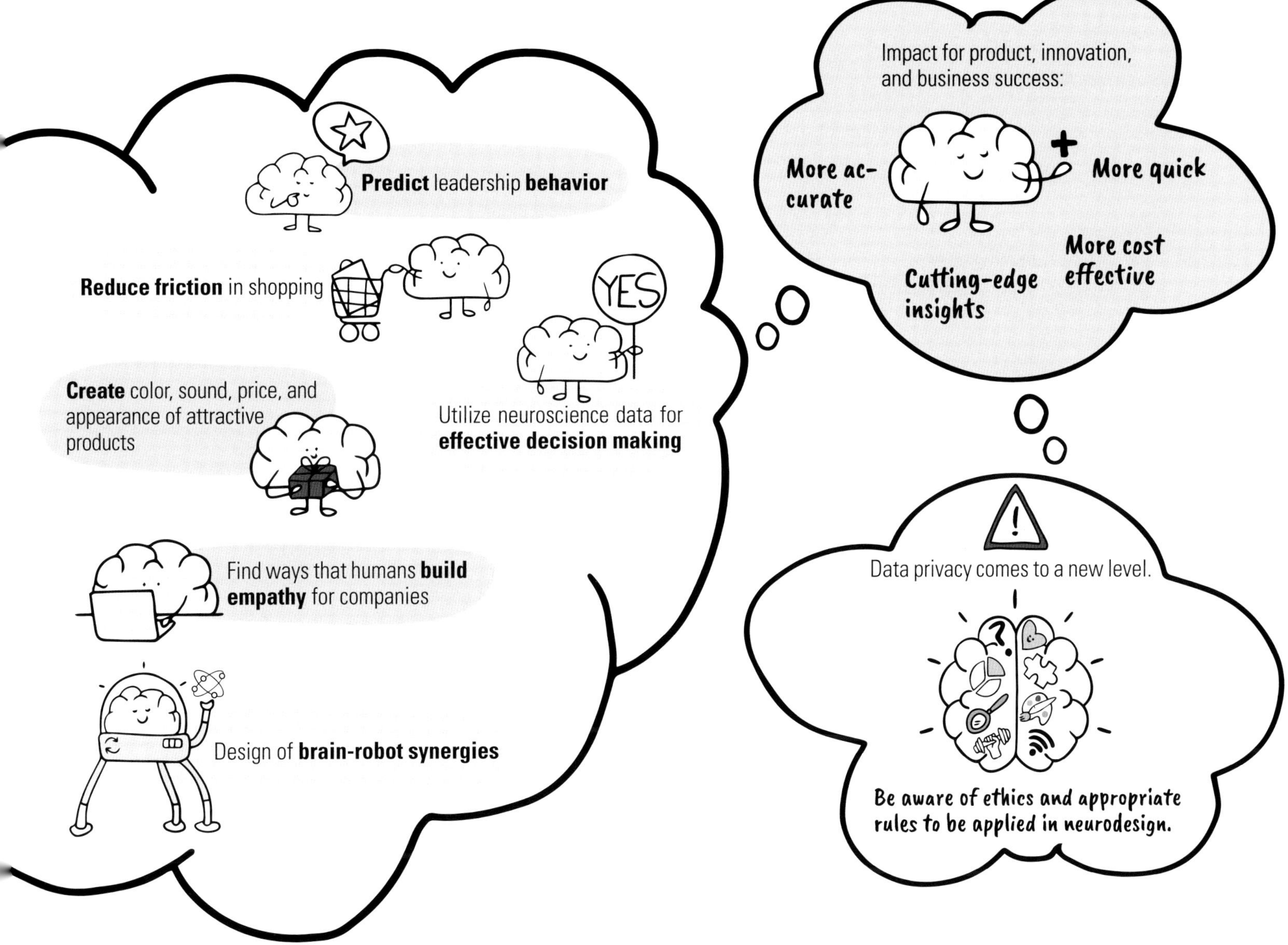
Predict leadership **behavior**
Reduce friction in shopping
YES
Create color, sound, price, and appearance of attractive products
Utilize neuroscience data for **effective decision making**
Find ways that humans **build empathy** for companies
Design of **brain-robot synergies**
Impact for product, innovation, and business success:
More accurate
More quick
More cost effective
Cutting-edge insights
Data privacy comes to a new level.
Be aware of ethics and appropriate rules to be applied in neurodesign.

Making What Seems Immeasurable Measurable

Cognitive neuroscience methods and tools are clustered by the location capabilities and timescale over which a particular measurement is obtained. The location capabilities refer to spatial sensitivity. While an MRI is able to detect, for example, large regions of the cortex, measurements performed by the patch clamp methods provide for example, information in milliseconds and low spatial resolution. Signals are recorded via patch, clamp, intracellular, extracellular, and mass unit recording as well as better-known techniques such as electroencephalography (EEG) and electrocorticography (ECoG). For operating in a more real environment, for example, measuring teams simultaneously on empathy hyperscanning is usually used.

To innovate far beyond our current imagination, smart tools are needed. These are similar to the tools created to overcome physical limitations. By understanding how the brain behaves when creating extraordinary ideas, we can find ways to expand the intellectual capabilities of our teams and potential customers/users through stimulation.

Different techniques and tools help throughout the design thinking cycle with insights and indicators about a certain behavior, reaction, or decision made by a potential customer/user. Especially interesting are also the future possibilities of measurements and metrics in interaction with neurodesign. Already today, many of the methods and tools are used to predict certain conditions. In particular, the most obvious differences can be identified when testing and comparing traditional methods, with single tools and methods from neurosciences as well as when applying a combined toolbox of all methods and paradigms.

There is a new discipline emerging which combines traditional needfinding and brainstorming with "neuro ideation."

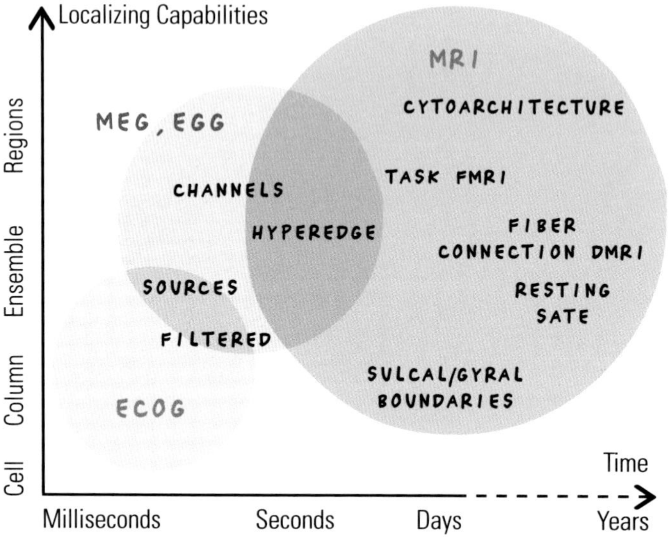
Localizing Capabilities
Regions
Ensemble
Column
Cell
MRI
CYTOARCHITECTURE
MEG, EGG
CHANNELS
TASK FMRI
HYPEREDGE
FIBER CONNECTION DMRI
SOURCES
RESTING SATE
FILTERED
SULCAL/GYRAL BOUNDARIES
ECOG
Time
Milliseconds
Seconds
Days
Years

Impact of neurodesign on product performance and experience

Example

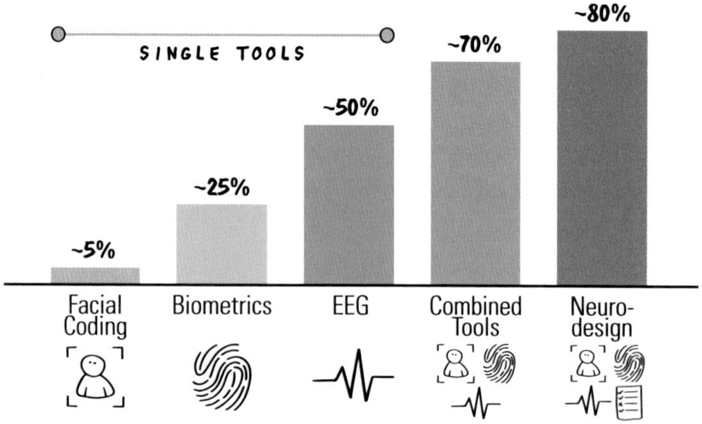
SINGLE TOOLS
~5%
~25%
~50%
~70%
~80%
Facial Coding
Biometrics
EEG
Combined Tools
Neuro-design

Creating the Perfect Symbiosis between Products, Services, and Experiences

Example

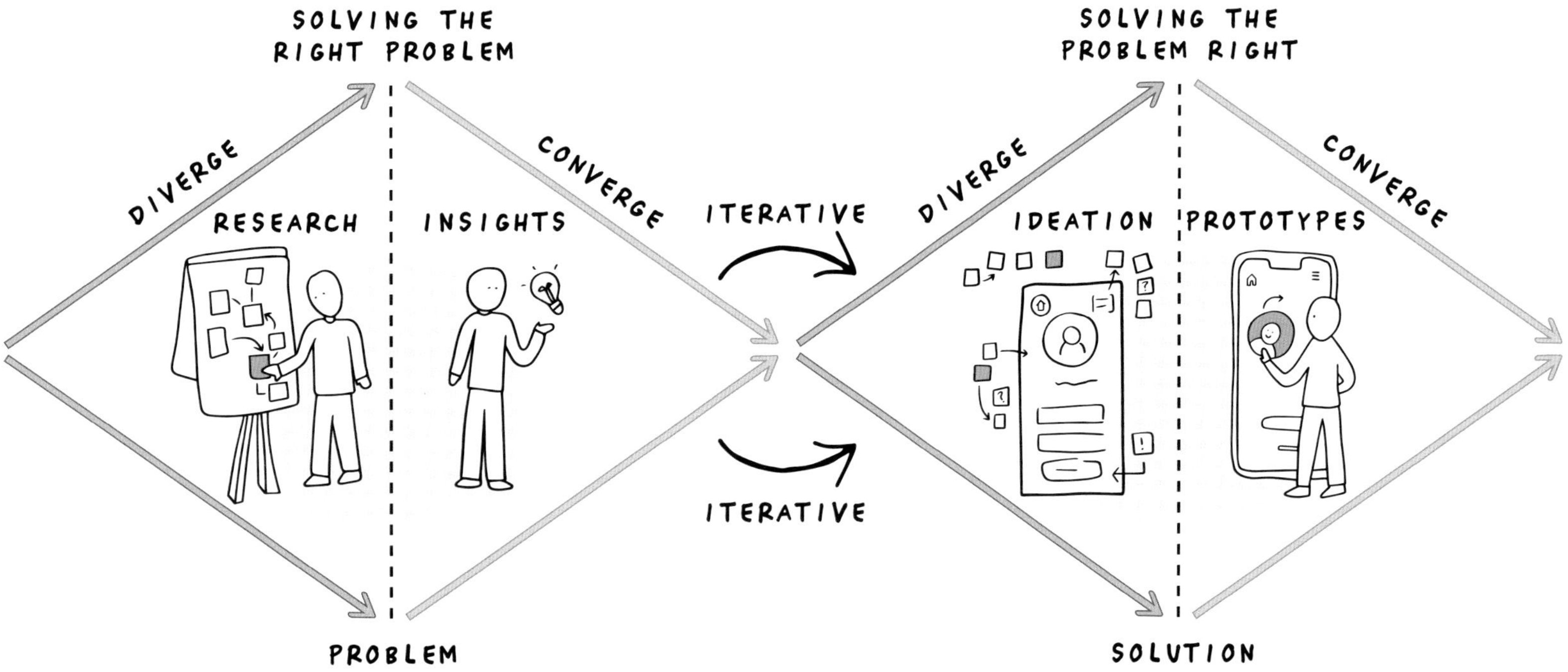
SOLVING THE RIGHT PROBLEM
SOLVING THE PROBLEM RIGHT
DIVERGE
CONVERGE
RESEARCH
INSIGHTS
ITERATIVE
DIVERGE
CONVERGE
IDEATION
PROTOTYPES
ITERATIVE
PROBLEM
SOLUTION

NEURO-TECHNOLOGIES

EEG
Multiple sensors to measure motivation, memory, and attention

BIOMETRICS
Skin conductance and heart rate to capture emotional journey

FACE CODING
Expressed emotions (positive, negative, neutral)

EYE TRACKING
Expressed interest (visual focus on content)

SELF-REPORT
Voice of the consumer

There is an opportunity, through neuroscience, to further understand the relationship between creative and analytical thinking, and their roles in the act of creating. However, bringing neurosciences and design thinking together has also created some challenges similar to the hybrid model (design thinking and data sciences), because there is an artistry in design which might clash culturally with the technical rationality of data and neuroscience. However, the combination has numerous advantages. Neurodesign, for example, enhances the design thinking activities from informing and diversifying the understand and observe phases to expanding the convergent-divergent design thinking Double Diamond with cognitive and affective neuroscience. For design thinking and innovation teams it is recommended to constantly observe what neurosciences explores to enhance creativity, innovation, and business success.

Understanding neuroscience is an integral part of the future of design thinking and innovation metrics. It will transform the way creative teams work and new products, services, and experiences are designed.

	Design Thinking	Neuroscience	Neurodesign
MISSION	Problem identification and problem solving by applying design thinking mindset, tools, and methods	Understanding the thinking of users and customers as well as the thinking of individuals and teams over the entire design cycle	Design, (re)design, and implementation of customer-led services, products, experiences, and processes based on the latest neuroscience research
STARTING POINT	Problem statement or design challenge	Hypotheses based on first observations or assumptions	Problem statement with findings of cognitive and affective neuroscience
REASONING	Abductive	Deductive	Combining abductive and deductive
THINKING MODE	High maturity of applications	Medium maturity of applications	Low maturity of applications (mainly research, sandbox experiments, and first initial implementations)
EXPERIMENTATION PERFORMED	Ideate, build prototype, and test	Controlled experiment	Neuroscience theories and tools integrated over the entire design cycle
MATERIAL & ENABLER	Design thinking tools and methodologies, material for prototypes	Scientific tools and methodologies with digital enabler technologies (machine learning and automation), application of brain imaging	Combination of physiological, neurological, and cognitive patterns with powerful design thinking tools

To the Point!

Neurodesign will be one of the key disciplines for the future to combine neuroscience and design thinking to promote human, economic, and ecological sustainability in an interdisciplinary setting.

It supports the work of design thinking and innovation teams by decoding needs, behaviors, and reactions of potential customers and users.

New technologies and analytical approaches, ranging from the molecular and cellular levels to the system and behavioral levels, open up new possibilities to apply neurodesign to create new measurements for creativity and design thinking activities, and, at the same time, apply neuroscience over the entire design cycle.

However, this not only requires a shift in how the work of teams is organized in the physical, hybrid, and virtual environment, but also in how the evidence is applied to build superior products, services, and experiences.

A FINAL WORD

Reflection and Outlook

The fact that design thinking is a powerful mindset for problem solving and value creation is demonstrated by the various approaches, examples, and combinations with other disciplines in this book. My personal observation over the last two decades confirms the perception that companies that have high performance in terms of the design thinking disciplines (focus on the customer problem, forming team-of-teams, continuous iteration, integration into a modern performance measurement system) also tend to realize the better outcomes in innovation, business results, and employee satisfaction.

The design thinking movement has been rapidly gaining traction over the last decades and it is better suited than ever to today's VUCA* world, where change has become a constant. The future need for design thinking skills is driven primarily by the demand for forward-thinking technologies and awareness of how design thinking can advance business objectives in complex environments.

Design thinking is able to deliver these types of results because it not only provides a mindset for reframing problems, finding solutions, and searching for better solutions, but it also provides a way around the many obstacles that can get in the way of creative thinking and problem solving, such as human bias, fear of failure, and lack of employee buy-in.

The magnitude of the challenge ahead should drive organizations to incorporate the all-important design thinking capabilities into their operations in a deliberate, structured, and dynamic way that will help redesign the appropriate vehicles for change so that decision makers and innovation teams can better manage complexity, make innovation measurable, and, ultimately, realize new market opportunities.

Structure-wise, exploratory design thinking activities and innovation projects require their own governance and new capabilities, separate from the efficiency-oriented exploitation activities. Unlike the mature business, which is trimmed for efficiency with predictable timelines, traditional financial metrics, and predefined "go" or "no go" decision criteria, the exploration portfolio uses an iterative testing and learning process, and evidence-based funding criteria.

* Volatility, Uncertainty, Complexity, and Ambiguity

Start Applying Design Thinking and Innovation Metrics

- ✓ Define a North Star and hold all relevant teams accountable for defining their objectives and key results.
- ✓ Introduce the measurements for exploring step by step and in iterations, but with the same professionalism as the measurements for exploit.
- ✓ Begin at the beginning with design thinking because changes on the final result are expensive to reserve or pivot.
- ✓ De-risk development by continually examining evidence across desirability, viability, and feasibility dimensions to evaluate highly uncertain, transformative innovation initiatives.
- ✓ Place the customer/user at the center of all activities, and break the internal silos and barriers to external partners to create winning value propositions.
- ✓ Expand the toolbox with data-driven tools, methods, and automation over the entire design cycle and establish new ways of measuring.
- ✓ Be aware that some of the signals and indicators that worked for measuring success in the past may not work for radical new products, services, and experiences.
- ✓ Gradually increase the organizational maturity in design thinking and innovation measurement systems, while aligning business/project objectives with efforts to change culture/behavior.
- ✓ And finally, don't forget that metrics used to evaluate the existing business (EXPLOIT) usually have no relevance to innovation initiatives (EXPLORE).

Building an effective innovation and performance measurement system is a journey. It requires changes in mindset that will make leaders, as well as teams, uncomfortable at first. As they develop an understanding of the differences between running and sustaining the mature business, and exploring the future, the pieces will slowly fall into place. Customer evidence will take the place of spreadsheets, metered funding will take the place of big bets, and asking questions will take the place of expressing opinions. Excitement will build as the appropriate problems are solved and the portfolio of many ideas is reduced to a few fully validated and scalable solutions.

In the future, companies, organizations, and teams should select and implement innovation and performance measurement systems with the presented methods and tools that fit the maturity level of the transformation and have the greatest impact at the given time. It is recommended to view the measurement system as a continuous cycle of four phases, starting with preparation, all the way to integration. Decision makers are encouraged to set and dynamically adapt the North Star and company objectives. All design thinking and innovation teams should be encouraged to apply the concept of minimum viable exploration metrics for EXPLORE and select the appropriate metrics for EXPLOIT. At the same time awareness is needed to moderate the transition from EXPLORE to EXPLOIT and sustain the business. The subsequent introduction phase is increasingly characterized by activities around collecting and validating data at all levels of measurement. During the activation phase, data is used for meaningful measurements, as well as establishing data-driven innovation as part of the innovation work. The integration phase aims at the operationalization of data and measurements and enables data-driven management decisions, up to the establishment of algorithms that support the design and innovation work. The respective reports and data feeds are shared more frequently and as needed, ensuring the necessary transparency to manage the activities. However, all the measurement activities should not overlook the importance of the teams and individuals, who ultimately have the responsibility of selecting the appropriate tools and methods throughout the design cycle, and understanding together where and why they have been used and have created the desired impact.

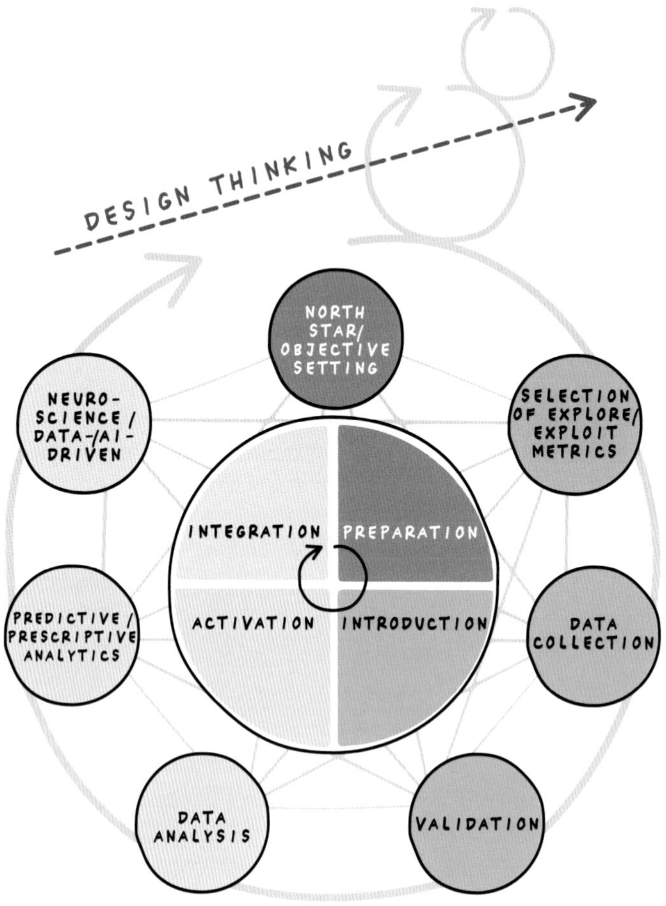
DESIGN THINKING
NORTH STAR/ OBJECTIVE SETTING
SELECTION OF EXPLORE/ EXPLOIT METRICS
DATA COLLECTION
VALIDATION
DATA ANALYSIS
PREDICTIVE / PRESCRIPTIVE ANALYTICS
NEURO-SCIENCE / DATA-/AI-DRIVEN
INTEGRATION
PREPARATION
ACTIVATION
INTRODUCTION

The continuous and step-by-step development of the innovation measurement system (IMS) drives evidence-based objectives and informs the portfolio on the realization of growth opportunities.

This means that companies, organizations, and innovation teams need talent with new capabilities and future skills to democratize the design thinking mindset by taking an integrative approach. Over the cycle of creating, adapting, and professionalizing the measuring systems, the future of managing innovation and business success takes shape. This includes real-time information processing based on in-depth interviews with potential customers/users. Design thinking teams, innovation teams, and decision makers will in the future receive custom-curated streams and stories, each tuned to the interests of a particular innovation topic or growth initiative. Instead of collecting insights from 50 potential customers manually, a machine will connect and interact with thousands of them at the same time and then summarize their insights, generate new points of view, and even test them in robot labs. Innovation teams will increasingly delegate innovation groundwork to machines, leaving the teams space to apply higher level of problem-solving competence, creativity, and innovativeness.

In a data-driven world, predictive analytics, along with leadership acumen, will help to identify and refine key strategic actions. The well-aligned and up-to-date North Star metrics lead to the better alignment of team behaviors with the respective objectives and key results. The iterative approach of creating a superior innovation measurement system will lead to better performance on business and innovation objectives. It is foreseeable that AI-driven systems will create new metrics and completely change the way measurement is done today.

Let's start collaborating across disciplines to discover new opportunity areas. Apply the full spectrum of statistics for forecasting, collecting, and clustering signals, connecting data points to uncover trajectories, and measuring what should be measured to reduce risks, and improve collaboration and the alignment between business and team performance!

Design thinking will continue to be of great importance in the future. It is the basis for the exploration of the actual problem up to the definition of AI projects.

START YOUR DESIGN THINKING AND INNOVATION METRICS JOURNEY. NOW!
AI

Practitioner Voices

Claudio Mirti, Sr. Advanced Analytics & AI Specialist EMEA, Microsoft

"The fusion of disciplines is becoming one of the most important competitive advantages for future business success. This book provides an excellent and inspiring guide on how to leverage best practices for executives, innovators, and business leaders."

Victor Schlegel, Director, Enterprise Analytics Consulting, EPAM Continuum

"Michael's new book gives an excellent outlook on what the future will look like. The combination of human-led and data-driven innovation forms a key pillar in 21st century sources of growth."

Omar Hatamleh, Author of the AI book *BetweenBrains*, ES2030

"AI is one of the most transformational technologies. The transversal nature of this powerful technology has the potential to change humanity on many different levels and create new waves of capabilities. This book offers tools to explore a future of unlimited possibilities."

Christian Hohmann, Program Lead Smart-up and Lecture, HSLU

"Startups and fast-growing companies often have no way of knowing if they are on the right path while innovating. The new *Design Thinking & Innovation Metrics* playbook helps us address the potential risks earlier."

Sabire Serap Keskin, Head of Branch, Allianz Technology

"A great read into the world of measuring EXPLORE AND EXPLOIT. Easy to understand, practical, and with engaging visualizations."

Markus Blatt, OKR-Trainer and Managing Director, neueBeratung

"Measuring the impact of your activities is crucial to objectives & key results. Thus, my favorite part of the book is the 201, where you can learn how to measure even the trickiest strategic impact!"

Patrick Link, Professor for Innovation Management, HSLU

"In university education, we focus on interdisciplinarity. Design thinking, measuring innovation, and data-driven innovation should be part of every professional innovation management education."

Roland Ringgenberg, Digital Architect, Digital Business Foundry

"Building a measurement system step by step while developing high-impact digital innovations is the new benchmark. This outstanding publication for innovation management provides the appropriate toolkit."

Karina Rempel, Head of Innovation Excellence, Siemens

"The understanding of how successful corporate venturing works has sharpened significantly in recent years. This book provides much inspiration for a customer-centric approach and how to measure the success of an innovation project. We will use this knowledge for our teams and customers."

We welcome more feedback and reviews on:

THANK YOU
FOR REVIEWING
AND PROVIDING
VALUABLE FEEDBACK
ON THE BOOK!

ARCHITECT | ILLUSTRATOR | MEDIA ARTIST

Rukaiya Karim is a talented designer with a great passion for design thinking. In 2022, she completed her master's, in International Media Architecture Master Studies, a double-degree program between Bauhaus-Universität, Weimar, Germany, and the State University of New York at Buffalo, USA, and now works as a freelancer for international art, architecture, and book projects. This deep immersion in the topic as well as working in teams are some of her most vital strengths in the field of art and design.

ILLUSTRATED BY RUKAIYA KARIM

LINKEDIN.COM/IN/RUKAIYA-KARIM

VISIT THE DESIGN THINKING TEMPLATE ONLINE SHOP

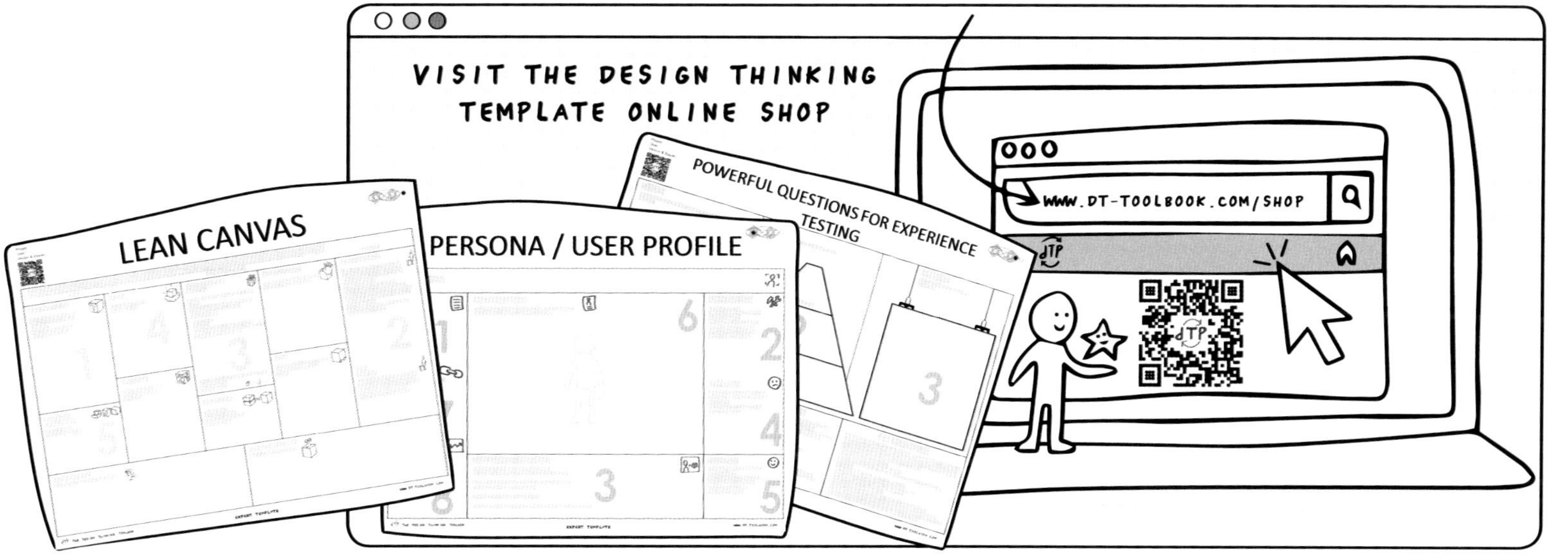
LEAN CANVAS
PERSONA / USER PROFILE
POWERFUL QUESTIONS FOR EXPERIENCE TESTING
WWW.DT-TOOLBOOK.COM/SHOP

INDEX AND SOURCES

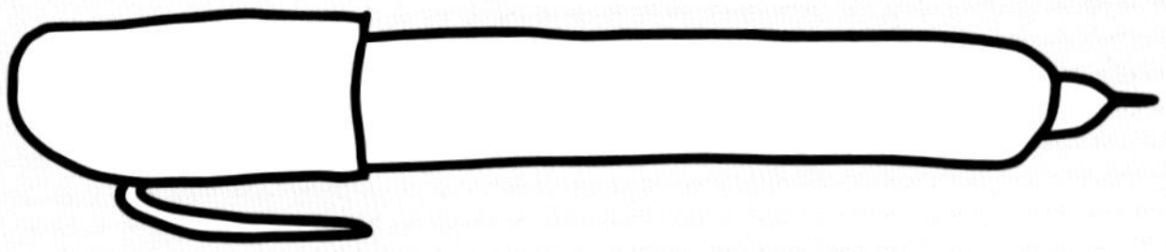

Index

The prepared index should significantly increase the usefulness of this book for our readers and design thinking practitioners.

Sources

- Abraham, A. (2018). *The Neuroscience of Creativity*. Cambridge, UK: Cambridge University Press.
- Acs, Z. J. and Audretsch, D. (1993). Analyzing innovation output indicators: The U.S. experience, in KLEINKNECHT Alfred and Donald BAIN (Eds.), *New Concepts in Innovation Output Measurement*, London: Macmillan, pp. 10–41.
- Aghera, A., Emery, M., Bounds, R., Bush, C., Stansfield, B., Gillett, B., & Santen, S. (2018). *A randomized trial of SMART goal enhanced debriefing after simulation to promote educational actions.* Western Journal of Emergency Medicine, 19(January), 112–120.
- Alford, L. P. (1929). Industry: Part 2-technical changes in manufacturing industries. In *Recent Economic Changes in the United States*, Vos. 1 and 2 (pp. 96–166).
- Amabile, T. M. (1983). The social psychology of creativity: A componential conceptualization. *Journal of Personality and Social Psychology* 45(2), 357–376.
- Amabile, T. M. (2013). Componential theory of creativity. In E. H. Kessler (Ed.), *Encyclopedia of Management Theory* (Vol. 1, pp. 135–139). Thousand Oaks: SAGE Publications, Ltd.
- Anticevic, A., Cole, M. W., Murray, J. D., Corlett, P. R., Wang, X. J., & Krystal, J. H. (2012). The role of default network deactivation in cognition and disease. *Trends in Cognitive Sciences* 16(12), 584–592.
- Archibugi, D., Cohendet, P., Kristensen, A., and Schaffer, K. A. (1997). "Evaluation of the community innovation survey. Report to the European Commission," Sprint/Eims Report, Luxembourg.
- Arden, R., Chavez, R.S., Grazioplene, R., Jung, R. E. (2010). Neuroimaging creativity: A psychometric review. *Behavioural Brain Research* 214, 143–156.
- Arnheim, R. (1969, neue Auflage 1997). *Visual Thinking*. Berkeley: University of California Press.
- Arundel, A., & Smith, K. (2013). History of the Community Innovation Survey. In F. Fault (Ed.), *Handbook of Innovation Indicators and Measurement. Cheltenham, UK:* Edward Elgar Publishing Ltd.
- Auernhammer J., Liu W., Ohashi T., Leifer L., Byler E., & Pan W. (2021) NeuroDesign: Embracing Neuroscience Instruments to Investigate Human Collaboration in Design. Human Interaction Emerging Technologies and Future Applications III. IHIET 2020. *Advances in Intelligent Systems and Computing* Vol. 1253. Springer Cham.
- Auernhammer, J., Sonalkar, N., & Saggar, M. (2021). NeuroDesign: From neuroscience research to design thinking practice. In Meinel and Leifer (Eds.), *Design Thinking Research*. New York: Springer.
- Auernhammer, J., Sonalkar, N., Xie, H., Monlux, K., Bruno, J., & Saggar, M. (2022). Examining the neuro-cognitive basis of applied creativity in entrepreneurs and managers. In Meinel and Leifer (Eds.), *Design Thinking Research*. New York: Springer.
- Augustine, A. A., & Hemenover, S. H. (2009). On the relative effectiveness of affect regulation strategies: A meta-analysis. *Cognition and Emotion* 23, 1181–1220.
- Autio, E., & Thomas, L. D. W. (2019). Value co-creation in ecosystems: Insights and research promise from three disciplinary persepctives. In S. Nambisan, K. Lyytinen, and Y. Yoo (Eds.), *Handbook of Digital Innovation. Cheltenham, UK:* Edward Elgar.
- Baars, J.E. (2018). Leading Design. München: Verlag Franz Vahlen GmbH.
- Baert, C., Meuleman, M., Debruyne, M., & Wright, M. (2016). Portfolio entrepreneurship and resource orchestration. Strategic Entrepreneurship Journal, 10(4), 346–370.
- Baker, E. (1933). *Displacement of Men by Machines—Effects of Technological Change in Commercial Printing*. New York: Columbia University Press.

- Baldwin, C. Y. (2012b). "Organization design for distributed innovation." Harvard Business School Working Paper (12-100), 1–12.
- Baldwin, C. Y., & Clark, K. B. (2000). *Design Rules: The Power of Modularity*. Cambridge, MA: MIT Press.
- Ballandonne, M. (2017). *On Geniuses and Heroes: Gilfillan, Schumpeter, and the Eugenic Approach to Inventors and Innovators*. Schumpeter. Paper ESSCA School of Management, April, 2017.
- Bansal, P., Kim, A., & Wood, M. O. (2018). Hidden in plain sight: The importance of scale in organizations' attention to issues. *Academy of Management Review*, 43(2), 217–241.
- Barbot, B. (2018). The dynamics of creative ideation: Introducing a new assessment paradigm. *Frontiers in Psychology* 9, 2529.
- Barnett, M. L. (2008). An attention-based view of real options reasoning. *Academy of Management Review*, 33(3), 606–628.
- Beaty, R. E., Benedek, M., Silvia, P. J., & Schacter, D. L. (2016). Creative cognition and brain network dynamics. *Trends in Cognitive Sciences* 20(2), 87–95.
- Beaty, R. E., Thakral, P. P., Madore, K. P., Benedek, M., & Schacter, D. L. (2018). Core network contributions to remembering the past, imagining the future, and thinking creatively. *Journal of Cognitive Neuroscience* 30(12), 1939–1951.
- Beaty, R.E., Seli, P., & Schacter, D. L. (2019). Network neuroscience of creative cognition: Mapping cognitive mechanisms and individual differences in the creative brain. *Current Opinion in Behavioral Sciences* 27, 22–30.
- Benedek, M., & Fink, A. (2019). Toward a neurocognitive framework of creative cognition: The role of memory, attention, and cognitive control. *Current Opinion in Behavioral Sciences* 27, 116–122.
- Benedek, M., Christensen, A. P., Fink, A., & Beaty, R. (2019). Creativity assessment in neuroscience research. *Psychology of Aesthetics, Creativity, and the Arts* 13, 218–226.
- Bernard, G., & Andritsos, P. (2017). A Process Mining Based Model for Customer Journey Mapping. In Forum and Doctoral Consortium Papers Presented at the 29th International Conference on Advanced Information Systems Engineering, 1848, 49–56.
- Bland, D., & Osterwalder A. (2019). *Testing Business Ideas: A Field Guide for Rapid Experimentation*. New York: Wiley.
- Blank, S. G. (2013). Why the lean start-up changes everything. *Harvard Business Review* 91(5): S. 63–72.
- Blank, S. G., & Dorf, B. (2012). *The Start-up Owner's Manual: The Step-by-Step Guide for Building a Great Company*. Pescadero, CA: K&S Ranch.
- Boccia, M., Piccardi, L., Palermo, L., Nori, R., & Palmiero, M. (2015). Where do bright ideas occur in our brain? Meta-analytic evidence from neuroimaging studies of domain-specific creativity. *Frontiers in Psychology* 6, 1195
- Bronwyn H., & Jaffe, A. (2012). Measuring science, technology, and innovation: A review. Report prepared for the Panel on Developing Science, Technology, and Innovation Indicators for the Future.
- Brouwer, E., & Kleinknecht, A. (1996). Firm size, small business presence and sales of innovativeproducts: A micro-econometric analysis. *Small Business Economics* 8(3): 189–201.
- Brouwer, E., & Kleinknecht, A. (1997). Measuring the unmeasurable: A country's non-R&D expenditure on product and service innovation. *Research Policy*, 25(8): 1235–1242.
- Brown, T., & Katz, B. (2009). *Change by design: How Design Thinking Transforms Organizations and Inspires Innovation*. New York: HarperCollins.
- Buchanan, R. (1992): Wicked problems in design thinking. *Design Issues* 8(2),S. 5–21
- Cao, L. (2017). Data science: A comprehensive overview. *ACM Computing Surveys* (CSUR) 50(3): 1–42.
- Chhabra, A., & Williams., S. (2019). Fusing Data and Design to Supercharge Innovation—in Products and Processes. *McKinsey Analytics*,

April 4. https://www.mckinsey.com/business-functions/mckinsey-analytics/our-insights/fusing-data-and-design-to-supercharge-innovation-in-products-and-processes.

- Carleton T., & Cockayne, W. (2013): *Playbook for Strategic Foresight & Innovation*. http://www.innovation.io.
- Carol et al. (2013). Innovation and intangible investment in Europe, Japan, and the United States. *Oxford Review of Economic Policy* 29(2): 261–286.
- Chesbrough, H. (2003). *Open Innovation. The New Imperative for Creating and Profiting from Technology*. Boston: Harvard Business School Press.
- Christensen, C. et al (2016). *The Innovator's Dilemma: When New Technologies Cause Great Firms to Fail*. New York: HarperBusiness.
- Cohen, W. M., & Levin, R. (1989). Empirical studies of innovation and market structure. In R. Schmalensee, et al. (Eds.), *Handbook of iIndustrial Organization*. New York: Elsevier 1059–1107.
- Colzato, L. S., Szapora, A., Pannekoek, J. N., & Hommel, B. (2013). The impact of physical exercise on convergent and divergent thinking. *Frontiers in Human Neuroscience* 7, 824.
- Coombs, R., Narandren, P., & Richards, A. (1996). A literature-based innovation output indicator. *Research Policy* 25(3): 403–413.
- Cornia, M., Baraldi, L., Serra, G., & Cucchiara, R. (2018). Predicting human eye fixations via an LSTM-based saliency attentive model. *IEEE Transactions on Image Processing* 27(10): 5142–5154.
- Corrado, C. A., & Hulten, C. R. (2014). Innovation accounting. In D. Jorgenson, et al. (Eds), *Measuring Economic Sustainability and Progress*. Chicago: University of Chicago Press, 595–628.
- Corsten, N., and J. Prick. (2019). The Service Design Maturity Model. *Touchpoint* 10(3): 72–77.
- Costa, N., Patrício, L., & Morelli, N. (2018). A designerly-way of conducting qualitative research in design studies. In *Service Design Proof of Concept*. Proceedings of the ServDes.2018 Conference, edited by A. Meroni, A. M. O. Medina, and B. Villari, 164–176. Linköpings: Linköping University Electronic Press.
- Crépon, Bruno, Duguet, E., & Mairesse, J. (1998). Research, innovation, and productivity: An econometric analysis at the firm level. *Economics of Innovation and New Technology*, 7, 115–158.
- Crocker, J. (2019). Measurement matters: The relationship between methods of scoring the Alternate Uses Task and brain activation. *Current Opinion in Behavioral Sciences* 27, 109–115.
- Cross, N. (2011). *Design Thinking*. Oxford: Berg Publishers.
- Csikszentmihalyi, M. (1997). *Creativity: Flow and the Psychology of Discovery and Invention*. New York: HarperCollins.
- Curedale, R. (2016): *Design Thinking – Process & Methods Guide*. 3rd ed. Los Angeles: Design Community College Inc.
- D'Ignazio, C. (2017). Creative data literacy: Bridging the gap between the data-haves and data-have-nots. *Information Design Journal* 23(1): 6–18.
- Davenport, T. (2014). *Big data @ work: Chancen erkennen, Risiken verstehen*. Vahlen Verlag.
- Davenport, T. H., & Patil, D. J. (2012). Data scientist: The sexiest job of the 21st century. In *Harvard Business Review*. October 2012, https://hbr.org/2012/10/data-scientist-the-sexiest-job-of-the-21st-century/.
- Day, T., & Tosey, P. (2011). Beyond SMART? A new framework for goal setting. *Curriculum Journal*, 22(4), 515–534.
- Doerr, J. (2018a). How Intel won the microprocessor wars. Retrieved April 10, 2019, from https://www.managementtoday.co.uk/intel-won-microprocessor-wars/any-otherbusiness/article/1464473.
- Doerr, J. (2018b) *Measure What Matters: How Google, Bono, and the Gates Foundation Rock the World with OKRs*. New York: Penguin Random House LLC.

- Doerr, J. (2019). OKR Cycle. www.whatmatters.com. Retrieved May 15, 2019 from https://assets.ctfassets.net/cn6v7tcah9c0/4ZKbGhsH-QI2MsGSMY6gASu/66fcfd4541502821dae 93221a6474105/A_Typical_OKR_Cycle_V1JS.pd.
- Doorley, S., Witthoft, S., & Hasso Plattner Institute of Design at Stanford (2012). *Make Space: How to Set the Stage for Creative Collaboration.* Hoboken, NJ: Wiley.
- Dorst, K. (2015). *Frame Innovation.* Cambridge (MA): MIT Press.
- Dorst, K. (2017). *Notes on Design – How Creative Practice Works.* Amsterdam: BIS publishers.
- Dorst, K., & Cross, N. (2001). Creativity in the design process: Co-evolution of problem–solution. *Design Studies*, 22(5), 425–437.
- Dow, S. P., Glassco, A., Kass, J., Schwarz, M., Schwartz, D. L., & Klemmer, S. R. (2010). Parallel prototyping leads to better design results, more divergence, and increased self-efficacy. *ACM Trans Comput Hum Interact* 17(4):18.
- Dow, S. P., & Klemmer, S. R. (2011). The efficacy of prototyping under time constraints. In *Design Thinking.* Springer, Heidelberg, pp. 111–128.
- Duschlbauer, T. (2018). *Der Querdenker. Zürich: Midas Management* Verlag AG.
- Dweck, C. (2006). *Mindset: The New Psychology of Success. New York:* Random House.
- Dweck, C. (2016). The remarkable reach of growth mind-sets. *Scientific American Mind*, 27(1), 36–41. Retrieved August 6, 2022, from https://www.jstor.org/stable/24945335.
- Dziallas, M, & Blind, K. (2018). Innovation indicators throughout the innovation process: An extensive literature analysis. *Technovation* 27.
- Fagerberg, J., and Verspagen, B. (2009). Innovation studies — The emerging structure of a newscientific field. *Research Policy* 38(2): pp. 218–233.
- Fagerberg, J., Mowery, D. C., & Nelson, R. R. (Eds.) (2005). *The Oxford Handbook of Innovation.* Oxford: Oxford University Press.
- Fagerberg, J. (2003). Schumpeter and the revival of evolutionary economics: An appraisal of the literature. *Journal of Evolutionary Economics* 13(2): 125–159.
- Fankhauser, G. (2019). *How to Establish Innovation Metrics.* Article, InnovationManagement.se.
- Freeman, C. (2007). A Schumpeterian renaissance? In H. Hanusch, Horst and A. Pyka (Eds.), Elgar Companion to Neo-*Schumpeterian* Economics. Cheltenham, UK: Edward Elgar Publishing, pp.130–147.
- Gilbreth. F. (1911). *Motion Study: A Method for Increasing the Efficiency of the Workman.* New York: David Van Nostrand Company.
- Gladwell, M. (2005). *Blink: The Power of Thinking Without Thinking.* New York: Back Bay Books.
- Godin, B. (2002). The rise of innovation surveys: Measuring a fuzzy concept. Canadian Science and Innovation Indicators Consortium, Project on the History and Sociology of S&T Statistics, Paper, 2002, 16.
- Godin, B. (2008). In the shadow of Schumpeter: W. Rupert Maclaurin and the study of technological innovation. *Minerva* 46, 343–360.
- Google. (2019). Google's OKR Playbook. Retrieved May 15, 2019 from: https://assets.ctfassets.net/cn6v7tcah9c0/4snZXJ821G6KYUoc-08masK/58ffcbc7c607d7c6056c7 da507727135/Google_OKR_Playbook_V1JS.pdf.
- Gray, D., Brown, S., & Macanufo, J. (2010). *Gamestorming.* Sebastopol, CA: O'Reilly Media Inc.
- Grimes, J., Fineman, A., Gillespie, B., Atvur, A., & Mager, B. (2018). Designing the future. *Touchpoint* 10(2): 88–93.
- Guilford, J. (1968). *Intelligence, Creativity, and Their Educational Implications.* San Diego, CA: Robert R. Knapp.

- Haefnera, N., Wincent, J., Parida, V., & Gassmann, O. (2021). Artificial intelligence and innovation management: A review, framework, and research agenda. *Journal of Technological Forecasting and Social Change.*
- Hall, B., & Rosenberg, N. (Eds.) (2010). *Handbook of the Economics of Innovation.* New York: Elsevier.
- Harbich, M., Bernard, G., Berkes, P., Garbinato, B., & Andritsos, P. (2017). Discovering customer journey maps using a mixture of Markov models. In P. Ceravolo, M. V. Keulen, and K. Stoffel (Eds.), *Proceedings of the 7th International Symposium on Data-Driven Process Discovery and Analysis.* 3–7. Neuchatel, Switzerland. ceur-ws.org.
- Herrmann, N. (1996). *The Whole Brain Business Book: Harnessing the Power of the Whole Brain Organization and the Whole Brain Individual.* New York: McGraw-Hill Professional.
- Hippel, E. V. (1986). Lead users. A source of novel product concepts. *Management Science* Vol. 32, S. 791–805.
- Hohmann, L. (2007). *Innovation Games.* Boston: Pearson Education Inc.
- IDEO. (2009). *Human-centered Design: Toolkit & Human-centered Design: Field Guide.* 2nd ed. [Beide verfügbar auf der IDEO Homepage oder unter: http://www.hcdtoolkit.com].
- Jerome, H. (1932). The measurement of productivity changes and the displacement of labor. *The American Economic Review* 22(1), 32–40.
- Kelly, T., & Littman, J. (2001). *The Art of Innovation: Lessons in Creativity from IDEO, America's Leading Design Firm* London: Profile Books.
- Kim, W., & Mauborgne, R. (2005). Der blaue Ozean als Strategie: wie man neue Märkte schafft, wo es keine Konkurrenz gibt. Hanser Verlag
- Kleinknecht, A., & Reijnen, J. (1993). Towards literature-based innovation outputindicators. *Structural Change and Economic Dynamics* 4(1): 199–207.
- Knight, J., Ross, E., Gibbons, C., & McEwan, T. (2020). Unlocking service flow fast and frugal digital healthcare design. In A. Woodcock, L. Moody, D. McDonagh, A. Jain, and L. C. Jain (Eds.), *Design of Assistive Technology for Ageing Populations*, 171–187. New York: Springer.
- Koch, J. (2017). Design Implications for Designing with a Collaborative AI. In AAAI 2017. Palo Alto, CA, USA: Spring Symposium Series.
- Köppen, E., Meinel, C., Rhinow, H., Schmiedgen, J., Spille, L. (2015). Measuring the impact of design thinking. In H. Plattnor, C. Meinel, and L. Leifler (Eds.), *Design Thinking Research. Switzerland:* Springer, pp. 157–170.
- Kunneman, Y. (2019). Data Science for Service Design: An Exploration of the Opportunities, Challenges and Methods for Data Mining to Support the Service Design Process. MSc thesis, University of Twente.
- Leifer, I. (1998). Design-team performance: metrics and the impact of technology. In *Evaluating Corporate Training*, pp. 297–0319.
- Leifer, L. (1998). Design-team performance: Metrics and the impact of technology. In *Evaluating Corporate Training*, pp. 297–319 (1998)
- Leifer, L., & Steiner, M. (2014). Dancing with ambiguity: Causality behavior, design thinking, and triple-loop-learning. In *Management of the Fuzzy Front End of Innovation*, pp. 141–158.
- Leifer, L., & Steinert, M. (2014). Dancing with ambiguity: Causality behavior, design thinking, and triple-loop-learning. In *Management of the Fuzzy Front End of Innovation*, pp. 141–158.
- Lewrick, M., & Link, P. (2015). *Hybride Management Modelle: Konvergenz von Design Thinking und Big Data.* IM+io Fachzeitschrift für Innovation, Organisation und Management (4), S. 68–71.
- Lewrick, M. (2014). Design Thinking – Ausbildung an Universitäten, S. 87–101. In Sauvonnet und Blatt (Hrsg). Wo ist das Problem? Neue Beratung.
- Lewrick, M. (2018). *Design Thinking: Radikale Innovationen in einer Digitalisierten Welt*, Beck Verlag; München.

- Lewrick, M. (2022). *Design Thinking for Business Growth*. New York: Wiley.
- Lewrick, M., Link, P., & Leifer, L. (2018). *The Design Thinking Playbook* New York: Wiley.
- Lewrick, M., Skribanowitz, P., & Huber, F. (2012). Nutzen von Design Thinking Programmen, 16. *Interdisziplinäre Jahreskonferenz zur Gründungsforschung* (G-Forum), Universität Potsdam.
- Liedtka, J. (2017). Evaluating the impact of design thinking in action. *Academy of Management Proceedings* (1).
- Lietka, J., & Ogilvie, T. (2011). *Designing for Growth*. New York: Columbia University Press Inc.
- Maclaurin, R. (1950). The process of technological innovation: The launching of a new scientific industry. *The American Economic Review*, 40(1): 90–112.
- Maeda, J. (2006). *The laws of Simplicity – Simplicity: Design, Tecnology, Business, Life*. Cambridge, London: MIT Press.
- Mairesse, J., & Mohnen, P. (2010). Using innovation surveys for econometric analysis. In B. H. Hall and N. Rosenberg (Eds.), *Handbook of the Economics of Innovation*. New York: Elsevier, pp.1129–1155.
- Maurya, A. (2013). *Running Lean - Das How-to für erfolgreiche Innovationen*.
- Miettinen, S. (2016). Introduction to industrial service design: What is industrial service design? In S. Miettinen (Ed.), *An Introduction to Industrial Service Design* 3–14. Abingdon: Routledge.
- Molaro, A., & White, L. (2015). *The Library Innovation Toolkit: Idea, Strategies, and Programs*. ALA edition.
- Murray, P. W., Agard, B., & Barajas, M. A. (2018). Forecast of individual customer's demand from a large and noisy dataset. *Computers & Industrial Engineering* 118: 33–43.
- Nagji, B., & Tuff, G. (2012). Managing your innovation portfolio. *Harvard Business Review*, 90(5), 66–74.
- Nilsson, P. (2012). *Four Ways to Measure Creativity. Sense and Sensation Writing on Education, Creativity, and Cognitive Science*. http://www.senseandsensation.com/2012/03/assessing-creativity.html.
- Nilsson, P. (2013). *Taxonomy of Creative Design. Sense and Sensation Writing on Education, Creativity, and Cognitive Science*.http://www.senseandsensation.com/2012/03/taxonomy-of-creative-design.html?view=magazine.
- Ohashi, T., Auernhammer, J., Liu, W., Pan, W., & Leifer, L. 2020. *NeuroDesignScience: Systematic Literature Review of Current Research on Design Using Neuroscience Techniques*.
- Osterwalder, A., Pigneur, Y., et al. (2015). *Value Proposition Design*. New York: Wiley
- Osterwalder, A., & Pigneur, Y. (2010). *Business Model Generation: A Handbook for Visionaries, Game Changers, and Challengers*. New York: Wiley.
- Osterwalder, A., Pigneur, Y., Smith, A., & Etiemble, F. (2020). *The Invincible Company: How to Constantly Reinvent Your Organization with Inspiration from the World's Best Business Models*. New York: Wiley.
- Patrício, L., Gustafsson, A., & Fisk, R. (2018). Upframing service design and innovation for research impact. *Journal of Service Research* 21(1): 3–16.
- Perchtold, C., Papousek, I., Koschutnig, K., Rominger, C., Weber, H., Weiss, E. M., & Fink, A. (2018). Affective creativity meets classic creativity in the scanner. *Human Brain Mapping* 39, 393–406.
- Plattner, H., Meinel, C., & Leifer, L. (2010). *Design Thinking: Understand Improve--Apply (understanding innovation)*. Heidelberg: Springer.
- Poria, S., Cambria, E., & Gelbukh, A. (2016). Aspect extraction for opinion mining with a deep convolutional neural network. *Knowledge-Based Systems* 108: 42–49.

- Prendiville, A., Gwilt, I., & Mitchell, V. (2018). Making sense of data through service design. In D. Sangiorgi and A. Prendiville (Eds.), *Designing for Service* 225–236. London: Bloomsbury.
- Rapp, K., & Stroup, C. (2016). How can organizations adopt and measure design thinking processes? Retrieved from Cornell University, ILR School site: http://digitalcommons.ilr.cornell.edu/student/138.
- Richtnée, A., et al. (2017). Creating better innovation measurement practices. *MIT Sloan Management Review.*
- Roth B. & Royalty, A. (2016). Developing design thinking metrics as a driver of creative innovation. In H. Plattner, C. Meinel, and L. Leifer (Eds.), *Design Thinking Research. Switzerland:* Springer, pp. 171–183.
- Sauvonnet, E., & Blatt, M. (2017). *Wo ist das Problem?* München: Verlag Franz Vahlen GmbH.
- Schmiedgen, J. et al. (2016). Measuring the impact of design thinking. In H. Plattner et al. (Eds.), *Design Thinking Research: Making Design Thinking Foundational* (pp. 157–170). Switzerland: Springer International Publishing.
- Schumpeter, Joseph A. (2017). *Theory of Economic Development. New York:*. Routledge (1st edition: 1934).
- Smithson, N. (2019). Google's Organizational Structure & Organizational Culture (An Analysis). Retrieved May 25, 2019, from http://panmore.com/google-organizational-structure-organizational-culture.
- Steiber, A. (2016). *The Google Model: Managing Continuous Innovation in a Rapidly Changing World.*
- Stetler, K. L., & Magnusson, M. (2014). Exploring the tension between clarity and ambiguity in goal setting for innovation. *Creativity and Innovation Management* 24(2), 231–246.
- Stickdorn, M., & Schneider, J. (2016). *This Is Service Design Thinking.* 6th ed. Amsterdam: BIS Publishers.
- Taylor, F. W. (1911). *The Principles of Scientific Management.* New York and London: Harper & Brothers.
- Toma, D., & Gons, E. (2022). *Innovation Accounting: A Practical Guide for Measuring Your Innovation Ecosystem's Performance. Amsterdam:* BisPublisher
- Töpfer, A. (2008). *Lean Six Sigma.* Heidelberg: Springer-Verlag GmbH.
- Tuarob, S., & C. S. Tucker. (2015). Automated discovery of lead users and latent product features by mining large-scale social media networks. *Journal of Mechanical Design* 137(7).
- Uebernickel, F., Brenner, W. et al. (2015). *Design Thinking – Das Handbuch.* Frankfurt am Main: Frankfurter Allgemeine Buch
- Ulrich, K. (2011). *Design Creation of Artifacts in Society.* Published by the University of Pennsylvania. http://www.ulrichbook.org/.
- Ulwick, A. (2016). *Jobs to Be Done.* Idea Bite Press
- van der Aalst, W. (2014). Data scientist: The engineer of the future. In *Enterprise Interoperability*, vol. 7, 13–26. Cham: Springer International Publishing.
- van der Pijl, P., Lokitz, J., & Solomon, L, K. (2016). *Design a Better Business.* New York: Wiley.
- von Thienen, J. P. A., Weinstein, T., Szymanski, C., Rahman, S., & Meinel, C. (2021). Design thinking, neurodesign und Curricula für das innovation engineering In C. Meinel and T. Krohn (Eds.), *Design Thinking in der Bildung. Innovation kann man lernen.* New York: Wiley.
- Weber, B., Koschutnig, K., Schwerdtfeger, A., Rominger, C., Papousek, I., Weiss, E. M., Tilp, M., & Fink, A. (2019). Learning unicycling evokes manifold changes in gray and white matter networks related to motor- and cognitive functions. *Scientific Reports* 9(1):4324.
- Weinberg, U. (2015). *Network Thinking.* Hamburg: Murmann Publishers GmbH.

- Wettersten, J., & D. Malmgren. (2018). What happens when data scientists and designers work together. *Harvard Busines Review*, March 05.
- Wodtke, C. (2016). *Introduction to OKRs.*
- Zuraik, A., & Kelly, L. (2019). The role of CEO transformational leadership and innovation climate in exploration and exploitation. *European Journal of Innovation Management* 22(1), 84–104.